Paint Flow and Pigment Dispersion

PAINT FLOW AND PIGMENT DISPERSION

A Rheological Approach to Coating and Ink Technology

Second Edition

TEMPLE C. PATTON

A Wiley-Interscience Publication

JOHN WILEY & SONS, New York • Chichester • Brisbane • Toronto

Library of Congress Cataloging in Publication Data:

Patton, Temple C.
Paint flow and pigment dispersion.

"A Wiley-Interscience publication."
Includes index.
1. Paint. 2. Pigments. I. Title.
TP936.P3 1979 667′.6 78-10774
ISBN 0-471-03272-7

Printed in the United States of America

10 9 8 7 6 5 4 3 2 1

Preface to Second Edition

Fifteen years have passed since the first edition of this book appeared. During this period many excellent papers have been published that have added greatly to our knowledge of paint flow and pigment dispersion. Prompted by this fact and also by the fine response to the first edition, I have prepared this updated version incorporating much of this new and important technical information.

As with the first edition, it has been necessary to exercise considerable selectivity in the choice of topics covered. Hence no claim is made that the coverage is exhaustive in either scope or depth. Many topics originally scheduled for discussion were reluctantly eliminated because of time and space considerations. However, it is hoped that this second edition will provide engineers in the coating, ink, and allied industries with a practical exposition of the more important aspects of the technology relating to pigment dispersion, the rheology of coating flow, and other topics related to these phenomena.

In keeping with the first edition, the objective has been to steer a middle course between a highly theoretical presentation and an unduly elementary treatment. In line with this objective problems have been worked throughout the text to demonstrate the practical use of equations derived from theoretical concepts.

Again I wish to express my appreciation for the help given to me by my wife Doris in typing and assembling the manuscript and for lending continuing encouragement during the whole writing effort.

TEMPLE C. PATTON

Brick Town, New Jersey
October 1978

Preface to Second Edition

Introductory Notes

SELECTED REFERENCE ABBREVIATIONS

Since four journals appear very frequently in the references, they have been assigned the following abbreviations to conserve space:

Off. Dig.	*Official Digest, Federation of Societies for Paint Technology* (before Vol. 321, No. 420, January, 1960, *Official Digest, Federation of Paint and Varnish Production Club*)
JPT	*Journal of Paint Technology*
JCT	*Journal of Coatings Technology*
JOCCA	*Journal of the Oil, Colour, and Chemists' Association*

LOGARITHM NOTATION

The notation "ln" is used to denote natural logarithms (to the base e), and "log" to denote logarithms to the base 10. Note that in algebraic equations (2.3026 log X) can be substituted for (ln X).

INSERTION OF PERCENTAGE VALUES IN EQUATIONS

Percentages are to be inserted as fractional values in all equations unless otherwise noted (32% PVC as 0.32; 0.12%w additive addition as 0.0012).

Contents

1 *Viscosity*

Rheology is defined as the science of flow and deformation. In this book only the rheology that involves coating and ink flow is considered. Even so, such coverage is extensive, since flow properties are of primary importance at every stage of paint or ink manufacture, from the time that liquid raw materials are first pumped into the plant (solvents, oils, resin solutions, pigment slurries) to the time that the prepared coating or ink is applied and converts to a solid film.

Subjectively, a viscous liquid might be described as thick, gluey, sticky, slow-flowing, or bodied. Conversely, a fluid liquid might be described as thin, watery, fast-flowing, or unbodied. Such terms are picturesque and helpful in roughly characterizing the flow properties of a liquid, but they are qualitative at best. Hence for technical work it becomes necessary to inquire more deeply into coating flow and to reduce coating rheological properties to a more rational and measurable basis.

Unfortunately, flow phenomena can become exceedingly complex. Even such a simple action as stirring paint in a can with a spatula involves a flow pattern that challenges exact mathematical analysis. However, simplifications and reasonable approximations can be introduced into coating rheology that permit the development of highly useful mathematical expressions. These in turn allow the ink or paint engineer to proceed with confidence in controlling and predicting the flow performance of inks or paint coatings.

VISCOSITY DEFINED

At the very outset it is important to develop a clear and exact concept of viscosity (the resistance of liquid to flow) in terms of measurable quantitities. To do

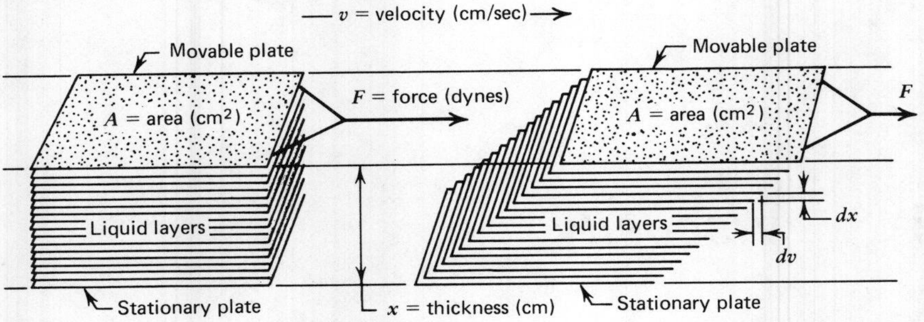

τ = shear stress = F/A (dynes / cm^2)
D = shear rate = v/x (sec^{-1})
η = viscosity = shear stress / shear rate = τ/D (dyne sec/cm^2) or (poise)

Fig. 1-1 Schematic diagram of a parallel plate arrangement illustrating the relationship of the quantities involved in simple (Newtonian) flow.

this, consider a model situation in which a liquid is confined between two parallel plates as shown schematically in Fig. 1-1. One plate is movable and one is held stationary, and they are separated by a distance x. Let a force F act on the top movable plate of area A in a tangential direction, so that the plate slides sidewise with velocity v relative to the stationary bottom plate. When it so moves, layers of liquid between the two plates are also moved in a sidewise direction, as depicted in the right-hand section of Fig. 1-1. The top layer of liquid moves with the greatest velocity, whereas the bottom liquid layer moves with the smallest (zero) velocity. Intermediate layers move with intermediate velocities. However, it should be noted that the velocity gradient dv/dx (differential or incremental change in velocity dv corresponding to a differential or incremental change in thickness dx) for any section of the liquid is constant. This velocity gradient D is referred to as the shear rate. In the given model situation the shear rate (velocity gradient) is uniform from top to bottom. Hence dv/dx is also equal to v/x, and both are equal to D as given by Eq. 1.

$$\text{D (shear rate)} = \frac{dv}{dx} = \frac{v\text{ (velocity)}}{x\text{ (thickness)}} \tag{1}$$

Shear velocity is conventionally expressed in cm/sec and thickness in cm. Hence shear rate D has the dimensions of reciprocal seconds (sec^{-1}), since the cemtimeter unit (present in both the numerator and the denominator of the expression v/x) is canceled, leaving only the unit of sec in the denominator to express the shear rate quantity:

$$\frac{\cancel{\text{cm}}}{\text{sec}} \cdot \frac{1}{\cancel{\text{cm}}} = \frac{1}{\text{sec}} = \text{sec}^{-1} \text{ (unit of shear rate)}$$

The total force acting tangentially on the top plate of area A in Fig. 1-1 is F. The force acting on a *unit* of top plate area is then F/A. This force per unit area is called the shear stress τ.

$$\tau \text{ (shear stress)} = \frac{F \text{ (force)}}{A \text{ (area)}} \tag{2}$$

Shear force F is conventionally expressed in dynes (1.0 g weight = 980 dynes), and the area A over which it acts in cm^2. Hence shear stress has the dimensions of $dynes/cm^2$.

Now that the two quantities of shear rate and shear stress have been developed, viscosity can be defined in quantitative terms. Viscosity η is simply the ratio of shear stress to shear rate.

$$\eta \text{ (viscosity)} = \frac{\tau \text{ (shear stress)}}{\text{D (shear rate)}} \tag{3}$$

In view of its importance, viscosity has been assigned a special unit, the poise. Viscosity in poises is automatically obtained when the shear stress is expressed as given above in $dyne/cm^2$ and the shear rate in sec^{-1}. From this it follows that the poise has the dimensions of $dyne\text{-}sec/cm^2$. It is the measure of absolute viscosity. Since a dyne in fundamental cgs units is expressed as $g\text{-}cm/sec^2$, the poise in turn has the dimensions of g/cm-sec in cgs units.

The three equations that have been developed (Eqs. 1, 2, and 3) underlie the science of flow and should be clearly understood for proper comprehension of the equations derived through their use.

The arrangement of the liquid between the two parallel plates as shown in Fig. 1-1 is obviously idealistic. Even so, this set of conditions is closely approached when a coating is applied to a surface by brushing or by a straight-edge spreading applicator. With a little imagination the similarity between the two can be readily visualized.

PROBLEM 1-1

A 6-in. brush applies a coating of 2.0-poise viscosity to a flat substrate to give a wet film thickness of 3.0 mils. The width of the brush that is flattened against the substrate is 1.0 in. when the brushing-stroke velocity is 4.0 ft/sec. Calculate the rate of shear and the drag on the brush under these application conditions.

Solution. If the wet film applied is 3.0 mils, the average gap during application is approximately twice this value (6.0 mils = 0.015 cm), since the coating is split between the brush and the substrate in about equal amounts. The information available is then:

$$\eta \text{ (viscosity)} = 2.0 \text{ poise}$$

$$x \text{ (thickness)} = 0.015 \text{ cm}$$

$$v \text{ (velocity)} = 122 \text{ cm/sec } (= 4.0 \cdot 30.5)$$

$$A \text{ (area)} = 39 \text{ cm}^2 \ (= 6.0 \cdot 1.0 \cdot 6.5)$$

(Note the conversion of all values into a consistent set of units before any substitution is made in Eqs. 1, 2, and 3.)

$$\tau \text{ (shear stress)} = \frac{F}{A} = \frac{F}{39} = 0.0256F \text{ dyne/cm}^2$$

$$\text{D (shear rate)} = \frac{v}{x} = \frac{122}{0.015} = 8130 \text{ sec}^{-1}$$

$$\eta \text{ (viscosity)} = \frac{\tau}{\text{D}} = 2.0 = \frac{0.0256F}{8130} \text{ poise}$$

$$F \text{ (force)} = \frac{2.0 \cdot 8130}{0.0256} = 635{,}000 \text{ dyne}$$

The shear rate is 8130 sec^{-1}, and the drag on the brush is 635,000 dynes or 1.4 lb [= 635,000/(980 · 454)].

At 68 F (20 C) water has a viscosity of about 1.0 centipoise (0.01 poise), linseed oil a viscosity of about 50 cP (0.50 poise), and castor oil a viscosity of about 1000 cP (10 poises). These are useful reference liquids for visualizing viscosity in terms of centipoise and poise units. With experience, viscosities in the range of about 0.5 to 50 poises can be estimated fairly closely by visual observation. However for lower (<0.1 poise) or higher (>200 poise) viscosities, judging values becomes most difficult, since the flow is either too rapid or too sluggish to distinguish differences.

Kinematic Viscosity

The poise, a measure of absolute viscosity η, is related to the stoke, a measure of kinematic viscosity ν, by Eq. 4, where ρ is the fluid density in g/cm^3.

$$\nu \text{ (stoke)} = \frac{\eta \text{ (poise)}}{\rho \text{ (g/cm}^3\text{)}} \tag{4}$$

The stoke is a useful unit when working with so-called kinematic viscometers (such as the orifice and bubble types) that depend on gravitational fall for their operation.

PROBLEM 1-2

Calculate the kinematic viscosity of linseed oil ($\rho = 0.93$ g/cm^3) at 20 C in centistokes.

Solution: Substitute the appropriate data in Eq. 4.

$$\nu = \frac{\eta}{\rho} = \frac{50}{0.93} = 54 \text{ centistokes}$$

NEWTONIAN FLOW

The coating of Problem 1-1 was assigned a viscosity of 2.0 poises for the application conditions. Does this imply that this coating under all conditions will exhibit this viscosity? The answer is no! Changes in temperature, in shear stress, in shear rate, or even in time may or may not alter a specified viscosity value. However, it is useful to conceive of an ideal liquid that has a constant viscosity over a broad shear rate range at any given temperature. Such a liquid is said to be Newtonian in its flow behavior, and liquids which approach this ideal are called Newtonian liquids. Many raw materials for the paint and ink industries, such as water, solvents, mineral oils, and some resin solutions, are Newtonian in nature. However, finished commercial coatings or inks are rarely Newtonian.

NON-NEWTONIAN FLOW

When a liquid system exhibits a changing viscosity with changing shear stress, it is said to be non-Newtonian in nature (viscosity no longer constant). Pigmented coatings in the paint industry are notoriously non-Newtonian in their flow behavior. This is shown dramatically by reference to Table 1-1, which gives the viscosity profile for a (non-Newtonian) pigmented latex coating. To help in visualizing the tremendous spread in the viscosity profile data, the four graphs shown in Fig. 1-2 plot the same information as is given in Table 1-1 in different ways. Two sets of uniform scales and two sets of logarithmic scales have been used in constructing these four graphs. The two top graphs of Fig. 1-2 have vertical scales carrying shear rate values and horizontal scales carrying shear stress values. These two graphs differ in that the scales of graph *a* are uniform, whereas those of graph *b* are logarithmic. The viscosity profile of graph *a* has the advantage of being a straight line. However, only a small portion of the data in Table 1-1 is covered or can be read from the graph. On the other hand, the

Table 1-1 Viscosity Profile for Non-Newtonian Latex Paint Coating (η = 2.0 + 10/D) or (τ = 2D + 10)

Shear Rate D (sec^{-1})	Viscosity η (poise)	Shear Stress τ ($dyne/cm^2$)
0.000	∞	10.000
0.001	10,002.0	10.002
0.01	1,002.0	10.02
0.10	102.0	10.2
0.20	52.0	10.4
0.50	22.0	11.0
1.0	12.0	12.0
2.0	7.0	14.0
5.0	4.0	20.0
10.0	3.0	30.0
50.0	2.5	110.0
100.0	2.1	210.0
1,000.0	2.01	2,010.0
10,000.0	2.001	20,010.0
∞	2.000	∞

viscosity profile curve of graph *b* covers most of the data listed in Table 1-1, and this information can be read from the graph with reasonable and consistent accuracy.

The two bottom graphs of Fig. 1-2 have vertical scales carrying viscosity values and horizontal scales carrying shear rate values. Again they differ in that the scales of graph *c* are uniform, whereas those of graph *d* are logarithmic. As before, only the logarithmic type of chart is capable of providing a satisfactory rendering of substantially the entire viscosity profile. Long experience has convinced this author that, of the four types of graphical presentation, graph *d* is most feasible for illustrating the viscosity profiles of coatings. On the *d* graph any Newtonian liquid is represented by a horizontal line running across the graph; any non-Newtonian liquid is represented by either a slanting straight line or a curve. Except for representing a small selected section of the viscosity profile, the use of uniform scales is technically inadequate and can often lead to misinterpretation of the rheology of paint systems. The *d*-type graph (log viscosity vs. log shear rate) will be used throughout this book.

Some of the terms that are used to describe and define the different types of non-Newtonian flow are now taken up in turn.

Pseudoplastic and Dilatant Flow

Coatings that decrease in viscosity with increased shear stress (shear-thinning) are described as *pseudoplastic.* Conversely, coatings that increase in viscosity with

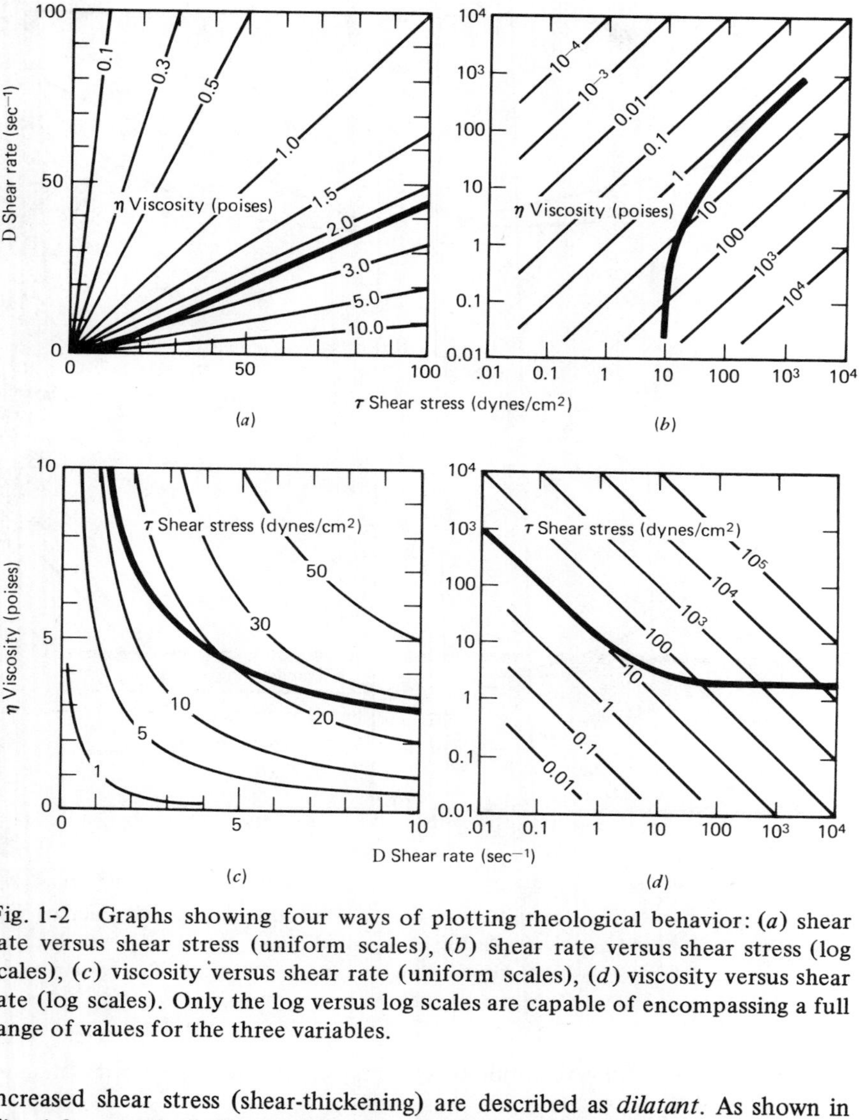

Fig. 1-2 Graphs showing four ways of plotting rheological behavior: (*a*) shear rate versus shear stress (uniform scales), (*b*) shear rate versus shear stress (log scales), (*c*) viscosity versus shear rate (uniform scales), (*d*) viscosity versus shear rate (log scales). Only the log versus log scales are capable of encompassing a full range of values for the three variables.

increased shear stress (shear-thickening) are described as *dilatant*. As shown in Fig. 1-3*a*, pseudoplastic systems give viscosity profiles that are either slanting straight lines or curves both with a negative slope (sweep downward from left to right); dilatant systems (Fig. 1-3*b*) give viscosity profiles that are either slanting straight lines or curves both with a positive slope (sweep upward from left to right).

Most finished coatings are pseudoplastic; dilatant coatings are seldom, if

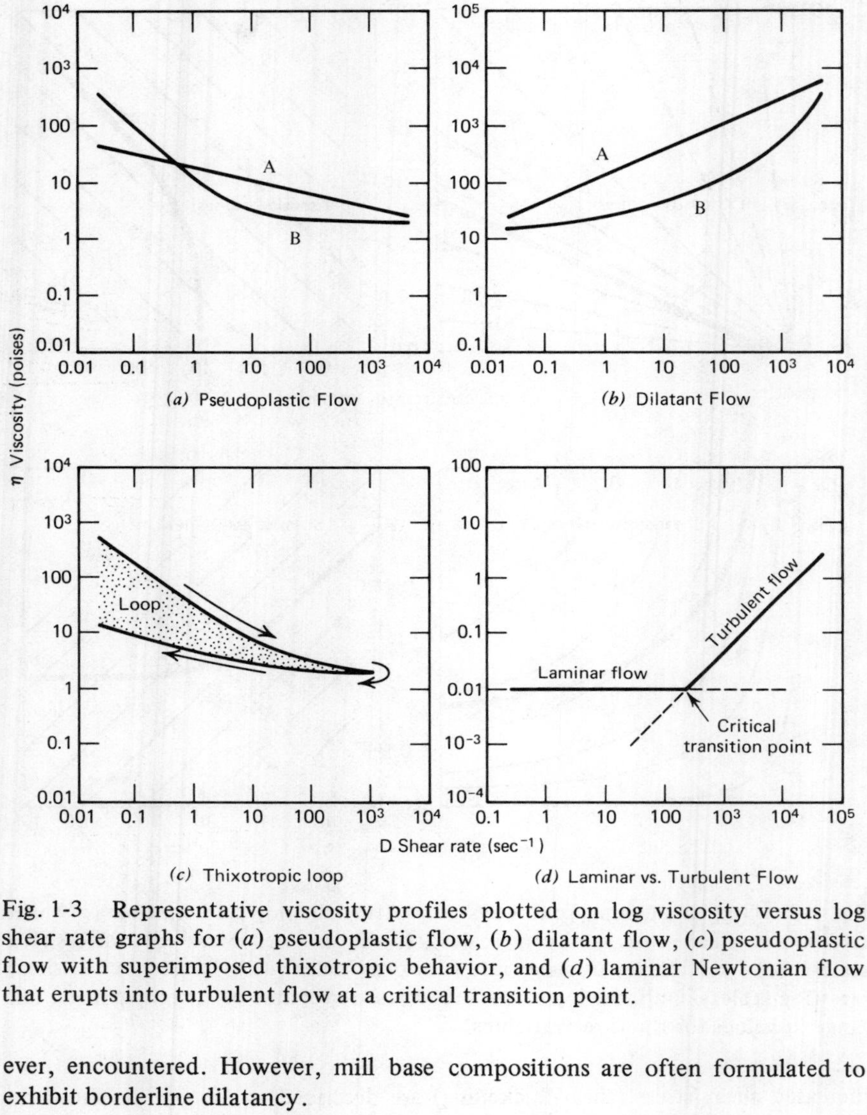

Fig. 1-3 Representative viscosity profiles plotted on log viscosity versus log shear rate graphs for (*a*) pseudoplastic flow, (*b*) dilatant flow, (*c*) pseudoplastic flow with superimposed thixotropic behavior, and (*d*) laminar Newtonian flow that erupts into turbulent flow at a critical transition point.

ever, encountered. However, mill base compositions are often formulated to exhibit borderline dilatancy.

Yield Value

Yield value refers to a certain minimum shear stress that must be exceeded before flow takes place. Below this yield value the coating acts as an elastic solid;

it may deform but it does not flow. Coating systems showing this type of behavior are often referred to as Bingham bodies. Once the shear stress exceeds the yield value, flow begins, and this may be variously referred to as pseuodoplastic flow with a yield value, as dilatant flow with a yield value, or under certain conditions as simply plastic flow.

A yield value is manifested only by two-phase systems such as foams, emulsions, slurries, or pigment dispersions where bubbles or particles are dispersed in a continuous phase. It is postulated that the mutual attraction among the entities of the discontinuous phase sets up a weak yet rigid structure that blocks flow, leading to a yield value.

Apparent Viscosity

The term *apparent viscosity* is used to indicate that the viscosity is that of a non-Newtonian liquid. The adjective *apparent* is not meant to imply that the measured viscosity is an illusory value but rather that the viscosity pertains to only one shear rate condition. Since a non-Newtonian coating has as many apparent viscosities as there are shear rates, it becomes obvious that only a viscosity profile is capable of expressing this wide-ranging and varying viscosity behavior.

Thixotropy

The viscosity relationships shown in Figs. 1-3*a* and *b* are either straight lines or curves. Hence for any given shear stress (corresponding to a point on the viscosity profile curve) there is only one associated shear rate and viscosity. Presumably no difficulty should be encountered in reproducing this set of measurements. One might assume that it would make no difference whether the test liquid was, or was not, stirred vigorously before the test, or whether the viscosity measurements were made in a descending or an ascending order to establish the viscosity profile. However, to add another rheological complication, flow behavior may very often be dependent on just such variables, especially at low shear rates. Both the history of the system before testing and the exact pattern of testing can be of critical importance. This is not always true, but with coatings and inks it is so more often than not.

The word *thixotropy*, derived from two Greek words, means literally "change by touch." Flow behavior in which the viscosity is reduced by touch (agitation, stirring) is called thixotropic. Thixotropy in coating systems is generally due to the breakdown of some loosely knit structure that is built up during periods of rest (as during storage conditions) and torn down during periods of stress.

Thixotropic structure is commonly deliberately built into a coating system by introducing a paint additive that imparts a thixotropic effect. The thixotropy so produced is a valuable asset, since the lower viscosity at high shear rates (during application) facilitates paint flow and ease of application, whereas the higher viscosity at low shear rates (before and after application) prevents settling and sagging.

The technology of thixotropic behavior is quite complex. How can it be measured when, just by touching, rheological conditions are changed? One approach to measuring the amount of thixotropy present is to construct a so-called thixotropic loop, as shown in Fig. 1-3*c*. Such a loop is obtained by imposing an ascending series of shear stresses on the test coating and measuring its rheological properties. The process is then reversed, usually at once or after a brief holding period, and a descending shear stress series is imposed to provide a second set of rheological data. A plot of this overall information (log viscosity vs. log shear rate) gives the required thixotropic loop, which lies between the ascending and the descending data. The area enclosed by this loop as obtained by this arbitrary procedure is taken as an index to the thixotropy resident in the test coating.

It is always somewhat disconcerting to take a viscosity reading under a constant shear rate and find that the viscosity drifts downward under this steady-shear-rate condition. With time a minimum viscosity value is observed. On cessation of the test and given a rest period, the structural viscosity rebuilds, although the recovery may never be complete, even over prolonged periods of time (days). This type of flow behavior is typical of thixotropic systems.

Rheopexy

Rheopexy is the opposite of thixotropy in that under a steady shear rate the viscosity increases and approaches some maximum value. Rheopexy has been observed in some dilatant systems. However, rheopectic behavior is of no special concern to the paint industry and is mentioned solely for its academic interest.

The relationships of the four terms used to describe these types of non-Newtonian flow are outlined in Table 1-2.

LAMINAR VERSUS TURBULENT FLOW

When the shear stress producing Newtonian flow is sufficiently low, the layers of liquid slide over each other in an orderly fashion to give the *laminar* flow schematically depicted in Fig. 1-1. This Newtonian flow is also called *viscous* or *streamline* flow. If the shear rate is gradually raised, however, and if the in-

Table 1-2 Relationships of Four Terms Used to Describe Non-Newtonian Behavior

Imposed Condition	Observed Change in Viscosity: Decrease	Observed Change in Viscosity: Increase
Increased shear rate	Pseudoplastic (shear thinning)	Dilatant (shear thickening)
Increased shearing time (same shear rate)	Thixotropic	Rheopectic

crease is sufficient, a critical point is reached where the flow suddenly becomes chaotic. The former orderly flow is disrupted, and in its place there appears a swirling chaos of eddies and vortices that is referred to as *turbulent* flow. Such turbulent flow offers much greater resistance to flow than does viscous Newtonian flow, as shown in Fig. 1-3*d*. The onset of this turbulence can be fairly well predicted for many situations, as will be discussed in subsequent chapters.

By definition, shear rate is proportional to shear stress for Newtonian flow (constant viscosity). However, for turbulent flow the shear rate becomes approximately proportional to the square root of the shear stress. Hence, when a Newtonian liquid transfers from the viscous to the turbulent regime, it no longer exhibits a fixed or Newtonian viscosity but rather manifests an increasing viscosity with increasing shear stress (or shear rate). Turbulence should not be confused with rheopexy, where the viscosity increase takes place under conditions of a constant shear rate.

GENERAL COMMENT

Even a cursory reading of this chapter reveals that flow phenomena are exceedingly complex. The task of the rheologist—to understand flow and reduce its vagaries to concise mathematical expressions—has not been easy. In this chapter several major flow patterns have been briefly discussed and characterized.

Figure 1-4 represents an effort to relate roughly several more common terms describing flow behavior. Because such terms are highly subjective in nature, their placement on the chart should be considered as highly approximate. Also noted on the chart are liquid systems that illustrate the particular flow behavior, also at approximate locations.

The enormous span of values for viscosity are all of interest to the coatings or ink technologist. In Problem 1-1 it was assumed that the coating viscosity was 2.0 poises under brush application conditions. However, under settling or sagging conditions, the viscosity may well be several hundred times this value or even

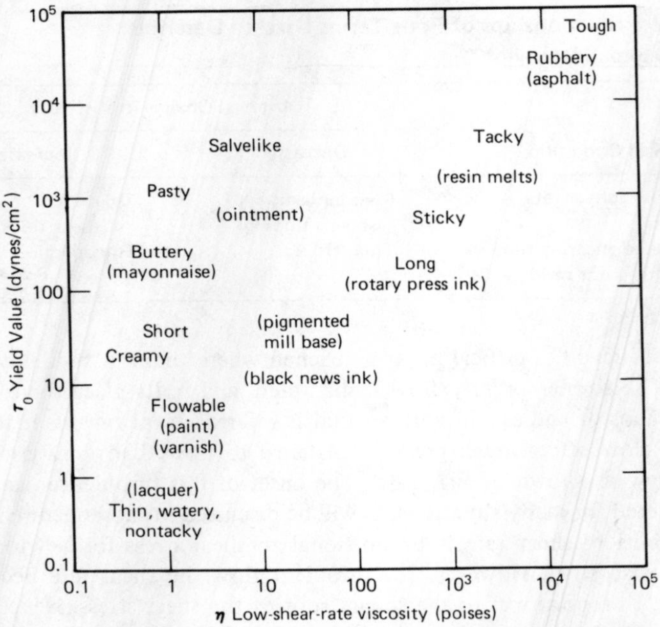

Fig. 1-4 Some common terms used to describe flow behavior related to yield value and low-shear-rate viscosity. The posting of terms is highly approximate, as are the systems used to illustrate the flow behavior.

Table 1-3 Approximate Shear Rate Ranges for Several Common Coating Operations

	Range (sec^{-1})							
Operation	10^{-2}	10^{-1}	10^0	10^1	10^2	10^3	10^4	10^5
Sagging	←	→						
Leveling	←	→						
Dipping				←→				
Flow coating				←→				
Pumping				←	→			
Mixing				←	→			
Dispersion						←	—	→
Spraying						←	—	→
Roller coating						←	—	→
Brushing							←→	

more. Shear rate values for different coating operations are also wide ranging, as indicated in Table 1-3.

In view of the bewildering diversity in coating flow behavior and the wide range of values assumed by the rheological variables (viscosity, shear stress, shear rate), it is understandable that, in the past, paint and ink rheology has developed largely along practical rather than theoretical lines. Today this state is rapidly changing, and the fundamental factors affecting the flow of coatings are now fairly well established.

BIBLIOGRAPHY

Brush, S., "Theories of Liquid Viscosity," *Chem. Rev.*, **62**, 513 (1962).

Patton, Temple C., "Fundamentals of Paint Rheology," *JPT*, **40**, No. 522, 301–307, July (1968).

Van Wazer, J. R. et al., "Viscosity and Flow Measurement," Wiley-Interscience, New York, (1963).

2 Viscometers: Theory and Reduction to Practice

THEORETICAL CONSIDERATIONS

Theoretically, two parallel plates (one movable and one stationary) with a liquid inserted in between, as depicted schematically in the model situation of Fig. 1-1, should serve as an ideal device for measuring viscosity. Practically, the arrangement is difficult to devise. Two problems immediately come to mind: complete confinement of the liquid to prevent it from spilling out the sides, and a practical method for continually moving the top plate further and further away from the bottom stationary plate. Even so, this configuration has been adapted, in modified forms, to the construction of commercial viscometers (band, concentric cylinder, and cone/plate types). But the ingenuity of man does not stop there. Just about every type of flow that can be conceived of has been made the basis for the construction of a viscometer: a liquid flow in a tube (capillary viscometer), a ball bearing dropping through a liquid (falling sphere viscometer), or a liquid flowing out of a hole at the bottom of a container (orifice viscometer).

The purpose of this chapter is to inquire into the background of the viscometers commonly used in the coating and ink industries. In some cases the geometric design of the viscometer lends itself to the precise determination of absolute viscosity. In other designs the geometric configuration is such that only relative values are given, and these may be only casually related to viscosity.

For purposes of discussion, viscometers can be conveniently grouped into two major classes: those limited essentially to the measurement of viscosities of Newtonian systems (constant viscosity at all shear rates), and those suitable for

Table 2-1 Classification of Viscometers Used in the Coatings and Ink Industries Based on Accuracy of Measurement and Suitability for Newtonian and Non-Newtonian Systems

Accuracy of Measurement	Suitable Essentially for Newtonian Systems Only (Constant Viscosity)	Suitable for Both Newtonian and Non-Newtonian Systems (Either Constant or Changing Viscosity with Shear Rate Change)
Accurate	Capillary tube Falling sphere Concentric cylinders (wide gap clearance) Single cylinder	Band Concentric cylinders (narrow gap clearance) 1. Radial motion 2. Axial motion Cone/Plate
Approximate	Capillary orifice Bubble Rotating disk	
Casually related	Hole orifice(s) Paddle Mixers	

the measurement of viscosities of both Newtonian and non-Newtonian systems (see listing in Table 2-1).

Among the first (constant viscosity) type is the capillary viscometer, which is capable of measuring the viscosity of a Newtonian system to a high degree of accuracy. However, all too often this viscometer is made to undertake the task of measuring the apparent viscosities of a non-Newtonian system, resulting in data of dubious or useless value. A mathematically inclined rheologist might object to the viscometer limitation specified above on the grounds that any non-Newtonian flow can be expressed by a power series expression, and that this in turn can be used to derive complex equations that express the flow taking place at different points in the capillary viscometer. However, the formidable mathematics required for this non-Newtonian programming and the sophisticated computer assistance that would be almost essential to make it technically feasible render this approach of borderline interest to the practicing engineer. Hence, from a pragmatic standpoint, the proposed breakdown into two major viscometer classes, as listed in Table 2-1, is believed to be reasonable.

The second class of viscometer (suitable for both constant and variable viscosity situations) embraces those types that are capable of directly measuring apparent viscosities at a number of specific shear rates or at least within very narrowly defined shear rate ranges. These viscometers are eminently suitable for measuring the viscosities of either Newtonian or non-Newtonian systems and hence are ideally equipped for establishing a viscosity profile over a wide shear rate range. Such viscometers include the band, narrow gap concentric cylinder, and cone/plate types, listed in Table 2-1.

Except where noted, the theoretical expressions that are derived in this chapter are based on the assumption of Newtonian flow.

From the voluminous body of literature dealing with viscometry, three reference sources (1-3) are especially recommended for further study of this subject.

Although the following discussion is fairly comprehensive, it is not exhaustive, since no attempt has been made to cover all makes of viscometers, nor are all models featured by any one supplier or manufacturer discussed. Selection has been more or less arbitrary with the idea of presenting in some depth representative viscometer instrumentations that are closely allied to the coatings and ink industries.

CAPILLARY VISCOMETER (4)

This extremely accurate method for measuring viscosity was experimentally discovered and reported in great detail in 1842 by Poiseuille. Later the mathematical expression for fluid flow in capillaries (Poiseuille's law) was derived theoretically.

Theory

Consider a fine-bore uniform cylindrical tube (a capillary) of inside radius R. Let a Newtonian liquid be maintained in uniform viscous flow through this capillary by means of a pressure drop p acting along length L of the tube (see Fig. 2-1). Consider a cylindrical shell of the liquid, concentric with the tube, having radius r, thickness dr, and length L. From the symmetrical arrangement of this thin shell, it is evident that each particle of liquid at the outer surface of the shell moves with velocity v, and each particle at its inner surface with a differentially greater velocity $v + dv$. Hence the velocity gradient (shear rate) across the shell thickness dr is $-dv/dr$. The minus sign indicates that the velocity decreases with an increase of radius.

The force F pushing on the cylinder of liquid enclosed by the shell equals the pressure drop p acting along the shell length multiplied by the cross-sectional area of the cylindrical shell πr^2. This force is distributed as a shearing stress over the inner surface of the shell, an area equal to $2\pi rL$. The shearing stress τ along the inner shell surface is then given by Eq. 1.

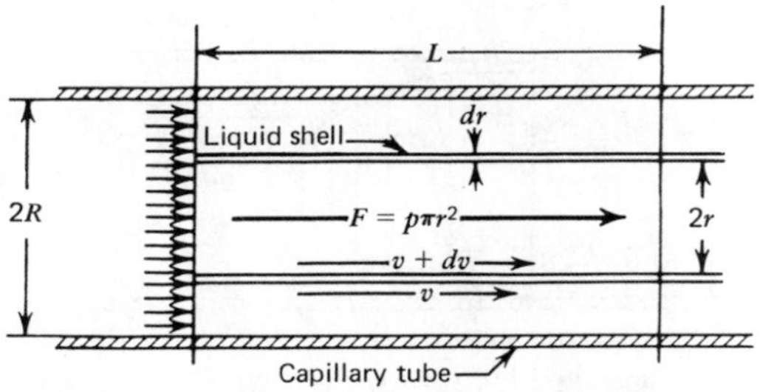

Fig. 2-1 Schematic flow pattern for a Newtonian liquid in viscous flow through a cylindrical capillary tube.

$$\tau = \frac{p\pi r^2}{2\pi r L} = \frac{pr}{2L} \tag{1}$$

By definition, viscosity η is the ratio of shear stress to shear rate. Hence

$$\eta = \frac{pr/2L}{-dv/dr}$$

Solving for dv results in Eq. 2.

$$dv = \frac{-pr}{2\eta L}\, dr \tag{2}$$

Equation 2 gives the differential velocity across a thickness of liquid dr at a radial distance r from the axis of the capillary tube. Integrating this expression results in Eq. 3.

$$v = \frac{-pr^2}{4\eta L} + K \tag{3}$$

where K = the constant of integration.

Since $v = 0$ when $r = R$, the constant of integration $K = pR^2/4\eta L$. Hence the velocity of flow at any radial distance r is given by Eq. 4.

$$v = \frac{p}{4\eta L}(R^2 - r^2) \tag{4}$$

To obtain the rate of volume flow through the tube, it is necessary to sum up the flow contributions of all the concentric shells in the tube ranging in radius from $r = 0$ to $r = R$. This can be done by noting that the volume of liquid that flows across a given cross section of the tube in unit time has the shape of a paraboloid of revolution, as shown in Fig. 2-2. For purposes of summation (integration), consider the paraboloid of revolution as being cut crosswise into

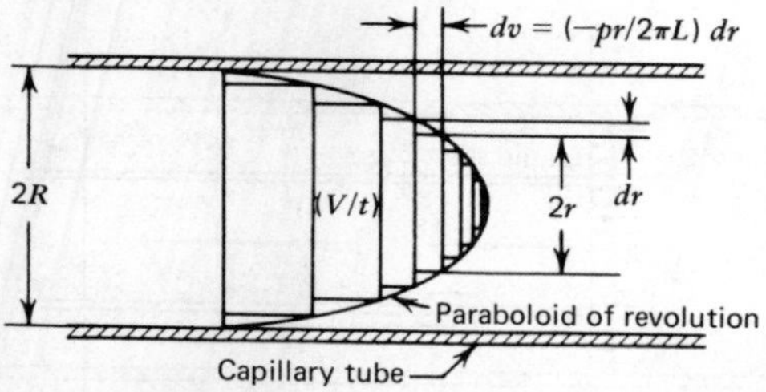

Fig. 2-2 Schematic diagram illustrating the variable flow rate across a cylindrical capillary tube for a Newtonian liquid in viscous flow.

thin circular disks, each of thickness dv and area πr^2. The volume of this paraboloid of revolution is then given by Eq. 5.

$$\frac{V}{t} = \int_{r=0}^{r=R} \pi r^2 \, dv \tag{5}$$

Substituting the value for dv from Eq. 2 in Eq. 5 gives Eq. 6.

$$\frac{V}{t} = \pi \int_{r=0}^{r=R} \frac{r^2 pr}{2\eta L} \, dr = \frac{\pi p}{2\eta L} \int_{r=0}^{r=R} r^3 \, dr \tag{6}$$

Integrating Eq. 6 between the given limits results in Eq. 7 for the rate of volume flow through a capillary tube.

$$\frac{V}{t} = \frac{\pi p R^4}{8\eta L} \tag{7}$$

Solving explicitly for viscosity η gives Eq. 8.

$$\eta = \frac{\pi p R^4 t}{8LV} \tag{8}$$

Kinetic Energy Correction Factor. In deriving Poiseuille's law, the assumption was made that all the energy expended went entirely into overcoming the viscous resistance of the liquid. In many industrial viscometers, however, an appreciable part of the energy input may be diverted into getting the fluid into motion through the capillary, that is, into kinetic energy. Hence a kinetic energy (KE) correction factor is frequently a necessary appendage of any equation used for converting time in sec (for a given volume of flow through a capillary viscometer) into poises. This correction is especially important in the case of low-viscosity liquids with high flow rates.

The KE correction factor can be computed by deriving an equation for the kinetic energy of the fluid passing through a cross section of the viscometer capillary in unit time. Since KE = $\frac{1}{2}mv^2$, where m is the liquid mass, the expression for the KE of the liquid that flows through the capillary tube of a viscometer per unit of time is given by Eq. 9.

$$\mathrm{KE} = \int_{r=0}^{r=R} \underbrace{\overbrace{2\pi r\, dr\, v}^{V}\,\rho}_{m} \frac{v^2}{2} = \pi\rho \int_{r=0}^{r=R} r\, dr v^3 \tag{9}$$

But v at any radial distance r from the axis of the capillary tube has already been determined as equal to $v = (p/4\eta L)(R^2 - r^2)$ from Eq. 4.

Note that, in Eq. 4, p represents the portion of the pressure devoted solely to overcoming the frictional resistance of the liquid. Substituting this value of v in the kinetic energy equation, Eq. 9, gives Eq. 10.

$$\mathrm{KE} = \frac{\pi\rho p^3}{64\eta^3 L^3} \int_{r=R}^{r=R} (R^2 - r^2)^3 r\, dr$$

$$= \frac{\pi\rho p^3}{64\eta^3 L^3} \int_{r=0}^{r=R} (R^6 r - 3R^4 r^3 + 3R^2 r^5 - r^7)\, dr \tag{10}$$

Integrating Eq. 10 gives Eq. 11.

$$\mathrm{KE} = \frac{\pi\rho p^3}{64\eta^3 L^3} \Big|_{r=0}^{r=R} \left[\frac{1}{2}R^6 r^2 - \frac{3}{4}R^4 r^4 + \frac{3}{6}R^2 r^6 - \frac{r^8}{8}\right] \tag{11}$$

When $r = R$, Eq. 11 reduces to Eq. 12 for the kinetic energy.

$$\mathrm{KE} = \frac{\pi\rho p^3}{64\eta^3 L^3} \frac{R^8}{8} \tag{12}$$

Now the average velocity u of the liquid flowing through the capillary tube must equal the rate of volume flow divided by the capillary cross section. Since the rate of volume flow has previously been determined from Eq. 7 as $V/t = \pi p R^4/8\eta L$, the average velocity u can be expressed as Eq. 13.

$$u = \frac{V/t}{\pi R^2} = \frac{\pi p R^4}{8\eta L \pi R^2} = \frac{pR^2}{8\eta L} \tag{13}$$

The cube of this equation is used for substituting in the work that follows. Cubing and rearranging results in Eq. 14 for p^3.

$$p^3 = \frac{8^3\eta^3L^3u^3}{R^6} \tag{14}$$

Substituting this value for p^3 in the integrated KE equation (Eq. 12) gives a simplified expression for the rate of kinetic energy input in terms of average fluid velocity u.

$$\text{KE} = \pi\rho R^2u^3 \tag{15}$$

This kinetic energy is generated by the action of p_{KE}, a portion of the total pressure acting on the volume πR^2u of the liquid flowing through the capillary. From this standpoint the energy input is equal to $\text{KE} = p_{\text{KE}}\pi R^2u$. But this energy input for KE must also be equal to the KE computed by Eq. 15, leading to Eq. 16.

$$p_{\text{KE}}\,\pi R^2u = \text{KE} = \pi\rho R^2u^3 \tag{16}$$

Let p_T be the total pressure required to maintain the flow of liquid through the capillary. Then the total energy input is equal to $p_T\pi R^2u$. The energy input that goes into overcoming frictional resistance is the difference between the total and KE inputs.

$$p\pi R^2u = \underbrace{p_T\pi R^2u}_{\substack{\text{(total}\\ \text{energy)}}} - \underbrace{\pi\rho R^2u^3}_{\substack{\text{(kinetic}\\ \text{energy)}}}$$

$$p = p_T - \rho u^2 \tag{17}$$

Equation 17 tells us that, if a true viscosity value is to be obtained from Poiseuille's equation, the imposed total pressure p_T must be diminished by the correction factor ρu^2 to compensate for the KE input. Introducing this correction in Eq. 7 (the equation giving the rate of volume flow) results in Eq. 18.

$$\frac{V}{t} = \frac{\pi(p_T - \rho u^2)R^4}{8\eta L} \tag{18}$$

Solving explicitly for V and replacing u by its equivalent $V/\pi R^2t$ gives Eq. 19.

$$V = \frac{\pi p_T R^4 t}{8\eta L} - \frac{\rho V^2}{8\pi\eta L t} \tag{19}$$

Finally, solving for viscosity η yields Eq. 20, an expression for determining the viscosity of a liquid that embodies a KE correction factor.

$$\eta = \frac{\pi p_T R^4 t}{8VL} - \frac{\rho V}{8\pi L t} \tag{20}$$

Reduction to Practice

In practice a pressure drop p acting along a capillary length L can be simply obtained by arranging for a hydrostatic head of the test liquid above the entrance to the capillary tube. Two such arrangements for the construction of the so-called kinematic type of capillary viscometer are shown in Fig. 2-3. In each case the effective pressure p acting along the capillary length L is equal to $h\rho g$, where h is the vertical height (difference in elevation) between the two free surfaces of the test liquid, ρ is the liquid density, and g is the gravitational constant of acceleration, 980 cm/sec^2.

$$p \text{ (hydrostatic pressure)} = h\rho g \tag{21}$$

As liquid flows through the capillary, it is obvious that the hydrostatic head becomes reduced. It can be shown that for purposes of calculation, a logarithmic mean can be used for the average effective hydrostatic head.

PROBLEM 2-1

Let the hydrostatic head h for the kinematic viscometer A of Fig. 2-3*a* be initially h_0. After volume V of the test liquid has flowed through the capillary, let the hydrostatic head be reduced to h_t. Neglecting KE effects, compute a value for the mean effective head h_m during this volume of flow.

Solution. The differential volume of flow dV through a capillary in differential time dt can be derived from Eq. 7.

$$\frac{dV}{dt} = \frac{\pi p R^4}{8\eta L} \tag{22}$$

For a kinematic viscometer $p = h\rho g$. Also, h is related to volume V by the expression $V = Ah$, where A is the uniform cross-sectional area of the container that is located above the capillary entrance. For a container of this cross-sectional area there can be derived the differential expression $dV = A\ dh$. Sub-

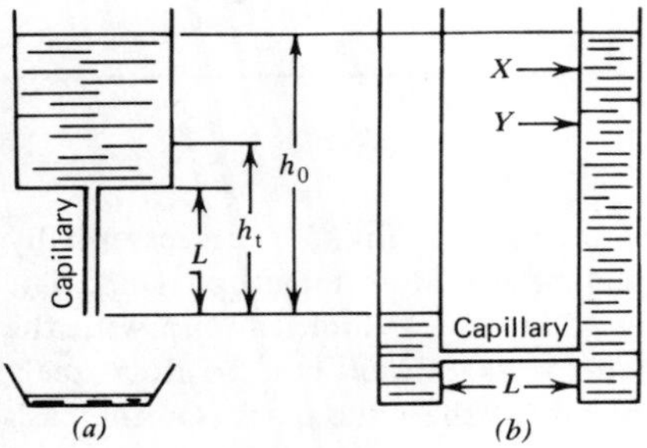

Fig. 2-3 Schematic diagram of hydrostatic capillary viscometers.

stituting this and the expression for p (as given in Eq. 21) in Eq. 22, we obtain Eq. 23.

$$\frac{dh}{h} = \frac{-\pi \rho g R^4}{8\eta L A}\, dt \tag{23}$$

Integrating this expression and computing a constant of integration from the fact that the hydrostatic head is h_0 when $t = 0$, we obtain Eq. 24, where h_t is the hydrostatic head at time t.

$$2.3 \log \frac{h_0}{h_t} = \frac{\pi g R^4}{8\eta L A}\, t \tag{24}$$

Returning for the moment to Eq. 7, note that the volume of flow V in time t is equal to $V = A(h_0 - h_t)$. Also, the mean effective pressure p_m during the given volume of flow V (for time period t) equals $h_m \rho g$, where h_m is the mean effective liquid head. If h_m could be conveniently obtained, it would facilitate the numerical solution of problems involving Eqs. 7 and 8. Substituting $A(h_0 - h_t)$ for V and $h_m \rho g$ for p in Eq. 8 and solving for the expression $(h_0 - h_t)/h_m$, we obtain Eq. 25.

$$\frac{h_0 - h_t}{h_m} = \frac{\pi \rho g R^4}{8\eta L A}\, t \tag{25}$$

Comparison of Eqs. 24 and 25 shows both right-hand terms to be identical. Hence the left-hand terms of these two equations must be equal. Setting them equal to each other and solving explicitly for h_m results in an expression for calculating the mean effective hydrostatic head.

$$h_m = \frac{h_0 - h_t}{2.3 \log (h_0/h_t)} \tag{26}$$

Saybolt Viscometer. Resort is made to Eq. 26 for determining the mean hydrostatic head for a Saybolt viscometer.

PROBLEM 2-2

The dimensions of the kinematic viscometer given in Fig. 2-4 are essentially those for a Saybolt viscometer as specified in ASTM test method D88. The viscosity of a test liquid is determined by filling the cylindrical cup with the liquid and then measuring the time required for exactly 60 cm^3 to flow freely through the capillary in the plug at the bottom of the viscometer. On the basis

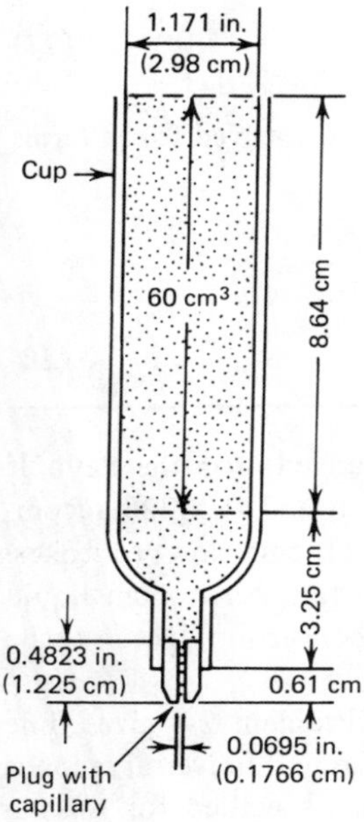

Fig. 2-4 Schematic diagram giving dimensions for the Saybolt Universal Viscometer (see ASTM D88).

of these data, compute a theoretical expression for converting time in sec (for 60 cm^3 of flow) to kinematic viscosity for this Saybolt instrument. Consider only the case of liquids of relatively high viscosity, where the flow rate is sufficiently low to make a kinetic energy correction unnecessary.

Solution. The following dimensions as read from the diagram of Fig. 2-4 are necessary for the solution of the problem:

$$h_0 = 12.5 \text{ cm}$$

$$h_t = 3.86 \text{ cm}$$

$$L = 1.225 \text{ cm}$$

$$R = 0.0883 \text{ cm}$$

The mean effective hydrostatic head for the problem conditions can be calculated from Eq. 26.

$$h_m = \frac{12.50 - 3.86}{2.3 \log (12.50/3.86)} = \frac{8.64}{2.3\,(0.51)} = 7.38 \text{ cm}$$

For kinematic viscometers, Eq. 8 can be more conveniently expressed in a slightly altered form as Eq. 27.

$$\frac{\eta}{\rho} = \nu = \frac{\pi h_m g R^4}{8LV}\, t \tag{27}$$

The kinematic viscosity for the Saybolt viscometer is now solved for in terms of its dimensional constants.

$$\nu = \frac{3.14 \cdot 7.38 \cdot 980 \cdot 0.0883^4}{8 \cdot 1.225 \cdot 60}\, t$$

$$\nu \text{ (stokes)} = 0.00235t \text{ (sec)} \tag{28}$$

This conversion factor of 0.00235 (theoretically derived) for the Saybolt viscometer compares favorably with factors of 0.00216 and 0.00226 recommended in the literature for converting time in sec (for a 60-cm^3 flow) to stokes. For actual measurements a calibration based on a standard Newtonian liquid should be made to determine precisely the specific conversion factor to be used with any given instrument.

The capillary orifice for the Saybolt viscometer of Problem 2-2 gives time values (for 60-cm^3 volume flow) expressed in terms of Saybolt universal seconds (SUS). An alternative and larger capillary orifice is also specified for this instrument; it results in 10 times the flow rate. The time period in this case (still for a 60-cm^3 flow volume) is expressed in terms of Saybolt Furol seconds. (Furol is a contraction of the term "fuel and road oils.")

PROBLEM 2-3

Calculate a kinetic energy correction factor for the conversion equation (Eq. 28) that is used for changing Saybolt seconds to stokes. The volume flow applying to a Saybolt viscometer is 60 cm^3.

Solution. By substituting $h_m \rho g$ for p_T in Eq. 20 and solving explicitly for kinematic viscosity ($\nu = \eta/\rho$), there results Eq. 29.

$$\frac{\eta}{\rho} = \nu = \frac{\pi h_m g R^4}{8VL}\, t - \frac{V}{8\pi L t} \tag{29}$$

The required correction factor is then computed from the term $V/8\pi L t$ of the above equation.

$$\frac{60}{8\pi \cdot 1.225t} = \frac{1.95}{t}$$

From this, the generalized conversion equation for converting Saybolt seconds to stokes is obtained.

$$\nu \text{ (stokes)} = 0.00235t \text{ (sec)} - \frac{1.95}{t} \text{ (sec)} \tag{30}$$

For the more viscous liquids (longer flow times t), the term $1.95/t$ becomes negligible in comparison with the term $0.00235t$. Again, Eq. 30 is based on theory. Hence specific numerical values for this equation should preferably be based on a calibration of any given Saybolt instrument, using liquids of known or certified viscosity.

ASTM test method D445-61 describes a common procedure for measuring viscosity, based on liquid flow through a capillary tube. Although the arrangements of the test apparatus parts vary widely among viscometers of this type, the essential construction is shown in the kinematic viscometer of Fig. 2-3.

The test procedure consists in accurately measuring the time required for a given volume of the test liquid (under its own hydrostatic head) to flow through the length of capillary L. In the typical capillary viscometer shown in Fig. 2-5, the hydrostatic head and the exact volume of flow are gauged by the markings above and below the bulb container located above the entrance to the capillary tube.

Although viscosity can be computed theoretically by using the geometric configuration of the capillary viscometer (as was done with the Saybolt viscometer of Problem 2-2), for routine industrial work viscosities are invariably calculated with a so-called viscometer constant. This constant is established through

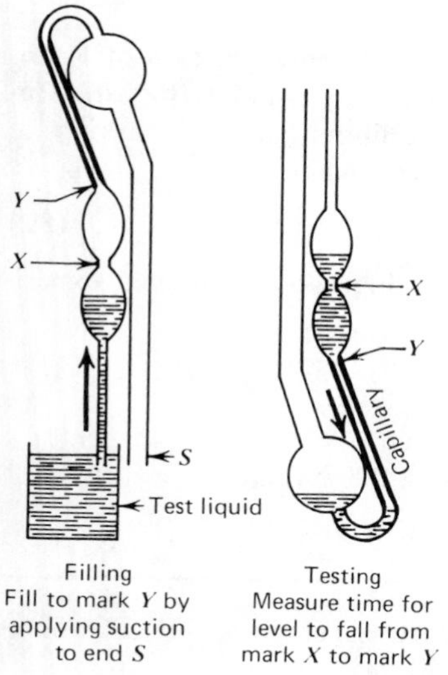

Fig. 2-5 Cannon-Fenske capillary viscometer (see ASTM D445-61).

the use of a liquid of known or certified viscosity. These ideas are more fully developed in Problem 2-4.

PROBLEM 2-4

Distilled water at 68 F is introduced up to the Y marking of the capillary viscometer of Fig. 2-5 by applying suction to end S. The viscometer is then turned upright to a vertical position. It is found that a time period of 500 sec is required for the water level to fall from the X to the Y marking. The viscometer is emptied and cleaned, and the procedure exactly repeated with butyl acetate, at 68 F. The time required for the butyl acetate level to drop from the X to the Y marking is 420 sec. Calculate the viscosity of the butyl acetate and the viscometer instrument constant.

Solution. The differential flow of liquid dV through a capillary in differential time dt is derived from Eq. 7.

$$\frac{dV}{dt} = \frac{\pi p R^4}{8\eta L}$$

Since the pressure causing the flow in the capillary viscometer is the hydrostatic head of the test liquid, p can be replaced by its equivalent $h\rho g$, where h is the head of liquid at any given time (the difference in elevation between the free surfaces of the test liquid). Introducing this substitution and solving explicitly for kinematic viscosity ν $(= \eta/\rho)$ gives Eq. 31.

$$\nu = \frac{\pi h g R^4}{8L} \frac{dt}{dV} \tag{31}$$

The term $\pi h g R^4/8L$ of Eq. 31 can be expressed as some function of V. Introducing this functional expression, we can write Eq. 31 as $f(V)\,dV = dt/\nu$. Integrating this equation between two fixed volume limits (such as markings X and Y), we obtain Eq. 32, where k is an instrument constant.

$$\nu = kt \tag{32}$$

The constant k for the capillary viscometer of Fig. 2-5 is obtained experimentally from the data given in the problem.

$$k = \frac{\nu}{t} = \frac{0.010\text{ (stoke)}}{500\text{ (sec)}} = 0.000020$$

The viscosity of the butyl acetate is calculated using this k value.

$$\nu = 0.000020 \cdot 420\text{ (sec)} = 0.0084\text{ stoke}$$

From this calculation it is seen that for capillary viscometers (Fig. 2-3*a* or *b*) flow times are proportional to kinematic (not absolute) viscosities.

$$\frac{\nu}{t}\text{ (liquid } A) = k = \frac{\nu}{t}\text{ (liquid } B)$$

For this reason they are called kinematic viscometers.

An inspection of Fig. 2-2 reveals capillary flow to vary enormously in its behavior, depending on the location of liquid in the capillary space. Thus at the center of the capillary (along the capillary axis) the shear stress is zero and the shear rate is also zero, but the velocity is at a maximum. Conversely, at the capillary wall the shear stress is at a maximum and so also is the shear rate, whereas the velocity is zero. In the following problem these extremes in the flow variables are calculated for a typical flow pattern in the Saybolt viscometer.

PROBLEM 2-5

The viscosity of a test liquid as measured by a Saybolt viscometer is 100 SUS. If the density of the test liquid is 1.0 g/cm^3, calculate the shear stress, shear rate, and velocity of the liquid at both the center of the capillary and the wall of the capillary.

Solution. Equations for shear stress, shear rate, and liquid velocity at a radial distance r from the axis of a capillary tube are given by Eqs. 1, 2, and 4, respectively.

$$\tau\text{ (shear stress)} = \frac{pr}{2L}$$

$$\text{D (shear rate)} = \frac{dv}{dr} = \frac{pr}{2\eta L}$$

$$v\text{ (velocity)} = \frac{p(R^2 - r^2)}{4\eta L}$$

In all three equations the quantity p can be replaced by its equivalent, $h\rho g$ (from Eq. 21). For the Saybolt viscometer the length of the capillary is 1.225 cm, r is a maximum (= 0.0883 cm) at the capillary wall, and the effective or mean head of liquid during the 60-cm^3 volume flow is 7.38 cm (see Problem 2-2). The kinematic viscosity of the test liquid in stokes can be calculated from the given 100-SUS viscosity value by the use of conversion Eq. 30.

$$\nu\text{ (test liquid)} = 0.00235\ (100) - \frac{1.95}{100} = 0.216\text{ stoke}$$

Since the fluid density is 1.0 g/cm^3, the absolute viscosity in poise ($\eta = \rho\nu$) is also numerically equal to 0.216. A full set of values has now been developed for substitution in the equations for shear stress, shear rate, and velocity.

For $r = 0$ (center of capillary):

$$\tau = 0$$

$$D = 0$$

$$v = \frac{7.38 \cdot 1.0 \cdot 980 \cdot 0.0883^2}{4 \cdot 0.216 \cdot 1.225} = 53.2 \text{ cm/sec}$$

For $r = 0.0883$ (at wall of capillary):

$$\tau = \frac{7.38 \cdot 1.0 \cdot 980 \cdot 0.0883}{2 \cdot 1.225} = 260 \text{ dynes/cm}^2$$

$$D = \frac{7.38 \cdot 1.0 \cdot 980 \cdot 0.0883}{2 \cdot 0.216 \cdot 1.225} = 1200 \text{ sec}^{-1}$$

$$v = 0$$

These computed values are summarized in Table 2-2.

The maximum velocity of 53.2 cm/sec as calculated above is more than twice the average velocity of 24.5 cm/sec for the liquid flow through the capillary.

ASTM Capillary Viscometers. The capillary viscometers specified by ASTM test method D445-61 (see Fig. 2-5) are manufactured to cover a complete range of viscosities. Thus the common Cannon-Fenske viscometer (one of 16 approved ASTM designs) is supplied in a number of sizes ranging from No. 25 (0.31 ± 0.02 mm inside capillary diameter; 0.004 to 0.016 stoke range) to No. 600 (4.00 ± 0.05 mm inside capillary diameter; 40 to 160 stoke range). The Ubbelohde viscometer, with an alternative design, covers a somewhat greater span of viscosities ranging from No. 0 (0.24 ± 0.01 mm inside capillary diameter; 0.003 to 0.010 stoke range) to No. 5 (6.7 ± 0.1 mm inside capillary diameter; 200 to 1000 stoke range).

The tremendous viscosity range covered by these viscometers is achieved solely by varying the capillary diameter (length and other dimensions remain

Table 2-2 Rheological Values Calculated from Problem 2-5 for Saybolt Viscometer

Parameter	Value	
	At Center of Capillary	At Wall of Capillary
Shear stress (dynes/cm^2)	0	260
Shear rate (sec^{-1})	0	1200
Velocity (cm/sec)	53.2	0
Viscosity (poise)	0.216	0.216

substantially constant). Inspection of Eq. 8 shows why this is possible: viscosity varies with the *fourth* power of the capillary radius. Hence a 10-fold increase in the capillary bore dimension corresponds to a 10,000-fold increase in the viscosity value that can be measured.

PROBLEM 2-6

A set of capillary viscometer tubes is designed to cover a range of viscosities from 0.005 to 200 stokes. What variation in capillary radii can be expected for the set?

Solution. The ratio of the limiting viscosities for the specified set of capillary tubes is 40,000 (= 200/0.005). In turn, the ratio of the limiting radii for the tubes will be approximately 14.1 (= $\sqrt[4]{40,000}$).

Table 2-3 Viscosity Standards[a]; Approximate Kinematic Viscosity in Stokes (approximate absolute viscosity in poises).

Coding for Viscosity Standard[b]	Temperature					
	−40 C −40 F	20 C 68 F	25 C 77 F	40 C 104 F	50 C 122 F	100 C 212 F
S3 (D)	0.80	0.046	0.040	0.029	—	0.012
	—	(0.039)	(0.033)	(0.024)	—	(0.009)
S6 (H, I)	—	0.11	0.089	0.057	—	0.018
	—	(0.094)	(0.076)	(0.048)	—	(0.014)
S20 (J, K)	—	0.44	0.34	0.18	—	0.039
	—	(0.38)	(0.29)	(0.15)	—	(0.031)
S60 (L, SB)	—	1.70	1.20	0.54	—	0.072
	—	(1.50)	(1.10)	(0.46)	—	(0.059)
S200 (M)	—	6.40	4.50	1.80	—	0.17
	—	(5.60)	(3.90)	(1.50)	—	(0.14)
S600 (N, SF)	—	24.0	16.0	5.2	2.8	0.32
	—	(21.0)	(14.0)	(4.6)	(2.4)	(0.28)
S2000	—	87.0	56.0	17.0	—	0.75
	—	(76.0)	(49.0)	(15.0)	—	(0.62)
S8000 (OB)	—	370.0	230.0	67.0	—	—
	—	(330.0)	(200.0)	(59.0)	—	—
S30000 (P)	—	—	810.0	230.0	110.0	—
	—	—	(720.0)	(200.0)	(96.0)	—

[a] Cannon Instrument Company, State College, PA 16801.
[b] The letter coding in parentheses refers to the discontinued NBS coded standards, which have been replaced by the equivalent listed S number series.

Liquids of known viscosity for use in calibrating viscometers can be either common liquids, like water or castor oil (for which viscosity values have been established and recorded in the literature), or liquids with certified viscosities supplied from a reliable laboratory. Table 2-3 lists a sampling of commercially available liquids having certified viscosities at specified temperatures (the National Bureau of Standards has discontinued supplying these standards). They are based on the National Bureau of Standards value of 0.01002 poise for water at 20 C. The viscosity standards are supplied with precise values of viscosity, kinematic viscosity, and density at the temperatures listed. Prices range from \$20 per pint sample for S3 through S2000 to \$30 per pint sample for S8000 and S30000.

CAPILLARY PLASTOMETERS

The capillary viscometers so far discussed have depended on a relatively small hydrostatic head to force the liquid through the capillary. However, high external pressures can also be employed to drive a liquid or polymer melt through a capillary bore. This can be accomplished by applying either a vacuum or a positive pressure to one end of the capillary tube. Such viscometers, sometimes referred to as plastometers, are necessarily more complex and expensive. Furthermore, although they provide positive and measurable rates of shear, the inherent geometry is such that a viscosity profile curve versus shear rate for a non-Newtonian system can probably never be obtained with any degree of certainty.

***Instron Capillary Rheometer* (5).** One of the capillary viscometers designed for studying the flow behavior of viscous materials (thermoplastics, polymer melts) is the Instron Capillary Rheometer (Model 3211), based on a design originating with the Monsanto Chemical Company. This apparatus provides a measured mechanical driving force that pushes a sample of material via an overhead plunger through a tungsten carbide or stainless steel capillary bore. Capillaries can be supplied from stock in diameters from 0.03 to 0.06 in. and in lengths from 1.0 to 4.0 in. A synchronous motor with six pushbutton speeds and a selection of change gears allows a choice of speeds through a ratio range from 1 to 333. From this, shear rates from $1.4 \cdot 10^4$ sec^{-1} (20 in./min with 0.03-in. diameter capillary) to $5.2 \cdot 10^{-1}$ sec^{-1} (0.006 in./min with 0.06-in. diameter capillary) can be scheduled. Loading up to 5000 lb (5.0 in./min speed) provides a maximum stress of $4.5 \cdot 10^7$ $dynes/cm^2$. This viscometer, which can be operated at temperatures from 40 to 499 C, can readily simulate shear rates for compression molding (1 to 10 sec^{-1}), milling and calendering (10 to 100 sec^{-1}), extrusion (100 to 1000 sec^{-1}), and injection molding (1000 to 10,000 sec^{-1}). The weight of the complete unit is 850 lb.

***Burrell-Severs Extrusion Rheometer* (6).** Another capillary viscometer (plastometer) designed for measuring the viscosity of very viscous materials is the Burrell-Severs Extrusion Rheometer, which uses gas pressure for driving the test material through a stainless steel bore (choice from 0.075- to 0.30-cm radius by 5.0-cm length). Two models are available, depending on the pressure and temperature requirements (Model A-120: room temperature, 0 to 100 psi, 0 to 10,000 poises, price ~\$360 as of 1977; Model A-250: 0 to 300 C, 0 to 500 psi, 0 to 10,000 poises, price ~\$3325 as of 1977).

For both the research investigation and routine laboratory testing of Newtonian systems, capillary viscometers based on a hydrostatic head provide accurate and useful information. For highly viscous systems (such as polymer melts) resort to external pressure becomes necessary. In any case, if the system is non-Newtonian in character, the results tend to be of marginal value except for purposes of manufacturing control or for establishing a rough assessment of the viscosity behavior of the non-Newtonian material.

ROTATING CONCENTRIC CYLINDER VISCOMETER

The rotating coaxial cylinder (or rotational) type of viscometer (see Fig. 2-6) can be thought of as an adaptation of the parallel plate arrangement of Fig. 1-1 in which the plates have been wrapped completely around an axis to give two cylindrically curved surfaces moving (rotating) relative to each other around a common axis. This practical arrangement lends itself to the establishment of a precise method for the measurement of viscosity. Unlike the capillary viscometer, in which the flow pattern varies enormously across the capillary cross section, a narrow gap rotational viscometer can be designed to give a flow pattern within a narrow range of flow variables. When evaluating non-Newtonian liquids, this feature represents an enormous advantage. In fact, the measurement of the viscosity of non-Newtonian liquids with a capillary viscometer is a questionable procedure.

Theory

The rotational viscometer consists essentially of two concentric cylinders with the space (annulus) between them filled with the liquid under investigation. One of the cylinders is made to rotate with a constant angular velocity ω. By measuring the torque M necessary to maintain this rotation and from the dimensions of the viscometer cylinders, a viscosity value for the test liquid can be calculated.

Consider a section of length L for two concentric cylinders disposed in this manner (see Fig. 2-6). Let a be the radius of the inner cylinder (it can be visualized as an axle), and b the inside radius of the outer cylinder (it can be visualized as a bearing). Let the annular space between these two cylindrical surfaces be filled with the test liquid of unknown viscosity η.

Consider the case where the inner cylinder (bob) is acted on by a torque M which rotates the inner cylinder about its axis with constant angular velocity ω_a (the outer cylinder or cup is held stationary). Consider a very thin cylindrical shell of the test liquid of radius r and thickness dr, which is concentric with the two cylinders and located between them. Let F equal the shearing force acting on the inner surface of this shell, which in terms of the applied torque is equal to M/r. This force acts over the entire face of the shell (of area $2\pi rL$), so that the shearing stress (shearing force per unit of area) is $F/2\pi rL$, or in terms of the applied torque is equal to $M/2\pi r^2 L$. The rate of shear D across the shell is equal to the differential change in velocity dv over the shell thickness dr, or $-dv/dr$. In terms of angular velocity this rate of shear is equal to $-r d\omega/dr$.

By definition, viscosity is the ratio of shear stress to shear rate. Setting up this ratio for the values derived above results in Eq. 33.

$$\eta = \frac{M/2\pi r^2 L}{-r\, d\omega/dr} = \frac{-M\, dr}{2\pi r^3 L\, d\omega} \tag{33}$$

Rearranging and solving explicitly for $d\omega$ gives Eq. 34.

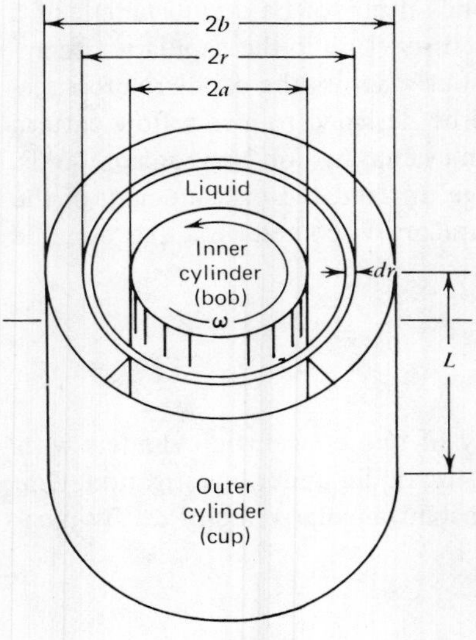

Fig. 2-6 Simplified schematic detail of a rotating concentric cylinder viscometer.

$$d\omega = \frac{-M}{2\pi L\eta} \frac{dr}{r^3} \tag{34}$$

On integrating this equation and establishing the constant of integration from the fact that $\omega = 0$ when $r = b$ (since the liquid next to the face of the outer cylinder is held stationary), there results Eq. 35 for the angular velocity at radius r.

$$\omega = \frac{M}{4\pi L\eta} \left(\frac{1}{r^2} - \frac{1}{b^2} \right) \tag{35}$$

Finally, when $r = a$, the angular velocity ω_a is the same as the angular velocity of the rotating inner cylinder with which the liquid is in contact. Hence, in terms of radii a and b, the angular velocity is given by Eq. 36.

$$\omega_a = \frac{M}{4\pi L\eta} \left(\frac{1}{a^2} - \frac{1}{b^2} \right) \tag{36}$$

Solving explicitly for viscosity results in Eq. 37.

$$\eta = \frac{M}{4\pi L\omega_a} \left(\frac{1}{a^2} - \frac{1}{b^2} \right) \tag{37}$$

For some calculations Eq. 37 is more conveniently expressed in the alternative form given by Eq. 38.

$$\eta = \frac{M}{4\pi L\omega_a} \frac{(b^2 - a^2)}{a^2 b^2} \tag{38}$$

As the gap between the bob and the cup narrows (radius a approaches radius b in value), the opposing faces of the viscometer become more nearly parallel. In turn the average shear stress between the two wall surfaces approaches the value $M/2\pi\,[(a + b)/2]^2 L$, and the shear rate across the gap approaches the value $v/(b - a)$, where v is the velocity of the rotating bob surface ($= \omega_a a$). Hence the test liquid viscosity can be expressed approximately by Eq. 39.

$$\eta = \frac{M}{4\pi L\omega_a} \frac{8(b - a)}{a(a + b)^2} \tag{39}$$

This approximate equation is applicable only for a narrow gap between bob and cup. The extent of the error introduced by Eq. 39 is illustrated by Problem 2-7.

Reduction to Practice

A viscosity determination with a rotational viscometer invariably requires a measurement of both rotational velocity and torque M. The viscometer design may call for either the inner or the outer cylinder to be rotated (the other cylinder being held stationary). The amount of torque imparted is commonly determined by measuring the compression (windup) of a helical spring or the twisting of a wire, either of which is arranged to oppose the impressed torque. Manufacturers' literature shows the ingenious ways in which these ideas have been adapted to the manufacture of industrial coaxial cylinder rotational viscometers.

PROBLEM 2-7

The radii for the bob (inner cylinder) and cup (outer cylinder) for an industrial rotational viscometer are 1.30 and 1.40 cm, respectively. When operating, the length of bob immersed in the test liquid is 5.0 cm (Fig. 2-7). If the torque necessary to maintain the bob rotation at 100 revolutions per minute (rpm) is 24,000 dyne-cm, calculate the viscosity of the test liquid.

Solution. The angular velocity must be expressed in radians/sec before substitution in Eqs. 38 and 39. The given rotational speed of 100 rpm is equal to 10.5 radians/sec (= $100 \cdot 2\pi/60$). This and the other given values are substituted in Eqs. 38 and 39 in turn.

$$\eta \text{ (exact)} = \frac{24{,}000}{4 \cdot 3.14 \cdot 5.0 \cdot 10.5} \frac{1.4^2 - 1.3^2}{1.3^2 \cdot 1.4^2}$$

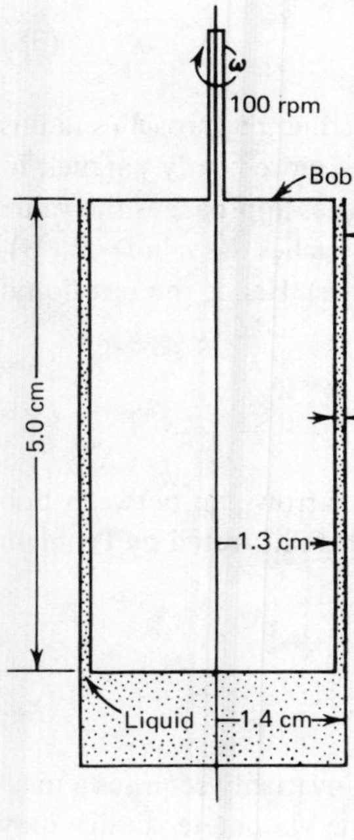

Fig. 2-7 Typical dimensions for a rotating concentric cylinder viscometer.

$$= \frac{24{,}000}{660} \frac{1.96 - 1.69}{1.69 \cdot 1.96}$$

$$= 2.96 \text{ poises}$$

$$\eta \text{ (approx.)} = \frac{24{,}000}{660} \frac{8(1.4 - 1.3)}{1.3 \cdot 2.7^2}$$

$$= 36.25 \frac{0.8}{1.3 \cdot 7.29} = 3.06 \text{ poises}$$

The viscosity given by the approximate Eq. 39 is 3.4% too high. However, if the bob and cup radii had been 1.39 and 1.40 cm, respectively, the error would have been reduced accordingly by about one-tenth (i.e., about 0.3%). A study of the derivation of Eqs. 38 and 39 serves to point up the similarity between the parallel plate arrangement of Fig. 1-1 and the coaxial cylinder arrangement of Fig. 2-6.

In the practical operation of rotational viscometers end effects must be taken into consideration. In certain designs the geometric arrangement may be such that the drag on the bob due to end effects may be neglected in comparison with the gap drag. In other designs the configuration of the bob end may be such that a hypothetical length can be added to the length proper to compensate for end influences. In still other rotational viscometers independently operated guard rings may be introduced to avoid the necessity for any end effect correction. It is also possible to arrange for different immersion depths and calculate the end drag (a constant) by extrapolation to a hypothetical zero immersion. Whatever the means of correction or compensation, the end treatment for any given instrument should be observed when it is called for in the calculations.

The outstanding advantage of the rotational coaxial cylinder viscometer (narrow gap clearance) over the capillary viscometer is its ability to provide a viscosity measurement based on a narrow band of shear stresses and shear rates. The span of such variables for a typical run is illustrated by the calculations of Problem 2-8.

PROBLEM 2-8

The viscometer described in Problem 2-7 is used to measure the viscosity of a test liquid. If a torque of 24,000 dyne-cm is required to produce a bob rotational velocity of 100 rpm, calculate the shear stress range, the shear rate range, and the velocity for the test liquid under these conditions.

Solution. Equations for shear stress, shear rate, and liquid velocity at radial distance r from the axis of a coaxial rotating cylinder viscometer (Eqs. 40, 41, and 42) are derived from expressions previously developed in this part of the text.

$$\tau = \frac{M}{2\pi r^2 L} \tag{40}$$

$$\text{D} = \frac{\omega_a 2a^2 b^2}{r^2(b^2 - a^2)} \tag{41}$$

$$v = \omega_a r \frac{a^2 b^2 - a^2 r^2}{r^2 b^2 - a^2 r^2} \tag{42}$$

The required range of values for τ, D, and v can be compuated by substituting the limiting values for r of 1.3 and 1.4 cm in turn in these three equations. The angular velocity ω_a for the bob in radians/sec is 10.5 (= $100 \cdot 2\pi/60$).

For $r = 1.3$ (bob radius):

$$\tau = \frac{24{,}000}{2 \cdot 3.14 \cdot 1.3^2 \cdot 5.0} = 452 \text{ dynes/cm}^2$$

$$\text{D} = \frac{10.5 \cdot 2 \cdot 1.3^2 \cdot 1.4^2}{1.3^2(1.4^2 - 1.3^2)} = 153 \text{ sec}^{-1}$$

$$v = (10.5 \cdot 1.3) \frac{1.3^2 1.4^2 - 1.3^2 1.3^2}{1.3^2 \cdot 1.4^2 - 1.3^2 \cdot 1.3^2}$$

$$= 13.7 \text{ cm/sec}$$

(This is also the bob surface velocity.)

For $r = 1.4$ (cup radius):

$$\tau = \frac{24{,}000}{2 \cdot 3.14 \cdot 1.4^2 \cdot 5.0} = 390 \text{ dynes/cm}^2$$

$$\text{D} = \frac{10.5 \cdot 2 \cdot 1.3^2 \cdot 1.4^2}{1.4^2(1.4^2 - 1.3^2)} = 132 \text{ sec}^{-1}$$

$$v = 0$$

Velocity $v = 0$ since the term $1.3^2 \cdot 1.4^2 - 1.3^2 \cdot 1.4^2$ in Eq. 42 is equal to zero. These computed values have been summarized in Table 2-4.

Let Eq. 38 for viscosity be rewritten as Eq. 43, where the dimensions for the viscometer bob and cup have been grouped together in a separate fraction.

$$\eta = \frac{M}{\omega_a} \frac{(b^2 - a^2)}{4\pi L a^2 b^2} \tag{43}$$

Table 2-4

Parameter	Value	
	At Bob Wall Surface	At Cup Wall Surface
Shear stress (dynes/cm^2)	452	390
Shear rate (sec^{-1})	153	132
Velocity (cm/sec)	13.7	0
Viscosity (poises)	2.96	2.96

Since the dimensions of a viscometer are not subject to change, the numerical value of this fraction $(b^2 - a^2)/4\pi La^2b^2$ can be considered as an instrument constant k for any given viscometer.

$$\eta = \frac{kM}{\omega_a} \tag{44}$$

PROBLEM 2-9

Calculate the instrument constant for the viscometer of Problem 2-7.

Solution. Substitute appropriate values for a, b, and L in the expression $(b^2 - a^2)/4\pi La^2b^2$ of Eq. 43.

$$k = \frac{1.4^2 - 1.3^2}{4\pi \cdot 5.0 \cdot 1.3^2 \cdot 1.4^2} = 0.0013$$

PROBLEM 2-10

A viscometer designed for the measurement of viscosity at high rates of shear (the Brushometer) consists essentially of a cylindrical shaft (0.500-in. diameter) rotating at 1800 rpm within a cylindrical collar as shown in Fig. 2-8 (7). The length of contact L between shaft and collar is 1.00 in. and the collar is restrained from rotating with the shaft by a helical spring. Calculate the gap seperation which must be provided to give a shear rate of 20,000 sec^{-1} and the torque exerted on the collar when a liquid having a viscosity of 4.0 poises is introduced between the rotating shaft and the stationary collar.

Solution. The gap separation required to provide a shear rate of 20,000 sec^{-1} can be computed by substituting in Eq. 41 the data given in the problem. The rotational velocity ω in radians/sec is 189 (= $1800 \cdot 2\pi/60$). For purposes of calculation let $r = a$, thus simplifying Eq. 41 to $D = 2\omega b^2/(b^2 - a^2)$.

$$20{,}000 = \frac{2 \cdot 189b^2}{b^2 - 0.635^2}; \qquad b = 0.641$$

The necessary gap spacing is then 0.006 cm (= 0.641 - 0.635).

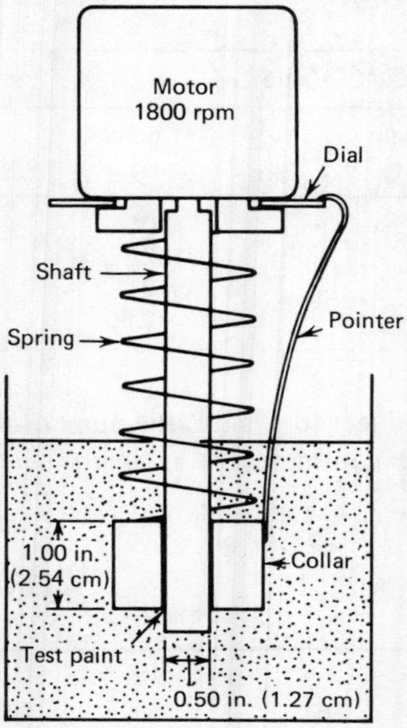

Fig. 2-8 Schematic diagram illustrating the arrangement of parts for the Brushometer viscometer.

In view of this low value, the gap distance could alternatively have been computed by simply dividing the peripheral velocity of the shaft by the required rate of shear (from the relation $\mathrm{D} = v/x$ or $v/\mathrm{D} = x = b - a$).

$$\frac{189 \cdot 0.635}{20{,}000} = 0.006 \text{ cm (gap spacing)}$$

The torque M exerted on the collar is computed from Eq. 45, where $2\pi rL$ is the surface area involved.

$$M = \tau(2\pi rL)r = \eta \mathrm{D}(2\pi rL)r \qquad (45)$$

$$M = 4.0 \cdot 20{,}000(2 \cdot 3.14 \cdot 0.635 \cdot 2.54) \cdot 0.635 = 515{,}000 \text{ dyne-cm}$$

Extension of Theory to Plastic Viscosity. The shear stress acting on a liquid in a rotational viscometer is given by Eq. 40. For a Newtonian liquid this entire stress is expended in overcoming viscous resistance. However, with a Bingham liquid (a liquid exhibiting plastic flow with a definite yield value), part of the shear stress must be diverted to overcoming yield value forces, the remainder being expended in inducing fluid flow. If τ is the total applied stress and τ_0 is the opposing yield value stress, then $\tau - \tau_0$ is the leftover shear stress which is available for overcoming viscous resistance. As before, the total applied stress is equal to $M/2\pi r^2 L$. Hence the stress τ_v available for inducing fluid flow is given by Eq. 46.

$$\tau_v = \frac{M}{2\pi r^2 L} - \tau_0 \tag{46}$$

The shear rate (as for a Newtonian liquid) is equal to $r\,d\omega/dr$. Hence the plastic viscosity η' for a Bingham liquid can now be expressed by Eq. 47.

$$\eta' = \frac{\tau - \tau_0}{r\,d\omega/dr} \tag{47}$$

Replacing τ by its equivalent expression, $M/2\pi r^2 L$, and solving explicitly for $d\omega$ results in Eq. 48.

$$d\omega = \left(\frac{M}{\eta' \cdot 2\pi r^3 L} - \frac{\tau_0}{\eta' r}\right) dr \tag{48}$$

Integrating this expression results in Eq. 49.

$$\omega = \frac{-M}{\eta' \cdot 4\pi r^2 L} - \frac{\tau_0 \ln r}{\eta'} + K \tag{49}$$

Consider a rotational viscometer where the outer cylinder is rotated. For these conditions $\omega = 0$ when $r = a$. Using this fact, we can establish a constant of integration K, leading to Eq. 50.

$$\omega\eta' = \frac{M(r^2 - a^2)}{r^2 a^2 \cdot 4\pi L} - \left(\tau_0 \ln \frac{r}{a}\right) \tag{50}$$

This equation is valid (experimentally) only when the liquid is in laminar flow at all points (i.e., the yield value is overcome throughout the liquid). Assuming this situation, we find that the angular velocity at radius b (the outer cylinder velocity) is given by Eq. 51.

$$\omega_b\eta' = \frac{M(b^2 - a^2)}{b^2 a^2 \cdot 4\pi L} - \left(\tau_0 \ln \frac{b}{a}\right) \tag{51}$$

When the outer cylinder velocity is reduced to zero, Eq. 51 simplifies theoretically to Eq. 52, where M_0 is the applied torque value.

$$\tau_0 = M_0 \frac{b^2 - a^2}{b^2 a^2 \cdot 4\pi L \cdot \ln (b/a)} \tag{52}$$

Substituting this value for τ_0 in Eq. 51 and solving for η' gives Eq. 53.

$$\eta' = \frac{M - M_0}{\omega_b} \frac{b^2 - a^2}{b^2 a^2 4\pi L} \tag{53}$$

Comparison of Eq. 53 with Eq. 54 (an alternative way of writing Eq. 38 for Newtonian flow) reveals these equations to be quite alike.

$$\eta = \frac{M}{\omega_a} \frac{b^2 - a^2}{b^2 a^2 \cdot 4\pi L} \tag{54}$$

They differ only in the torque value to be used, and even here they are essentially similar; the torque value to be entered into each equation is the portion expended in overcoming viscous resistance (the total torque in the case of a Newtonian liquid; part of the torque in the case of a Bingham liquid).

As can be seen from a study of Eq. 53, a plot of applied torque M versus angular velocity ω_b will give a straight line intersecting the torque scale at $M = M_0$. Although this intercept value is a theoretical point, it is valid for calculating a true yield point value from Eq. 52. In extrapolating the straight line back to the torque axis, only experimental points in the middle to upper range should be used (i.e., points above a torque value of $M = 2\pi b^2 L \tau_0$).

PROBLEM 2-11

A TiO_2 paste dispersion is introduced into a rotational viscometer (rotating cup, stationary bob) having the following dimensions: bob radius $a = 1.8$ cm, bob length $L = 5.0$ cm, cup radius $b = 2.0$ cm. The torques (M values) required to produce angular velocities of 5.0, 10.0, and 20.0 radians/sec are found to be 391,500, 700,000, and 1,317,000 dyne-cm, respectively. If end effects can be considered negligible, calculate the plastic viscosity and the yield value for this TiO_2 dispersion.

Solution. A plot of the given M versus ω values results in a straight line intersecting the torque axis at a value of 83,000 dyne-cm. This is the value of M_0, the theoretical torque for zero angular velocity. This torque value could also have been obtained by substituting two sets of M and ω values in Eq. 53 and solving simultaneously for M_0.

The yield value and plastic viscosity are now obtained by substituting in Eqs. 52 and 53, M_0 and the other given values.

$$\tau_0 = \frac{83{,}000(2.0^2 - 1.8^2)}{2.0^2 \cdot 1.8^2 \cdot 4 \cdot 3.14 \cdot 5.0 \cdot 2.3 \log(2.0/1.8)}$$

$$= 740 \text{ dynes/cm}^2 \text{ (yield value)}$$

$$\eta' = \frac{700{,}000 - 83{,}000}{10} \; \frac{2.0^2 - 1.8^2}{2.0^2 \cdot 1.8^2 \cdot 4 \cdot 3.14 \cdot 5.0}$$

$$= 58 \text{ poises (plastic viscosity)}$$

The Brushometer viscometer considered in Problem 2-10 (see also Fig. 2-8) is a relatively inexpensive, special purpose type designed to cover a shear rate range corresponding to paint brushing and roller coating conditions. Many other narrow gap, concentric cylinder viscometers, both special purpose and general purpose, are commercially available, ranging from relatively simple designs to highly sophisticated instrumentations.

Narrow versus Wide Gap Clearances. Although the borderline region between narrow and wide gap clearances has never been clearly defined, it would seem reasonable that, to qualify for narrow gap clearance, the radii of the two cylinders (outer bob face, inner cup face) should be such as to give shear rates that deviate by less than ±10% from an average value. The plus or minus maximum deviation from an average shear rate can be calculated from the expression $(r_b^2 - r_a^2)/(r_b^2 + r_a^2)$, where r_b and r_a refer to the outer and inner radii involved (see Fig. 2-6). It can be shown that a 10% deviation corresponds to a radii ratio (r_a/r_b) of 0.905. Hence, when the r_a/r_b ratio becomes less than about 0.9, wide gap clearance conditions must be assumed. Actually, specifying a narrower gap clearance, corresponding to an r_a/r_b ratio of 0.95 and providing only a 5% deviation from the average for the minimum and maximum shear rates, would be more realistic. However, this stricter criterion tends to automatically eliminate from the narrow gap category a good proportion of the stator and rotor combinations supplied by manufacturers of concentric cylinder viscometers. Lowering the cutoff point from the stricter r_a/r_b ratio of 0.95 to the more permissive 0.90 value is helpful in bringing more of the stator/rotor combinations into the narrow gap class. This more liberal differentiation between narrow and wide gap clearances will be used later in commenting on some industrial concentric cylinder viscometers.

In considering a viscometer for serious research study or development work on non-Newtonian systems, the paint engineer should be well aware that, by resorting to a wide gap, concentric cylinder, he or she must be prepared to handle complex equations that may or may not be applicable to the non-Newtonian system being tested. Computer assistance is almost mandatory. It is far preferable to initially schedule the use of narrow gap concentric cylinders that are applicable to the non-Newtonian system under investigation. This automatically avoids the complicated mathematics and possibly uncertain results obtained through the use of large gap clearances. Another alternative is to shift to a cone/plate type of viscometer.

Effect of Temperature on Viscosity Measurement. Since viscosity is highly dependent on temperature, the question arises of how much temperature increase is generated during a test run as a result of the work input during the viscosity measurement. It can be shown that, per unit of cylinder surface, the rate of energy dissipated as heat due to the viscous flow between the two cylinders of the concentric cylinder viscometer must equal the product of the shear stress times the velocity. The generation of heat is substantial, but so is the removal of this heat by forced convection (warm liquid moves out of the annular space as cooler liquid moves in) and, more importantly, by the rapid conduction of the heat from the test sample to the metal walls of the concentric cylinders. This situation has been treated mathematically in cases where an equilibrium temperature is calculated based on a balance between the heat input due to the dissipation of mechanical energy and the rate of heat conduction from the sample in the annulus to the massive heat sink provided by the metal cylinders (3). Such calculations indicate that the temperature rise in the sample in the concentric cylinder gap space is essentially negligible for viscosities below 100 poises at boundary velocities below 50 cm/sec. Since the testing of most coating compositions involves lower values, the temperature rise due to mechanical heat generation can generally be neglected as a contributing influence to the viscosity measurement. Practical experience also bears out this conclusion.

Effect of Pigment Particle Size on Viscosity Measurement. When measuring the viscosities of pigmented coatings, slurries, and the like, consideration should be given to the size of the particles in the subject composition relative to the gap clearance. As a safe general rule, the gap size should preferably be at least 10 times the size of the largest particles (3). However, this stipulation is not always observed in practice. For example, a Hegman grind reading of 2H (admittedly a rough grind for a coating composition) indicates the presence of particles up to 3 mils (0.0076 cm) in diameter. This means that the largest particle size is of the same order as the gap clearance in either the Brushometer viscometer (0.006 cm) or the CRGI/Glidden miniviscometer (0.0051 to 0.0076 cm). Since both these viscometers are special purpose (high-shear) instruments designed expressively for the paint industry, there are evidently not enough of these oversize particles in a Hegman 2H grind to seriously jeopardize the viscosity measurement. Furthermore, it is only fair to consider the other extreme. The white pigment titanium dioxide usually varies only slightly from an average diameter of 0.30 μ (0.00003 cm). Hence, for this most common pigment, 200 TiO_2 particles could be stretched across the narrow gap clearance of either of the aforementioned viscometers. However, the possible interference of large particles, due to an excessive number becoming jammed in the narrow gap of a concentric cylinder viscometer, should always be kept in mind when measuring the viscosity of coarse compositions.

Comments on Some Commercial Concentric Cylinder Viscometers. The measuring head of many of the commercial concentric cylinder viscometers can also be used with cone/plate sensors, as discussed later. The following thumbnail sketches serve to illustrate the wide choice of instrumentations, from simple portable viscometers to highly sophisticated designs with extensive ancillary equipment for precise research studies.

Brushometer (7). This viscometer is a special purpose instrument designed for routinely measuring the brushability of paints. It has been made the subject of Problem 2-10 and is schematically diagrammed in Fig. 2-8 (3) (~\$765 as of 1977).

CRGI/Glidden Miniviscometer (7-9). This viscometer, which is competitive with the Brushometer, is actually an accessory to the conventional paddle-type Stormer viscometer. The Stormer instrument is discussed later in this chapter. Instead of rotating a paddle, the driving force of the Stormer apparatus (a falling weight) is converted to rotating the inner cylinder (0.635-cm radius) of a concentric cylinder miniviscometer. The gap clearance is on the order of 0.0065 cm, making this instrument a narrow gap type ($r_a/r_b = 0.99$). The Stormer instrument is adjusted to provide a rotational speed of 10 rev/sec. The average shear rate developed is accordingly 6170 $\sec^{-1}$ ($= 10 \cdot 2\pi \cdot 0.638/0.0065$). This shear rate corresponds to the one developed when a painter uses a brush or roller coater. This special purpose viscometer is capable of measuring viscosities between 0.2 and 10 poises, the range within which most trade sales coatings fall at the indicated shear rate. It is one of the least expensive concentric cylinder viscometers on the market (~\$270 as of 1977).

Brookfield UL Adapter (10). Another of the less expensive concentric cylinder viscometers designed specifically for the ultralow viscosity range represents an adaptation of the conventional Brookfield viscometer (normally a single-cylinder type) to narrow gap conditions. The driving and indicating head of the standard Brookfield instrument (see next section) is used to rotate a cylindrical spindle within a concentric cylindrical tube (radii ratio 0.91). Viscositites in the ultralow range from 0 to 0.1 poise can be measured with an accuracy of 0.0002 poise. The maximum viscosity range is from 0 to 20 poises.

In contrast to the special purpose concentric cylinder viscometers with their limited ranges of operation, the general purpose types span a tremendous range of rheological variables. For this reason and because of the intense technical effort that has been expended to design highly refined and sophisticated instruments, they are generally more expensive. These general purpose viscometers invariably offer a selection of concentric cylinders. When supplied with accessory mechanisms to automatically program instructions and record the rheological data on strip charts, they can become quite costly. However, in evaluating these viscometers for a given use, the distinction between narrow and wide gap geometries should still be kept in mind when non-Newtonian systems are being investigated.

Brookfield Small Sample Adapter (11). This general purpose viscometer is an adaptation of the conventional Brookfield viscometer (normally a single-cylinder type) to concentric cylinder conditions. The driving and indicating head of the standard Brookfield instrument (see next section) is used to rotate any one of several cylindrical spindles within a choice of several jacketed concentric tubes. Although a range from 0.05 to 200.00 poises is covered by the many spindle/tube combinations, only one or two can be considered as falling in the narrow gap classification. "Small sample" refers here to test sample volumes from 2.0 to 16 cm^3. Simplicity of design and elimination of ancillary equipment (except the Brookfield head) make this a most reasonably priced, general purpose viscometer (~\$238 as of 1977 for a single spindle, concentric tubular chamber, and necessary coupling facility).

Ferranti (12). This portable (weight <5 lb), general purpose concentric cylinder viscometer has been available to industry since 1949. It consists of a rotating outer cylinder (rotor) driven by a small, high-torque synchronous motor and a second cylinder (stator) located coaxially within it that is free to twist and assume an equilibrium position against a calibrated spring. The open ends of the two cylinders are exposed downward so that they can be directly immersed in a test liquid by lowering the viscometer assembly from above. A reading is given by a pointer that changes position with reference to a fixed dial at the top of the unit. This is converted to absolute viscosity by an appropriate multiplying factor. Three models (VL, VM, and VH) are available with a selection of three inner cylinders that permit the assessment of viscosity through a range from 0.02 to 200,000 poises (shear rates from 0.2 to 1200 sec^{-1}). However, of these three models only two of the three rotor/stator combinations supplied with Model VM and one of three combinations of Model VL can be considered as providing narrow gap conditions for non-Newtonian systems. Prices range from ~\$3492 for Model VL to ~\$3910 for Model VH as of 1977.

Fann (13). This general purpose, concentric cylinder viscometer was originally designed by the Socony Mobil Oil Company to measure the viscosities and yield values of oil well drilling muds. It is now claimed that these viscometers are the most widely used in the world. As in the Ferranti, the bob and cup are open at the bottom, with the sample introduced from below. Several models are available, ranging from Model HC34A, which is cranked by hand (~\$781/1977), to Model 50C, which incorporates such features as electronic controls, programmable heat adjustment to 500 F, pressures to 1000 psi, and dual pen recording on *X-Y* strip charts (~\$15,780/1977). Intermediate, nonautomated Model 35A (weight 35 lb) is possibly best suited for the paint industry (~\$1400/1977). A gear train driven by a synchronous motor provides speeds from 3 to 600 rpm, and these can be reduced by a factor of 0.3 by an accessory gear box. There is a selection of eight torsion springs, three rotors (outside cups) and four stators (inside bobs). However, of the 24 rotor/stator combinations only 2 (R1/B1 and R2/B1) qualify for the narrow gap category. Only the R1/B1 combination with an F1 torsion spring is furnished as standard equipment with each viscometer.

Rotovisco (14). This general purpose RV3 viscometer, produced in West Germany, is a versatile instrument that can be adapted to either concentric cylinder or cone/plate sensor systems. Speed can be selected from 40 fixed speeds (0.1 to 724 rpm) or can be made continuous (0.1 to 1000 rpm) by external programming. Flow curves are recorded on an *X-Y* recorder, and a thermostatically controlled bath can be supplied to maintain the temperature of the test sample at a preselected level during the test program. By a suitable selection of concentric cylinders, the RV3 Rotovisco viscometer can be made to cover shear rates from 10^{-3} to $4 \cdot 10^4$ $\sec^{-1}$, shear stresses from 2 to 10^7 dynes/cm^2, and viscosities from 0.02 to 10^8 poises. For the Model MV or T system, three concentric cylinder combinations are available (one outside stator cup and three inside rotor bobs). One of these qualifies as narrow gap clearance, and another just misses with an r_a/r_b ratio of 0.88. An alternative Model HS system of concentric cylinders provides very narrow gap clearances (r_a/r_b ratios of 0.990 and 0.997). Still another sensor system (Model NV) is characterized by a unique double-gap design (a bell-shaped rotor rotating within a narrow, double-walled annulus). Other Rotovisco models are supplied to industry for both special and general purpose applications.

Rheotron (15). This general purpose viscometer (weight ~18 lb) is a relatively new entry that has been designed and constructed to comply with the latest scientific findings. It can be fitted with either a rotating outside cylinder or a rotating plate that impresses a shear flow on either an inside stator cylinder or cone, respectively. Fixed rotational speeds can be selected from a keyboard (32 choices), or a continuous speed adjustment can be programmed. Three measuring springs give torque ranges of 100, 500, and 5000 g-cm. An *X-Y* recorder is available that works in conjunction with the Rheotron programmer. Viscosity measurements can be made at temperatures from −30 to 300 C, using a thermostatically controlled circulation bath. Three cups (outside rotors) and six bobs (inside stators) are supplied as standard concentric cylinders. Of the six recommended rotor/stator combinations, four fall within the narrow gap classification, one is just outside (r_a/r_b) = 0.88), and one is in the wide gap class. The Rheotron is a modern, wide-ranging viscometer that covers shear rates from $5 \cdot 10^{-2}$ to $2 \cdot 10^4$ $\sec^{-1}$, shear stresses from $1 \cdot 10^3$ to $3 \cdot 10^6$ dynes/cm^2, and viscosities from 0.005 to 10^7 poises (~$14,000/1977).

Contraves (16). This general purpose viscometer (Rheomat RM30), designed to meet the most demanding requirements of continuous shear rheometry, features 30 rotational speeds in geometric progression from 0.47 to 350 rpm (the slowest speed corresponds to one revolution every 21.3 min). Rotational speeds are set manually on the Rheomat control or continuously changed by an external programmer. Four torque ranges are available from 50 to 500 g-cm. Both concentric cylinder and cone/plate viscosity sensing systems can be attached to the measuring head. Temperature control is maintained by a thermostatically stabilizing bath. As of 1977 the Rheomat 30 with adjustable support stand cost ~$7250, the *X-Y* recorder/programmer ~$2400, the MSA/E (general use) concentric cylinder sensing system ~$780, and the cone/plate sensing system ~$860. Other concentric cylinder combinations are supplied, such as the MS-O (a high-sensitivity system for low-viscosity materials), so that the Rheomat 30 viscometer is capable of covering a complete span of rheological variables as encountered in research and industrial work.

Rheogoniometer (17). Probably the most intricately designed viscometer is the Weissenberg Rheogoniometer R19, which features, among other capabilities, the ability to generate an independent sine wave oscillation movement that can be imposed on a continuous rotational movement. A choice can be made between either a concentric cylinder and a cone/plate sensor system. Four models are available, ranging from the basic Model A (Rheo-Viscometer) to the comprehensive and versatile Model D (Weissenberg Rheogoniometer). The measuring head of Model D is provided with sophisticated facilities to measure continuous shear, oscillation effects, and normal stress force (normal to direction of shear). The range of rheological variables is extensive (viscosity from $1 \cdot 10^{-3}$ to $5 \cdot 10^{7}$ poises, shear rate from $7 \cdot 10^{-4}$ to $9 \cdot 10^{3}$ sec^{-1}, shear stress from $9.5 \cdot 10^{-4}$ to $1.2 \cdot 10^{7}$ dynes/cm^2). The rotational speed can be varied from 1.0 radian/5.6 hr to 39 radians/sec, and the ambient temperature from −50 C (with fluid chamber) to 400 C (with oven). All models are supplied with rotary switches for control of forward or reverse rotation, oscillatory movement, and braking by an electromagnetic clutch. Readout for Model D includes gap setting, rotational torque, oscillation torque input, and normal force. Ancillary equipment is available, including a recorder of either the single- or the twin-channel chart pen type, the latter for recording both tangential and normal stress.

ROTATING CYLINDER IN CUP OF INFINITE RADIUS

Theory

A single cylinder (bob) rotating in a cup (container) of infinite radius corresponds to a gap clearance also of infinity, since $b = \infty$ makes $(b - a) = \infty$. When the value of infinity is substituted for b in Eqs. 37 and 41, there result two reduced theoretical expressions, Eqs. 55 and 56, for the viscosity and shear rate, which apply to a Newtonian system as measured by a single-cylinder viscometer (theoretical shear stress and velocity at any radius are calculated as before by Eqs. 40 and 42).

$$\eta = \frac{M}{4\pi L \omega_a a^2} \tag{55}$$

$$\mathrm{D} = \frac{2\omega_a a^2}{r^2} \tag{56}$$

When the radius r equals a (the bob radius), $\mathrm{D} = 2\omega_a$, $\tau = M/2\pi a^2 L$, and $v = \omega_a a$. At the outside radius ($b = \infty$), D, τ, and v all become zero. The shear rate within the single-cylinder viscometer is thus seen to vary from twice the angular velocity to a value of zero. This is of no concern for a Newtonian system, since this variability in shear rate has been taken into account in the derivation of the several equations. However, for non-Newtonian systems, a much more complicated body of mathematics is required to establish the viscosity profile of a test material (3). If a yield value is present, even mathematical manipulation appears unable to handle the condition.

Reduction to Practice

Despite the serious limitations for non-Newtonian systems, the single-cylinder viscometer is probably the most widely used in industry, partly because of such features as its uncluttered design, long life, and simplicity in operation, as well as the low initial investment required. Correction for deviation from theory for Newtonian liquids can be easily made by calibration with viscosity standards. The literature is also replete with articles dealing with mathematical systems that, it is claimed, render non-Newtonian systems susceptible to measurement by the single-cylinder viscometer or at least make the instrument provide relevant rheological data (18-20).

Brookfield SynchroLectric Viscometer (11). Undoubtedly the most widely used commercial single-cylinder viscometer is the Brookfield SynchroLectric, which has been marketed for more than 40 years. It is portable (weight 3.5 lb), reasonably priced (~\$570 to \$655 per model/1977), and rugged in construction, and scale values as indicated by a pointer are easily readable on a 13.5-in. circular scale (guaranteed accuracy 1%). A viscosity determination, which takes less than 30 sec, consists simply in inserting the cylindrical spindle in the test sample (for calibration a 600-cc Griffin beaker is specified) and activating the rotating mechanism at one of the several speeds provided. The reading is converted to viscosity by an appropriate multiplying factor.

The viscometer comes with a number of spindles; some are true cylinders, but most are composites of cylinders and disks. Because of the wide range of geometric configurations available and also because of the fact that the four models (LV, light viscosity; RV, medium viscosity; HA, heavy viscosity; HB, very heavy viscosity) are supplied with different spring tensions and a range of speed selections, the Brookfield instrumentations cover a full range of rheological variables (0 to 640,000 poises). However, the inherent limitations of the single-cylinder viscometer should be kept in mind when working with a non-Newtonian system so that the scrambled data that are generated can be properly interpreted. The popularity of the Brookfield SynchroLectric viscometer is reflected by its identification with at least a dozen ASTM test standards.

PROBLEM 2-12

Calculate a theoretical length for the portion of a cylindrical rodlike spindle (radius 0.159 cm) which, when submerged in a viscous Newtonian liquid (1000 poises) in a large container and rotated at 6 rpm (0.1 rev/sec), exerts a torque of 673 dyne-cm (corresponding to a scale reading of 100 on a LVF or LVT Brookfield SynchroLectric viscometer). What shear rates are involved during the spindle rotation?

Solution. Substitute the given data in Eq. 55, which applies to a rotating cylinder in a bath of infinite extent, and solve for length L.

$$1000 = \frac{673}{4\pi L \cdot 2\pi \cdot 0.1 \cdot 0.159^2}; \qquad L = 3.37 \text{ cm}$$

Equation 55 shows that, for a rotating cylinder in a large container, the shear rate ranges from a maximum of $D = 2\omega_a$ at the bob to a value of zero at the container walls. Hence the shear rate range in the given problem is from 1.26 ($= 2 \cdot 2\pi \cdot 0.1$) to 0 sec^{-1}.

Comment. The calculated theoretical length of 3.37 cm for the portion of the spindle that must be submerged compares favorably with the actual working length of 3.10 cm (3.40 cm when corrected for end effects) specified for Spindle LV4, which is supplied with the Brookfield viscometer for the conditions outlined in the problem.

CONCENTRIC CYLINDER WITH AXIAL (TELESCOPIC) MOTION

As opposed to rotational concentric cylinder viscometers with their radial motion, axial viscometers have concentric cylinders which slide lengthwise relative to each other (axial motion). In both types the test liquid is located in the annular space between the cylinders.

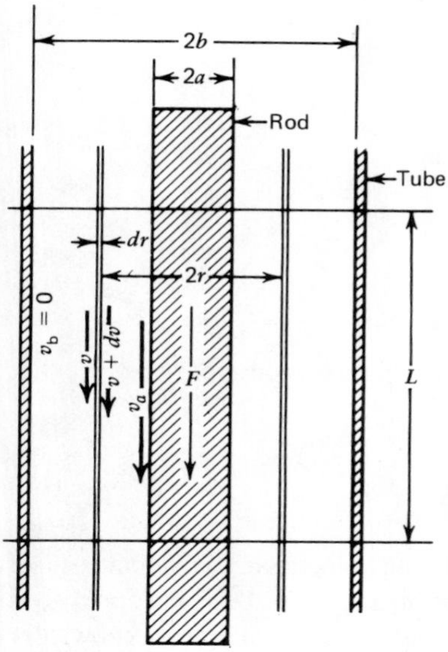

Fig. 2-9 Flow pattern for a Newtonian liquid in viscous flow in a concentric cylinder viscometer with axial (telescopic) motion.

Theory

Consider a volume of Newtonian fluid confined between the two coaxial cylinders shown in Fig. 2-9. The inner cylinder (a rod) has a radius a, and the outer cylinder (a tube) has a radius b. Consider a length L of this concentric arrangement, and let the outside cylinder be maintained stationary ($v_b = 0$). Let an axial force F act over length L of the inner cylinder so that it moves axially relative to the outer cylinder with velocity v_a. The shear stress τ_a at the surface of the inner cylinder is then given by Eq. 57.

$$\tau_a = \frac{F}{2\pi a L} \tag{57}$$

Consider a cylindrical shell of length L, radius r, and differential thickness dr concentrically located between the coaxial cylinders. The axial shear stress τ_r on this shell is $(a/r)\,\tau_a$, since the shear stress in moving toward the outer cylinder is attenuated in inverse proportion to the area over which it acts.

$$\tau_r = \frac{a}{r}\tau_a = \frac{F}{2\pi r L} \tag{58}$$

The shear rate D at radius r is $-dv/dr$, the negative sign indicating that the velocity decreases as the radius increases. Since a Newtonian liquid (η = constant) is being considered, Eq. 59 can be established immediately for the shear stress/shear rate relationship.

$$\eta = \frac{\tau_r}{D} = \frac{a\tau_a/r}{-dv/dr} \tag{59}$$

Solving for dv gives Eq. 60.

$$dv = \frac{-a\tau_a}{\eta}\frac{dr}{r} \tag{60}$$

Integrating and then establishing a constant of integration from the fact that $v = 0$ when $r = b$ results in Eq. 61.

$$v = \frac{a\tau_a}{\eta} \ln \frac{b}{r} = \frac{a\tau_a}{\eta} \cdot 2.31 \log \frac{b}{r} \tag{61}$$

At the surface of the inner cylinder ($r = a$) the velocity is the velocity v_a of the moving cylinder. Substituting these values in Eq. 61 and solving for η gives Eq. 62, which can be used to compute the viscosity of a liquid from a knowledge of the radii a and b, the shear stress τ_a, and the axial velocity v_a, all of which are readily subject to measurement.

$$\eta = \frac{\tau_a}{v_a} a \cdot 2.31 \log \frac{b}{a} = \frac{F}{2v_a \pi L} \cdot 2.31 \log \frac{b}{a} \tag{62}$$

For narrow gap spacings between inner and outer cylinders, the cylinder walls can be considered as essentially parallel. Under these conditions an approximate "parallel plate" type of expression can be derived (Eq. 63).

$$\eta = \frac{\tau_a}{v_a}(b - a) \tag{63}$$

The error introduced by this approximate Eq. 63 is usually negligible, as will be illustrated by Problem 2-13.

By substituting the expression for viscosity given by Eq. 62 in Eq. 61 we obtain Eq. 64 or the liquid velocity at any radial distance r.

$$v_r = v_a \frac{\log b - \log r}{\log b - \log a} \tag{64}$$

Differentiation of Eq. 64 to obtain shear rate ($= dv/dr$) results in Eq. 65.

$$D = \frac{dv}{dr} = \frac{v_a}{r} \frac{1}{2.31 \log (b/a)} \tag{65}$$

Reduction to Practice

A practical concentric cylinder viscometer (axial type) of simple design for measuring the viscosity of viscous liquids (say a heavy-bodied ink) is shown in Fig. 2-10 (21). The accuracy of this instrument depends primarily on the precision of the machining of the rod and tube diameters. For industrial use the clearance between the two cylindrical surfaces ($= b - a$) is usually held to a few thousandths of a centimeter. Hence the accuracy with which this gap can be controlled and measured determines the precision of the viscometer.

Laray Viscometer. Problem 2-13 illustrates the use of the Laray viscometer in a practical application.

PROBLEM 2-13

The viscometer of Fig. 2-10 (concentric cylinders with axial motion) has the following specifications:

Radius of rod	0.4000 cm	Weight of rod	140 g
Radius of tube	0.4060 cm	Added weights	100 g
Gap spacing	0.0060 cm	Added weights	200 g
Length of rod	25.0 cm	Added weights	400 g
Length of tube	4.00 cm		

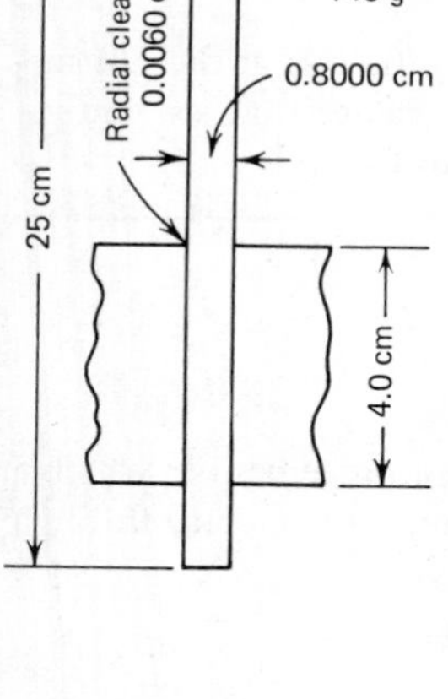

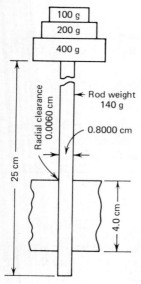

Fig. 2-10 Representative dimensions for a commercial concentric cylinder viscometer with axial (telescopic) motion.

The rod is smeared with a test liquid (heavy-bodied ink) and then allowed to drop through the tube so that only the viscous resistance of the ink lodged between the walls of the rod and tube impedes its free fall. When the rod is loaded with 700 g of added weight, the rod drops a distance of 15.0 cm with uniform velocity in 10.0 sec. Calculate the shear rate, the shear stress, and the viscosity of the ink for these conditions.

Solution. The rod weight together with the added weights totals 840 g. The shear stress at the rod surface is then computed from Eq. 57.

$$\tau_a = \frac{840 \cdot 980}{2 \cdot 3.14 \cdot 0.400 \cdot 4.00} = 81{,}800 \text{ dynes/cm}^2$$

The rod velocity is 1.5 cm/sec (= 15.0/10.0).

The ink viscosity is calculated by substituting these values for τ_a and v_a in turn in Eqs. 62 and 63.

$$\eta = \frac{81{,}800}{1.5} \cdot 2.31 \cdot 0.400 \cdot \log \frac{0.406}{0.400}$$

$$= 50{,}450 \cdot 0.00647 = 326 \text{ poises (accurate value)}$$

$$\eta = \frac{81{,}800}{1.5} \cdot 0.0060 = 327 \text{ poises (approximate value)}$$

The shear rate is calculated from the expression $\text{D} = v/x$, where $v = v_a$ and $x = b - a$.

$$\text{D} = \frac{1.5}{0.0060} = 250 \text{ sec}^{-1}$$

This is an approximate overall shear rate. The shear rate at any given radius r can be computed from Eq. 65. For the problem conditions the shear rate at radius r is equal to $100.6/r$.

$$\text{D} = \frac{1.5}{r}\left[\frac{1}{2.31 \log (0.406/0.400)}\right] = \frac{100.6}{r}$$

The shear rate then varies between 252 sec^{-1} (= 100.6/0.400) at the rod surface and 248 sec^{-1} (= 100.6/0.406) at the tube surface, values that are equivalent to an essentially constant shear rate throughout the test liquid.

BAND VISCOMETER (22)

The band viscometer is substantially a dual parallel plate arrangement in which a thin, flat band (extended movable plate) is pulled with constant velocity through a rectangular slit with parallel side walls (stationary plates).

Theory

Consider a band of indefinite length that moves with velocity v through a slit with parallel walls, as shown in Fig. 2-11. Let the force acting on the band and causing it to move with velocity v be F. The effective band area that takes part in the shearing action at any given moment is $2BL$, where B is the band width and L the slit length. Let the gap spacing between the band and the wall surface be x (this is for either side, since symmetrical conditions are assumed). Then the shear stress, shear rate, and viscosity for a test liquid (located between the band and the slit walls) are given by Eqs. 66, 67, and 68, respectively.

$$\tau = \frac{F}{2BL} \tag{66}$$

$$\mathrm{D} = \frac{v}{x} \tag{67}$$

$$\eta = \frac{F/2BL}{v/x} = \frac{F}{v}\frac{x}{2BL} \tag{68}$$

Reduction to Practice

In practice, the band (say a vinyl tape) is made as thin as possible to minimize edge effects. From a practical standpoint the use of the viscometer is generally

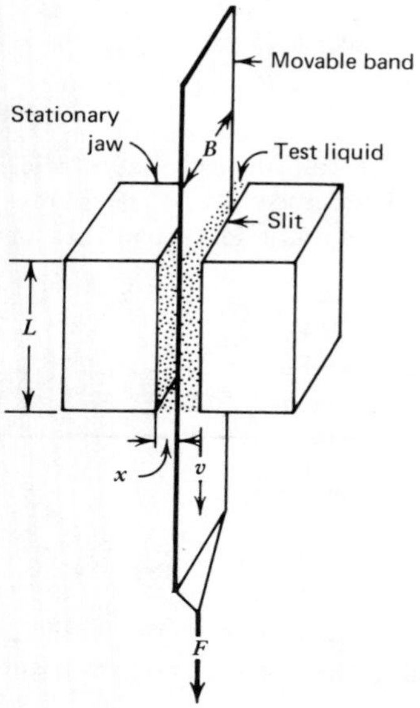

Fig. 2-11 Schematic diagram illustrating the arrangement of parts for a band viscometer.

limited to fairly viscous liquids. Moreover, the design is such that leakage to the side is prevented and a reservoir of test liquid is provided at the top to feed liquid into the slit continuously as the band is fed between the slit walls.

PROBLEM 2-14

A 4-mil vinyl tape (1.00 in. wide) is pulled through a rectangular slit by a 200-g weight (attached to the lower end of the tape). The slit gap is 0.020 in. across, and the parallel faces of the slit walls measure 2.00 in. in width by 1.50 in. in length. A viscous test dispersion is continuously fed into the top of the slit area as the tape is allowed to pass freely through the slit under the pull of the 200-g weight. If a 21.0-in. length of tape moves past the slit opening with uniform velocity during a measured 12.0-sec period under these conditions, calculate the test dispersion viscosity. Neglect edge effects and the weight of the tape.

Solution. The data as presented in the problem are listed below, along with conversion to consistent units in parentheses.

$$B = 1.00 \text{ in. } (2.54 \text{ cm})$$

$$L = 1.50 \text{ in. } (3.81 \text{ cm})$$

$$\text{Gap spacing} = 0.020 \text{ in. } (0.0508 \text{ cm})$$

$$\text{Tape thickness} = 0.004 \text{ in. } (0.0102 \text{ cm})$$

$$v = 21.0 \text{ in.}/12.0 \text{ sec} = 1.75 \text{ in./sec } (4.45 \text{ cm/sec})$$

$$F = 200 \cdot 980 \text{ dynes} = 196{,}000 \text{ dynes}$$

The distance x between the band surface and the slit wall surfaces that parallel it is equal to 0.0203 cm [= (0.0508 − 0.0102)/2]. The viscosity of the test dispersion is calculated by substituting the appropriate numerical quantities in Eq. 68.

$$\eta = \frac{196{,}000}{4.45} \; \frac{0.0203}{2 \cdot 2.54 \cdot 3.81} = 46.2 \text{ poises}$$

CONE/PLATE VISCOMETER (ROTATIONAL TECHNIQUE)

The cone and plate viscometer, as its name implies, consists of a cone with its vertex in point contact with a horizontal flat plate. The normal design for this viscometer calls for a very wide vertical angle for the cone, which in turn makes the angle between the cone and flat plate extremely small (on the order of 0.2 to 3.0°). The circular wedgelike space between the cone and the plate is filled with the test fluid. One of the confining surfaces is held stationary, while the other is made to rotate (around the cone axis). A model cone and plate viscometer arrangement is shown in Fig. 2-12.

Theory

Consider a differential volume of fluid of area $r\,d\theta\,dr$ and thickness αr positioned between the stationary and rotating faces of the viscometer of Fig. 2-12. Here the thickness expression αr is an approximation to the true thickness value $\sin \alpha \cdot r$, but for small angles the approximation is almost exact (e.g., for 2° angle α is equal to 0.03491 radian, whereas $\sin \alpha = 0.03490$). The top surface of this element of liquid is in contact with the rotating face of the viscometer and hence is moving with velocity ωr. The other surface is stationary. Therefore the shear rate for this differential volume of fluid is given by Eq. 69.

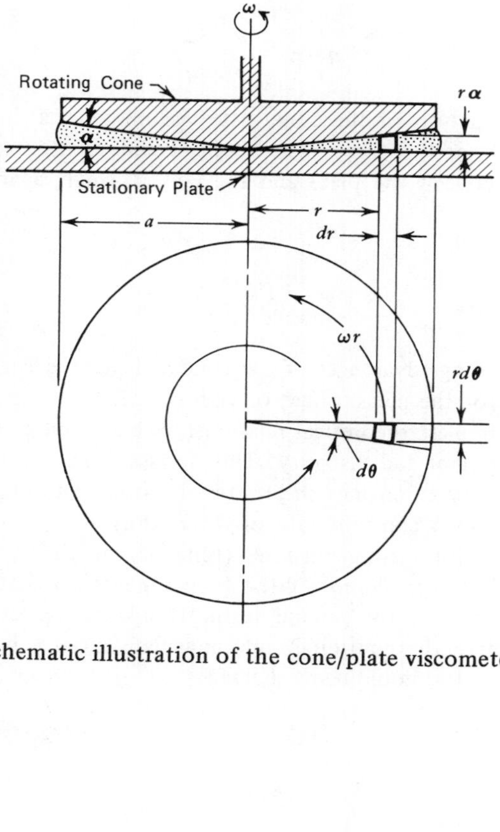

Fig. 2-12 Schematic illustration of the cone/plate viscometer geometry.

$$D = \frac{\omega r}{\alpha r} = \frac{\omega}{\alpha} \tag{69}$$

Since this ratio is fixed by the cone/plate angle α and the angular velocity ω only (the ratio is independent of r), it follows that the shear rate is constant over the entire volume of the test liquid. It also follows that the shear stress must be constant for any given angular velocity ω.

Let τ be the shear stress at the surface of the rotating cone. The total torque M is then equal to the cumulative torque contributions from the various differential areas ($r\,d\theta\,dr$) as given by Eq. 70.

$$M = \int_0^{2\pi} \int_0^{a} \tau r^2 \, dr \, d\theta = \tau \frac{2\pi a^3}{3} \tag{70}$$

The shear stress/shear rate relationship can now be established.

$$\tau = M \frac{3}{2\pi a^3} \tag{71}$$

$$D = \frac{\omega}{\alpha}$$

$$\eta = \frac{\tau}{D} = \frac{M}{\omega} \frac{3\alpha}{2\pi a^3} \tag{72}$$

A more rigorous analysis of the cone/plate viscometer that has been reported demonstrates that inertial forces can be neglected relative to viscous forces when the angle between the plate and the cone is small, as in commercial instruments (23).

Reduction to Practice

The many outstanding advantages of the cone and plate viscometer make it an ideal instrument for the meásurment of viscosity. The geometry of the system is such that the simplest theoretical relationships exist among the quantities of shear stress, shear rate, and viscosity. This derives from the fact that for any given angular velocity a constant shear stress is imposed throughout the liquid, which in turn means a constant rate of shear. Only a minimal amount of test sample is required for testing purposes (generally on the order of 0.1 cm^3). Furthermore, it has been demonstrated (experimentally) that peripheral edge effects can be neglected for general testing purposes. Stress-induced heat is quickly dissipated to the confining cone and plate surfaces because of the extremely thin film of test liquid being sheared.

As presently designed, the cone and plate viscometer is easy to set up, fill, run, and clean. It has recently been adapted to automatic flow-curve recording.

Special truncated cones are supplied for measuring the viscosity of pigment dispersions and slurries in cases where the largest particles might conceivably become jammed in the apex region between the cone and the plate. The necessary truncation, which is adjusted to the largest particle size, is usually so trivial as to have little or no effect on the viscosity measurement. As in the case of the concentric cylinder viscometer, the temperature increase due to the conversion of mechanical energy into heat during the test run can generally be safely ignored as a contributing factor to the viscosity determination.

Both special and general purpose cone/plate (C/P) viscometers are commercially available.

ICI Cone and Plate Viscometer (24). This special purpose C/P viscometer, designed specifically to measure the brushability, roller coatability, and sprayability of coatings, requires less than 1.0 cm^3 of test sample. The approximate shear rate is 10,000 sec^{-1}, and models are available covering viscosity ranges from 0 to 5, 0 to 10, and 0 to 40 poises. Model VR-4000 has a single temperature setting (25 C), but other models (VR-4006 and VR-4008) provide temperature settings in steps from 25 to 200 C. All the cones are less than 1.0 in. in diameter with cone angles from 0.5 to 2.0°.

The test sample is brought to a temperature of 25 C in less than 15 sec and then held there during the test run by a thermostat control. Insertion of the sample is quickly accomplished, and its removal is equally rapid (a wipe-off with a paper tissue). Readout is given directly by a pointer on a viscosity scale. It is claimed that a run of a new test sample can be made every minute with excellent accuracy. In 1977 these ICI cone/plate viscometers ranged in price from \$1350 to \$1550 (total weight ~35 lb).

Ferranti-Shirley Cone/Plate Viscometer (25). The first general purpose C/P viscometer for research and industrial use was developed around 1950 by Ferranti Ltd., based on an earlier research cone/plate configuration in use at the Shirley Institute (both establishments located in Manchester, England). Several hundred Ferranti-Shirley viscometers have now been in use for up to 20 years. The instrumentation of this C/P viscometer is highly sophisticated, including solid state circuitry that controls a programmable rotational speed, sensitive yet rugged torque sensors (spring deflection measured by a precision potentiometer), electrical sensing of the cone/plate positioning (cone/plate adjustment at apex <0.0001 in.) and subsequent automatic gap adjustment that compensates for changes in ambient temperature, thermostatically controlled temperature by water circulation through the plate, digital temperature indication via thermocouples in the plate, and automatic recording of the shear rate versus shear stress displayed on an X–Y recorder chart. The standard cone angle is

0.30°, although other cone angles can be provided (a standard set of cones 1.0, 2.0, and 3.5 cm in radii is supplied with the unit). Truncated cones (with a center pin for positioning) can be supplied that accommodate particles in pigment dispersions with no loss of accuracy (the degree of truncation depends on the particle sizes involved, from 5 up to 400 μ). Only a small sample is required (0.1 to 1.0 cm^3), and productivity is high (up and down, 0- to 10-rpm curves are printed out in less than 10 min). The viscosity range of the standard model runs from 300,000 poises (1.8 sec^{-1}) to ~0.2 poise (18,000 sec^{-1}). A shear relaxation adapter is provided that allows the measurement of viscosities in the ultralow-shear-rate range, using the relaxation technique discussed later. The price of the Ferranti-Shirley viscometer, excluding *X-Y* recorder, digital temperature indicator, and thermostatic circulation bath, was ~$22,000 as of 1977 (operating temperature to 200 C).

Wells-Brookfield Microviscometer (26). This lightweight viscometer (weight 3.5 lb) was developed originally as a special purpose type for investigating the rheological properties of biologic fluids through a low viscosity range. It has since been expanded to include four C/P models (LVT, RVP, HAT, and HBT) that bring it within the general purpose class. Each model can be supplied with a choice of three cone angles (0.8, 1.565, and 3.0°) and up to eight rotational speeds with ratios of 100, 50, 20, 10, 5, 2, 5, 1, and 0.5. The plate is located within a jacketed cup through which water or other liquid can be circulated to maintain a constant temperature (±0.01 C°) at preselected temperatures in the range from -30 to 180 C. Clearance between the plate and cone apex is set to within 0.0001 in. by a unique mechanical design. Test sample volumes are generally from 0.5 to 2.0 cm^3. The viscosity is obtained by multiplying the reading given by a pointer on a circular scale by an appropriate factor. By a proper selection of model, cone angle, and motor speed, viscosities can be measured, with 1% accuracy of the range employed, from 0.1 poise (300 sec^{-1}) to 500 poises (4.0 sec^{-1}). Testing is readily carried out, the test sample being rapidly inserted and removed to permit good productivity. This C/P viscometer is ideally suited for the measurement of viscosities in the ultralow-shear-rate range by the relaxation technique, as discussed in the next section. The Wells-Brookfield Microviscometer was priced from $827 to $1065 per model as of 1977.

Other Adaptations. Several of the general purpose concentric cylinder viscometers discussed previously in this chapter are adaptable to cone/plate sensor systems. Thus the Contraves, Rotovisco, Rheotron, and Rheogoniometer instrumentations can function as either concentric cylinder or C/P viscometers. The driving, programming, and recording systems are identical. Only the sensing geometrical configurations are altered.

PROBLEM 2-15

The viscosity of an alkyd paint is measured with a C/P viscometer ($\alpha = 1°$; $a = 2.0$ cm). When a constant torque M of 100,000 dyne-cm is imposed on the cone, an immediate angular velocity of 40 rpm is recorded. Ten seconds later the angular velocity increases to 160 rpm, the torque remaining constant. Calculate the viscosities for these two conditions, and explain the discrepancy between them.

Solution. The shear stress, which is maintained constant, is calculated from Eq. 71.

$$\tau = 100{,}000 \frac{3}{2 \cdot 3.14 \cdot 2^3} = 5970 \text{ dynes/cm}^2$$

The initital and 10-sec shear rates are calculated from Eq. 69. Note that the 1° cone/plate angle in radians is equal to 0.01745 (= 1° · 2 · 3.14/360) and that the two angular velocities in radians/sec are equal to 4.2 (= 40 · 2 · 3.14/60) and to 16.8 (= 160 · 2 · 3.14/60), respectively.

$$D \text{ (initial)} = \frac{4.2}{0.01745} = 241 \text{ sec}^{-1}$$

$$D \text{ (10 sec)} = \frac{16.8}{0.01745} = 963 \text{ sec}^{-1}$$

In turn the two viscosities are calculated from Eq. 72.

$$\eta \text{ (initial)} = \frac{5970}{241} = 24.8 \text{ poises}$$

$$\eta \text{ (10 sec)} = \frac{5970}{963} = 6.2 \text{ poises}$$

The reason for the drastic drop in viscosity is the thixotropic nature of the alkyd paint. Apparently this paint initially exhibited a high degree of structural viscosity, which was then broken down by the shearing action of the viscometer surfaces.

CONE/PLATE VISCOMETER (RELAXATION TECHNIQUE)

Measurement of the viscosities of non-Newtonian systems at ultralow shear rates (shear rate range from about 0.001 to 1.0 sec^{-1}) poses many technical difficulties. However, by resorting to the so-called relaxation technique, conventional fundamental viscometers (cone/plate viscometer, concentric cylinder viscometer) can be adapted to the measurement of viscosities in this problem area.

For example, the normal procedure for running the C/P viscometer calls for the imposition of constant rotational velocity, where the viscous resistance offered by the test liquid confined between the cone and the plate is just counterbalanced by a partially wound-up helical spring. However, in the relaxation technique the helical spring is wound up and then allowed to relax (no other force acts on the liquid). Since the spring relaxation is being opposed only by the viscous resistance of the liquid, several minutes may be required for the spring to unwind to an equilibrium position. Such long time periods mean that very low shear rates are involved. Mathematical details of the relaxation technique were developed and reported by Patton in 1966 (27), although the general concept had earlier been outlined qualitatively by Krieder (28).

One of the attractive features of the relaxation method is that a plot of a single relaxation curve (log scale reading versus time) permits the calculation of viscosities for any number of shear rates imposed by the helical spring as it unwinds. This calculation is done simply by multiplying the slope of the curve at any selected shear stress by the instrument constant. The relaxation technique is now routinely used for establishing the low-shear-rate range for the viscosity profiles of coating systems (29, 30).

It should be pointed out that the relaxation technique applies only to conditions where the inertial component of the cone is negligible in comparison to the viscous component during the relaxation period. In the case of the Wells-Brookfield RVT cone/plate microvisometer used by Patton (27) this condition prevails when the viscosity is above about 5.0 poises. Since the relaxation technique was developed specifically to measure viscosities, many orders higher than this (tens to thousands of poises), corresponding to the high viscosities exhibited by coating compositions in the ultralow-shear-rate range, it is apparent that here the inertial component can be completely ignored.

ROTATING DISK VISCOMETER

The rotating disk type of viscometer presently supplied by a number of manufacturers provides a convenient method for rapidly obtaining a viscosity value for a test liquid. Unfortunately, the geometry of the typical industrial disk viscometer does not lend itself to precise theoretical treatment. Hence the reading obtained with a disk viscometer must be based on a scale calibrated against reference liquids of known viscosity. Viscosity readings obtained for Newtonian liquids with an industrial disk viscometer can be considered as relatively accurate. With non-Newtonian liquids, however, the viscosity reading represents a sort of overall average for a moderate range of viscosities which are being recorded simultaneously, corresponding to the indeterminate but wide span of shear stresses set up by the instrument. Despite the objections that the rotating disk viscometer of commercial design neither is a truly fundamental instrument nor is able to pinpoint a viscosity in a narrow stress range, it has become popular and useful.

Theory

Consider a cylinder of liquid of height x confined by two parallel disks, each of radius R as shown in Fig. 2-13. Let the upper disk revolve with a constant angular velocity of ω radians/sec under an imposed torque M while the lower disk is held stationary. The layer of liquid in contact with the upper disk revolves with angular velocity ω also. The layer of liquid in contact with the bottom disk remains stationary. Between these two boundary layers lie intermediate disklike layers of liquid rotating with intermediate angular velocities.

Consider next a cylindrical shell of the liquid having a radius r and differential thickness dr, concentric with the axis of revolution of the revolving upper disk. The velocity of the topmost layer of liquid in this shell is ωr, and the velocity of the bottom layer is zero. Since the shell height is x, the velocity gradient (shear rate) D down through the shell height is $\omega r/x$.

$$\mathrm{D} = \frac{\omega r}{x} \tag{73}$$

Let the differential torque acting on the annular ring of the shell (of area $2\pi r\, dr$) be dM. The force acting on the angular ring is then dM/r. The shear stress τ in turn is $(dM/r)/(2\pi r\, dr)$.

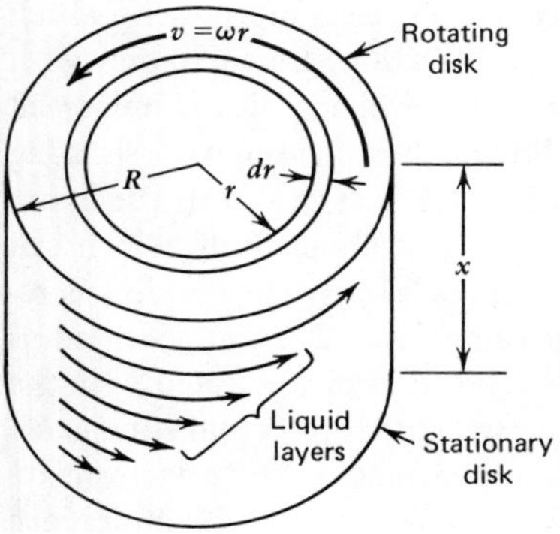

Fig. 2-13 Pattern for viscous flow in a model disk viscometer.

$$\tau = \frac{dM/r}{2\pi r\, dr} = \frac{dM}{2\pi r^2\, dr} \tag{74}$$

Since viscosity is the ratio of shear stress to shear rate, an expression for η can be established in terms of their equivalents for D and τ, given by Eqs. 73 and 74.

$$\begin{aligned} \eta &= \frac{dM/2\pi r^2\, dr}{\omega r/x} \\ &= \frac{dMx}{2\pi\omega r^3\, dr} \end{aligned} \tag{75}$$

Solving for dM gives Eq. 76.

$$dM = \frac{2\pi\eta\omega r^3\, dr}{x} \tag{76}$$

Integrating this expression between the limits of $r = 0$ and $r = R$ results in Eq. 77.

$$M = \frac{\pi\eta\omega R^4}{2x} \tag{77}$$

Solving explicitly for viscosity gives Eq. 78.

$$\eta = \frac{M}{\omega}\,\frac{2x}{\pi R^4} \tag{78}$$

Reduction to Practice

The viscosity given by Eq. 78 is a theoretical value for a hypothetical set of idealized conditions (see Fig. 2-13). In practice, edge effects are present which exert an influence and must be taken into consideration. Furthermore, with the common type of portable disk viscometer, the revolving disk is immersed a set distance below the surface of the test liquid, thus creating two shearing zones, one above and one below the disk (see Fig. 2-14). The top of the cylinder of liquid above the disk is the surface of the test liquid itself, which can hardly be considered a completely stationary liquid layer when the disk is revolving. Below the disk the cylinder of liquid being sheared is not specified as to a height dimension. In view of these varied conditions encountered with the portable type of disk viscometer, it can be understood that resort must be made to liquids of known viscosity to calibrate these instruments. Efforts to relate exact mathematical expressions to embrace all the variables met with in practice

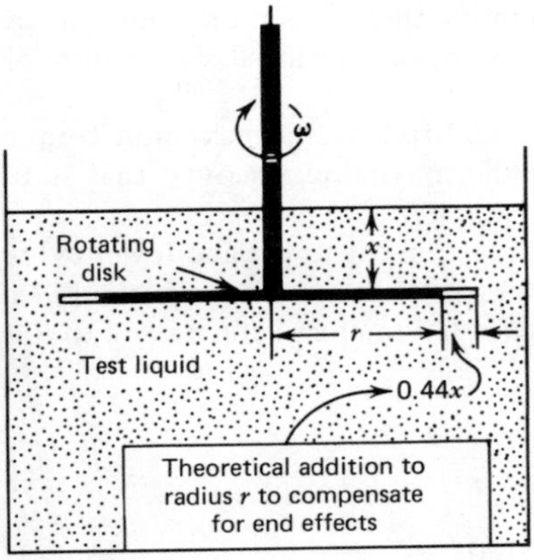

Fig. 2-14 Schematic illustration of a commercial disk viscometer in a test liquid.

lead only to hopelessly involved equations which are of questionable practical use.

However, it has been deduced (31) that edge effects can be approximately compensated for by increasing the actual radius of the disk by a hypothetical addition equivalent to $(2 \ln 2/\pi)x$ or $0.44x$, where x is the depth of immersion of the disk. This adjustment allows a closer approximation to the actual conditions encountered in industrial work and appears to work out fairly well in practice.

With the proper calibration, and by following the instructions of the manufacturer, these disk viscometers represent handy instruments for obtaining viscosities within the areas they are meant to serve.

PROBLEM 2-16

Specify the approximate radius for a thin, flat disk to be used for measuring the viscosity of liquids in the range from 0 to 40 poises. The disk will be immersed 3.0 cm below the surface of the test liquid, and the spindle of the disk will be adapted to a motorized head capable of (*a*) rotating the disk at a steady angular velocity of 10 rpm and (*b*) concurrently measuring the torque necessary to maintain this velocity (at the full-scale reading the acting torque is 7200 dyne-cm).

Solution. Two assumptions are made in computing a radius to fulfill the problem conditions: (*a*) that the viscous resistance due to end effects can be compensated for by increasing the actual radius by a theoretical addition factor equal to 1.32 ($= 0.44 \cdot 3.0$), and (*b*) that the two layers of liquid parallel to the disk and at a distance x ($= 3.0$) above and below it, respectively, remain stationary as the disk revolves. Essentially the liquid cylinder below the disk is similar to that given in Fig. 2-13, whereas the cylinder of liquid above the disk is the reverse of Fig. 2-13. The actual setup is shown in Fig. 2-14. The assumptions made represent attempts to simplify the calculations and will of necessity have to be checked for plausibility by an eventual calibration of the disk against liquids of known viscosities.

To take advantage of the full torque scale, bracket the maximum torque reading at full scale (7200 dyne-cm) with the maximum viscosity that is to be measured (40.0 poises).

The torque acting will be equally divided between the upper and lower cylinders of liquid. Hence, halve the maximum torque when substituting in Eq. 78. Also, the value of R to be entered into the calculation must be increased to $R + 1.32$ to compensate for end or edge effects.

$$40.0 = \frac{3600}{10 \cdot 2 \cdot \pi/60} \cdot \frac{2 \cdot 3.00}{\pi(R + 1.32)^4}$$

$$(R + 1.32)^4 = \frac{3600 \cdot 6.0 \cdot 6.0}{40 \cdot 2 \cdot \pi^2} = 164$$

$$R + 1.32 = 3.58; \qquad R = 2.26 \text{ cm}$$

Comment. The above set of conditions and the given and calculated dimensions closely approximate the No. 2 spindle of the Brookfield RVF viscometer.

FALLING SPHERE VISCOMETER

The falling sphere viscometer measures the viscosity of a liquid based on the rate of fall of a sphere through the test liquid. The test is simple and can be made to yield accurate viscosity data for Newtonian systems by resorting to a proper mathematical treatment.

Theory

Equation 79 is a theoretical expression for computing the viscous resistance encountered by a sphere when moving through a Newtonian liquid with uniform velocity (Stokes's law).

$$F = 6\pi r \eta v \tag{79}$$

where F = viscous resistance (dynes)
r = radius of sphere (cm)
η = viscosity (poises)
v = velocity (cm/sec)

In deriving this expression, Stokes assumed the sphere to reside in a liquid bath of infinite extent. By using a modified form of this equation, it is possible to work out a method for measuring the viscosity of a liquid based on the free fall of a sphere in a container of *finite* dimensions, as shown in Fig. 2-15.

Consider first the derivation of the basic equation for the falling sphere viscometer. Let a sphere have radius r and density ρ_s. The volume of the sphere, V, is then equal to $\frac{4}{3}\pi r^3$. When the sphere is placed in a liquid of density ρ, the weight of the sphere (gravitational pull) acts to pull it down through the liquid, whereas the buoyant force of the liquid acts to push the sphere upward. The end effect is a net force equal in value to Eq. 80.

$$F = \frac{4}{3}\pi r^3(\rho_s - \rho)g \tag{80}$$

When a sphere is allowed to fall freely through a liquid under the influence of gravity, the active force (computed by Eq. 80) is opposed by a viscous resisting force (computed by Eq. 79). When these two forces are exactly balanced, the sphere falls with a constant velocity v. Setting these two forces equal to each other for this constant velocity condition results in Eq. 81.

$$6\pi r\eta v = \frac{4}{3}\pi r^3(\rho_s - \rho)g \tag{81}$$

Solving for η gives Eq. 82 for the liquid viscosity.

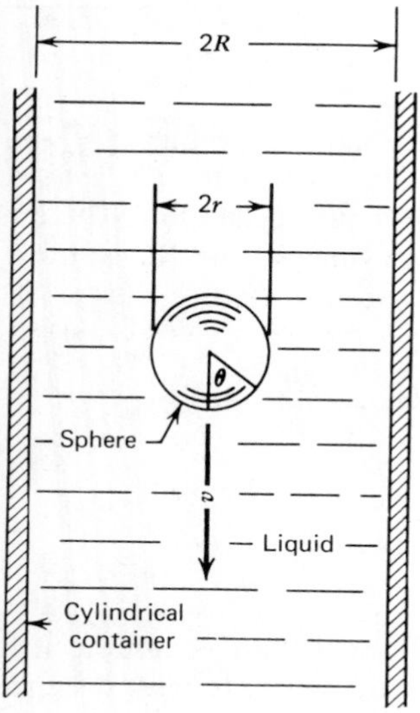

Fig. 2-15 Schematic drawing of the falling sphere viscometer (see ASTM D1343).

$$\eta = \frac{1}{v}\frac{2}{9}r^2(\rho_s - \rho)g$$

$$= \frac{1}{v}\,218r^2(\rho_s - \rho) \qquad (82)$$

This equation holds for the free fall of the sphere in a liquid bath of infinite extent. For a free fall in a cylindrical container of finite dimensions (radius R), the viscosity as calculated by Eq. 82 must be multiplied by a wall correction factor given by Eq. 83.

$$\text{Wall correction factor} = 1 - 2.104\,\frac{r}{R} + 2.09\left(\frac{r}{R}\right)^3 - 0.95\left(\frac{r}{R}\right)^5 \qquad (83)$$

where r = radius of falling sphere (cm)
R = inside radius of cylindrical container (cm)

Reduction to Practice

The falling sphere viscometer has the advantages of simplicity and inexpensiveness. Provided that it is used to measure the viscosity of a Newtonian liquid, good results can be obtained which are expressible in fundamental poise units. Employment of the falling sphere viscometer for non-Newtonian liquids is not recommended.

PROBLEM 2-17

A commercial adaptation of the falling sphere viscometer specifies a $\frac{5}{16}$-in. (0.794-cm) diameter steel ball bearing (density 7.78 g/cm^3) for the sphere, which is allowed to fall freely through a solution of nitrocellulose (NC) contained in a 1.00-in. (2.54-cm) diameter (ID) glass cylinder. The NC solution has the following composition:

Component	Parts by Weight
Nitrocellulose (20 sec)	12.2
Ethyl acetate	17.5
Ethanol	22.0
Toluene	48.3

The density of this NC solution is 0.91 g/cm^3. Under these conditions it is determined that a period of 20.0 sec is required for the sphere to fall a distance of 10.0 in. with uniform velocity through the NC solution. Calculate the viscosity of this 12.2% solids NC composition.

Solution. The viscosity of the NC solution is calculated by substituting the given data in Eq. 82. The value for the velocity v is 1.27 cm/sec (= 20.0 · 2.54/20).

$$\eta = \frac{1}{1.27} \cdot 218 \cdot 0.397^2(7.78 - 0.91) = 186 \text{ poises}$$

This is the viscosity free from wall effects. Correction for the confining action of the cylinder walls is carried out by multiplying by a theoretical wall correction factor (Eq. 83). The r/R ratio is 0.312 (= 0.397/1.27).

$$\eta = 186(1 - 2.104 \cdot 0.312 + 2.09 \cdot 0.312^3 - 0.95 \cdot 0.312^5)$$

$$= 186(1 - 0.656 + 0.0635 - 0.0028) = 186 \cdot 0.405 = 75.3 \text{ poises}$$

It has been established by experiment that each second of fall through 10.0 in. of this apparatus for the given conditions corresponds to 377 cP of viscosity. With this relation as a basis for calculation the NC solution has a viscosity of 75.4 poises (= 3.77 · 20). Thus the theoretical value (75.3 poises) and the experimental value (75.4 poises) are seen to agree remarkably well, provided that the wall correction factor is included in the theoretical viscosity calculation.

ASTM Test Method. The following problem illustrates a simple application of the falling ball principle.

PROBLEM 2-18

ASTM test method D1343-56 specifies a falling sphere viscometer that consists essentially of a bottle (inside diameter 6.4 cm; height 6.70 cm) and a steel ball bearing ($\frac{3}{32}$-in. diameter). The test procedure consists in dropping the steel ball through a hole in the center of the bottle cap and accurately measuring the time it takes the ball to fall freely a distance of 2.00 in. with uniform velocity through the test liquid filling the bottle.

The density of the steel ball is 7.68 g/cm^3. The density of the NC solution is 0.97 g/cm^3. If it takes 36.0 sec for the steel ball to drop a distance of 2.00 in. through the solution at 25 C, calculate the viscosity of the NC solution at this temperature.

Solution. The viscosity of the NC solution is calculated by substituting the given data in Eq. 82. The value to be used for the velocity is 0.141 cm/sec (= 2.00 · 2.54/36.0). The radius of the ball is $\frac{3}{64}$ in. (0.119 cm).

$$\eta = \frac{1}{0.141}[218 \cdot 0.119^2(7.68 - 0.97)] = 147 \text{ poises}$$

This is the viscosity uncorrected for wall effects. Correction for side wall confinement is carried out by multiplying by the wall correction factor given by Eq. 83, where $r/R = 0.0372$ (= 0.119/3.20).

$$\eta = 147(1 - 2.104 \cdot 0.0372 + 2.09 \cdot 0.0372^3 - 0.95 \cdot 0.0372^5)$$

$$\eta = 147(1 - 0.0782 + 0.00011 - 0.00000007) = 49 \cdot 0.922 = 135.6 \text{ poises}$$

ASTM test method D1343-56 lists sphere sizes suitable for several ranges of viscosity (for a 2-in. fall). An adapted form of this tabulation is given in Table 2-5.

A drop period of between 20 and 100 sec usually provides adequate accuracy for this type of falling sphere viscometer.

Table 2-5 ASTM Sphere Specifications for Falling Sphere Viscometer

Diameter of Sphere (in.)	Radius of Sphere (cm)	Metal	Density (g/cm^3)	Weight (g)	Suggested Range of Viscosity (poises)
$\frac{1}{16}$	0.0794	Aluminum	2.82	0.00591	10–50
$\frac{1}{16}$	0.0794	Stainless steel	7.66	0.01605	35–150
$\frac{3}{32}$	0.1190	Stainless steel	7.66	0.0542	76–300
$\frac{1}{8}$	0.1585	Stainless steel	7.66	0.1277	125–600
$\frac{7}{32}$	0.2775	Stainless steel	7.66	0.6897	250–1800

***Hoeppler Precision Viscometer* (32).** Falling sphere viscometers of more sophisticated design are commercially available that permit the precise measurement of the viscosities of Newtonian systems (gases and liquids) through a range from 10^{-4} to 10^4 poises with a precision of ±0.1%. The Hoeppler viscometer is a highly refined instrument in which glass or metal balls are allowed to fall freely (actually, roll freely) down a 1.6-cm inside diameter tube that is tilted at an angle of 10° from the vertical. Both the tube and the balls are machined to a tolerance of 0.0005 mm. In this instrument the balls are made only slightly smaller than the tube bore (ratios from ~0.745 to 0.995). The time of traverse down the tube through a distance of 10.00 cm runs from 30 to 500 sec (balls are selected so that at least a minimum 30-sec requirement is met). The test fluid temperature is closely controlled by a surrounding water jacket that automatically holds any selected temperature constant to ±0.02 C°. The viscosity of the test fluid is calculated by multiplying the time required for the descent of the ball through the fluid (between two etched marks on the tube) by the difference between the density of the fluid and that of the ball and by the ball constant. The ball constant is obtained by calibration with fluids of standardized Newtonian viscosities. An equation relating the fluid viscosity to the instrument dimensions

ASTM cups failed to achieve any popularity. In the United States the most important surviving efflux viscometers are the Saybolt viscometer and the Ford, Shell, and Zahn cups. The Shell cup is not actually a survivor since it is a relatively new entry, but it appears destined to become solidly entrenched in the field.

In the 1964 edition of this book a nomograph was submitted that permitted the conversion of efflux times from one cup to equivalent efflux times for any one of nearly 20 other cups. In this edition only conversions among the Ford, Shell, and Zahn cups will be considered, as shown in Figs. 3-1*a* and *b*, since these are now the workhorse efflux cups of the U.S. coatings and ink industries (2-4).

In Chapter 2 it was pointed out that an appreciable part of the energy involved in the flow of a fluid through an orifice or capillary is diverted into getting the fluid into motion in the first place, during and after which it encounters viscous resistance. Hence any equation representing liquid flow through an efflux viscometer should recognize both the kinetic and viscous components of the flow. Such an expression takes the form of Eq. 1 (see Eq. 30 of Chapter 2), where ν is the kinematic viscosity in stokes (or centistokes) and t is the drain time in sec.

$$\nu = At - \frac{B}{t} \qquad (1)$$

Although Eq. 1 has sound theoretical significance, it takes second place in actual usage to the simpler relationship given by Eq. 2.

$$\nu = A(t - B) \qquad (2)$$

In both equations the constant B serves as a correction factor for the kinetic component of the efflux flow. In Eq. 2 it is notable that practical experience shows the value of B to decrease with an increase in liquid viscosity. This occurs because a slower flowing viscous material possesses a lower kinetic energy. As a matter of fact, with very viscous liquids Eq. 2 reduces to Eq. 3.

$$\nu = At \qquad (3)$$

Table 3-1 lists equations in the form of Eq. 2 for the Ford, Shell, and Zahn-type efflux cups that are presently marketed in the United States. From an inspection of these equations two conclusions can be drawn; (*a*) there is confirmation that the constant B is smaller for more viscous systems, and (*b*) for a given viscosity range the constant B becomes relatively smaller in progressing from an orifice to a capillary type of discharge. Note that the intercepts of the

Table 3-1 Efflux Viscosity Cups

Cup Designation	Radius of Orifice or Capillary (cm)	Equation for Converting Drain Time (sec) to Kinematic Viscosity (centistokes)	Kinematic Viscosity for Given Drain Time Values (centistokes) t (sec) 30	60	90
Ford (Stubby Capillary), ~100-cm^3 Capacity					
2	0.126	$\nu = 1.60\ (t - 23.5)$	10.4	58	106
3	0.170	$\nu = 2.31\ (t - 6.6)$	54	123	193
4	0.206	$\nu = 3.85\ (t - 4.5)$	98	214	329
Shell (Long Capillary), ~23-cm^3 Capacity					
S-1	0.089	$\nu = 0.226\ (t - 13)$	3.8	10.6	17.4
S-2	0.119	$\nu = 0.576\ (t - 5)$	14.4	31.7	49
S-3	0.152	$\nu = 1.51\ (t - 2)$	42	88	133
S-4	0.191	$\nu = 3.45\ (t - 1)$	100	204	307
S-5	0.229	$\nu = 6.5\ (t - 1)$	189	384	579
S-6	0.290	$\nu = 16.2\ (t - 0.5)$	478	964	1450
Zahn Type (Orifice Only), ~44-cm^3 Capacity					
1	0.100	$\nu = 1.1\ (t - 29)$	—	34	67
2	0.137	$\nu = 3.5\ (t - 14)$	56	161	266
3	0.188	$\nu = 11.7\ (t - 7.5)$	263	614	965
4	0.213	$\nu = 14.8\ (t - 5.0)$	370	814	1258
5	0.264	$\nu = 23t$	690	1380	2070

plotted cup lines with the bottom scale of Figs. 3-1*a* and *b* correspond to *B* constant values.

Both theory and practical experience indicate that the long capillary outlet of the Shell cup is preferable to the stubby capillary of the Ford cup or the orifice opening of the Zahn cup. The Shell cup capillary is more sensitive to viscosity change, exhibits a cleaner drain time break (no dribbling), and is more accurate (3). Both the Shell and the Zahn cup have a hemispherical bottom below a short cylindrical tube (1.58-cm radius for Shell cup; 1.64-cm radius for Zahn-type cup). They differ in that the Shell cup drains through a 1.0-in. capillary, whereas the Zahn-type cup drains directly through a hole (orifice). The Ford cup is an intermediate design, a conical bottom discharging to a stubby capillary (see Fig. 2-18).

Conversion of drain time t from one type of efflux cup to another can be carried out graphically or numerically. Equation 4, where subscripts 1 and 2 refer to the two cups in question, is suitable for carrying out a numerical conversion.

$$t_2 = \frac{A_1(t_1 - B_1)}{A_2} + B_2 \tag{4}$$

A graphical conversion can be carried out by resorting to the graphs given in Figs. 3-1*a* and *b*.

PROBLEM 3-1

A Ford No. 4 cup gives a drain time of 77 sec for a given coating composition. It is desired to shift to a Shell S-5 cup to reduce the time period for the test while maintaining equivalent accuracy. What drain time can be expected for the Shell S-5 cup, and what is the kinematic viscosity of the given composition?

Solution. Numerical: Substitute the given data in Eq. 4, taking appropriate constants from the listing in Table 3-1.

$$t_2 = \frac{3.85\ (77 - 4.5)}{6.5} + 1 = 44 \text{ sec}$$

Graphical: Enter the graph of Fig. 3-1*a* at a drain time *t* of 77 sec, proceed vertically upward (see dashed lines) to intersect Ford No. 4 cup line, proceed

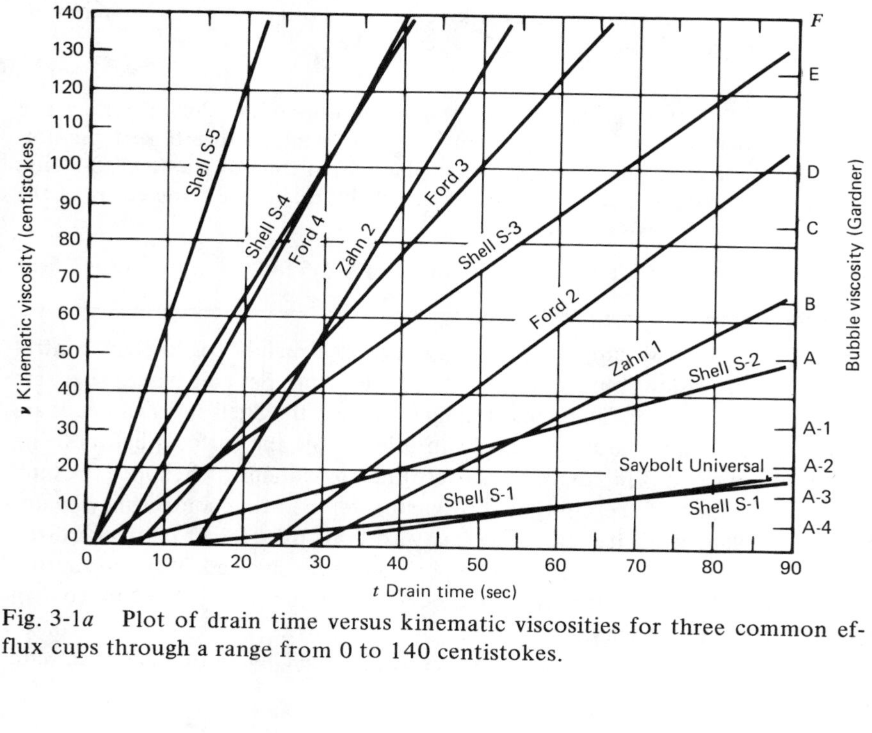

Fig. 3-1*a* Plot of drain time versus kinematic viscosities for three common efflux cups through a range from 0 to 140 centistokes.

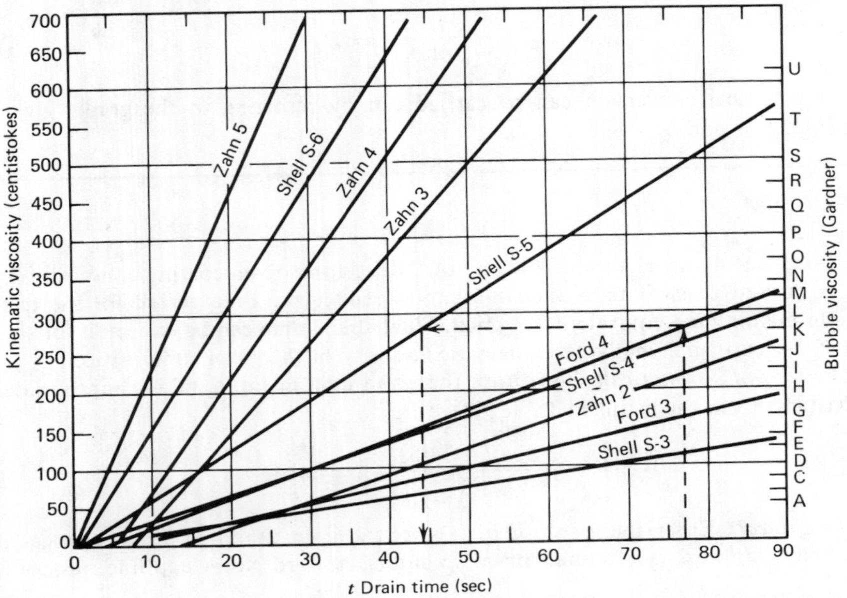

Fig. 3-1*b* Plot of drain time versus kinematic viscosity for three common efflux cups through a range from 0 to 700 centistokes.

horizontally to left from this intersection to intersect the Shell S-5 line, and proceed vertically downward from the intersection point to intersect the bottom scale at a value of 44 sec, the equivalent drain time of the Shell S-5 cup. The kinematic viscosity of 279 centistokes can be calculated from either of the efflux equations or read off the graph.

$\nu = 3.85\ (77 - 4.5) = 279$ (Ford 4) or $\nu = 6.5\ (44 - 1) = 279$ (Shell S-5)

The efflux or flow cup viscometers are intended mainly for in-plant control and field inspection purposes. The Ford cup and Saybolt viscometer are mounted on stands for filling and draining, whereas the Shell and Zahn cups are dipped directly into the test liquid (2 to 3 in. below the surface) before being quickly lifted (2 to 3 in. above the surface) to allow draining back into the bulk liquid. In the case of the dip-type efflux cups, temperature control may be difficult. For plant work it may be necessary to draw up a target or standardized viscosity/temperature plot for the composition being checked. The temperature of the efflux liquid is then used in checking its viscosity against this control viscosity/temperature curve.

The range of acceptable efflux times is governed by both theoretical and practical considerations. If the efflux drain time is too short, the flow period cannot be measured with precision. Furthermore, the correction factor B becomes more significant relative to the time t, which is also undesirable. Conversely, if the efflux time is unduly long, the procedure becomes impractical from a time consumption standpoint. As a compromise between these two extremes, practical experience indicates that drain times from 30 to 90 sec are suitable for the Ford, Shell, and Zahn cups. Hence, in selecting an efflux viscometer, it is suggested that a check be made of the centistoke ranges listed in Table 3-1 (see viscosities for the 30-, 60-, and 90-sec drain times).

CONVERSION OF BUBBLE RISE TIME TO KINEMATIC UNITS

The standard travel distance in the ASTM bubble viscometer is set at 7.30 cm, so that in most cases (viscous liquids) the time for a bubble to traverse this distance is also automatically the kinematic viscosity (inside tube diameter 1.065 cm). Although this relation holds true for kinematic viscosities above about 2.5 stokes, it fails to hold for viscosities below this level. The reason for this difference is that in the light viscosity range the kinetic energy factor can no longer be neglected as compared with the viscous resistance. As expected, the departure from equivalency between rise time and kinematic viscosity becomes more pronounced as lower viscosities are encountered. The trend of this deviation is shown in Fig. 3-2, where bubble rise time is plotted versus kinematic viscosity. Only the lower portion of the plot is shown, corresponding to the region where deviations arise, since above 3.0 stokes or 3.0-sec rise time (upper right corner of graph) a straight-line relationship exists (extension of the straight line given on graph).

The circles on the graph correspond to values as given in the literature (5). The solid curve represents the locus of Eq. 5, which was developed by the present author to represent the relationship between kinematic viscosity ν and ASTM bubble rise time t.

$$\nu = t - \frac{0.3}{t^2} \qquad (5)$$

PROBLEM 3-2

The bubble rise time for a test liquid in an ASTM bubble tube is 1.4 sec. Determine its kinematic viscosity.

Solution. Numerical: Substitute the given value of 1.4 sec in Eq. 5, and solve for the kinematic viscosity.

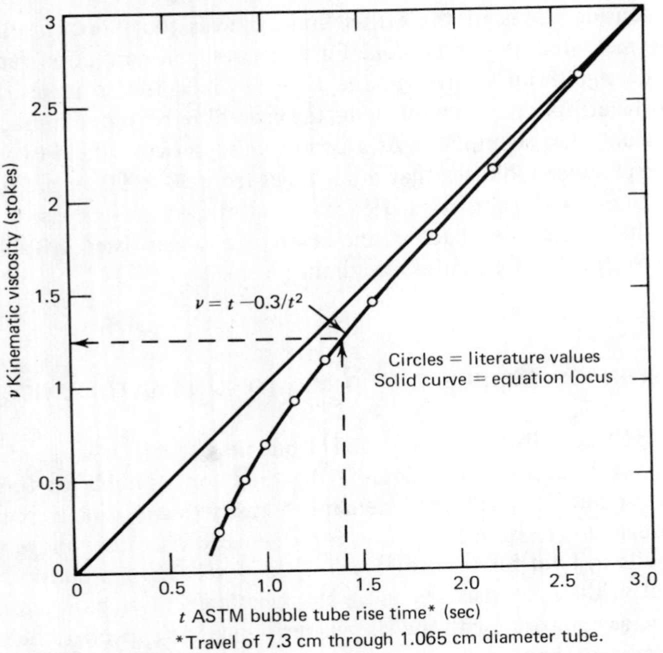

Fig. 3-2 Plot of bubble rise time versus kinematic viscosity for light-viscosity liquids (< 3.0 stokes).

$$\nu = 1.4 - \frac{0.3}{1.4^2} = 1.25 \text{ stokes}$$

Graphical: Enter the graph of Fig. 3-2 at a value of 1.4 sec; and, using the plotted curve, read off the corresponding value of 1.25 stokes (see dashed lines) on the kinematic viscosity scale.

As with any viscosity measurement, bubble rise time is very sensitive to any temperature change (~10% change/1.0 C). Departure from verticality is also critical (~10% change/5° slant).

CONVERSION OF WEIGHT LOADING OF STORMER PADDLE VISCOMETER TO KREBS UNITS AND POISES

From the standpoint of viscometer design, the Stormer paddle viscometer completely lacks the geometrical configuration demanded of a fundamental viscometer instrument. Despite this objection it is a viscometer that is still firmly entrenched in the coatings industry.

Results from a Stormer viscometer are given in terms of the weight loading required to rotate a paddle of specified dimensions (see Fig. 2-19) at a speed of 200 rpm in the test material. In turn this loading is converted to arbitrary Krebs units (KU) as shown in the graph of Fig. 3-3*a*. The curve relating Krebs units

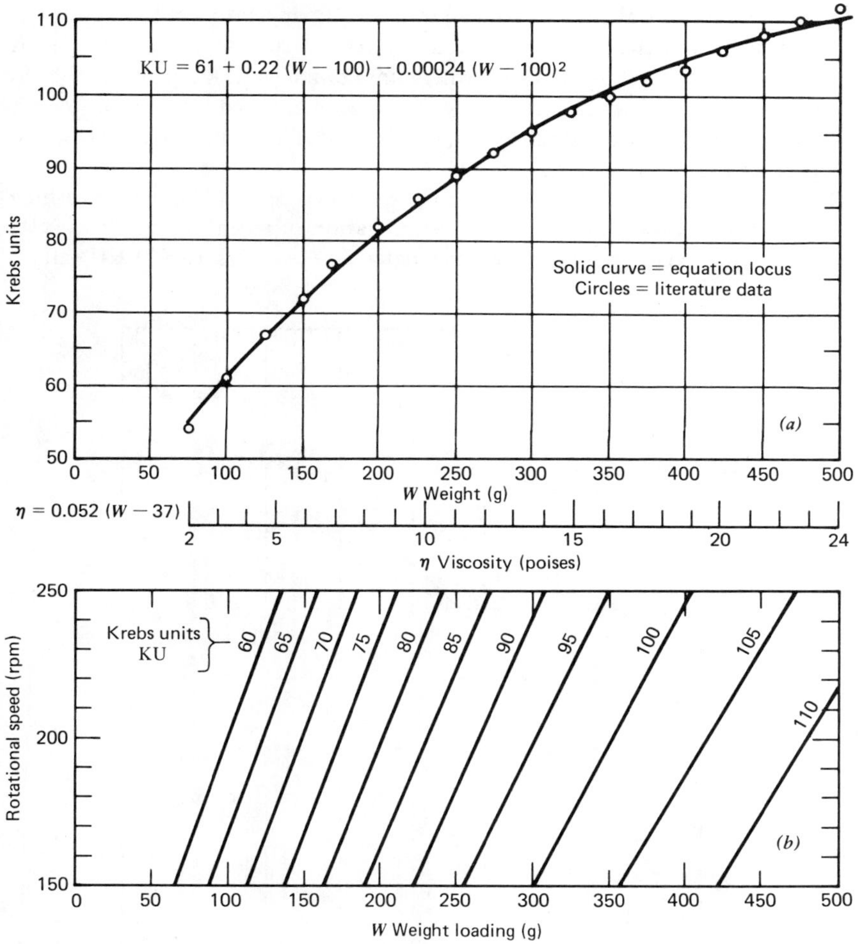

Fig. 3-3 (*a*) Plot of Stormer viscometer loading weight versus Krebs units (KU) for a speed of 200 rpm. (*b*) Lines showing constant Kreb unit values, plotted on a graph of weight loading versus rotational speed for the Stormer paddle viscometer.

with the weight loading W(g) at 200 rpm is based on Eq. 6. This expression, developed by the present author, covers the weight loading range from 50 to 500 g but not beyond.

$$KU = 61 + 0.22\,(W - 100) - 0.00024\,(W - 100)^2 \qquad (6)$$

Since an exact adjustment of the loading weights to achieve a 200-rpm speed can be time consuming, there has been developed a numerical compensating system wherein equivalent KU values are given, taking into account both the loading weight and the rotational speed (6). The nature of this adjustment is illustrated by the graph of Fig. 3-3*b*, where lines of constant KU value have been entered on a chart of weight loading versus rotational speed. Inspection of this graph shows that for a constant KU value higher loadings must be accompanied by higher speeds and vice versa.

Originally a KU value of 100 was supposed to correspond to good brushing properties for a house paint. However, examination of recent literature suggests that trade sales coatings today are formulated to lower (75 to 95) KU values.

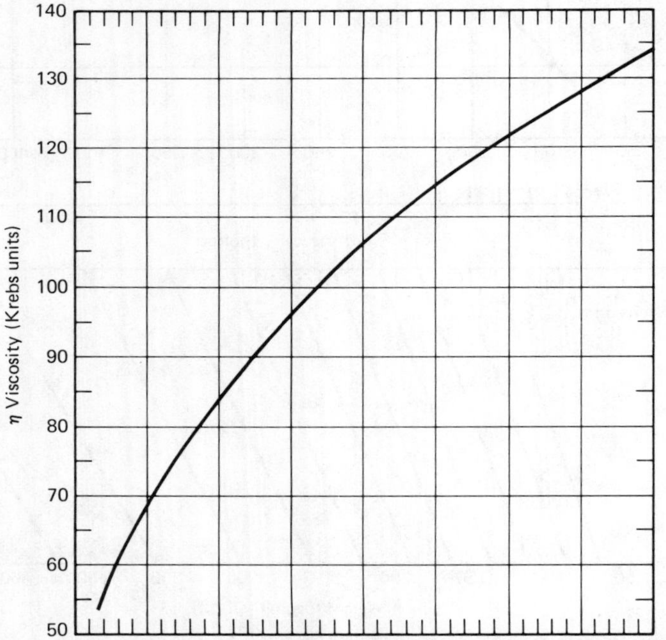

Fig. 3-4 Graph giving an approximate relationship between Krebs units (KU values) and viscosity in poises at a shear rate of about 400 sec^{-1}.

The two scales at the bottom of Fig. 3-3*a* provide a visual correlation of poises versus weight loading based on Eq. 7.

$$\eta \text{ (poises)} = 0.52\,(W - 37) \tag{7}$$

However, results provided by this relationship should be treated as rough approximations at best, since the equation assumes Newtonian flow and the pigmented coatings measured by the Stormer viscometer are seldom Newtonian in behavior.

Attempts to correlate KU units with true viscosity are somewhat questionable, since it has been demonstrated that the conversion depends markedly on the type of coating being measured (enamel, house paint, latex paint). Nevertheless, conversion charts relating the two viscosity measurements are widely reported in the technical literature; and an overall average-value graph, based on several of these conversion systems, has been worked out and is given in Fig. 3-4. This can be used to convert from KU units to poises or vice versa, with the understanding that errors of 10% or more can be involved.

The Stormer paddle viscosmeter is a venerable instrument that has served well in the past and is presently specified in some dozen official standards. However, it should now be retired and replaced by modern, fundamental viscometers that are specifically designed to furnish the basic rheological information demanded by present-day technology.

REFERENCES

1. Dallas Paint and Varnish Production Club, "Correlation of Viscosity Measurements," *Off. Dig.*, **24,** No. 329, 378 (1952).
2. Catalog C, "Ford Viscosity Cups," Gardner Laboratory, Inc., 5521 Landy Lane, Bethesda, Md. 20014.
3. PC-70-1, "Shell Cup Viscometers," Norcross Corporation, 255 Newtonville Avenue, Newton, Mass. 02158.
4. Brochure, "Zahn Computerized Viscosity Cups," Paul N. Gardner Company, Fort Lauderdale, Fla. 33316.
5. Catalog C, "Air Bubble Rheology," Gardner Laboratory, Inc., 5521 Landy Lane, Bethesda, Md. 20014.
6. ASTM Special Technical Publication 500, *Paint Testing Manual*, American Society for Testing and Materials, 1916 Race Street, Philadelphia, Pa. 19103, 1972, p. 189.

4 *Effect of Temperature, Binder (Polymer) Concentration, Solvent Viscosity, and Molecular Size on Viscosity*

The concepts developed in this chapter are applicable, in general, only to Newtonian systems, such as solvents and solvent solutions. Extension to non-Newtonian systems should be made with considerable reservation.

EFFECT OF TEMPERATURE ON VISCOSITY

Since even minor changes in temperature have a significant effect on viscosity, it is helpful to be able to compute this effect. Two situations will be considered. In the first the liquid viscosity is known for only one temperature. In the second the liquid viscosity is known for two (or more) temperatures.

Viscosity of Liquid Known for One Temperature Only

When the viscosity information is limited to a single viscosity value (one temperature only), the calculation of viscosity for some other temperature is subject to considerable uncertainty. Provided that some backlog of information is available, a viscosity/temperature correspondence with a liquid similar to the test liquid can be attempted. In the absence of such correlation with a related liquid, an approximate viscosity value can be estimated by the following graphical procedure.

It has been found experimentally that for any given viscosity η the change in viscosity $d\eta$ produced by a change in temperature dT is substantially the same for most liquids. Furthermore, the function $f(\eta)$ of Eq. 1 depends primarily on the magnitude of the viscosity only (it does not depend appreciably on the nature of the liquid).

$$\frac{d\eta}{dT} = f(\eta) \tag{1}$$

On the basis of this empirical finding a so-called universal viscosity curve has been established. By means of this standard relation, it is possible to estimate the viscosity of a liquid at some required temperature from its known viscosity at some other temperature. As generally submitted, this universal viscosity curve is a plot of the logarithm of viscosity versus temperature difference. However, since the use of a temperature difference involves an additional calculation step, Fig. 4-1 (a nomographic modification of the conventional graph) involves temperatures directly and thus eliminates the need for a temperature difference calculation.

The use of the nomograph is straightforward. The known viscosity and temperature values are located on their respective scales and then connected by a straight line to intersect the intermediate pivot line. To obtain a viscosity at some other specified temperature, a second straight line is drawn from the second temperature (as located on the temperature scale) through the intersection point of the first line with the pivot line. An extension of this second line intersects the viscosity scale at the required value.

PROBLEM 4-1

The viscosity of diethylene glycol ethyl ether at 25 C is 0.038 poise. Compute its viscosity at 60 C.

Solution. Locate the given values of 25 C and 0.038 poise on the temperature and viscosity scales of Fig. 4-1, and connect them by a straight line to intersect the pivot line. Next draw a straight line from the specified temperature of 60 C through the intersection point of the first line with the pivot line. An extension of this second line intersects the viscosity scale at the required value of 0.0175 poise (the literature value is reported as 0.0177 poise). The lines drawn across the graph correspond to the solution to this problem.

PROBLEM 4-2

The viscosity of heavy-gravity glycerol (99.0% pure) at 68 F is 15.0 poises. To what temperature must it be heated to reduce its viscosity to 1.0 poise?

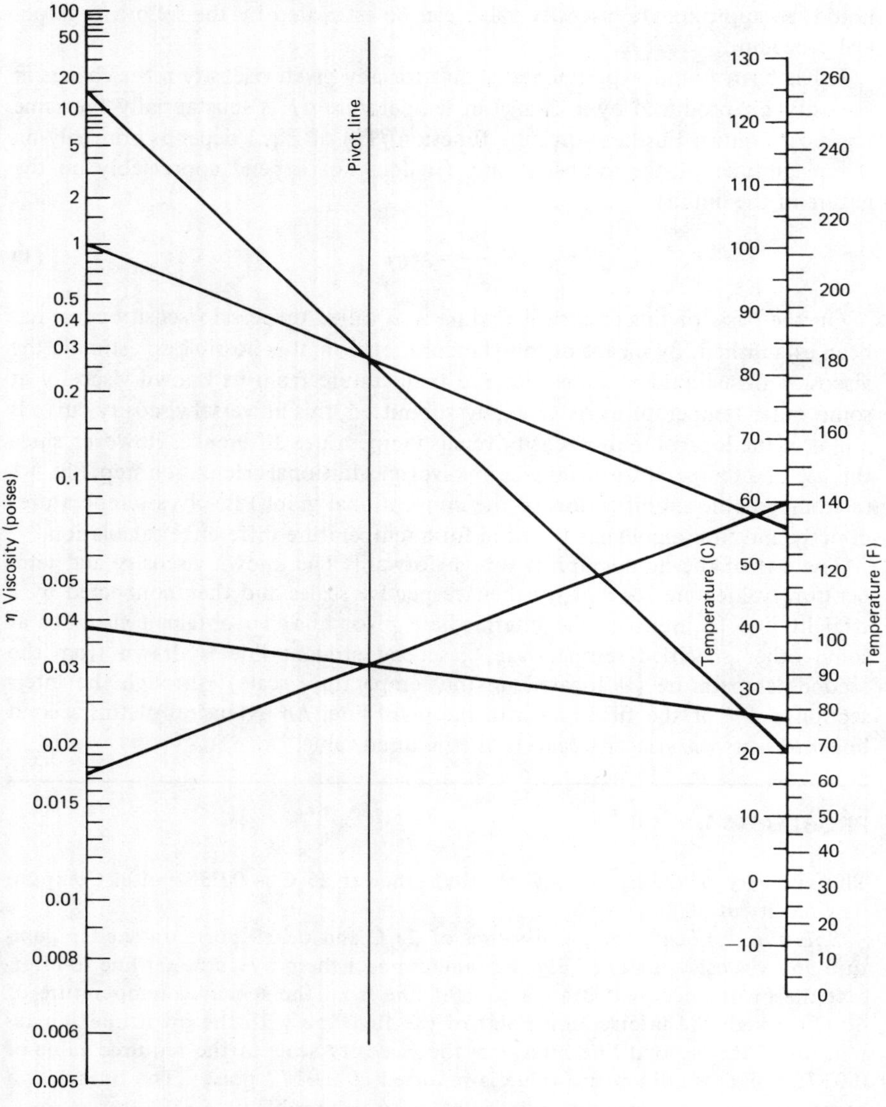

Fig. 4-1 Nomograph for determining the viscosity of a liquid at some required temperature, given the viscosity of the liquid at some other known temperature.

Solution. Connect the values of 68 F and 15.0 poises on their respective scales of Fig. 4-1 by a straight line, and mark the point of intersection of this line with the pivot line. Next draw a straight line from the value of 1.0 poise on the viscosity scale through the marked point of intersection (of the first line), and extend to intersect the temperature scale at the required value of 132 F.

A result obtained using the universal viscosity curve relationship serves simply as an approximation to an actual or true viscosity. However, for quickly establishing a rough estimate of the viscosity at some required temperature (given the known viscosity at some other temperature) the nomograph is a valuable graphical device. It is rather interesting that, whereas the temperature level determines the absolute viscosity of a liquid, it is the viscosity itself that largely determines the change in viscosity that occurs because of a change in temperature.

Viscosity of Liquid Known for Two Temperatures

When the viscosity of a given liquid is known for two different temperatures that are reasonably well separated from each other, the viscosity for a third temperature can be computed with good to excellent accuracy for interpolated values and with fair to good accuracy for extrapolated values.

Of the many equations that have been proposed for relating viscosity to temperature, one appears to represent the viscosity/temperature relationship most accurately. It is commonly referred to as Andrade's equation (Eq. 2).

$$\eta = A \cdot 10^{B/T} \tag{2}$$

Equation 2 can be expressed alternatively in logarithmic form as Eq. 3.

$$\log \eta = \log A + \frac{B}{T} \tag{3}$$

Temperature T must be expressed in absolute units (K = 273 + C or R = 460 + F), and A and B are constants for the liquid in question.

If subscripts 1 and 2 are used to denote the conditions for two different temperatures, it can be readily shown (by subtraction) that the two conditions are related by Eq. 4.

$$\log \frac{\eta_1}{\eta_2} = B\left(\frac{1}{T_1} - \frac{1}{T_2}\right) \tag{4}$$

PROBLEM 4-3

The viscosity of linseed oil is 0.60 poise at 10 C (283 K) and 0.18 poise at 50 C (323 K). Calculate its viscosity at 30 C and 90 C.

Solution. Using Eq. 4 and the given information, compute a value for the constant B that applies to linseed oil.

$$\log \frac{0.60}{0.18} = B\left(\frac{1}{283} - \frac{1}{323}\right)$$

$$0.5229 = B(0.003534 - 0.003096); \qquad B = 1194$$

Substitute this calculated value for B in Eq. 3, and calculate a value for the constant $\log A$.

$$\log 0.60 = \log A + \frac{1194}{283}$$

$$-0.2218 = \log A + 4.219; \qquad \log A = -4.4408$$

The relationship between viscosity and temperature for linseed oil is then given by Eq. 5.

$$\log \eta \text{ (linseed oil)} = \frac{1194}{T} - 4.4408 \tag{5}$$

From Eq. 5 calculate the required viscosities for linseed oil at 30 C (303 K) and 90 C (363 K).

$$\log \eta = \frac{1194}{303} - 4.4408 = -0.5002; \qquad \eta = 0.316 \text{ poise (30 C)}$$

$$\log \eta = \frac{1194}{363} - 4.4408 = -1.151; \qquad \eta = 0.071 \text{ poise (90 C)}$$

The form of Eq. 3 is such that in practice a plot of $\log \eta$ versus $1/T$ results in a straight line over a considerable range of experimental values. This relationship also permits the construction of a nomograph such as that given in Fig. 4-2, where one scale carries $\log \eta$ values and the other $1/T$ values. This nomograph permits a graphical solution to the above type of problem, namely, the determination of the viscosity at some requested temperature, given the viscosities at two other temperatures. Thus Problem 4-3 is solved graphically as follows:

1. Draw two straight lines across the chart connecting the two known paired temperature and viscosity values as located on their temperature and viscosity scales. The point of intersection of these two lines is marked and serves as a pivot point for subsequent lines that are drawn across the diagram. (As an example, the solid lines drawn across the nomogram of Fig. 4-2 represent the given information of Problem 4-3.)
2. Using the intersection point determined in step 1, pass straight lines through this point, starting either with a temperature value to obtain a viscosity value or with a viscosity value to obtain a temperature value (the two dotted lines drawn across the nomogram represent answers to Problem 4-3).

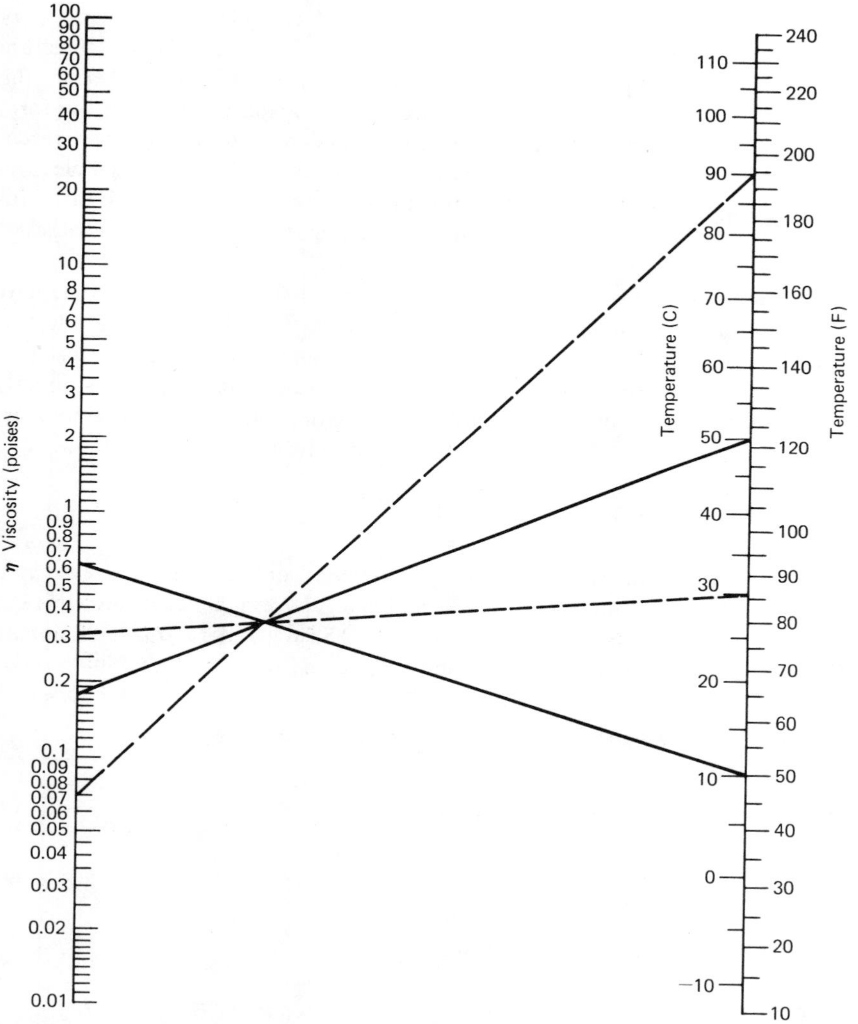

Fig. 4-2 Nomograph for determining the viscosity of a liquid at some required temperature, given the viscosities of the liquid at two other known temperatures.

Incidentally the graph of Fig. 4-1 can also be used in the same way as outlined for Fig. 4-2. In the case of Fig. 4-1, the intersection point of the two lines connecting the known viscosity/temperature pairs may or may not fall on the pivot line. This is not unexpected since the pivot line location simply represents the locus of the universal *average* for all intersection points.

Graphical Solutions. To avoid the mathematical computations demanded by Eqs. 3 and 4, to speed up the calculation process, and to establish a visual picture of the viscosity/temperature relationship, resort can be made either to the nomographs of Figs. 4-1 and 4-2 or to coordinate graph paper. Thus the form of Eq. 3 is such that, if a plot of viscosity versus temperature is made on coordinate paper that carries logarithmic and reciprocal temperature scales, respectively, a straight line results. Unfortunately, commercial graph papers with log cycles (ordinate scale) and reciprocal temperature values (abscissa scale) are no longer obtainable.

However, there are available a number of excellent graph papers that have been developed by workers in the petroleum industry for the purpose of plotting precise viscosity/temperature relationships through a wide range of temperatures. Although developed for petroleum products, these graphs are directly applicable to the viscosity problems of the coatings industry. The graphs are based on complex viscosity/temperature relationships of the form of Eq. 6.

$$\log \log (\nu_2 + 0.8) = n \log \frac{T_2}{T_1} + \log (\nu_1 + 0.8) \tag{6}$$

Hence, if improved plotting accuracy is required or if a temperature range is wider than can be conveniently handled by regular graph paper, recourse can be made to the charts that are described in ASTM standard D341, "Standard Viscosity-Temperature Charts for Liquid Petroleum Products" (see Table 4-1).

Table 4-1 Commercial Viscosity/Temperature Graph Paper

Chart Number and Coding	Kinematic Viscosity Range (centistokes)	Temperature Range	Size (in.)
(I) 12-403411-12	0.3 to 20,000,000	−70 to 370 C	$26\frac{3}{4} \times 32\frac{1}{2}$
(III) 12-403413-12	0.3 to 20,000,000	−70 to 370 C	$8\frac{1}{2} \times 11$
(IV) 12-403415-12	0.3 to 20,000,000	−100 to 700 F	$26\frac{3}{4} \times 32\frac{1}{4}$
(II) 12-403412-12	0.18 to 6.5	−70 to 370 C	$20\frac{1}{2} \times 32\frac{1}{4}$
(IV) 12-403414-12	0.18 to 6.5	−70 to 370 C	$8\frac{1}{2} \times 11$
(VI) 12-403416-12	0.18 to 3.0	−100 to 700 F	$20\frac{1}{2} \times 32\frac{1}{4}$

ASTM Standard D341, American Society for Testing and Materials, 1916 Race Street, Philadelphia, PA 19103.

By using the ASTM type of chart, excellent prediction of a third viscosity can be made based on two known viscosity values (a reasonable separation in temperatures for the two viscosities is assumed). Even extrapolation to a pour point (about 2000 poises) is not unreasonable when using two widely spaced higher temperatures. As shown by Problem 4-4, the projection of Andrade's equation (Eq. 3) into regions of very high viscosity (>1000 poises) is questionable.

PROBLEM 4-4

Castor oil is reported to have a viscosity of 6.5 poises at 25 C and 0.17 poise at 100 C. Compute its pour point both from Eq. 3 and from ASTM Chart I, and compare with its experimentally determined pour point of -10 F (-23 C).

Solution. Assume for both the numerical and graphical computation that the pour point viscosity for castor oil is 2000 poises.

Numerical Computation Using Eq. 3: Substitute appropriate given data in Eq. 4, and solve for the constant B.

$$\log \frac{6.5}{0.17} = B \left(\frac{1}{298} - \frac{1}{373} \right); \qquad B = 2344$$

Substitute this value for B in Eq. 3 to obtain a value for $\log A$.

$$\log 6.5 = \log A + \frac{2344}{298}; \qquad \log A = -7.053$$

The relationship between viscosity and temperature for castor oil is then given by Eq. 7.

$$\log \eta \text{ (castor oil)} = \frac{2344}{T} - 7.053 \tag{7}$$

From Eq. 7 calculate the pour point temperature of castor oil corresponding to a viscosity of 2000 poises.

$$\log 2000 = 3.301 = \frac{2344}{T} - 7.053$$

$$T = \frac{2344}{10.35} = 226 \text{ K } (-47 \text{ C})$$

Graphical Computation Using ASTM Chart I: Read centistokes on the chart as centipoises. Connect the viscosity value of 650 centistokes at 25 C (77 F) with the value of 17 centistokes at 212 F (100 C) by a straight line. Extend this straight line to intersect the 2000-poise (200,000-centistoke) line, and read off the pour point of castor oil: -23 C (-10 F).

Comment. The graphically determined pour point for castor oil is the same as the value determined experimentally (−23 C). The numerically calculated pour point (−47 C) is in error by 24 C.

In general, for coating problems involving viscosity/temperature relationships coordinate graph paper designed specifically for this type of calculation can be used to advantage. Results are obtained in a fraction of the time required for a numerical solution, the relationship expressed by the graph is usually highly accurate for most Newtonian liquids, the chance of error is minimized, and the visual portrayal of the relationship is helpful in understanding the interplay of the variables involved.

EFFECT OF BINDER (POLYMER) CONCENTRATION ON VISCOSITY OF ORGANIC SOLVENT SOLUTIONS

A common viscosity problem calls for calculating the change in solution viscosity produced by a change in polymer concentration. In this discussion the term "polymer" is widely interpreted to include both oligomers and resins.

Equations Relating Viscosity to Polymer Concentration

Any equation relating viscosity to polymer concentration should ideally cover a complete range of concentrations yet involve a minimum number of variables. One expedient in approaching this ideal is to avoid the use of densities, say by specifying the test composition in terms of the weight fractions of the two components (solvent and polymer solute). As a second expedient, viscosity ratios rather than viscosities can be used to reduce the number of equation parameters. The symbols listed below will be used in the following discussion, which relates solution viscosity to polymer concentration:

w = weight fraction of solute (polymer concentration)
$1 - w$ = weight fraction of solvent
η_0 = viscosity of solvent ($w = 0$)
η = viscosity of solution ($w = w$)
η_p = viscosity of polymer (oligomer) ($w = 1.00$)
$\eta_r = \eta/\eta_0$ = relative viscosity (viscosity ratio)

The most widely quoted and simplest relationship is Einstein's classic equation, given as Eq. 8, where ϕ is the volume fraction of solute (1). Unfortunately, this equation is limited to very dilute solutions of hard spherical particles and hence is unsuitable for industrial compositions, where the trend is to ever-higher solids content and hydrated molecules.

$$\eta_r = \frac{\eta}{\eta_0} = 1 + 2.5\phi \tag{8}$$

Equation 8 has been expanded by numerous investigators in attempts to cover higher concentrations. These modified equations generally involve higher powers of ϕ and frequently introduce the concept of a hydrodynamic volume (a spherical entity corresponding in effective volume to the solvated or swollen polymer molecule) (2). This whole approach tends to be highly research oriented, is theoretically quite complex, and at the present time appears to be of questionable technical value to the practicing coatings engineer.

In the 1964 edition of this book a number of equations were discussed for relating viscosity to polymer concentration. However, this author is convinced that they should all defer to Eq. 9, which is of general utility and appears to be the one best suited for practical coating work.

$$\log \eta_r = \frac{w}{k_a - k_b w} \tag{9}$$

Equation 9 is fairly simple and directly involves the two key variables η_r and w. The two constants k_a and k_b can be readily obtained either by calculation or by graphical plotting. In the case of oligomers, Eq. 9 is capable of covering a complete range of viscosities from 0 to 100% solute content (3). For higher molecular weight polymers, however, viscosities obtained by extrapolation from lower to higher weight fractions should be considered as becoming more uncertain in moving away from the highest measured viscosity value. In fact, any calculated 100% polymer viscosity should be regarded strictly as an equation constant rather than an actual viscosity for the high-molecular-weight polymer in question.

Equation 9 can also be rewritten as Eq. 10.

$$\frac{1}{\log \eta_r} = \frac{k_a}{w} - k_b \tag{10}$$

If subscripts 1 and 2 are assigned to identify two different weight concentrations of a polymer, then, by setting up two equations for these two conditions in the form of Eq. 10 and subtracting, the constant k_b is eliminated, leading to Eq. 11.

$$k_a = \frac{1/\log \eta_{r_1} - 1/\log \eta_{r_2}}{1/w_1 - 1/w_2} \tag{11}$$

Hence the constant k_a can be determined from a knowledge of the viscosities for two different concentrations (a reasonably wide spread between them is assumed). In turn, k_b can be calculated by inserting this k_a value in Eq. 10. From these data a complex viscosity concentration relationship can be established for a given composition.

PROBLEM 4-5

A polyester resin dissolved in styrene ($\eta_0 = 0.0088$ poise) has a viscosity of 0.531 poise at a 50%w concentration and a viscosity of 24.7 poises at an 80%w concentration. Calculate the viscosity of this polyester in styrene at 60%w and 100%w concentrations.

Solution. Using the given data, calculate a value for the constant k_a from Eq. 11.

$$\eta_r\,(w = 0.50) = \frac{0.531}{0.0088} = 60.3; \qquad \log 60.3 = 1.780$$

$$\eta_r\,(w = 0.80) = \frac{24.7}{0.0088} = 2810; \qquad \log 2810 = 3.447$$

$$k_a = \frac{1/1.780 - 1/3.447}{1/0.50 - 1/0.80} = 0.3623$$

Substitute this value for k_a in Eq. 10 to obtain the value for k_b.

$$k_b = \frac{0.3623}{0.80} - \frac{1}{\log 2810}; \qquad k_b = 0.1629$$

The relative viscosity of the polyester in styrene for any weight fraction w is then given by Eq. 12.

$$\log \eta_r = \frac{w}{0.3623 - 0.1629w} \tag{12}$$

Substitute in turn the weight fractions of 0.060 and 1.00 in Eq. 12 to obtain relative viscosities, and from these data calculate the required solution viscosities for the two specified concentrations.

For $w = 0.60$:

$$\log \eta_r = \frac{0.60}{0.3623 - 0.1629 \cdot 0.60}$$

$$= 2.2678; \qquad \eta_r = 185.2; \qquad \eta = 1.63 \text{ poises}$$

For $w = 1.00$:

$$\log \eta_r = \frac{1.00}{0.3623 - 0.1629}$$

$$= 5.051; \qquad \eta_r = 103{,}500; \qquad \eta = 911 \text{ poises}$$

Examination of Eq. 9 reveals some interesting interrelationships. As expected, if only solvent is present ($w = 0$), $\log \eta_r = 0$ and $\eta_r = 1.0$, making $\eta_{w=0} = \eta_0$. When only polymer is present ($w = 1.00$), Eq. 9 reduces to Eq. 13.

$$\log \eta_r = \log \frac{\eta_p}{\eta_0} = \frac{1}{k_a - k_b}$$

or

$$k_a - k_b = \frac{1}{\log (\eta_p/\eta_0)} \tag{13}$$

Equation 13 tells us that, of the innumberable theoretical curves that could be drawn relating viscosity to concentration between two given terminal conditions ($w = 0$ and $w = 1.00$), the difference between k_a and k_b remains constant, since η_p and η_0 are constant for any given system. If k_b happens to be zero, a straight line results from a plot of viscosity (log scale) versus concentration (linear scale) on semilog graph paper. However, a positive value for k_b results in a bending of the curve downward toward the concentration scale in proportion to the magnitude of k_b; a negative value for k_b, an upward bending. This effect is illustrated in Fig. 4-3*a* for a hypothetical case where the solvent viscosity is 0.01 poise and the polymer (oligomer) viscosity is 1000 poises (range in η_r from 1.00 to 100,000). Note that the difference $(k_a - k_b)$ remains constant and equal to the expression $1/\log (\eta_p/\eta_0)$. A value can be selected for either k_a or k_b but not both, since a value assigned to one determines the value of the other. In the given example, values of 0.1, 0.2, 0.3, 0.4, 0.7, and 1.2 were assigned to k_a. Corresponding values for k_b are less by a value of 0.2.

Turning to Eq. 10 (an alternative form of Eq. 9), let both sides be multiplied by w to give Eq. 14.

$$\frac{w}{\log \eta_r} = k_a - k_b w \tag{14}$$

Now, as w approaches zero, the term $k_b w$ becomes negligibly small in comparison with k_a. However, the term $w/\log \eta_r$ does not become vanishingly small, since $\log \eta_r$ is also approaching zero. As a result, as the polymer solution becomes ever more dilute ($w \to 0$), the ratio $w/\log \eta_r$ approaches a value of k_a as a final value (when $w = 0$).

A similar situation arises in the case of so-called intrinsic viscosity, which is discussed more fully later. Intrinsic viscosity $[\eta]$ is defined by the expression $(\ln \eta_r/c)_{c \to 0}$, where the concentration c is expressed in g/dl (gram per deciliter). Its value is obtained by extrapolation to a zero value. A comparison of k_a with $1/[\eta]$ shows that they have much in common *at infinite dilution.* In fact, they differ only in the choice of the logarithm base (10 vs. e) and in the units chosen for expressing the concentration (g/g vs. g/dl).

$$k_a = \frac{w}{\log \eta_r}; \qquad \frac{1}{[\eta]} = \frac{c}{\ln \eta_r}$$

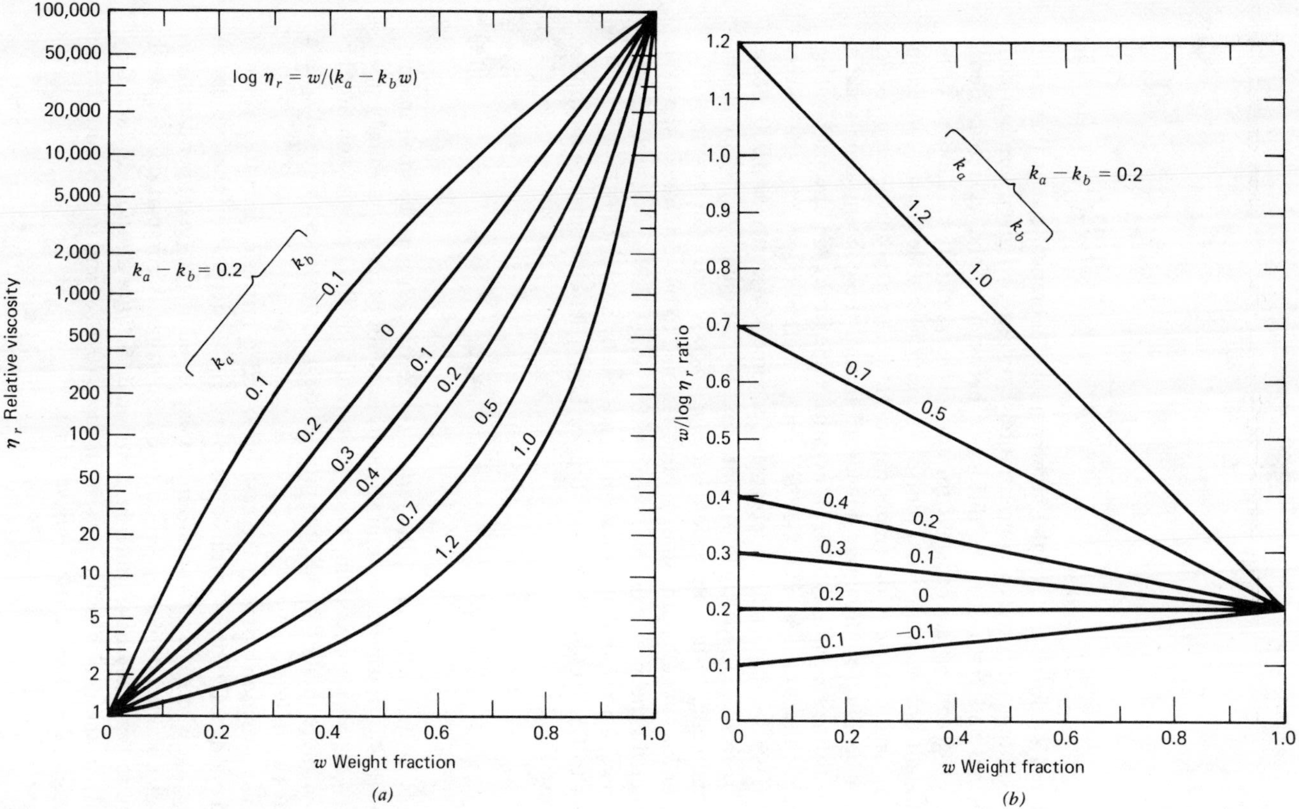

Fig. 4-3 (*a*) Plot of relative viscosity versus weight fraction of dissolved oligomer for hypothetical case where the solvent viscosity is 0.01 poise and the oligometer viscosity is 1000 poises. (*b*) Plot the ratio $w/\log \eta_r$ versus the weight fraction w of dissolved oligomer where the solvent viscosity is 0.1 poise and the oligomer viscosity is 1000 poises.

If the solvent density is taken as ρ, then, at nearly infinite dilution, this density must be essentially the solution density also. Hence Eq. 15 linking $[\eta]$ and k_a applies.

$$[\eta] = \frac{0.02303}{\rho k_a} \tag{15}$$

Graphical Plotting of Viscosity/Concentration Data

The form of Eq. 14 is such that, if $w/\log \eta_r$ is plotted against w, a straight line results. This is shown in Fig. 4-3*b*, where the curves of Fig. 4-3*a* have been replotted in this alternative manner. The intercept of any given line with the $w/\log \eta_r$ scale gives its k_a value. If the line is horizontal, (no slope), $k_b = 0$ and $k_a = 1/\log(\eta_p/\eta_0)$. If the line is sloped downward to the right, k_b has a positive value that is equal to $k_a - 1/\log(\eta_p/\eta_0)$. This value is also equal to the difference between the two $w/\log \eta_r$ values as read off the extreme left- and right-hand scales corresponding to $w = 0$ and $w = 1.00$. From practical experience it has been found that a plot of most experimental data in the form of $w/\log \eta_r$ versus w almost always gives a reasonably straight line that can be extended to intersect the outside scales. From such a plot an expression for the viscosity/concentration relationship can be immediately established.

PROBLEM 4-6

Solution viscosities were determined for an acrylic oligomer in diacetone alcohol ($\eta_0 = 0.031$ poise, $\rho = 0.94$ g/cm^3) at the weight concentrations listed in Table 4-2. Derive an expression for the viscosity/concentration relationship for this system, and from this expression calculate a theoretical viscosity for the oligomer and the intrinsic viscosity of the oligomer in diacetone alcohol.

Solution. From the given data calculate the relative viscosities, and from these the $w/\log \eta_r$ ratios for the given oligomer concentrations (see Table 4-2). Plot values of $w/\log \eta_r$ versus w on rectilinear coordinate paper, connect the plotted data by a straight line, extend the line to either side to intersect the outside scales, and read off values for $w/\log \eta_r$ at the points of intersection. This type of plotting has been carried out in Fig. 4-4. The intercept value of 0.375 at $w = 0$ corresponds to k_a, and the difference between the two intercept values of 0.275 (= 0.375 − 0.100) corresponds to k_b. The required viscosity/concentration relationship is then given by Eq. 16.

$$\log \eta_r = \log \frac{\eta}{\eta_0} = \frac{w}{0.375 - 0.275w} \tag{16}$$

The viscosity of the oligomer is obtained by substituting $w = 1.00$ in Eq. 16 and solving for η_p.

Table 4-2 Calculating Values for Problem 4-6

Given		Calculated		
w	η	η_r (= η/η_0)	$\log \eta_r$	$w/\log \eta_r$
0.1	0.060	1.93	0.285	0.351
0.2	0.132	4.25	0.628	0.318
0.3	0.341	11.0	1.041	0.288
0.4	0.983	31.7	1.501	0.266
0.5	3.91	126.0	2.10	0.238
0.6	21.5	693.0	2.84	0.211
0.7	187.0	6040.0	3.78	0.185

$$\log \frac{\eta_p}{0.031} = \frac{1}{0.375 - 0.275} = 10.0; \qquad \eta_p = 3.1 \cdot 10^8 \text{ poises}$$

The intrinsic viscosity is obtained by substituting the value of $k_a = 0.375$ in Eq. 15.

$$[\eta] = \frac{0.02303}{0.375 \cdot 0.94} = 0.065$$

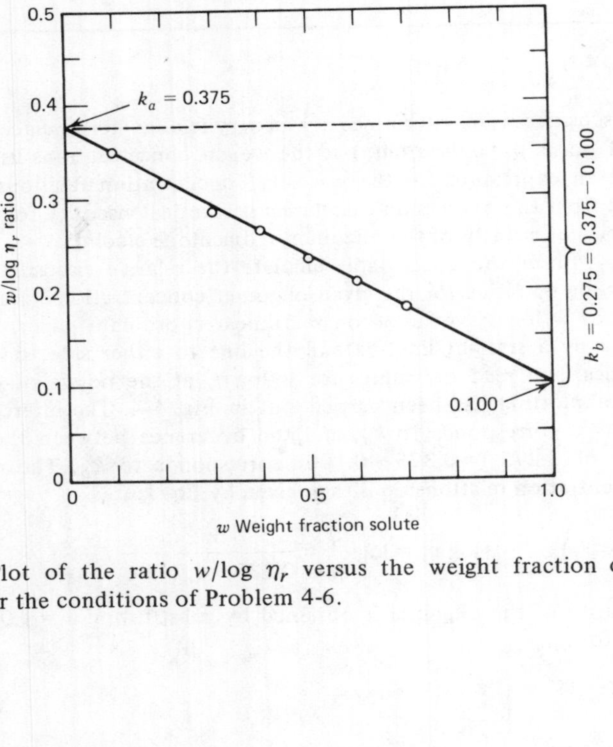

Fig. 4-4 Plot of the ratio $w/\log \eta_r$ versus the weight fraction of dissolved oligomer for the conditions of Problem 4-6.

Approximate Viscosity/Concentration Equation

If only one viscosity is known (at a single concentration), an approximate determination of viscosity/concentration relationship can still be attempted by ignoring the modifying term $k_b w$ of Eq. 9 (assume $k_b = 0$). This reduces Eq. 9 to the simpler but much less accurate Eq. 17.

$$\log \eta_r = \frac{w}{k_a} \tag{17}$$

For the case of a 100% polymer ($w = 1.0$), k_a becomes equal to $1/\log \eta_r = 1/\log (\eta_p/\eta_0)$. Substituting this value for k_a in Eq. 17 gives Eq. 18.

$$\log \eta_r = \log \frac{\eta}{\eta_0} = w \cdot \log \frac{\eta_p}{\eta_0}$$

or

$$\log \eta = w \cdot \log \eta_p + (1 - w) \cdot \log \eta_0 \tag{18}$$

PROBLEM 4-7

The viscosity of a long-oil alkyd in odorless mineral spirits ($\eta_0 = 0.012$ poise) at 40% NV (nonvolatile content by weight) is 2.5 poises. Calculate its viscosity at 30% and 50% NV content.

Solution. Substitute the given information in Eq. 18, and solve for η_p.

$\log 2.5 = 0.40 \log \eta_p + 0.60 \log 0.012$; $\log \eta_p = 3.88$; $\eta_p = 7540$ poises

Use this value of η_p to calculate the required viscosities at 30% and 50% NV contents.

For 30% NV:

$$\log \eta = 0.3 \cdot 3.876 + 0.7 \cdot (-1.921) = -0.182; \qquad \eta = 0.658 \text{ poise}$$

For 50% NV:

$$\log \eta = 0.5 \cdot 3.876 + 0.5 \cdot (-1.921) = 0.978; \qquad \eta = 9.5 \text{ poise}$$

Results obtained using Eq. 18 are very approximate because of the simplifying assumption made in its derivation ($k_b = 0$). Despite this disadvantage the fact that it yields straight lines on semilog paper makes it useful for roughly delineating viscosity/concentration information on graph paper. For example, the solution to Problem 4-7 could have been obtained without calculation by simply drawing a straight line between points located on semilog graph paper corresponding to the information submitted by the problem conditions (as indicated in Fig. 4-5). From the location of appropriate points on this line, the information required by the problem is obtained.

Viscosity of Cross Blends

Often in the literature, and occasionally in company brochures (4), the semilog type of graph is submitted for quickly ascertaining the viscosity resulting from blending two solutions of differing viscosity. The viscosities of the two initial solutions are first located on the logarithmic scales to either side of the chart. These points are then joined by a straight line. The viscosities of the blends can be read off at appropriate points on this line corresponding to any required weight fraction (concentration) as located on the bottom scale.

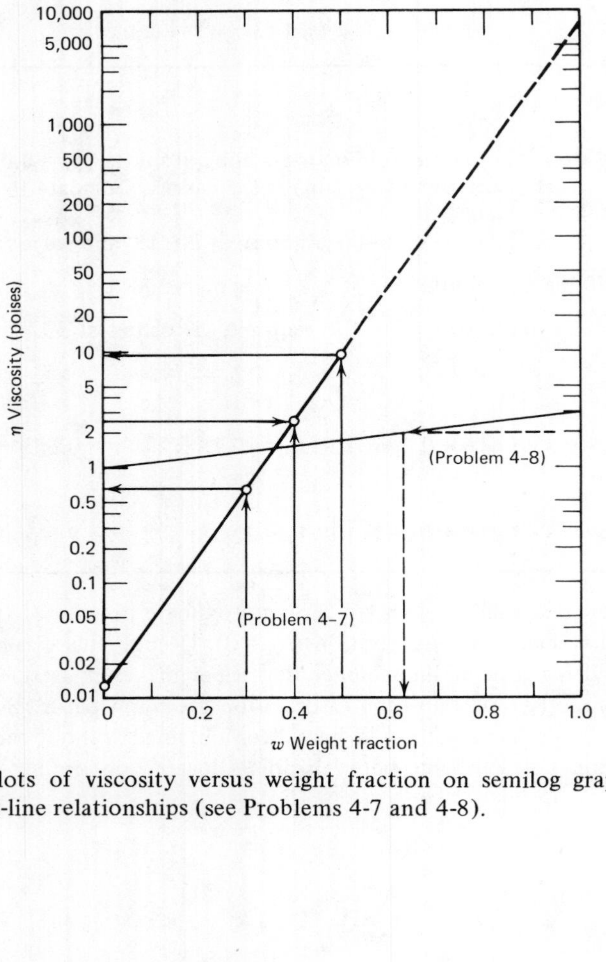

Fig. 4-5 Plots of viscosity versus weight fraction on semilog graph paper to give straight-line relationships (see Problems 4-7 and 4-8).

PROBLEM 4-8

A 4%w solution of Type 100 ethyl cellulose in a solvent mixture of 80% toluene and 20% ethanol has a viscosity of 1.0 poise. A similar solution of Type 300 ethyl cellulose has a viscosity of 3.0 poises. What proportions of the two solutions should be used to obtain a 2.0-poise viscosity blend?

Solution. Connect the 1.0-poise value on the left-hand scale of Fig. 4-5 (representing 100% of Type 100 ethyl cellulose) with the 3.0-poise value on the right-hand scale (representing 100% of Type 300 ethyl cellulose) by a straight line. At the intersection of this line with the horizontal 2.0-poise cross-line, draw a line vertically downward to intersect the bottom scale and read off the value of 0.63 for the weight fraction of the Type 300 ethyl cellulose solution to be used with the 0.37 (= 1.00 - 0.63) weight fraction of the Type 100 ethyl cellulose solution to give the required 2.0-poise viscosity for the mixutre (see lines drawn in on Fig. 4-5 for this problem solution).

EFFECT OF POLYMER CONCENTRATION ON VISCOSITY OF WATER-SOLUBLE SYSTEMS

The change in viscosity with polymer concentration for a water-soluble system is quite different from that for an organic solvent system. Solutions of polymers in organic solvents are mostly Newtonian in behavior, and changes in viscosity versus polymer concentration can be expressed by relatively simple mathematical expressions (such as Eq. 9). On the other hand, water-soluble systems tend to be non-Newtonian in behavior and the relationship between viscosity and polymer concentration is somewhat complex and not, as yet, too well understood.

Since water-soluble coatings are presently capturing an ever-larger share of the industrial coatings market, largely because of the economical, nonpolluting, and nonflammable nature of water as a solvent, more attention is being given to the rheological properties of these systems (5).

Many water-soluble systems are based on polymers with carboxyl groups that are converted to amine salts to achieve solubility either in water alone or, more commonly, in a water cosolvent blend. At certain intermediate polymer concentrations the polymer solution normally converts to a dispersion of micelles that may or may not be visually evident by a hazy appearance. However, the occurrence of this micelle development always has a significant effect on solution viscosity.

Water-soluble polymers are normally prepared as cosolvent-rich solutions that are let down to application viscosities with additional water. The polymers themselves tend to be hydrophobic, since the hydrophilic salt groups are present in minor amount. Hence on reduction with water these polymer molecules contract, engulf less water, and as a result solution viscosity tends to be reduced more than might be expected from simple dilution alone. However, there are two strong influences that more than counteract this reduction. In the first place an increased dissociation of the salt groups occurs as water is added to the polymer solution, leading to an increased repulsion among the ions attached to the same molecule, a consequently bulkier macromolecule, and in turn an increased viscosity. In the second place, and more important, the macromolecules, which are initially relatively immobile in the cosolvent-rich solution, become less stable on dilution and tend to form aggregates (micelles) that are sufficiently swollen with organic cosolvent to strongly interfere with solution flow. The net result is a dilution curve that exhibits a plateau or even an intermediate maximum for the middle range of dilution (say between about 30 and 50%w of polymer). Both organic solvent and water systems start with a downtrend in viscosity with dilution. However, whereas this trend is smooth and continuous with organic solvent systems, it is halted or reversed in a water-soluble system at an intermediate stage. After passing through the holding or intermediate maximum stage between about 30 and 50%w polymer concentration, further dilution of the water-soluble system (below about 30%w) results in a rather precipitous drop in viscosity. Hence the coatings engineer should not be too surprised to find only nominal changes in viscosity in the midrange of the water dilution process but rather drastic changes at further dilution (into the lower ranges of application viscosities).

Micelle formation is reduced by lower molecular weight, by an initially higher percentage of cosolvent, and by increased carboxyl content. Pigmentation acts to shift the characteristic water-soluble curve, but the general shape remains about the same.

Since the water/cosolvent ratio is an important factor controlling micelle formation, it can be used indirectly to control viscosity behavior (through micelle control). Thus the use of two cosolvents can be advantageous, in that one, by being more volatile, can reduce sagging by its rapid evaporation rate, and the other, by being less volatile, can remain to assure leveling and satisfactory film formation without "popping." Note, however, that any cosolvent that evaporates more slowly than the water is likely to impart an undesirable "delayed sagging."

EFFECT OF SOLVENT VISCOSITY ON SOLUTION VISCOSITY

The solvent component of a polymer solution controls the solution viscosity in two ways: (*a*) by its polymer dispersing or solvating ability, and (*b*) by its own viscosity. Although both strongly influence solution viscosity, all too often the viscosity of the solvent is overlooked or disregarded as a contributing factor.

It is rather striking that a fractional centipoise difference in solvent viscosity can result in a difference of hundreds to thousands of centipoises in solution viscosity. Let Eq. 17 be rewritten in terms of intrinsic viscosity instead of the constant k_a to give Eq. 19.

$$\log \frac{\eta}{\eta_0} = \frac{w[\eta]\,\rho}{0.023} \tag{19}$$

Now, if a polymer has essentially the same intrinsic viscosity in two solvents (which vary in viscosity but not in density), then, according to Eq. 19, for any given weight fraction of polymer the ratio (η/η_0) is a constant. Hence the solution viscosity for any given weight concentration is proportional to the solvent viscosity. For example, a difference of 0.002 poise between the viscosities of two solvents (0.012 and 0.010 poise) may appear trivial but can be reflected by substantial differences in solution viscosity; this could be a 20-poise difference at a weight fraction of 50% (120 and 100 poises).

To further illustrate this effect with some actual systems, consider the viscosities of three common hydrocarbon solvents that fall in the same molecular weight range, as tabulated in Table 4-3. Inspection of the data suggests that, from the standpoint of dispersing power, toluene with a KB value of 105 and a mixed aniline point of 52 should outperform both methylcyclohexane and heptane (poorest in solvating power). However, from the standpoint of minimal solution viscosity, heptane is the solvent of choice, since it possess the lowest solvent viscosity. It is true that the differences in the cited solvent viscosities are

Table 4-3 Properties of Three Hydrocarbon Solvents Falling in the Same Molecular Weight Range

	Solvent Type		
Solvent:	(Paraffinic) *n*-Heptane	(Aromatic) Toluene	(Naphthenic) Methylcyclohexane
Structural formula:	$CH_3CH_2CH_2CH_2CH_2CH_2CH_3$	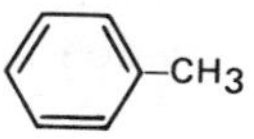	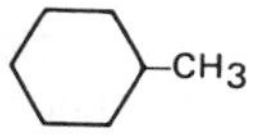
Molecular weight:	100	92	98
Viscosity at 77 F (cP)	0.39	0.56	0.69
Viscosity ratio	1.00	1.44	1.77
Boiling point (F)	209	231	214
Kauri butanol (KB) value	25	105	50
Aniline point, mixed (F)	158	52	130

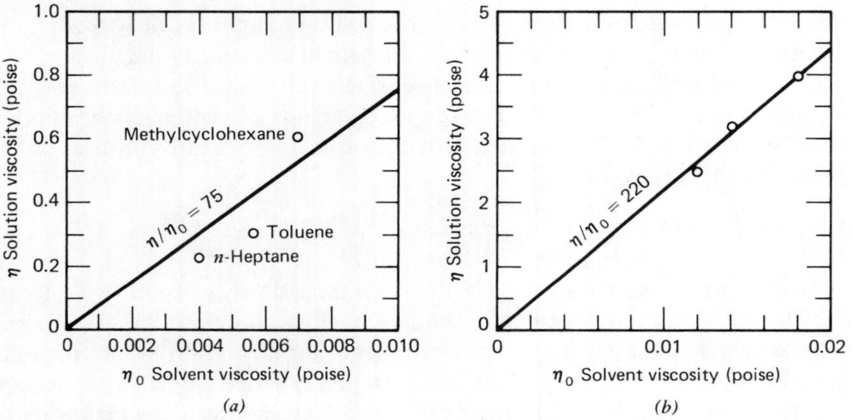

Fig. 4-6 (*a*) Plot of solution viscosity versus solvent viscosity for 50%w solutions of 5% limed rosin in three hydrocarbon solvents (50% NV). (*b*) Plot of solution viscosity versus solvent viscosity for 40%w solutions of a long-oil alkyd in three isoparaffinic (odorless) solvents (40% NV).

small (maximum difference 0.003 poise), but such arithmetical differences are not relevant here; rather, viscosity ratios dominate and control the buildup of viscosity in polymer solutions. Figure 4-6*a* gives a plot of the three solvent viscosities versus the viscosities of their limed rosin solutions (50% NV content). Note the controlling influence of solvent viscosity in determining solution viscosity, with dispersing ability playing a secondary role (see deviation for toluene).

If the solvent type is kept constant, an even closer tie between solvent and solution viscosity is generally obtained. For example, Fig. 4-6*b* gives a plot of the viscosities of three isoparaffinic solvents (essentially the same KB values) versus the viscosities of their 40%w solutions of a long-oil alkyd. In this instance the ratio of solution viscosity to solvent viscosity varies little from a value of 220. Therefore a reduction in solvent viscosity, say of 0.005 poise, can be expected to effect a reduction in the 40%w solution viscosity by a full 1.1 poise ($= 220 \cdot 0.005$).

In formulating any vehicle or paint to meet a viscosity specification, it is always essential to take into account the vital factor of solvent viscosity as well as solvent power.

Viscosity of Solvent Blends

The viscosity of an ideal (noninteracting) solvent blend can be closely calculated from Eq. 20, where η is the viscosity of the blend and w_i and η_i are the weight fraction and neat viscosity, respectively, of the ith component.

$$\log \eta = \sum (w \log \eta)_i \tag{20}$$

PROBLEM 4-9

Calculate the viscosity of a blend of 48%w of methyl ethyl ketone ($\eta = 0.41$ cP), 32%w of *n*-butyl acetate ($\eta = 0.68$ cP), and 20%w of toluene ($\eta = 0.55$ cP).

Solution. Since ketones, esters, and hydrocarbon solvents are relatively free of interactions when blended, substitute the given data in Eq. 20 and solve for the blend viscosity.

$$\log \eta = 0.48 \log 0.41 + 0.32 \log 0.68 + 0.20 \log 0.55; \qquad \eta = 0.51 \text{ cP}$$

Comment. The calculated value of 0.51 cP for the blend compares favorably with a measured value of 0.49 cP.

Unfortunately many solvent blends are far from ideal, especially when hydroxyl-type solvents are involved. Hence attention has been directed recently to modifying the quantities in Eq. 20 to encompass nonideal (interacting) solvent blends. Such an advance in calculation capability automatically means a corresponding improved capability in predicting the viscosity of resin solutions. As discussed in the preceding section, solution viscosity closely parallels solvent viscosity, and a knowledge of solvent blend viscosity behavior can be directly translated into a knowledge of solution viscosity behavior. It is always a much simpler task to measure solvent viscosity than solution viscosity.

To extend the use of Eq. 20 to cover the case of a nonideal solvent blend it has been proposed that, of two classes of interacting solvents, the members of one class be assigned so-called effective viscosity values that more accurately reflect their blending behavior when interaction occurs (6). These effective viscosity values are established from experimental data and, depending on the type of system, are applicable up to about 20 to 40% contents of the solvents with the assigned effective viscosity values. Two important blending systems between solvent classes have been investigated and reported (6). Some of these data are reproduced in Tables 4-4 and 4-5. Equation 20 is now modified in that effective viscosities are substituted for actual viscosities for one of the two classes of interacting solvents, and there is, of course, the reservation that the total weight fraction as noted in the tables should not be exceeded for the solvents with assigned effective viscosity values. Equation 21 applies to blends of hydrocarbon solvents (one class) with alcohol solvents (second class). Effective viscosities for solvents in the alcohol class are tabulated in Table 4-4.

$$\log \eta = \sum (w \log \eta_a)_i + \sum (w \log \eta_e)_j \tag{21}$$

Table 4-4 Actual and Effective Viscosities of Alcohol Solvents at 25 C When Blended up to 30 to 40% with Hydrocarbon Solvents

	Viscosity (cP)		
Alcohol Solvent	Actual η_a	Effective η_e	Ratio η_e/η_a
Ethanol	1.30	1.05	0.81
Methoxyethanol	1.60	1.20	0.75
Ethoxyethanol	1.90	1.20	0.63
Propanol	2.00	1.40	0.70
Isopropanol	2.40	1.10	0.46
Butanol	2.60	1.60	0.62
sec-Butanol	2.90	1.40	0.48
Butoxyethanol	2.90	2.00	0.69
Diacetone alcohol	2.90	2.00	0.69
Isobutanol	3.40	1.80	0.53
Methylisobutylcarbinol	3.80	1.80	0.47
Butoxyethoxyethanol	5.30	2.15	0.41
2-Ethylhexanol	7.78	3.30	0.42
Average			0.59

PROBLEM 4-10

Calculate the viscosity for a $\frac{70}{30}$ weight blend of toluene ($\eta = 0.55$ cP) and isopropanol.

Solution. Since the weight fraction of the alcohol (isopropanol) does not exceed the 40% limitation on its content, the blend viscosity can be calculated from Eq. 21, using the effective viscosity value for isopropanol given in Table 4-4.

$$\log \eta = 0.70 \log 0.55 + 0.30 \log 1.10; \qquad \eta = 0.68 \text{ cP}$$

Equation 21 is equally applicable to blends of oxygenated organic solvents with water. In this case the effective viscosity values are assigned to the oxygenated solvents, and Eq. 21 reduces to Eq. 22 (viscosity water 0.92 cP).

$$\log \eta = (w \log 0.92)_{H_2O} + \sum (w \log \eta_e)_j$$
$$\log \eta = (-0.0362w) + \sum (w \log \eta_e)_j \qquad (22)$$

It is notable that the effective/neat viscosity ratios for the alcohol/hydrocarbon solvent system do not vary greatly from an average value of 0.59. On the other hand, the ratios applying to the oxygenated solvent/water system vary tremendously (see Tables 4-4 and 4-5). Since water systems are becoming increasingly

Table 4-5 Actual and Effective Viscosities at 25 C of Oxygenated Organic Solvents When Blended up to 20 to 30% with Water

Oxygenated Solvent	Viscosity (cP) Actual η_a	Effective η_e	Ratio η_e/η_a
Acetone	0.31	6.06	19.5
Methyl ethyl ketone	0.41	9.8	23.9
Methoxyethanol	1.60	14.0	13.0
Ethoxyethanol	1.90	24.4	8.8
Isopropanol	2.10	59.3	28.2
Diacetone alcohol	2.90	22.2	7.7
Butoxyethanol	2.90	32.1	11.1
Methoxyethoxyethanol	3.80	17.3	4.6
Ethoxyethoxyethanol	4.00	26.1	6.5
tert-Butanol	4.50	116.4	25.9
Butoxyethoxyethanol	5.30	35.0	6.6
Ethylene glycol	17.4	10.0	0.57
Diethylene glycol	28.9	15.6	0.54
Hexylene glycol	29.8	65.0	2.2

important in coatings technology, it is convenient to be able to predict with reasonable accuracy the diverse viscosity behavior that can be anticipated for these water blends.

PROBLEM 4-11

Calculate the viscosity of a mixture of 10%w of isopropanol, 10%w of methoxyethanol, and 80%w of water.

Solution. Substitute appropriate effective viscosity data from Table 4-5 in Eq. 22, and solve for the blend viscosity.

$$\log \eta = (-0.0362 \cdot 0.80) + 0.10 \log 59.3 + 0.10 \log 14$$

$$\log \eta = -0.0290 + 0.1773 + 0.1146 = 0.263; \qquad \eta = 1.83 \text{ cP}$$

Comment. The calculated value of 1.83 cP for the water blend closely approximates a measured value of 1.81 cP.

Equations 21 and 22 are limited in their application to blend contents of less than 30 to 40% for the solvent class to which effective viscosity values are assigned. However, by resorting to a different type of equation this restriction can be removed. For example, for a binary water/alcohol blend this author recommends Eq. 23 as giving a reasonable approximation to experimentally determined values over the entire blend range (w = fractional weight of solvent).

$$\log \eta = (1 - w) \log \eta_{H_2O} + w^2 \log \eta_a + (w - w^2) \log \eta_e \qquad (23)$$

To illustrate the use of this equation both the calculated and the measured viscosities for water/isopropanol blends, through a complete range of concentrations, have been plotted in Fig. 4-7.

This type of equation can be expanded to cover multicomponent blends as shown by Eq. 24, where $w = \Sigma w_j$ and w_j is the fractional weight of the jth alcohol component.

$$\log \eta = (1 - w) \log \eta_{H_2O} + \sum w_j w \log \eta_{aj} + \sum w_j(1 - w) \log \eta_{ej} \qquad (24)$$

PROBLEM 4-12

Calculate the viscosity of a blend of 35%w of isopropanol, 25%w of ethylene glycol, and 40%w of water.

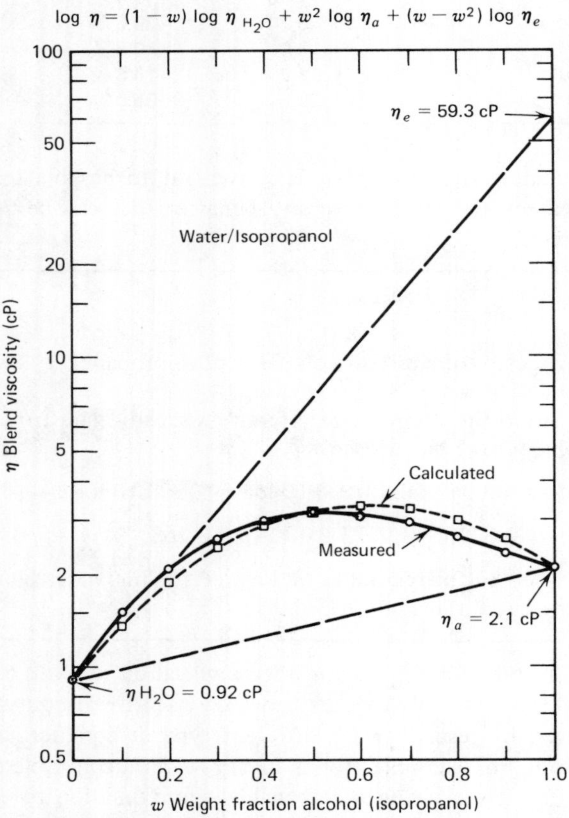

Fig. 4-7 Calculated and measured viscosities plotted versus weight fraction of isopropanol for isopropanol/water blends.

Solution. Substitute the given data in Eq. 24, noting that $w = 0.60$ (= 0.35 + 0.25).

$$\log \eta = 0.40 \log 0.92 + 0.35 \cdot 0.60 \log 2.1 + 0.25 \cdot 0.60 \log 17.4$$
$$+ 0.35 \cdot 0.40 \log 59.3 + 0.25 \cdot 0.40 \log 10$$

$$\log \eta = 0.588; \qquad \eta = 3.87 \text{ cP}$$

EFFECT OF MOLECULAR WEIGHT ON VISCOSITY

The viscosity of a polymer solution is profoundly affected by the molecular weight (M or MW) of the polymer (or resin) dissolved or dispersed in the solution solvent. In fact, probably the most important industrial method for characterizing polymer molecular weight is based on the measurement of polymer viscosity in dilute solutions (less than 1% concentration).

Number-Average and Weight-Average Molecular Weights

Since the paint chemist is working more and more with the adaptation of synthetic polymers to coating applications, he or she should become conversant with the special viscosity techniques and terminology employed by polymer chemists to describe and characterize these high-molecular-weight products. Whereas many methods for measuring molecular weight (melting-point depression, boiling-point elevation, osmotic pressure, and group analysis) furnish a number-average molecular weight (M_n) as defined by Eq. 25, a viscosity measurement differs by furnishing an index to the weight-average molecular weight (M_w) as defined by Eq. 26.

$$M_n = \frac{\sum N_x M_x}{\sum N_x} = \frac{W}{N} \tag{25}$$

$$M_w = \frac{\sum N_x M_x^2}{\sum N_x M_x} = \frac{\sum W_x M_x}{W} \tag{26}$$

Here N_x represents the number of moles in a polymer fraction of weight W_x, all having the same molecular weight M_x, and W and N represent the total weight and the total molecules, respectively, for the total polymer mixture.

PROBLEM 4-13

Consider an academic polymer consisting of five fractions, each of the same weight, which increase in molecular weight from 2000 to 10,000 by 2000

Table 4-6 Tabulation of Data Used in Computing M_n and M_w Values

$N_x M_x (= W_x)$	M_x	$N_x (= W_x/M_x)$	$N_x M_x^2 (= W_x M_x)$
0.2W	2,000	0.000100W	400W
0.2W	4,000	0.000050W	800W
0.2W	6,000	0.000033W	1200W
0.2W	8,000	0.000025W	1600W
0.2W	10,000	0.000020W	2000W
W		0.000228W(= N)	6000W(= $\Sigma W_x M_x$)

molecular weight increments. Calculate the M_n and M_w values for this polymer mixture

Solution. Let W be the total weight of the polymer mixture. Then each fraction x will contribute $0.2W$ to the total weight $W(N_x M_x = 0.2W)$. Calculate and tabulate $N_x M_x$, M_x, N_x, and $N_x M_x^2$ values for each x fraction as shown in Table 4-6. Use the totaled values of Table 4-6 to compute M_n and M_w from Eqs. 25 and 26.

$$M_n = \frac{W}{N} = \frac{W}{0.000228W} = 4380$$

$$M_w = \frac{\sum W_x M_x}{W} = \frac{6000W}{W} = 6000$$

Note that, when only one molecular weight is involved, M_n and M_w are identical. Moreover, for any actual molecular weight distribution, the M_w average is weighted in favor of the higher molecular weight fractions. For a random polymerization of a linear polymer M_w is normally twice M_n $(M_w = 2M_n)$.

Dilute Solution Viscometry; Intrinsic Viscosity

Since dilute solutions of polymeric materials can be made relatively low in viscosity by selecting a suitable solvent and a sufficiently high temperature, and since a high degree of accuracy is required for dilute solution viscometry, it is natural that recourse should be made to a capillary type of viscometer for evaluating polymer molecular weight. Although the polymer researcher indirectly measures solution viscosity, he or she rarely makes use of an actual viscosity per se. Rather, the calculations are carried out in terms of a viscosity ratio as defined by Eq. 27.

$$\eta_r(\text{viscosity ratio}) = \frac{\eta_{\text{solution}}}{\eta_{\text{solvent}}} \tag{27}$$

In terms of efflux times through the capillary viscometer, the viscosity ratio is also defined by Eq. 28, since efflux times are directly proportional to liquid viscosities (solvent and solution densities are essentially equivalent at dilute concentrations).

$$\eta_r = \frac{t_{\text{solution}}}{t_{\text{solvent}}} \tag{28}$$

At present a renaming of the terms used in dilute solution viscometry is under way in the interests of more exact description. Thus none of an intrinsic, an inherent, a specific, or a relative viscosity is an actual viscosity. Table 4-7 tabulates the older and the presently more common terminology versus the newer system proposed by the International Union of Pure and Applied Chemistry. Also given are equations defining these terms.

From inspection of the terms of Table 4-7 it is apparent that the polymer chemist does not work just with a viscosity ratio. Primarily this is so because a viscosity ratio fails to furnish a single number that can be conveniently related to a weight-average molecular weight. For one thing, the viscosity ratio varies with polymer concentration.

However, an arbitrary technique has been established that provides a limiting viscosity number applying to zero concentration, which in turn can be related to *M*. Thus, if the points of a plot of either the viscosity number or the logarithmic viscosity number versus the polymer concentration are connected by a straight line, the line will intersect the viscosity number scale at a limiting viscosity number (an intrinsic viscosity). This critical number for zero polymer concentration can in turn be related to a weight-average molecular weight.

Problem 4-14 serves to illustrate the method for obtaining a limiting viscosity number starting with viscosity ratio values obtained with a capillary viscometer.

Table 4-7 Older and Newer Terms Relating to Dilute Solution Viscometry

Symbols and Related Equations[a]	Proposed New Terms	Older Common Terms
η_r (Eqs. 27, 28)	Viscosity ratio	Relative viscosity
$\eta_{sp} = (\eta_r - 1)$		Specific viscosity
η_{sp}/c or $(\eta_r - 1)c$	Viscosity number	Reduced specific viscosity
$\ln \eta_r/c$ or $2.3 \log \eta_r/c$	Logarithmic viscosity number	Inherent viscosity
$[\eta] = (\eta_{sp}/c)_{c \to 0}$	Limiting viscosity number	Intrinsic viscosity
$[\eta] = (\ln \eta_r/c)_{c \to 0}$	Limiting logarithmic viscosity number	Intrinsic viscosity

[a] For dilute solution viscometry the polymer concentration *c* is expressed in grams of polymer per deciliter of solvent.

Table 4-8 Viscosity Data for Two Polymethylmethacrylate Polymers (Chloroform Solutions at 20 C)

Polymer Concentration c (g/dl)	Viscosity Ratio η_r	
	Medium High MW	Very High MW
0.20	1.095	1.288
0.40	1.198	1.632
0.60	1.306	2.026

PROBLEM 4-14

The viscosity ratios (at 20 C in chloroform) for solutions of a medium high and a very high molecular weight polymethylmethacrylate polymer, respectively, are given in Table 4-8. Calculate the limiting viscosity numbers (intrinsic viscosities) for these two polymers.

Solution. Determine the limiting viscosity numbers for the two polymers by computing and plotting their viscosity numbers (and as a check their logarithmic viscosity numbers) versus their corresponding concentrations on coordinate graph paper. Extension of the straight lines connecting the sets of plotted points to intersect the viscosity number scale will give the required limiting viscosity numbers at the points of intersection. Table 4-9 develops the information required for the plotting, and Fig. 4-8 gives the plot of these data. The limiting viscosity numbers for the medium high and very high polymethylmethacrylate polymers are read off the plot as 0.46 and 1.31, respectively.

The reason why the logarithmic viscosity number line meets the viscosity number line at zero concentration (see Fig. 4-8) can be explained by expressing the logarithm of the viscosity ratio in the logarithmic viscosity number expression as an algebraic series.

Table 4-9 Development of Data Required for Plotting Viscosity Numbers versus Polymer Concentration

	c	η_r	η_{sp}	η_{sp}/c	$\log \eta_r$	$2.3 \log \eta_r/c$
Medium high MW	0.20	1.095	0.095	0.475	0.0394	0.453
	0.40	1.198	0.198	0.495	0.0785	0.451
	0.60	1.306	0.306	0.510	0.1159	0.445
Very high MW	0.20	1.288	0.288	1.440	0.1101	1.265
	0.40	1.632	0.632	1.580	0.2127	1.225
	0.60	2.026	1.026	1.710	0.3066	1.175

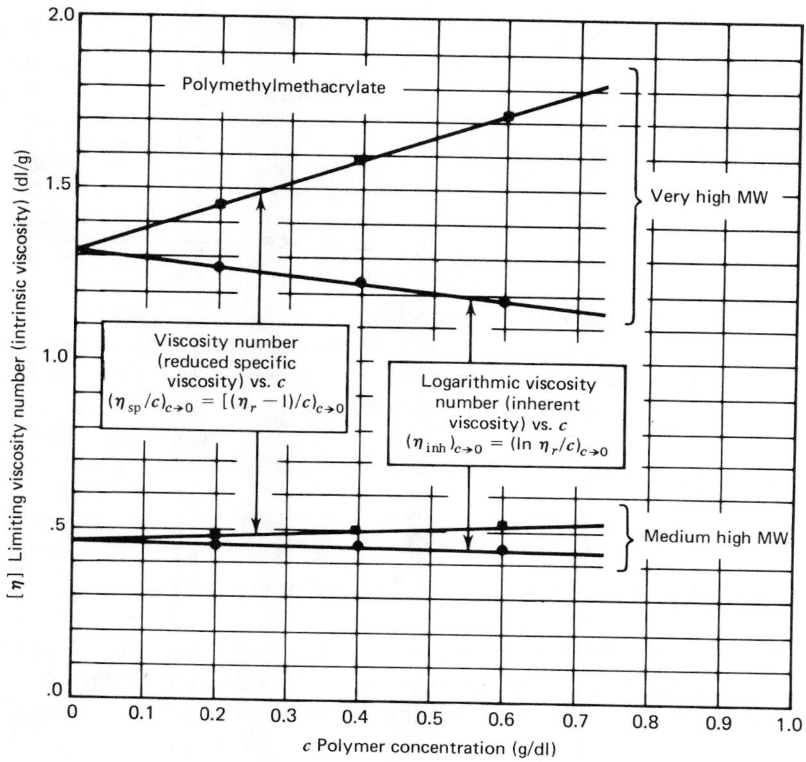

Fig. 4-8 Plot of viscosity number versus polymer concentration for polymethylmethacrylate polymers.

$$\frac{\ln \eta_r}{c} = \frac{(\eta_r - 1) - \frac{1}{2}(\eta_r - 1)^2 + \frac{1}{3}(\eta_r - 1)^3 - \cdots}{c} \tag{29}$$

As the solution becomes more dilute (approaches zero concentration), η_r approaches a value of 1.00. As a result the square and higher exponential terms become negligibly small compared with the first term $(\eta_r - 1)$ at infinite dilution, leading to Eq. 30.

$$\left(\frac{\ln \eta_r}{c}\right)_{c \to 0} = \left(\frac{\eta_r - 1}{c}\right)_{c \to 0} \tag{30}$$

The right-hand term of Eq. 30 is seen to be identical with the limiting viscosity number. Hence at zero concentration the two viscosity numbers become equal, a fact reflected graphically by the juncture of the lines at zero concentration.

Table 4-10 Conditions Specified for ASTM Dilute Solution Viscosity Measurements

ASTM Number	Type of Polymer	Solvents Specified	Temperature Specified (C)
D1243-60	Vinyl chloride	Cyclohexane	30
		Nitrobenzene	30
D1601-61	Ethylene	Decahydronaphthalene	130

The value of a polymer intrinsic viscosity $[\eta]$ is a function of both the choice of solvent and the temperature at which the viscosity measurements are made. Table 4-10 shows typical standard conditions for these two variables as specified by two ASTM procedures. Other solvents commonly employed are water, acetone, chloroform, benzene, and toluene. Temperatures of 20 and 25 C are also frequently used industrially for polymers that can be solubilized readily at room temperatures.

Calculation of Intrinsic Viscosity from a Single Viscosity Ratio

An inspection of the plotted data of Fig. 4-8 suggests an interesting slope relationship, namely, that the upward slope of the viscosity number line is about three times as steep as the downward slope of the logarithmic viscosity number line. Actually, this 3 : 1 slope relationship is found to be quite common when plotting viscosity numbers and has led to Eq. 31, which is recommended for computing the intrinsic viscosity (limiting viscosity number) from a single value for the viscosity ratio.

$$[\eta] = \frac{\eta_r - 1}{4c} + \frac{3 \ln \eta_r}{4c}$$

or

$$[\eta] = \frac{0.25(\eta_r - 1) + 1.725 \log \eta_r}{c} \tag{31}$$

The nomograph of Fig. 4-9 provides a graphical solution to Eq. 31.

PROBLEM 4-15

The viscosity ratio for a 0.5-g/dl solution of a polymethylmethacrylate polymer of very high molecular weight in chloroform at 20 C is 1.820. Calculate the polymer intrinsic viscosity.

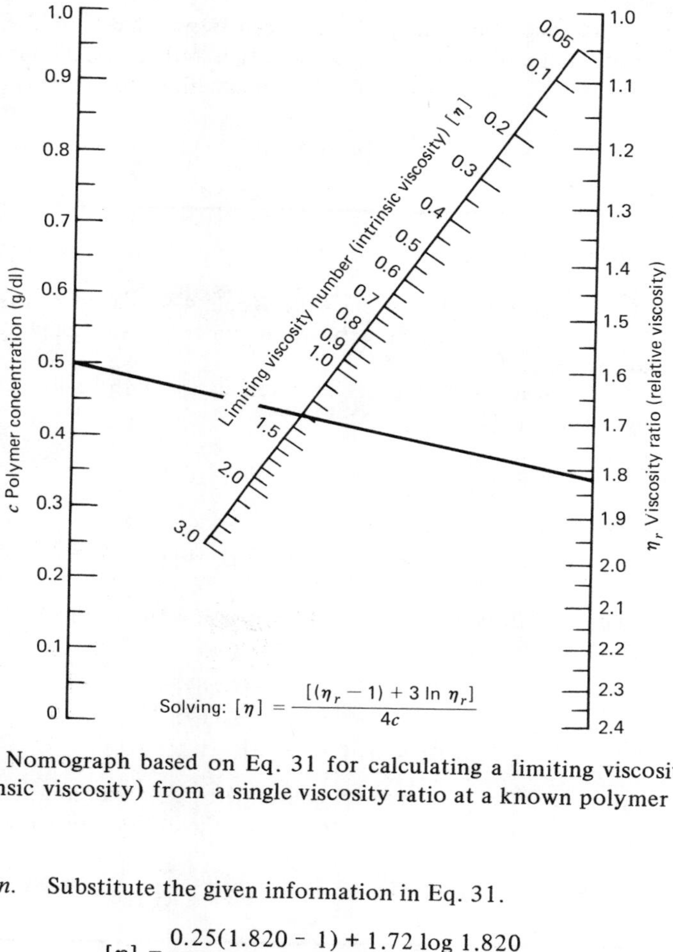

Fig. 4-9 Nomograph based on Eq. 31 for calculating a limiting viscosity number (intrinsic viscosity) from a single viscosity ratio at a known polymer concentration.

Solution. Substitute the given information in Eq. 31.

$$[\eta] = \frac{0.25(1.820 - 1) + 1.72 \log 1.820}{0.5}$$

$$[\eta] = \frac{0.25(0.820) + 1.725(0.260)}{0.5} = 1.31$$

This solution corresponds to the one given by the line crossing the nomograph of Fig. 4-9.

Huggins' Constant. A further inspection of Fig. 4-8 shows the slope of the viscosity number line to be proportionly steeper for the polymer with the higher intrinsic viscosity. This relationship is generally true, and from both experimental and theoretical studies it has been shown that Eq. 32 expressing this generalization holds quite well. In fact, it is rather remarkable that the constant k' of Eq. 32 is quite close to 0.38 for most dilute polymer solutions.

$$\frac{\eta_{sp}}{c} = [\eta] + k'[\eta]^2 c \tag{32}$$

PROBLEM 4-16

Using Eq. 32 and the data of Problem 4-14 (Table 4-9), calculate the intrinsic viscosity for the polymethylmethacrylate polymer of very high molecular weight. Also determine the value of k'.

Solution. Substitute two sets of the given data for the polymethylmethacrylate polymer in Eq. 32.

$$\frac{0.288}{0.2} = 1.44 = [\eta] + k'[\eta]^2(0.2)$$

$$\frac{1.026}{0.6} = 1.71 = [\eta] + k'[\eta]^2(0.6)$$

Solve for $[\eta]$ by subtracting the second expression from three times the first expression and simplifying.

$$4.32 = 3[\eta] + k'[\eta]^2(0.6)$$
$$1.71 = [\eta] + k'[\eta]^2(0.6)$$
$$2.61 = 2[\eta]; \qquad [\eta] = 1.31$$

Compute the value for k' (Huggins' constant) by substituting the calculated value of 1.31 for the limiting viscosity number in one of the above equations.

$$1.44 = [1.31] + k'[1.31]^2(0.2); \qquad k' = \frac{1.44 - 1.31}{0.343} = 0.38$$

If only a single viscosity ratio is known, an approximation to the limiting viscosity number can be obtained by assuming a value of 0.38 for k' and substituting in Eq. 33, which is an alternative way to expressing Eq. 32.

$$[\eta] = \frac{1}{2k'c}(\sqrt{1 + 4\eta_{sp}k'} - 1) \tag{33}$$

PROBLEM 4-17

The viscosity ratio for a 0.5-g/dl solution of a polymethylmethacrylate resin of very high molecular weight in chloroform at 20 C is 1.820. Compute its intrinsic viscosity.

Solution. Substitute the given information in Eq. 33.

$$[\eta] = \frac{1}{2 \cdot 0.38 \cdot 0.5}(\sqrt{1 + 4 \cdot 0.820 \cdot 0.38} - 1)$$

$$= 2.64(1.495 - 1) = 1.31$$

For most industrial purposes the intrinsic viscosity, of itself, is an adequate index of polymer molecular size. In fact, for exact work it may be preferable to resort to the intrinsic viscosity for measuring the size of a polymer rather than attempting to convert this value to a molecular weight figure. Certainly for routine production control, determination of the limiting viscosity number is a fully satisfactory guide to polymer quality. In general, for a polymer to display an acceptable degree of strength or film integrity, its intrinsic viscosity must have a value of at least 0.4 to 1.0, depending on the polymer and the solvent.

Viscosity-Average Molecular Weight

For research work it becomes necessary to probe more deeply into the relationship of intrinsic viscosity to molecular configuration, molecular weight, and molecular weight distribution. At present Eq. 34 is accepted as a reasonably valid expression for relating weight-average molecular weight M_w to intrinsic viscosity.

Table 4-11 Representative Values for the Constants *K* and *a* for Selected Polymers in Eq. 34 Relating Viscosity-Average Molecular Weight to Limiting Viscosity Number (Intrinsic Viscosity)

Polymer	Solvent	Temperature (C)	*K*	*a*
Cellulose acetate	Acetone	25	0.000019	1.03
Cellulose nitrate	Acetone	25	0.000038	1.00
Polystyrene	Benzene	25	0.00010	0.74
	Chloroform	25	0.00011	0.73
Polymethylmethacrylate	Benzene	25	0.000094	0.76
	Butanone	25	0.000068	0.72
	Acetone	25	0.000075	0.70
	Chloroform	20	0.000135	0.70
Polyisobutylene	Cyclohexane	30	0.00026	0.70
	Benzene	20	0.00036	0.64
Polyvinyl alcohol	Water	50	0.00059	0.67
Natural rubber	Toluene	25	0.00050	0.67
Polyvinyl acetate	Acetone	20	0.00028	0.66
Neoprene	Toluene	25	0.00050	0.62

$$[\eta] = KM_w^a \tag{34}$$

Since the weight-average molecular weight is being determined by a viscosity technique, it is often referred to as a viscosity-average molecular weight. Table 4-11 lists some typical values for the constants K and a of Eq. 34. Note that the solvent and temperature are still controlling influences and must be listed as part of the given information. In general, a limiting viscosity number changes by only a fraction of a percent for a 1 C change in solution temperature. On the other hand, a limiting viscosity number can vary by as much as a factor of 2 in changing from one solvent to another (each of which may be considered a good solvent for the polymer in question). Hence any given limiting viscosity number must be interpreted with both the polymer and the solvent in mind.

Theoretically, a value of 0.5 for the constant a in Eq. 34 corresponds to a completely flexible, randomly kinked polymer chain, whereas a value of 2.0 indicates a rigid, rod-shaped molecule. This theory is borne out by noting that the elastomeric polymers (rubbers) have lower a values than the stiffer cellulosic polymers.

PROBLEM 4-18

Compute the viscosity-average molecular weight for a cellulose nitrate polymer having an intrinsic viscosity of 2.10 (25 C in acetone).

Solution. Using the constants for K and a listed in Table 4-11, substitute the value of 2.10 in Eq. 34 and solve for M_w.

$$2.10 = 0.000038 M_w^{1.0}; \qquad M_w = 55{,}000$$

PROBLEM 4-19

Compute the viscosity-average molecular weight for a polymethylmethacrylate polymer having a limiting viscosity number of 0.45 (20 C in chloroform).

Solution. Using the constants for K and a listed in Table 4-11, substitute the given value of 0.45 for $[\eta]$ in Eq. 34 and solve for M_w.

$$0.45 = 0.000135 M_w^{0.70}; \qquad 0.70 \log M_w = 3.518$$

$$M_w = 108{,}000$$

REFERENCES

1. Mysels, K. J., *Introduction to Colloid Chemistry*, Interscience, New York, 1959, p. 280.
2. Rudin, A. and G. Strathdee, "Model for Predicting Polymer Solution Viscosities," *JPT*, **46,** No. 591, 33–43, April (1974).

3. Erickson, James R., "Viscosity of Oligomer Solutions for High Solids and UV Curable Coatings," *JPT*, **48,** No. 620, 58–67, September (1976).
4. Bulletin 800-48, "Ethyl Cellulose: Properties and Uses," Hercules, Inc., Coatings and Specialty Products Department, 910 Market Street, Wilmington, Del. 19899.
5. Hill, L. W., and L. B. Brandenburger, "Viscosity Variation and Solvent Balance in Water-Soluble Coatings," *Prog. Org. Coatings*, **3,** 361–379 (1975).
6. Rocklin, A. L. and G. D. Edwards, "Predicting Viscosities of Organic and of Aqueous Solvent Blends," *JCT*, **48,** No. 620, 68–74, September (1976).

5 *Pigment/Binder Geometry*

At an early period in the paint industry it was common practice to compare pigments in coating formulations by making weight-for-weight substitutions. However, by the mid-1950s this practice had largely been abandoned in favor of making substitutions on the more nearly correct volume-for-volume basis. This trend to a volume rather than a weight replacement was accelerated by research carried out by a number of workers in this field leading to the useful concepts of saturation value in 1947 (1) and critical pigment volume concentration (CPVC) in 1949 (2). Since the term "saturation value" (numerically equal to CPVC/PVC) soon became obsolete, the present book uses the well-recognized term "CPVC" in discussing pigment/binder geometry. More recently the ratio PVC/CPVC has been introduced as a so-called reduced pigment volume concentration and has been assigned the symbol lambda, Λ (3).

IDEALIZED CRITICAL PIGMENT VOLUME CONCENTRATION (CPVC)

At the outset it is instructive to consider what happens when an unlimited amount of binder is systematically added to a limited amount of dry pigment powder. This corresponds to progressing from a condition where only a bed of pigment is present (air is considered as filling the voids among the pigment particles) to a condition where only binder is present. These initial and final conditions represent two extremes on the pigment volume concentration scale, extending from all pigment (PVC = 100%) to all binder (PVC = 0%).

In going from one extreme to the other (all pigment to all binder), it is obvious that there is a first stage during which binder progressively displaces air from within the interstices of the packed bed of pigment. During this stage there exists a system of pigment, binder, and air. Finally a point is reached where the last trace of air is displaced, leaving the voids of the packed pigment bed completely filled with binder. This unique point is called the critical pigment volume concentration (CPVC). Further addition of binder now leads to a second stage where the pigment particles start to separate from each other (their earlier packed or contacting state is disrupted), and as binder addition continues the distance between two particles becomes increasingly remote. Finally an end state is reached where essentially only binder is present.

Of course, an idealized progression from one extreme condition to the other could have been made in the other direction by adding an unlimited amount of pigment to a limited amount of binder. Here again a critical intermediate point is reached where the limited binder content is just adequate to fill the interstices among the pigment particles as they finally all come into contact. Further introduction of pigment beyond this critical pigment volume concentration acts to produce voids (air pockets), and with further continuing pigment addition an essentially all-pigment system eventually results.

The main point here is that, below the CPVC, pigment particles lose contact, whereas, above the CPVC, binder is replaced by air. Certainly the value where this dramatic change to either side of this unique pigment volume concentration occurs deserves the designation of *critical* pigment volume concentration.

The discription of the changes in the nature of a pigment/binder system in progressing from all pigment to all binder (or vice versa) given above should be considered as idealized. Real systems have perturbing variables. These must be recognized and generally compensated for if the concept of a CPVC is to be useful to the paint engineer. For example, an admixture of solvent-type binder with pigment invariably involves adsorption of binder on the pigment surface. This tends to promote a more expanded arrangement of the pigment particles, the degree of added separation depending on the nature and extent of the adsorbed binder. Again, latices tend to resist penetration into the innermost voids of pigment particles or fail to adequately wet the pigment surface. All these effects lower the CPVC. On the other hand, high-intensity and prolonged mixing produce denser pigment packing, leading to higher CPVCs. These and other modifying influences bearing on the critical pigment volume concentration will be considered in some detail.

PIGMENT VOLUME CONCENTRATION (PVC) VERSUS PIGMENT PACKING FACTOR (ϕ)

Failure to differentiate between the pigment volume concentration (PVC) and the pigment packing factor (ϕ) can lead to unnecessary confusion. The PVC is based only on the solids content of a system, whereas the pigment packing factor is based on the solids content plus any voids (interstitial air spaces).

The pigment volume concentration (PVC) is defined as the fractional volume of pigment in a unit volume of a given pigment/binder mixture (only the solids content is considered).

The pigment packing factor (ϕ) is defined as the fractional volume of pigment in a unit volume of a pigment/binder/air mixture (air volume corresponds to interstitial void space, if any). If no air is present, PVC = ϕ.

Let V_p = volume of pigment
V_b = volume of binder (volatiles excluded)
V_a = volume of air (interstitial voids)

Then

$$\text{PVC} = \frac{V_p}{V_p + V_b} \qquad (1)$$

$$\phi = \frac{V_p}{V_p + V_b + V_a} \qquad (2)$$

Since void space (air) is absent below the CPVC, it follows that through this region (PVC < CPVC) PVC and ϕ are the same. However, above the CPVC, PVC and ϕ diverge. For this region (PVC > CPVC), ϕ undergoes little or no change, whereas PVC increases to an end-point value of 1.0 (100% pigment).

These and other relationships to be discussed are graphically depicted in Fig. 5-1, where the horizontal scale shows PVC values and the vertical scale ϕ values.

PROBLEM 5-1

The weight composition of a flat alkyd film is given below, together with applicable density data. The overall porosity of the film is 5.0% v, based on the pigmented porous film. Calculate the PVC, ϕ, CPVC, and Λ for this flat alkyd film, assuming idealized conditions.

Solution. First convert the weight composition of the flat alkyd to a volume composition.

	W (g)	ρ (g/cm^3)	V (cm^3)	
Alkyd (100% NV)	12.5	1.05	11.9	11.9 (alkyd)
TiO_2 (rutile)	40.0	3.95	10.1	
Talc	7.5	2.84	2.6	
$CaCO_3$ (whiting)	40.0	2.71	14.8	
	100.0			27.5 (pigment)

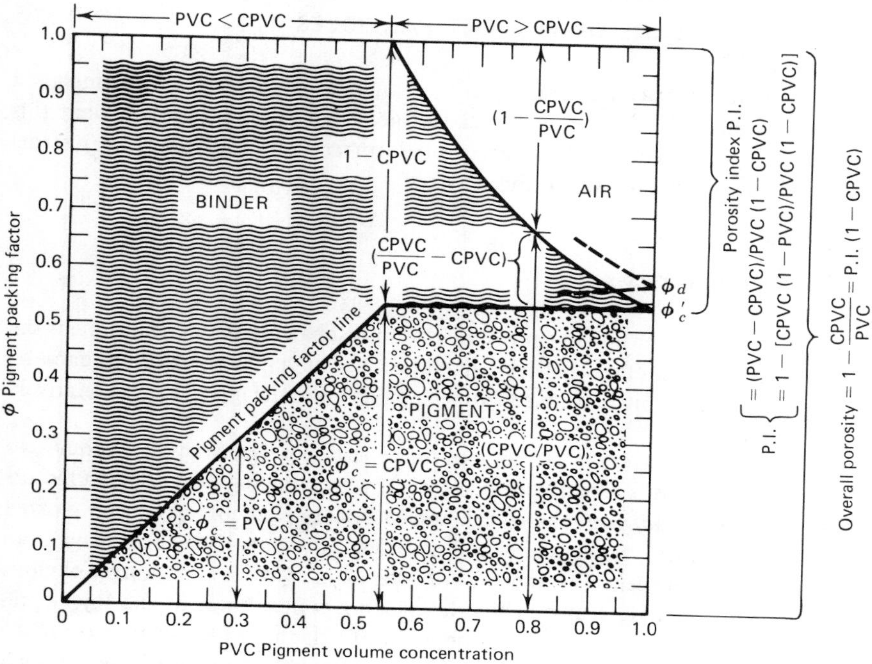

Fig. 5-1 Graph showing the relationships among the key quantities that define pigment/binder geometry.

Next calculate the alkyd PVC from Eq. 1.

$$\text{PVC} = \frac{27.5}{27.5 + 11.9} = 0.70\ (70.0\%)$$

Since the overall film porosity is 5.0% (0.05), the remaining 95.0% (0.95) of volume must be filled with pigmented alkyd, of which 70% (0.70) is pigment and 30% (0.30) is alkyd binder. This permits calculation of the overall volume composition.

	V (cm^3)
Air (void space)	0.050
Pigment	0.665 (= 0.70 · 0.95)
Alkyd binder	0.285 (= 0.30 · 0.95)
	1.000

The pigment packing factor (ϕ) is then 0.665 (66.5%). Also, since the film is porous, the condition PVC > CPVC prevails, and for this condition CPVC = ϕ. The reduced pigment volume concentration is 1.05 (= 0.700/0.665).

PACKING PATTERNS OF PIGMENT PARTICLES

Prediction of the packing arrangement of pigment particles into a compacted bed is complicated by many technical factors. Chief among these are what this author calls the seven S factors–size, shape, surface, spacing (geometric), structure, spread (size distribution), and stirring.

Size of Pigment Particles

Particle size is conventionally measured by some characteristic linear dimension, normally by the average particle diameter d. The pigment particle diameter sizes considered in this text range from very fine colloidal (~0.01 μ) to relatively coarse (~100 μ).

The smaller the average pigment particle, the more resistant it becomes to dense packing. This is illustrated in Fig. 5-2, where the pigment packing factors ϕ for a number of dry commercial calcium carbonate pigments have been plotted versus their average particle diameters. Although the data have been taken from various sources, the general trend toward a less dense (fluffier) packing is obvious as the particle size decreases. Thus the packing compactness of the finer (precipitated) calcium carbonates is on the order of half that of the coarser (ground) calcium carbonates. Similar results have been reported quantitatively for such diverse powdered products as gypsum, Portland cement, and chrome yellow pigment (4).

Shape of Pigment Particles

Pigment particles vary greatly in shape. They may be essentially perfect spheres, or they may take the form of flat, leafy plates or of long needles or spikes. Some of the terms used to describe nonspherical pigments are listed below in the order of their departure from sphericity.

Nodular (irregularly rounded in shape, approximately spherical)
Blocky (roughly rectangular, chunky, or boxlike)
Platy (leafy, lamelliform, flaky, sheetlike)
Acicular (needlelike, bristlelike)
Fibrous (threadlike)

This diversity in particle shape makes the prediction of particle packing patterns extremely difficult, and more so as the particle shape departs from true sphericity. Fortunately, the particles of many important commodity pigments are nodular or even spherical in shape. Even blocky pigments can be approximately related to the packing of spherical pigments.

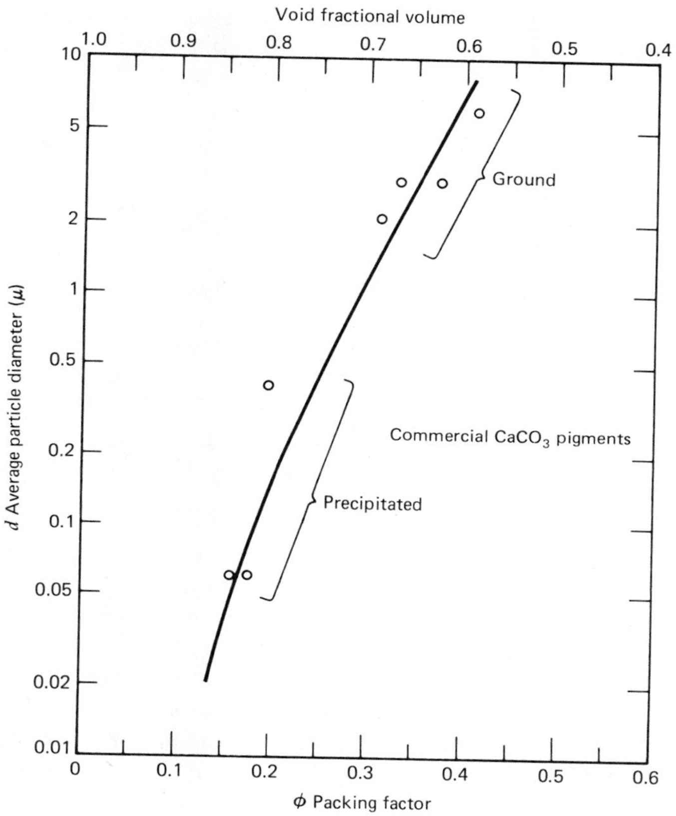

Fig. 5-2 Plot of average particle diameter versus packing factor for dry commercial calcium carbonate pigments.

One method for characterizing the degree of departure of a particle shape from sphericity is to assign it a so-called sphericity value. A sphericity value is defined (Eq. 3) as the ratio of the surface of a sphere, equal in volume to the nonspherical particle, to the surface area of the nonspherical particle (5).

$$\text{Sphericity value} = \frac{\text{surface area of sphere equal in volume to nonspherical particle}}{\text{surface area of nonspherical particle}} \tag{3}$$

PROBLEM 5-2

Calculate the sphericity value for a cubical particle.

Solution. Let x equal the length of each side of the cubical particle. The cube surface area is then $6x^2$, and the cube volume x^3. Let the diameter of a sphere d be such that its volume ($V = \pi d^3/6$) exactly equals the cube volume. From this equality, compute the sphere diameter in terms of x.

$$x^3 = \frac{\pi d^3}{6}; \qquad d = x\left(\frac{6}{\pi}\right)^{1/3}$$

Calculate the surface area of this sphere of equivalent volume.

$$\text{Sphere surface area} = \pi d^2 = \pi x^2 \left(\frac{6}{\pi}\right)^{2/3}$$

Determine the ratio of the area of this sphere (equal in volume to the cube) to the area of the cube.

$$\text{Sphericity value} = \frac{\pi x^2 (6/\pi)^{2/3}}{6x^2} = \left(\frac{\pi}{6}\right)^{1/3} = 0.806$$

Since a cube represents a very significant departure from a sphere, it follows that sphericity values less than 0.806 are indicative of quite irregular shapes (such as crushed glass or jagged flint sand, each with a value of about 0.65). On the other hand, a cylindrical particle with a height equal to its diameter has a sphericity value of 0.874; a cylinder with twice this height, a value of 0.860; and a cylinder with half this height (disk-shaped), a value of 0.827. Ottawa sand that is used in sand mill grinding has a sphericity value of 0.95, indicating a close approach to true roundness (5).

Surface Area of Pigment Particles

The specific particle surface, defined as the surface area of a unit weight of pigment, varies inversely with the particle diameter. This most important theoretical relationship must be well understood to appreciate some of the difficulties that arise in preparing satisfactory pigment dispersions. Thus, on holding the weight of pigment particles constant, halving the particle diameter acts to double the surface area.

Consider a unit weight (1.0 g) of uniform spherical particles of density ρ. The volume of particles in this 1-g weight is $1.0/\rho$. Let the diameter of the uniform spherical particles be d. Then the volume of each particle is $\pi d^3/6$, and the number of particles in the volume $1.0/\rho$ is $6/\pi d^3 \rho$. The area of each uniform spherical particle is πd^2. Therefore the total area of the spherical particles in a 1-g weight is given by Eq. 4.

$$S \text{ (specific surface area)} = \frac{6}{\pi d^3 \rho} \cdot \pi d^2$$

$$S = \frac{6}{d\rho} \qquad (4)$$

When the diameter in Eq. 4 is expressed in cm and the density in g/cm³, the units of S are cm²/g. However, particle diameters are generally given in μ units. If such micrometer units (10,000X smaller than a cm unit) are used in Eq. 4, S must be adjusted unitwise accordingly (made 10,000X larger) to give S units of m²/g. The specific surface areas of pigment particles, then, are generally expressed as m²/g, and these units are obtained when particle diameters, expressed in μ, are entered into Eq. 4.

PROBLEM 5-3

Calculate the approximate square feet of particle surface area for 1.0 lb of rutile titanium dioxide pigment (ρ = 4.16 g/cm³; average particle diameter d = 0.30 μ).

Solution. Substitute the given data in Eq. 4 to obtain the specific surface area of the TiO_2 particles in m²/g.

$$S = \frac{6}{0.30 \cdot 4.16} = 4.81 \text{ m}^2/\text{g}$$

The surface area of 1.0 lb (= 454 g) of TiO_2 pigment is then 2183 m² (= 4.81 · 454). This is equivalent to 23,500 ft² (= 2183 · 10.76) or more than 0.5 acre of surface.

PROBLEM 5-4

A medium color furnace black (pigment density 1.80 g/cm³) has a specific surface area of 190 m²/g. Calculate an approximate average particle size (diameter) for the carbon black particles.

Solution. Substitute the given data in Eq. 4.

$$190 = \frac{6}{1.80d}; \qquad d = 0.018\ \mu\ (= 180\ \text{Å})$$

Although Eq. 4 relating specific surface area S to particle diameter d and particle density ρ applies strictly only to uniform spherical particles, it still provides a simple and very useful approximate relationship for most of the commodity pigments encountered in practice. For more accurate work, equations relating S to d and ρ have been developed that also take into account both particle size distribution and departure from sphericity. However, the added data necessarily involve more complex equations (4, 5).

Since specific surface area varies inversely with particle size, it follows that phenomena associated with the surfaces of particles are accentuated in progressing from coarser to finer pigment grades. For example, a tenfold increase in surface adsorption can be anticipated for a tenfold reduction in particle size (as from 10 to 1.0 μ). Similarly, in comparison with the coarser pigment, the dispersion of the finer pigment in a vehicle requires the displacement of 10 times as much adsorbed gas and other impurities from the particle surface to effect the same acceptable dispersion.

Measurement of the specific surface of pigment particles is generally carried out using either permeametric techniques (measurement of the flow of gas through a bed of pigment particles) or adsorption techniques (measurement of the amount of gas or liguid adsorbed by a sample of pigment).

Specific Particle Surface Determined by Permeability Measurements. Since the flow of a liquid is quite slow through a bed of pigment particles, resort is almost always made to gaseous flow in determining specific surface area by permeametric measurements.

Two major types of gas flow are used (the transition stage between the two is also recognized as a hybrid type of flow but is purposely avoided in experimental work). Laminar (viscous) flow occurs with gases at or near atmospheric pressure; Knudsen (diffusional) flow, with gases under a high vacuum. The latter type of flow is most appropriate for accurately measuring the specific surface areas of pigment particles. However, viscous flow is also routinely used for determining the specific surface areas of many industrial powders. An excellent description of permeametric methodology is given by Orr and Tyree (6).

Specific Particle Surface Determined by Adsorption Measurements. The adsorption method for determining the specific surfaces of pigment particles is probably most widely used in research studies. Both gaseous and liquid adsorption are recognized procedures. Of the two, however, gaseous adsorption is generally preferred, since the experimental techniques for gas adsorption are well established and the procedure is relatively simple and straightforward. A comprehensive introduction to gaseous adsorption methodology is given by Ettre (7).

Spacing of Pigment Particles

In considering the spacing (geometric) patterns of pigment particles, it is helpful to consider first some symmetrical arrangements. For example, as schematically shown in Fig. 5-3*a*, uniform spherical particles (monosize spheres) can be visu-

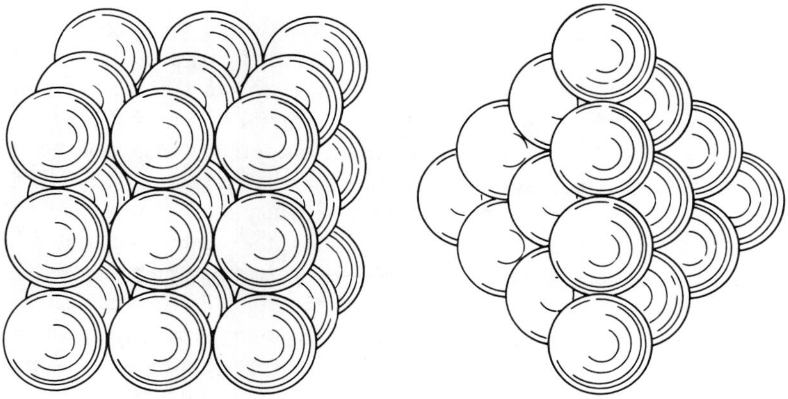

(*a*)

Fig. 5-3*a* Schematic illustration of cubical and tetrahedral packing patterns for uniform spherical particles.

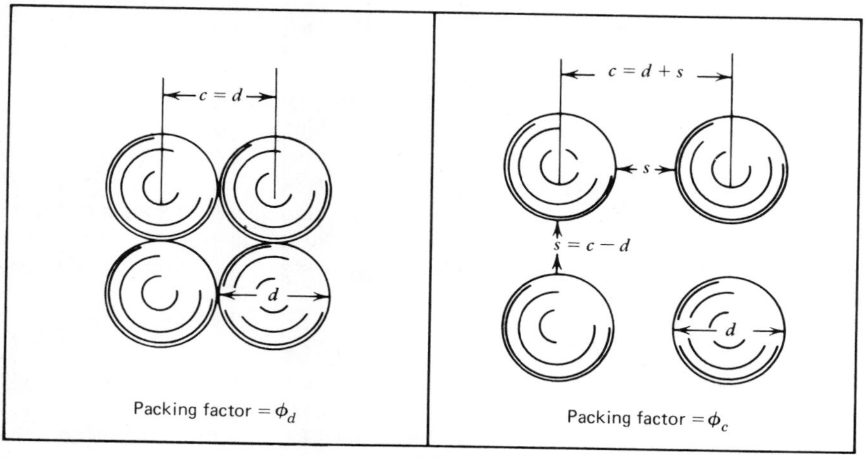

(*b*)

Fig. 5-3*b* Schematic diagram showing linear relationships for contacting and separated uniform spheres conforming to a cubical packing arrangement.

alized as spaced according to either a cubical (loose) packing arrangement ($\phi = 0.524 = \pi/6$) or a tetrahedral or rhombohedral (tight) packing arrangement ($\phi = 0.740 = \sqrt{2}\pi/6$). Experimental studies have shown that actual monosize spheres tend to pack randomly to give a packing factor of $\phi = 0.639$, a value close to the average of 0.632 for the limiting cubical and tetrahedral arrangements.

In Fig. 5-3*a* the monosize spheres are shown as touching each other, but what happens to the pigment packing factor when these spheres move away from each other in a uniform manner so that the center-to-center distance between any two spheres, originally equal to d, is increased to some value c? This type of separation is schematically shown in Fig. 5-3*b* for a cubical packing pattern. Let ϕ_d be the packing factor when the monosize spheres are in contact (center-to-center distance equal to particle diameter d), and ϕ_c be the packing factor when the particles have moved apart and their center-to-center distance is c. Then it can be shown that the simple relation given by Eq. 5 is valid (8).

$$\phi_c c^3 = \phi_d d^3 \tag{5}$$

Another useful spacing parameter is the stand-off distance s, which is defined as the distance separating the outside surfaces of two particles. When two particles are in contact the stand-off distance is zero ($s = 0$); when two particles have a center-to-center separation distance of c, their stand-off distance is $s = c - d$, as indicated in Fig. 5-3*b*.

PROBLEM 5-5

The CPVC of a zinc-rich epoxy primer is 0.80. Assuming particle-to-particle contact of the spherical zinc dust particles ($d = 5.0\ \mu$) at this CPVC, calculate the separation distance between particles (stand-off distance s) that can be expected if the pigment volume concentration is reduced to 0.60.

Solution. Since the zinc particles are originally in contact, the pigment packing factor ϕ equals the CPVC value ($\phi = 0.80$). Substitute this value and other given data in Eq. 5, and solve for c. Note that, for the condition PVC < CPVC, PVC and ϕ_c are identical.

$$0.60c^3 = 0.80 \cdot 5.0^3; \qquad c = 5.5\ \mu$$

The stand-off distance s for the PVC of 0.60 is equal to 0.5 μ (= 5.5 − 5.0).

Comment. A zinc-rich primer depends on contact among the zinc dust particles for its anticorrosion activity over ferrous substrates. Reducing the PVC from 80% (0.80) to 60% (0.60) probably introduces an intolerable separation distance and a consequent unsatisfactory anticorrosion behavior.

PROBLEM 5-6

The hiding power HP of rutile titanium dioxide pigment (expressed in units of ft^2/lb at 0.98 contrast ratio) has been empirically related to the ratio s/c by the expression HP = 370 (s/c), based on the assumption of a tetrahedral spacing pattern (8). Calculate the hiding power of a glossy white paint that has been pigmented with rutile TiO_2 to a PVC of 15% (0.15).

Solution. Let the average diameter of the TiO_2 particles be d, and calculate the center-to-center separation distance c for the particles in terms of d for the given PVC of 0.15, using Eq. 5. Note that, for the condition PVC < CPVC (which is certainly true for a glossy paint), ϕ_c = PVC. Also, for a contacting tetrahedral spacing, $\phi_d = 0.740$.

$$0.15 \cdot (d+s)^3 = 0.740 d^3$$

$$s = 0.702 d$$

$$c = d + s = 1.702 d$$

Substitute these derived data in the given expression HP = 370 (s/c).

$$\text{HP} = 370 \cdot \frac{0.702 d}{1.702 d} = 153 \text{ ft}^2/\text{lb}$$

Comment. Since s increases at a relatively faster rate than $(s + d)$, it follows that the ratio $s/c = s/(s + d)$ must increase with an increase in s. This means that the hiding power, expressed in ft^2/lb, increases as the particles become further separated. Conversely, bringing the TiO_2 particles closer together (by increasing the PVC) reduces the hiding power. Actually the introduction of an *excessive* concentration of titanium dioxide to achieve whiteness can be self-defeating.

Binder from most solvent-type vehicles is more or less tightly adsorbed on the surface of pigment particles, and this adsorbed layer is effective in preventing particle-to-particle contact. As a result the theoretical maximum packing factor ϕ that is possible for the bare pigment (corresponding to particle-to-particle contact) is generally never attained; rather, some lesser value ϕ_c', corresponding to the condition where the outside surfaces of the adsorbed layers just make contact, is reached. This effect becomes increasingly significant in progressing from coarser to finer particle sizes, since the adsorbed layer thickness, which remains essentially the same, plays a more important role as the particle diameter diminishes.

When a layer of adsorbed binder, of thickness a, envelopes the pigment particles, the closest center-to-center contact that can be effected corresponds to a c value of $(d + 2a)$. This results in a reduced pigment packing factor, as given by Eq. 6.

$$\phi_c'(d + 2a)^3 = \phi_d d^3$$

$$\phi_c' = \phi_d \left(\frac{d}{d + 2a} \right)^3 \tag{6}$$

Inspection of this equation shows that adsorbed binder gives a decreased packing factor.

The critical pigment volume concentration (CPVC) is represented best by ϕ_c' as calculated by Eq. 6, rather than ϕ_d for the bare particles, and will be considered so in this book.

PROBLEM 5-7

The adsorption of linseed oil on carbon black particles has been reported to provide an adsorbed layer of oil that is 0.0025 μ in thickness (9). Assuming that furnace blacks are essentially spherical and that their bare packing factor ϕ_d is independent of particle size, compare the CPVCs for two furnace blacks (average diameters of 0.017 and 0.070 μ, respectively) when dispersed in linseed oil.

Solution. Substitute the given data in Eq. 6, in turn, for the two carbon blacks.

Carbon black of 0.017-μ diameter:

$$\phi_c' = \text{CPVC} = \phi_d \left(\frac{0.017}{0.017 + 0.0050} \right)^3 = 0.461\phi_d$$

Carbon black of 0.070-μ diameter:

$$\phi_c' = \text{CPVC} = \phi_d \left(\frac{0.070}{0.070 + 0.0050} \right)^3 = 0.813\phi_d$$

A comparison of the calculated CPVCs shows the CPVC for the finer carbon black to be 57% of that for the coarser one (0.57 = 0.461/0.813).

Comment. Inspection of these results reveals the profound effect of adsorption on the CPVC values of pigments of very fine particle size. On the other hand, adsorption exerts only a very minor influence on coarse pigments. For example, for a 5.0-μ diameter particle (typical of an extender pigment) the CPVC value given by an adsorbed layer 0.0025 μ thick will be only 0.3% less than the theoretical ϕ_d value for the bare particles.

Structure of Pigment Particles

The term "structure" refers to the chain- or grapelike clusters of particles that form during the manufacture of certain very fine pigments. The prime examples of structured pigments are the carbon blacks, which generally exhibit low to high structure. Inspection of an electron micrograph of a high-structure carbon black reveals that the black structure is composed (at least in part) of fused chains of nodular particles. An increased carbon black structure assists dispersibility in polymer systems and contributes to viscosity. On the other hand, an increased black structure lowers the jetness and gloss of carbon black/polymer systems.

Pyrogenic silica is a second major example of a structured pigment. The primary particles of pyrogenic silica are spherical, and during manufacture these particles coagulate into chains. The silica particles are extremely fine, and even after the particle chains are disrupted by mechanical action they tend to regroup to form a chicken-wire type of structure.

Fine colored pigments also display some evidence of structure. Structure, as will be shown, is associated with pigment texture, a term that is used to describe whether a pigment disperses with ease (soft or good texture) or with difficulty (hard or bad texture).

Of the above pigments, only the carbon blacks have been routinely assigned structure ratings. Normal-structure blacks are given a structure index of 100. Low-structure blacks have ratings in the region from 50 to 100; high-structure blacks, in the region from 100 to 200. Unfortunately this rating system is presently being applied only to assess the performance of carbon blacks in rubber compounds.

Structure Rating for Paint Pigments. As one possible approach for establishing a structure rating for paint pigments, this author proposes that the packing factor ϕ_d for the bare pigment particles can usefully serve as an index of pigment structure. Such a packing factor can be calculated from a knowledge of the density ρ, the CPVC (as determined by an oil absorption value), and the specific surface area S (or the average particle diameter d) of the pigment in question. As part of the computation procedure it is assumed that a layer of linseed oil, 0.0025 μ thick, is tightly adsorbed on the surface of the pigment particles during the oil absorption test.

Let ϕ_d be the packing factor for the bare pigment particles, and CPVC the packing factor for the coated particles at the end of the oil absorption test. At this CPVC end point, the coating of adsorbed linseed oil effectively increases the pigment diameter from d to $(d + 0.0050)$. Entering these data into Eq. 5 gives Eq. 7.

$$\text{CPCV} \cdot (d + 0.0050)^3 = \phi_d d^3$$

$$\phi_d = \text{CPVC}\left(1.0 + \frac{0.0050}{d}\right)^3 \tag{7}$$

From Eq. 4, $d = 6/S\rho$, and substituting this in Eq. 7 gives Eq. 8.

$$\phi_d = \text{CPVC}\,(1 + 0.00083\,\rho S)^3 \tag{8}$$

A low value for this structure index corresponds to a highly structured particle system, as indicated by a considerable void volume. Conversely, a high value for ϕ_d corresponds to a dense pigment packing with little or no structure.

Texture Rating. It is suggested that pigment texture (inherent dispersibility of a pigment) can be related to its structure index and specific surface area by Eq. 9

$$\text{Texture rating} = S(\phi_d - 0.4) \tag{9}$$

This expression, empirically derived by the present author, is based on a considerable quantity of experimental data (10, 11). These data have been plotted in Fig. 5-4 where circles represent soft-textured (easy dispersing) pigments, squares represent normal-textured (normal dispersing) pigments, and triangles represent hard-textured (difficult dispersing) pigments. Also shown are texture curves calculated from Eq. 9. Inspection shows that the normal-textured pigments fall within a band bounded by texture ratings of 12 and 15, the soft-textured pigments are in a region below the 12 rating curve, and the hard-textured pigments lie above the 15 rating curve. This scheme for quantitatively assigning texture ratings (dispersibility values) to pigments appears to merit further investigation.

PROBLEM 5-8

A phthalocyanine green pigment has an average diameter of 0.040 μ, a density of 1.96 g/cm^3, and a CPVC value of 0.563 (calculated from an oil absorption

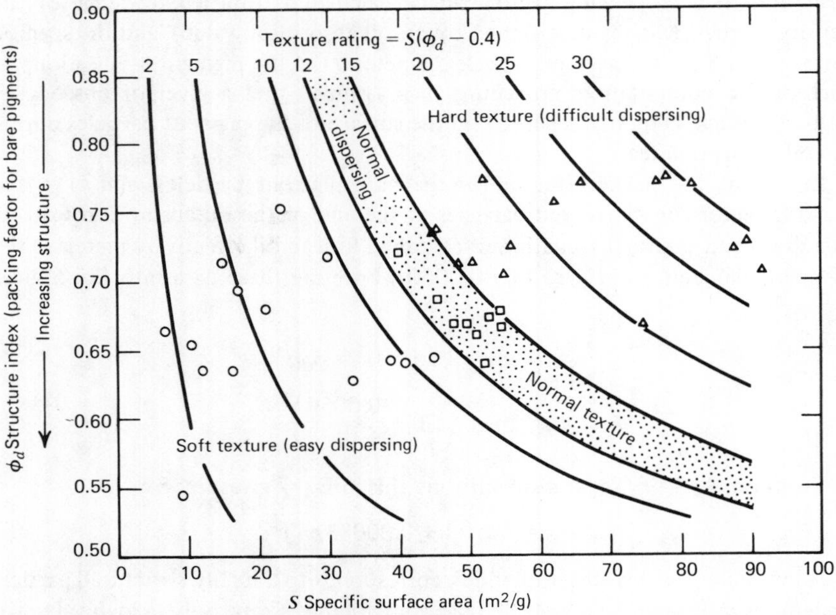

Fig. 5-4 Plot of texture rating as a function of structure index and specific surface area. Plotted points are literature values for soft-, normal-, and hard-texture pigments.

value of 37 g/100 g). From this information compute a texture rating for the green pigment, and from this rating estimate its inherent dispersibility.

Solution. Substitute the given CPVC and particle diameter values in Eq. 7 to obtain a packing factor for the bare particles.

$$\phi_d = 0.563\left(1.0 + \frac{0.0050}{0.040}\right)^3 = 0.802$$

Calculate the specific surface area for the phthalocyanine green pigment from Eq. 4.

$$S = \frac{6}{d\rho} = \frac{6}{0.040 \cdot 1.96} = 76.5\ \text{m}^2/\text{g}$$

Compute the texture rating from Eq. 9.

$$\text{Texture rating} = 76.5(0.802 - 0.4) = 31$$

The texture rating of 31 suggests that this green phthalocyanine pigment is very difficult to disperse (see location in Fig. 5-4) and should be processed accordingly.

Comment. The texture ratings for three carbon blacks having the same specific surface ($S = 80\ \text{m}^2/\text{g}$) and the same density ($\rho = 1.75\ \text{g/cm}^3$), but different oil absorption values (117, 75, and 37 g/100 g), were calculated as 3, 16, and 33, respectively. This rank order is in accord with the relative dispersibilities of these blacks as indicated by their values in the rubber structure index system (150, 100, 50), where a higher value represents easier dispersion.

The suggested texture rating system for paint and ink pigments is based on the experimental observation that finer and more compacted pigments are more difficult to disperse than coarser and less densely aggregated ones. Other factors play a role in determining dispersibility, but the proposed texture rating can serve as a benchmark value from which deviations can be considered to occur because of variables other than particle fineness and structure (packing density of bare particles).

Spread in Particle Size (Distribution Patterns)

Commercial pigments may be supplied within a narrow range of particle sizes (e.g., titanium dioxide, most carbon blacks, and many precipitated calcium carbonates) or over a wider range of particle sizes (e.g., many clays, zinc oxides, lead pigments, and ground calcium carbonates). A voluminous body of literature dealing with both the theoretical and the actual distribution patterns for particulate matter is available (4). Standard distribution functions (binomial, Gaussian, log-normal) and deviations from these laws have been utilized to characterize particle size distributions. Procedures for computing average particle size based on geometric mean weights, arithmetic mean weights, mean volumes, and specific surface areas are well established. From this vast mathematical background a few simple relationships have been extracted for discussion in this book.

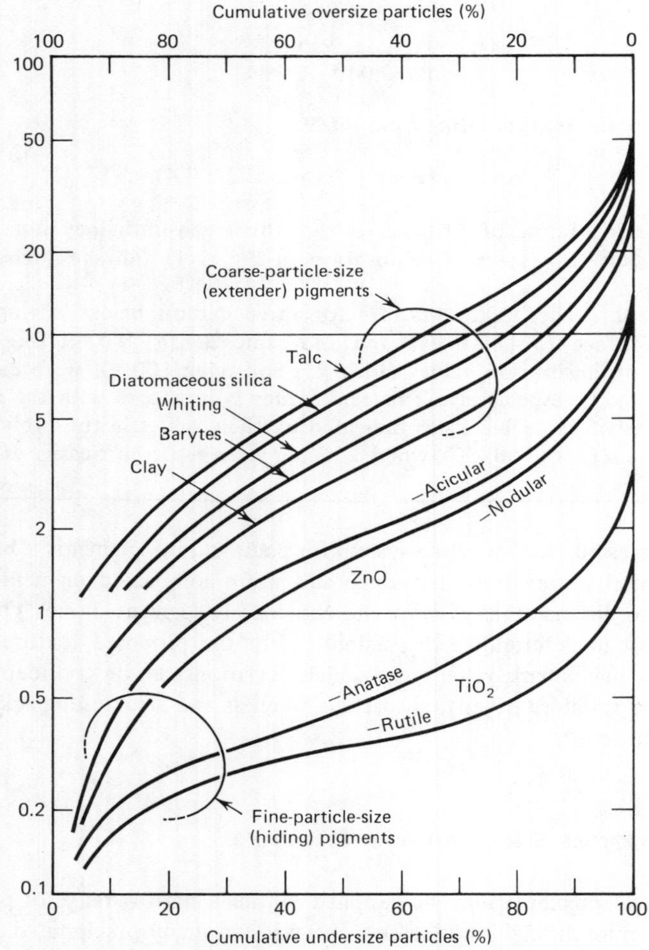

Fig. 5-5 Approximate distribution patterns for representative coarse extender and fine hiding pigments, plotted on a cumulative type of graph.

The following common graphical methods are used to present particle size distributions:

Histograms
Plot of the frequency of particle occurrence as a function of size range.
Size frequency curve
Smoothed out histograms; justified when many values are available for plotting.
Cumulative graphs
Plot of the fraction (or percent) of particles greater (or less) than a given particle size against particle size.

The cumulative type of graph as plotted on semilog paper is most helpful for visually portraying and understanding particle size distributions and is used throughout this book (see Fig. 5-5).

The packing patterns of pigment particles of different size constitute another area that has been studied over the years by many investigators. However, because of the complexity of this problem, it has been necessary to blend theoretical analysis with empirical findings to achieve even a minimal proficiency in calculating pigment packing factors for mixtures of different-size particles.

Packing of Binary Mixtures of Pigment Particles. There is considered first a simple binary (two size) mixture in which particles of a finer size are mixed with particles of a larger size (on the order of 10X larger in diameter). It is obvious that in such a combination the smaller or finer particles will tend to fill the voids formed by the larger or coarser particles. As a result a denser packing is obtained for the binary mixture than for either particle size alone.

Let the total solid volume of a pigment mixture (interstitial voids are excluded) equal unity (1.00), and let the solid fractional volume of the larger particles equal v. Then the solid fractional volume of the smaller particles equals, by difference, $(1.00 - v)$. Also let the pigment packing factor for the larger particles be ϕ_l, that for the smaller particles ϕ_s, and that for the binary mixture ϕ_m. These and other symbols used in the derivation to follow are listed below.

	Solid Volume	Packing Factor
Binary mixture		
In general	1.00	ϕ_m
For maximum packing	1.00	ϕ_{m*}
Large particles		
In general	v	ϕ_l
For maximum packing	v^*	
Small particles		
In general	$1 - v$	ϕ_s
For maximum packing	$1 - v^*$	—

Consider first a packed bed of larger particles, and let the smaller particles progressively fill the interstitial voids within these large particles, as shown in Fig. 5-6*a*. Before the start of the addition the packing factor is that for the larger particles alone (ϕ_l). However, as the smaller particles fill the large particle voids, the packing for the binary mixture becomes more dense, as indicated by Eq. 10. The geometric volume in this equation refers to the overall volume, including interstitial voids (air spaces); the total solid volume is 1.00.

$$\phi_m = \frac{\text{solid volume}}{\text{geometric volume}} = \frac{1}{v/\phi_l}$$

$$\phi_m = \frac{\phi_l}{v} \tag{10}$$

Note that during this stage the geometric volume of the binary mixture is the same as the geometric volume of the large particles alone. Eventually a point is reached where the voids of the large particles are completely filled with the finer pigment. This occurs when the condition expressed by Eq. 11 is reached.

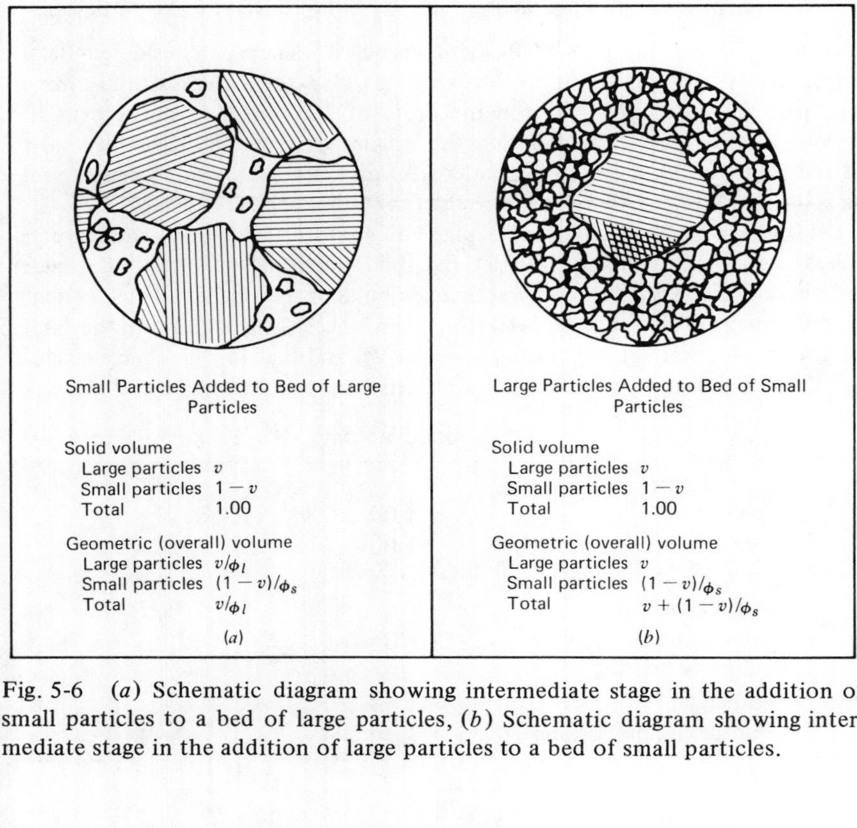

Small Particles Added to Bed of Large Particles
Solid volume
Large particles v
Small particles $1 - v$
Total 1.00
Geometric (overall) volume
Large particles v/ϕ_l
Small particles $(1 - v)/\phi_s$
Total v/ϕ_l
(*a*)
Large Particles Added to Bed of Small Particles
Solid volume
Large particles v
Small particles $1 - v$
Total 1.00
Geometric (overall) volume
Large particles v
Small particles $(1 - v)/\phi_s$
Total $v + (1 - v)/\phi_s$
(*b*)

Fig. 5-6 (*a*) Schematic diagram showing intermediate stage in the addition of small particles to a bed of large particles, (*b*) Schematic diagram showing intermediate stage in the addition of large particles to a bed of small particles.

$$\frac{1 - v^*}{\phi_s} = \frac{v^*}{\phi_l}(1 - \phi_l) \tag{11}$$

Solving for v^* gives Eq. 12.

$$v^* = \frac{\phi_l}{\phi_l + \phi_s - \phi_l\phi_s} \tag{12}$$

Substituting this v^* value in Eq. 10 yields Eq. 13 for the packing factor ϕ_{m^*}, the maximum packing that can be attained for the binary mixture.

$$\phi_{m^*} = \phi_l + \phi_s - \phi_l\phi_s \tag{13}$$

Of course, a reverse procedure could have been used in arriving at this filled condition, namely, starting with a bed of smaller particles to which are added the larger particles, as shown in Fig. 5-6*b*. Here the initial packing factor is that for the smaller particles alone (ϕ_s). However, as large particles are introduced, they progressively displace a proportion of the fine particles and their voids in an amount equal to the solid volume of the added large particles. This promotes a denser packing, as given by Eq. 14.

$$\phi_m = \frac{\text{solid volume}}{\text{geometric volume}} = \frac{1}{(1 - v)/\phi_s + v}$$

$$\phi_m = \frac{\phi_s}{1 - v(1 - \phi_s)} \tag{14}$$

Inspection of Eqs. 10 and 14, both giving ϕ_m values, shows that, regardless of finer particles are added to coarser ones (Eq. 10) or vice versa (Eq. 14), an increase in packing density occurs. Obviously, there must be some intermediate ϕ_{m^*} value where these two rising trends meet to give a common maximum packing factor peak for the binary mixture. The value of v^* for this maximum packing factor can be determined by equating the two values for ϕ_m, as given by Eqs. 10 and 14, and solving for v^*.

$$\phi_{m^*} = \frac{\phi_l}{v^*} = \frac{\phi_s}{1 - v^*(1 - \phi_s)}$$

$$v^* = \frac{\phi_l}{\phi_l + \phi_s - \phi_l\phi_s}$$

Not unexpectedly, this is the same expression as given by Eq. 12 for v^*.

To render these equations more meaningful Fig. 5-7 shows a graph where the packing factors for all possible binary mixtures are plotted against the fractional volumes of the larger particles in these mixtures (solids basis). The intersection of a finer particle packing factor curve ϕ_s with a coarser particle

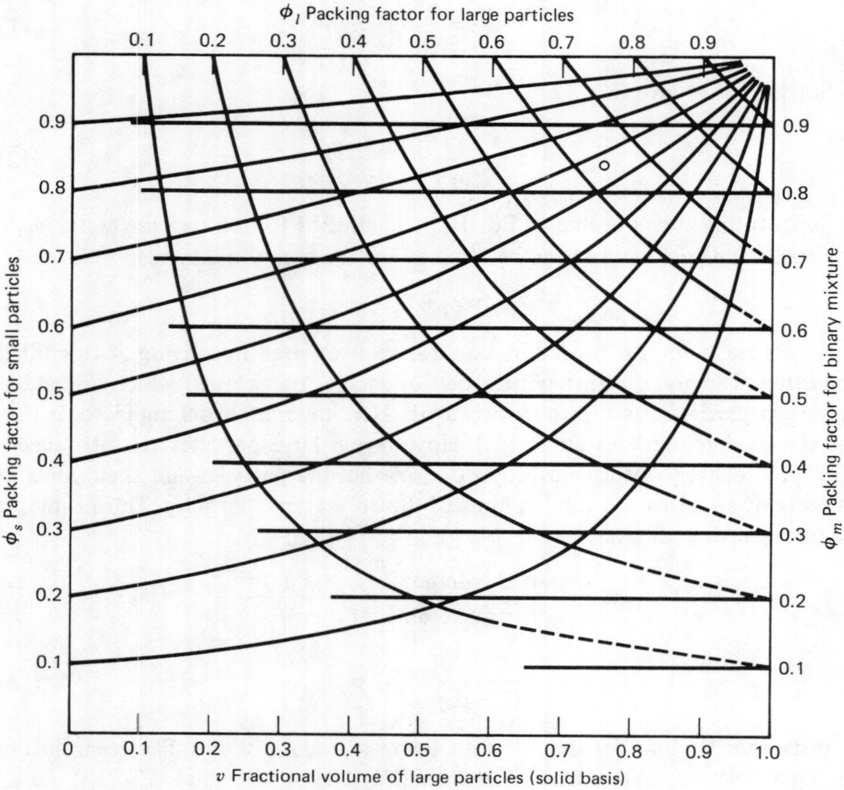

Fig. 5-7 Graph giving packing factors for binary mixtures of small and large particles, and fractional solid volume of the large particles, as a function of the individual packing factors for the small and large particles.

packing factor curve ϕ_l gives a peak point corresponding to the maximum packing factor ϕ_{m*} for the two ϕ_l and ϕ_s packing factors. Also given by the plot is the larger particle fractional volume for this maximum packing (v^*).

PROBLEM 5-9

A coarse ground calcium carbonate is mixed with a titanium dioxide pigment of fine particle size. If the packing factor for the calcium carbonate pigment is 0.64 and that for the titanium dioxide is 0.56, calculate the maximum packing density (maximum packing factor) that is possible with a combination of these two pigments, and the solid volume fraction of coarse pigment necessary to give this densest binary mixture.

Solution. This problem can be solved either graphically (see circled point in upper right section of Fig. 5-7) or by substitution in Eqs. 13 and 12.

$$\phi_{m*} = 0.64 + 0.56 - (0.64 \cdot 0.56) = 0.84\ (84\%)$$

$$v^* = \frac{0.64}{0.84} = 0.76\ (76\%)$$

Comment. The calculated result for v^* of 0.76 agrees reasonably well with the value for v^* of 0.74 reported for this binary mixture on the basis of laboratory testing. In general, such good agreement is obtained for v^* values. However, the calculated value for ϕ_{m*} of 0.84 is significantly higher than an experimentally determined value of 0.72 for this mixture (8). The reason for this discrepancy is discussed next.

The change in the packing factor that occurs when a fine pigment is added to a coarse one (or vice versa) can be traced on the graph of Fig. 5-7 by locating appropriate pigment packing curves to either side of the diagram (corresponding to the packing factors ϕ_l and ϕ_s for the coarse and fine particles) and then tracking inwardly along these curves to their intersection within the graph. This juncture is always characterized by a peak. However, the maximum packing factor given by this peak is invariably too high. Actually, this discrepancy might have been anticipated, since it is unrealistic to expect a binary mixture to behave with theoretical perfection at such a critical juncture. Rather than a sharp break, there is a transition region through which a gradual rounding-over from one curve to the other takes place. The gently rounded top that joins the two curves may be some 10 to 20% below the peak formed by the theoretically derived upsweeping curves. The rounded top reflects the existence of a confused intermixture of packing densities through this region, where either side intrudes on the other to smooth out the pigment packing factor to a lower value.

Packing of Fine and Coarse Particles in Practice. Flat wall paints exhibit many optimum properties when formulated at, or close to, their CPVC points, such as minimum change in hiding on drying and minimum color and sheen difference over substrates of varying porosity. It has been demonstrated repeatedly that, by formulating a flat paint at its CPVC value, color uniformity is assured. Prompted by this fact, a study was undertaken to develop a method for computing CPVC values for pigment mixtures based on the packing characteristics of the individual pigments composing the mixture. Results of this study were reported in 1959 and 1961 (12, 13). Data have been abstracted from this material and recast in a somewhat different manner to apply directly to the text presentation here.

The determination of the CPVC by calculation, as proposed in the studies cited, calls for a differentiation between fine hiding pigments as a group and coarse extender pigments as a group. As in the preceding theoretical section, it is assumed that the fine hiding pigments fill the voids of the coarser extender pigments as part of the pigment mixing process.

The first step in the proposed calculation procedure calls for the computation of an average CPVC. This is then corrected to the true CPVC by the application of an experimentally derived correction factor that accounts for the void filling action. The densities and packing factors (CPVC values) for the fine and coarse pigments used in this work are listed in Table 5-1. The approximate particle size distributions for these same pigments are indicated in Fig. 5-5.

The correction factor c that is read off the empirical correction factor graph, as given in Fig. 5-8, is based on the fractional solid volume of the fine pigment in the total composition. It is entered into Eq. 15 to give the corrected or true CPVC*.

$$\text{CPVC}^* \text{ (corrected)} = \frac{\text{CPVC}}{1.0 - c \cdot \text{CPVC}} \tag{15}$$

If more than one hiding pigment is present, correction factors are first obtained, based on the fractional solid volume of the combined fine pigments. These are then prorated on the basis of the percentage of each fine pigment in the fine pigment mixture to give a final composite correction factor. Problem 5-10 serves to show the general method of calculation based on a typical flat wall paint.

PROBLEM 5-10

Calculate the critical pigment volume concentration for a pigment mixture composed of 45.45%w of titanium dioxide (rutile), 9.10%w of talc, and 45.45%w of whiting (calcium carbonate). Use applicable data from Table 5-1.

Table 5-1 Critical Pigment Volume Concentrations and Densities for Selected Fine Hiding and Coarse Extender Pigments

Type of Pigment	CPVC Value	ρ Density (g/cm^3)
Fine Hiding		
Titanium dioxide		
Rutile TiO_2	0.578	4.16
Anatase TiO_2	0.472	3.90
Zinc oxide		
Acicular ZnO	0.429	5.58
Nodular ZnO	0.455	5.58
Coarse Extender		
Barytes ($BaSO_4$)	0.658	4.45
Clay	0.500	2.63
Silica, diatomaceous	0.282	2.20
Talc	0.495	2.70
Whiting ($CaCO_3$)	0.649	2.71

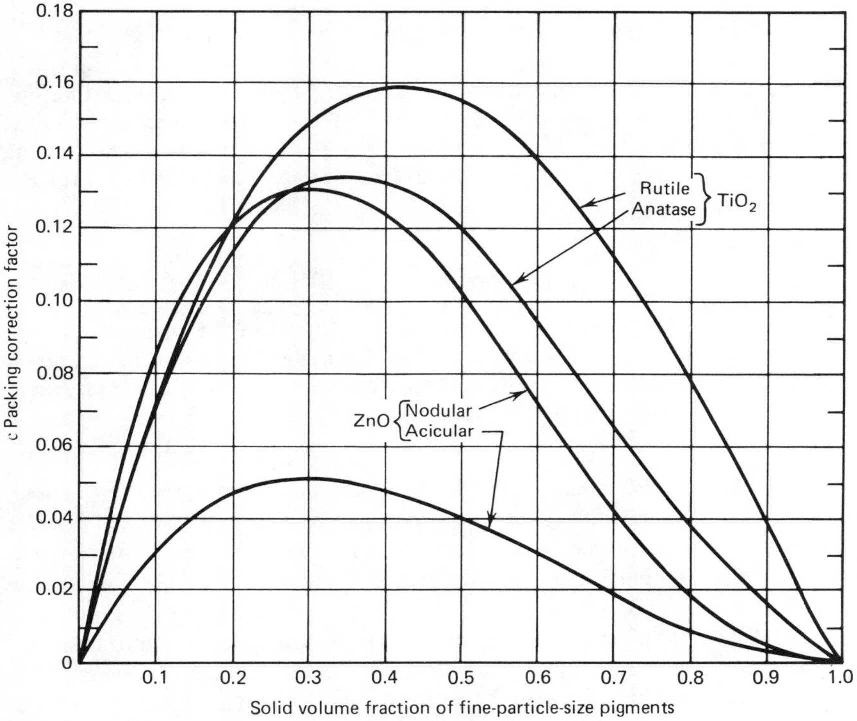

Fig. 5-8 Packing correction factor plotted as a function of the solid volume fraction of fine particles.

Solution. Determine first the volume composition of the pigment mixture, using the given density data, and then the volume of space occupied by the three pigments based on their CPVC values.

Pigment	W (g)	ρ (g/cm^3)	V (cm^3)	CPVC	V/CPVC (cm^3)
TiO_2	45.45	4.16	10.93	0.578	18.91
Talc	9.10	2.70	3.37	0.495	6.81
Whiting	45.45	2.71	16.77	0.649	25.84
	100.00		31.07		51.56

From these derived data calculate the fractional solid volume of the fine hiding TiO_2 and the average (uncorrected) CPVC for the pigment mixture.

$$\text{Fractional solid volume of } TiO_2 = \frac{10.93}{31.07} = 0.352$$

$$\text{Average CPVC (uncorrected)} = \frac{31.07}{51.56} = 0.603$$

Reference to the correction factor graph (Fig. 5-8) shows that at a fine pigment fractional solid volume of 0.352 the rutile TiO_2 correction factor is 0.155. Substitute this c value and the uncorrected average CPVC value of 0.603 in the correction expression (Eq. 15) to obtain the corrected and true CPVC for the pigment mixture.

$$\text{CPVC*} = \frac{0.603}{1.0 - 0.155 \cdot 0.603} = 0.665\ (66.5\%)$$

Comment. This calculated value for the true CPVC of the pigment mixture is the same as the CPVC value obtained experimentally by pigmenting an alkyd with varying proportions of this pigment mixture, and from this ladder ascertaining the PVC value that provides optimum optical properties (13).

Packing of Spherical Particles. Up to this point only binary mixtures characterized by a significant difference in particle size have been considered (on the order of a tenfold difference in diameter or more). However, no limitation was placed on the packing factor or the particle shape for either the fine or the coarse fraction.

Conversely, in this section both binary and multicomponent mixtures are considered with no restriction on the difference in particle size. However, each fraction is restricted to a content of uniform spherical particles, and the monosized spheres in each fraction must randomly pack to give a standard packing factor of 0.639.

Binary mixtures of spherical particles meeting these specifications have been extensively studied over the years, both theoretically and experimentally. A paper that summarizes much of this prior information also proposes a novel procedure for computing the packing factor for a multicomponent mixture of spherical particles (14). The computation method is premised on experimentally determined maximum packing factors for binary mixtures as reported in the literature. These data are shown in Fig. 5-9 (lower curve), where maximum packing factors are plotted against diameter ratios for the large and small spheres (d_l/d_s). From these empirical binary data, extension is made to cover tertiary and multicomponent systems.

An inspection of the empirically derived curve for the maximum packing factor for binary mixtures of uniform spherical particles shows that it starts with a packing factor of 0.639 (for randomly packed monosize spheres alone), rises rapidly to a value of 0.835 at a d_l/d_s ratio of 10, and then tapers off to approach asymptotically a theoretical value of 0.870, as given by Eq. 13.

$$\phi_{m*} = 0.639 + 0.639 - 0.639 \cdot 0.639 = 0.870$$

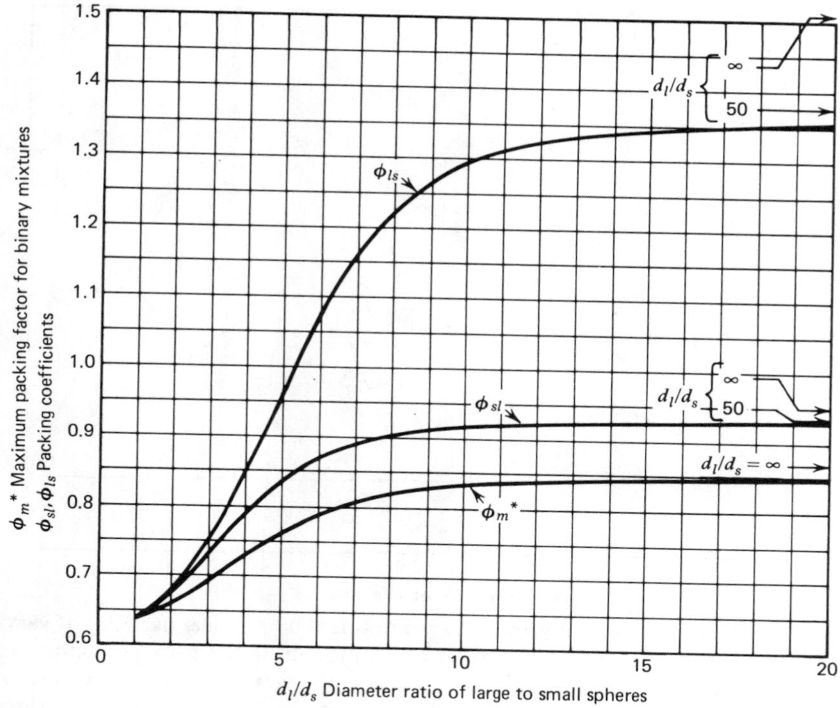

Fig. 5-9 Maximum packing factors for binary mixtures of spherical particles and derived packing coefficients plotted as a function of the diameter ratio of large to small spheres.

The fractional volume of the large spherical particles for this terminal binary condition is given by Eq. 12.

$$v^* = \frac{0.639}{0.870} = 0.735$$

From this experimental binary curve there can now be derived a series of packing coefficients that can be used to calculate the packing factors for multi-component systems. The manner of obtaining and using these coefficients starts with the graphs of Figs. 5-10 and 5-11. The experimental packing factors for binary mixtures of varying d_l/d_s values from Fig. 5-9 have been replotted in Fig. 5-10, where the horizontal bottom scale of the graph gives the solid fraction of the large particles, and the vertical scale the packing factor for the binary mixtures. In the original paper the binary compositions for maximum packing were assumed to be independent of the diameter ratio of the spherical particles

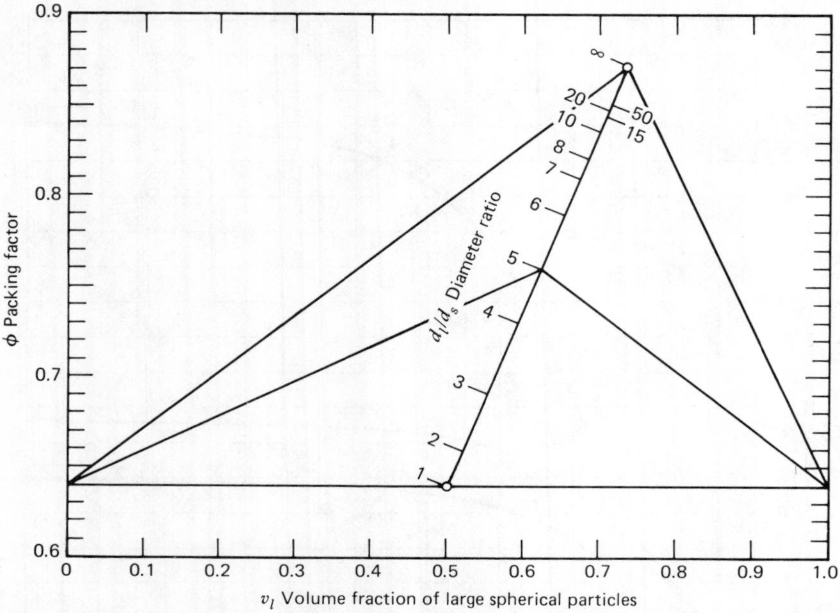

Fig. 5-10 Maximum packing for binary mixtures of spherical particles of varying large/small diameter ratios plotted as a function of the packing factor and the solid volume fraction of the large particles. Values on diameter ratio scale correspond to points of maximum packing.

(ϕ_{m*} fixed at a value of 0.735). However, this assumption seems somewhat unrealistic, and this author suggests that, intuitively, it is more nearly correct to assume that compositions of maximum diameter ratio progressively change from a value of 0.735 for an infinite diameter ratio to a value of 0.500 for a bed of monosize spherical particles ($d_l/d_s = 1.0$). The latter assumption has been made in constructing the graph of Fig. 5-10. Other than maximum packing compositions for binary mixtures are shown in Fig. 5-10 by straight lines drawn from the maximum packing factors (as found on the diameter ratio scale) to the value of 0.639 as located on the scales at either side of the graph (see illustrative lines drawn for d_l/d_s values of 5.0 and ∞). From this base plot, empirically derived packing factors for all binary mixtures of spherical particles can be obtained.

To extend this calculation scheme to tertiary and other multicomponent mixtures, it is necessary to develop packing coefficients. These are calculated by the use of Eqs. 16 and 17. The analytical geometry involved in their derivation is schematically outlined in Fig. 5-11.

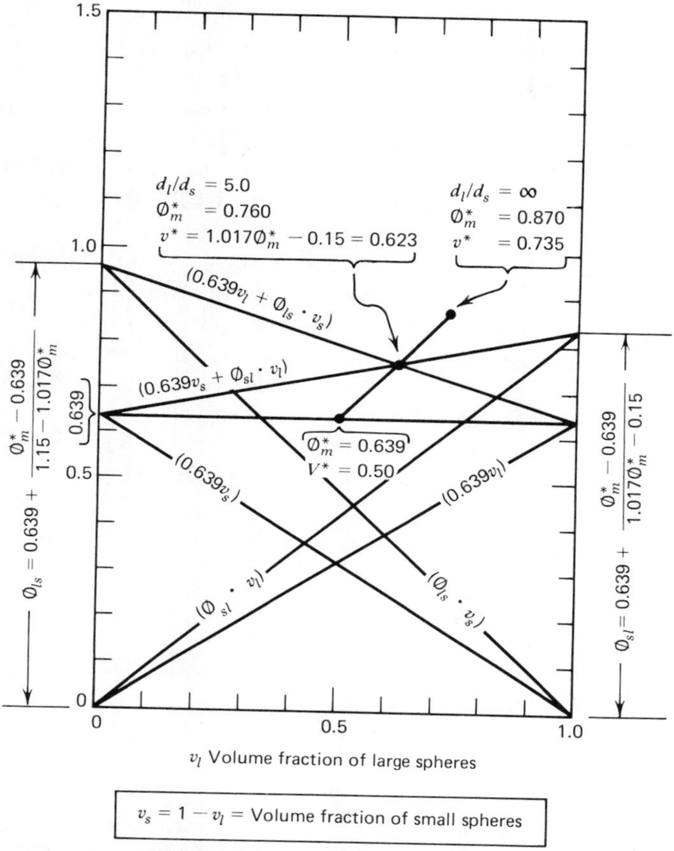

Fig. 5-11 Diagram illustrating the analytical geometry used in developing the packing coefficients from binary systems that can be applied to higher multi-component mixtures.

ϕ_{ls} = coefficient to be used in relating a larger to a smaller particle group
ϕ_{sl} = coefficient to be used in relating a smaller to a larger particle group

$$\phi_{ls} = 0.639 + \frac{\phi_{m*} - 0.639}{1.15 - 1.017\phi_{m*}} \tag{16}$$

$$\phi_{sl} = 0.639 + \frac{\phi_{m*} - 0.639}{1.017\phi_{m*} - 0.15} \tag{17}$$

Values for these coefficients have been plotted against ϕ_{m*} values on the graph given in Fig. 5-9 (upper two curves).

As an example of a packing coefficient calculation consider a 1:5 diameter ratio for the spherical particles. The maximum packing factor for this ratio is 0.760 (read off from Fig. 5-9). Substituting this value in Eqs. 16 and 17 gives packing coefficients of 0.833 for $\phi_{1:5}$ and 0.960 for $\phi_{5:1}$ (the same as can be read off the graph of Fig. 5-9). The computation for multicomponent mixtures uses these coefficients by systematically covering all possible packing factor combinations and selecting from the several results the one with the lowest value as the required packing factor.

Consider a tertiary mixture where the subscripts s (small), i (intermediate), and l (large) refer to the diameter classes (fractions) involved. Let v_s, v_i, and v_l be the respective volume fractions (noting that $v_s + v_i + v_l = 1.00$). Let $(\phi_m)_s$, $(\phi_m)_i$, and $(\phi_m)_l$ represent the packing factors for the three possible systems. Each starts with a value of 0.639, corresponding to a value of 1.00 for v_s, v_i, and v_l, respectively (only one class of sphere is initially present in each). The three possible packing factor combinations are given by Eqs. 18, 19, and 20. Note that v_l is inserted in these equations by its equivalent $(1 - v_s - v_i)$.

$$(\phi_m)_s = 0.639\,v_s + \phi_{si}v_i + \phi_{sl}(1 - v_s - v_i) \tag{18}$$

$$(\phi_m)_i = \phi_{is}v_s + 0.639v_i + \phi_{il}(1 - v_s - v_i) \tag{19}$$

$$(\phi_m)_l = \phi_{ls}v_s + \phi_{li}v_i + 0.639(1 - v_s - v_i) \tag{20}$$

The application of these equations to a tertiary system is illustrated in the next two problems.

PROBLEM 5-11

Calculate the packing factor for a mixture of spherical particles consisting of 30%v of 0.40-μ diameter spheres, 40%v of 2.0-μ spheres, and 30%v of 10.0-μ spheres.

Solution. The diameter ratios for the three pigment classes are 1:5:25 (0.40:2.0:10.0). Determine the necessary packing coefficients for these ratios, either by calculation of by reading them off the graph of Fig. 9.

Packing Coefficients

ϕ_{sl} Type (small to large size direction)	ϕ_{ls} Type (large to small size direction)
$\phi_{1:5}$ or $\phi_{5:25} = 0.833$	$\phi_{5:1}$ or $\phi_{25:5} = 0.960$
$\phi_{1:25} = 0.930$	$\phi_{25:1} = 1.360$

Substitute these packing coefficient values and the given values of $v_s = 0.30$, $v_i = 0.40$, and $v_l = 0.30$ $(= 1 - v_s - v_i)$ in Eqs. 18, 19, and 20, and calculate the three packing factors.

$$(\phi_m)_s = (0.639 \cdot 0.30) + (0.833 \cdot 0.40) + (0.930 \cdot 0.30) = 0.804$$

$$(\phi_m)_i = (0.960 \cdot 0.30) + (0.639 \cdot 0.40) + (0.833 \cdot 0.30) = 0.794$$

$$(\phi_m)_l = (1.360 \cdot 0.30) + (0.960 \cdot 0.40) + (0.639 \cdot 0.30) = 0.984$$

The lowest of these three values is the correct packing factor for the given volume composition ($\phi_m = 0.794$).

The next and more complex problem calls for the volume composition that gives a maximum packing factor.

PROBLEM 5-12

Calculate the fraction volume composition that provides maximum packing for a tertiary mixture of spherical particles that have diameter ratios of 1 : 5 : 25.

Solution. Since these ratios are similar to those of Problem 5-11, the same packing coefficients apply. Substitute these values in Eqs. 18, 19, and 20.

$$(\phi_m)_s = 0.639v_s + 0.833v_i + 0.930(1 - v_s - v_i)$$

$$(\phi_m)_i = 0.960v_s + 0.639v_i + 0.833(1 - v_s - v_i)$$

$$(\phi_m)_l = 1.360v_s + 0.960v_i + 0.639(1 - v_s - v_i)$$

At the maximum packing composition, ϕ_{m*}, there exist the following equalities: $(\phi_m)_s = (\phi_m)_i = (\phi_m)_l$. Hence, by setting the equations equal to each other and solving for the volume fractions, data are obtained that can be used to compute a maximum packing factor.

$$v_s = 0.254; \qquad v_i = 0.082; \qquad v_l = 0.664$$

$$\phi_{m*} = 0.849$$

Comment. Anyone who ventured to carry through the tedious but necessary calculations required for the solution to this problem can appreciate that, in working with even higher multicomponent systems, it becomes almost necessary to enlist the aid of computer technology. Otherwise the numerical task becomes overwhelming. Incidentally, the following results were obtained for the above tertiary system, using a fixed value of 0.760 for calculating the packing coefficients as was done in the original article: $v_s = 0.216$; $v_i = 0.092$; $v_l = 0.692$; $\phi_{m*} = 0.850$.

With the aid of computer programming, packing factors for idealized stepwise particle size distributions (10 different volume fractions of spherical particles) have been calculated for rectangular, symmetrical, positively skewed, and negatively skewed distributions (14). The results of these computations have been plotted in Fig. 5-12, together with a reference value of 0.639 for the random packing factor for uniform spherical particles. These curves may prove useful in situations where it is necessary to estimate (by comparison) a packing factor for an actual size distribution of nodular or spherical particles not too fine in size.

Recently a mathematical model has been proposed, based on the foregoing

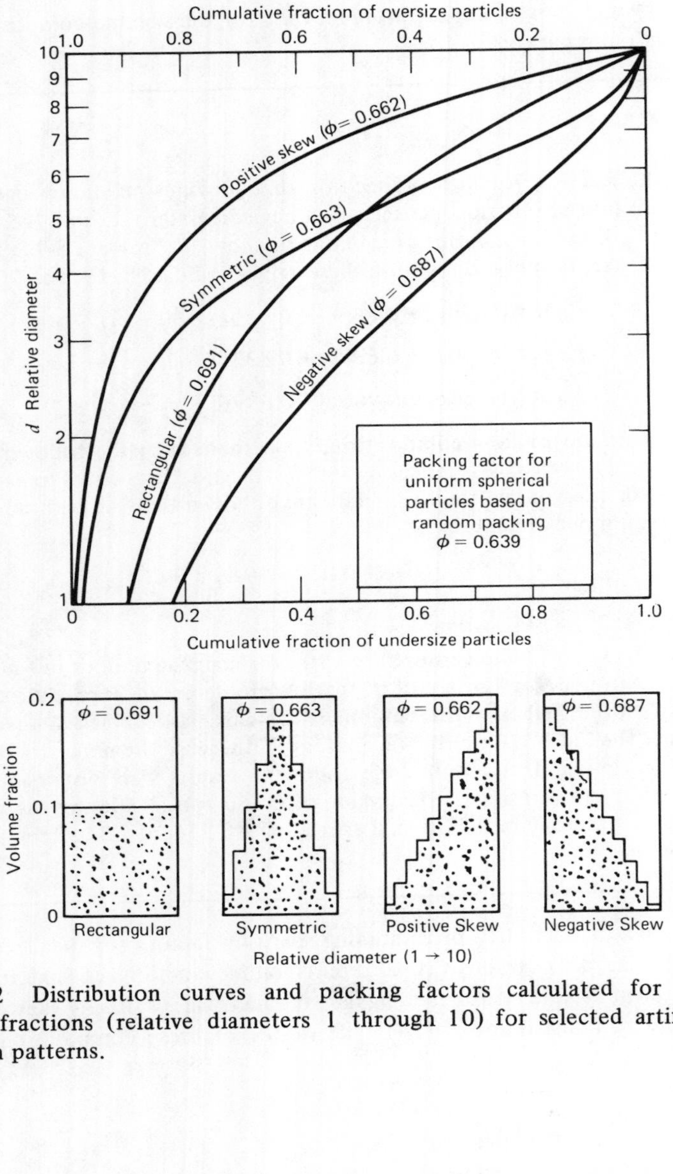

Fig. 5-12 Distribution curves and packing factors calculated for idealized 10-step fractions (relative diameters 1 through 10) for selected artificial distribution patterns.

concepts, that permits direct calculation of the CPVC of a paint, using particle size distribution data (corrected for resin absorption on the particle surface) (15). However, since a computer (such as an IBM 1130) is necessary to carry out the columinous calculations, and since the total mathematical procedure is quite involved, it appears that this approach for determining the CPVC is probably infeasible for most paint engineers. Furthermore, there is no certainty that a CPVC value obtained in this manner is any more accurate than one determined directly by simple, routine laboratory procedures.

Stirring of Pigment Particles

Stirring implies a disturbance of the relative positions of pigment particles, especially by agitation such as vibratory motion (shaking) or continuous circular movement (milling, grinding, dispersing).

When dry powdered pigments are loosely packed, stirring promotes their rearrangement to a denser packing because of their weight (a compacting gravitational force). In the packaging of industrial powders, use is made of vibrating (shaking) devices to achieve compaction and conserve space. Average values for loose and dense packing factors for the random packing of uniform spheres (lead shot, steel balls, glass spheres, nylon spheres) have been reported in the literature as 0.589 and 0.639, respectively. These values indicate that a 5% reduction in occupied space occurs in progressing from a loose to a dense packing condition (0.05 = 0.639 - 0.589).

Stirring pigment particles in a vehicle also acts to promote denser packing. The correct determination of an oil absorption value, to be presently discussed, depends on the vigorous movement of the pigment particles in linseed oil over a prolonged period to obtain a maximum packing effect.

POROSITY

Porosity (presence of voids or air pockets), as existing in all paint films above the CPVC, can be considered in terms of either an overall porosity or a porosity index (P.I.). Overall porosity considers the film as a whole, including the pigment solid volume. On the other hand, the porosity index considers only the volume represented by the void space within the pigment particles that may be filled partly with air and partly with binder. The portion (fraction or percentage) of the interstitial space occupied by air is defined as the porosity index (P.I.). By difference the fraction or percentage filled with binder is equal to (1.00 - P.I.). The relationships of the overall porosity to the porosity index, and of these two quantities in turn to the quantities PVC, CPVC, and the pigment packing factors ϕ_c, ϕ'_c, and ϕ_d, are shown in Fig. 5-1. Some of these relationships are repeated below as numbered equations for future reference.

$$\text{Overall porosity} \begin{cases} = 1 - \dfrac{\text{CPVC}}{\text{PVC}} & (21) \\ = \text{P.I.}\,(1 - \text{CPVC}) & (22) \end{cases}$$

$$\text{Porosity index (P.I.)} \begin{cases} = \dfrac{\text{PVC} - \text{CPVC}}{\text{PVC}(1 - \text{CPVC})} & (23) \\ = 1 - \dfrac{\text{CPVC}(1 - \text{PVC})}{\text{PVC}(1 - \text{CPVC})} & (24) \\ = \dfrac{\text{overall porosity}}{1 - \text{CPVC}} & (25) \end{cases}$$

Inspection of Fig. 5-1 clearly reveals the change from a solid to a porous condition that occurs at the CPVC point as air is introduced into the space between the pigment particles.

The nomograph given in Fig. 5-13 provides a graphical solution to Eqs. 23 and 24, which relate the PVC and the CPVC of a paint film to the porosity index.

PROBLEM 5-13

A paint film has a pigment volume concentration of 58%. If the CPVC of the paint film is 54%, what is its P.I. value?

Solution. Substitute the given PVC and CPVC values in either Eq. 23 or Eq. 24, and solve for the porosity index.

$$\text{Eq. 23:} \quad \text{P.I.} = \frac{0.58 - 0.54}{0.58(1 - 0.54)} = 0.15\ (15\%)$$

$$\text{Eq. 24:} \quad \text{P.I.} = 1 - \frac{0.54(1 - 0.58)}{0.58(1 - 0.54)} = 0.15\ (15\%)$$

The same result is also obtained by use of the nomograph of Fig. 5-13 (see index line on nomograph).

PROBLEM 5-14

What is the overall porosity of the paint in Problem 5-13?

Solution. Substitute the appropriate data in either Eq. 21 or Eq. 22, and solve for the overall porosity.

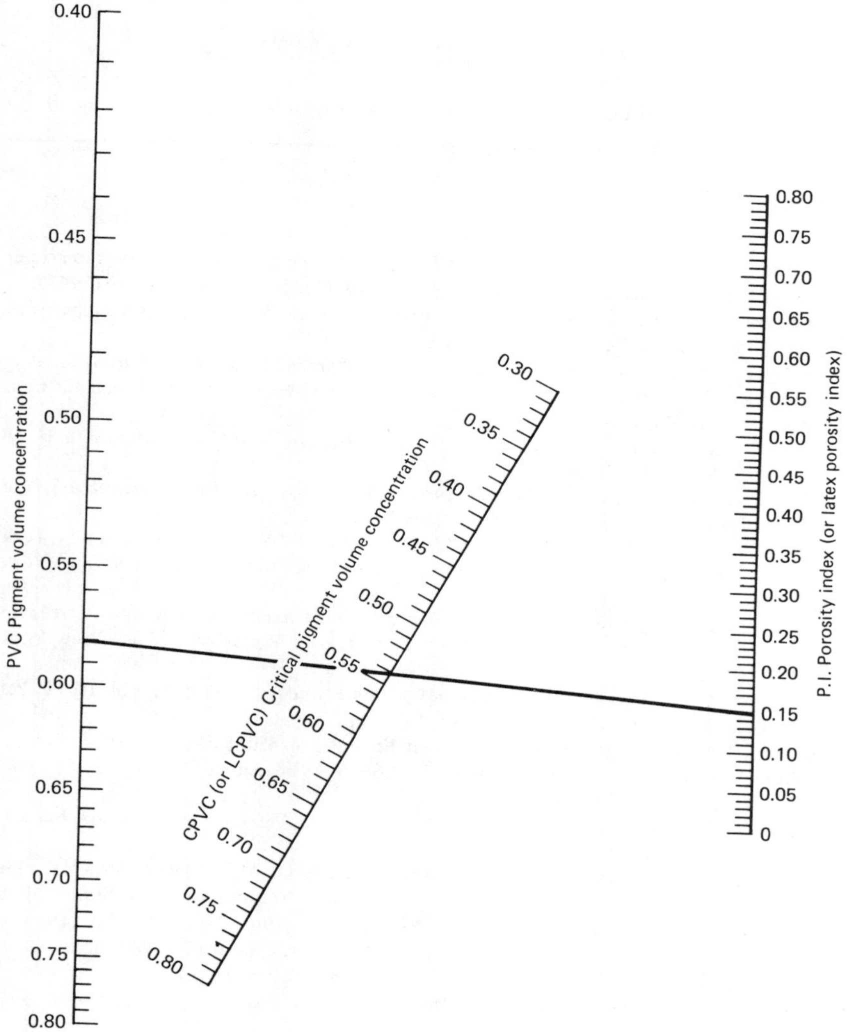

Fig. 5-13 Nomograph relating pigment volume concentration PVC and critical pigment volume concentration CPVC (or latex critical pigment volume concentration LCPVC) to porosity index (or latex porosity index).

Eq. 21: Overall porosity $= 1 - \dfrac{0.54}{0.58} = 0.069\ (6.9\%)$

Eq. 22: Overall porosity $= 0.15(1 - 0.54) = 0.069\ (6.9\%)$.

REFERENCES

1. Armstrong, W. G. and W. H. Madson, "The Effect of Pigment Variation on the Properties of Flat and Semi-Gloss Finishes," *Off. Dig.*, **19**, No. 269, 321–335, June (1947).
2. Asbeck, W. K. and Maurice Van Loo, "Critical Pigment Volume Relationships," *Ind. Eng. Chem.*, **41**, No. 7, 1470–1475, July (1949).
3. Bierwagen, G. P. and T. K. Hay, "The Reduced Pigment Volume Concentration as an Important Parameter in Interpreting and Predicting the Properties of Organic Coatings," *Prog. Org. Coatings*, **3**, 281–302 (1975).
4. Irani, Riyad R. and C. F. Callis, *Particle Size: Measurement, Interpretation, and Application*, Wiley, New York, 1963.
5. Orr, Clyde, Jr., "Converting Particle Surface Area Data," in *Pigment Handbook*, Vol. III (T. Patton, Ed.), Wiley-Interscience, New York, 1973, p. 129.
6. Orr, Clyde, Jr., and R. W. Tyree, "Characterization of Pigments: Permeability Techniques," in *Pigment Handbook*, Vol. III (T. Patton, Ed.), Wiley-Interscience, New York, 1973, pp. 117–126.
7. Ettre, L. S., "Pigment Surfaces: Measurement and Characterization by Gas Adsorption," in *Pigment Handbook*, Vol. III (T. Patton, Ed.), Wiley-Interscience, New York, 1973, pp. 139–155.
8. Stieg, Fred B., "Pigment/Binder Geometry," in *Pigment Handbook*, Vol. III (T. Patton, Ed.), Wiley-Interscience, New York, 1973, pp. 203–217.
9. Patton, Temple C., "Reflections of a Paint Engineer on Paint Flow, Interface Physics, and Pigment Dispersion," *JPT*, **42**, No. 551, 666–694, December (1970).
10. Carr, W. J., *JOCCA*, **49**, 831 (1966).
11. Carr, W. J., "Texture and Dispersibility of Organic Pigments and Carbon Blacks," *JOCCA*, **50**, No. 12, 1115 (1967).
12. Philadelphia Paint and Varnish Production Club (R. G. Alexander, Chairman), "Predicting the Oil Absorption and the Critical Pigment Volume Concentration of Multicomponent Pigment Systems," *Off. Dig.*, **31**, No. 418, 1490–1530, November (1959).
13. Philadelphia Society for Paint Technology (D. Engler, Chairman), "Determination of CPVC by Calculation," *Off. Dig.*, **33**, 1437–1452, November (1961).
14. Lee, Do Ik, "Packing of Spheres and Its Effect on the Viscosity of Suspensions," *JPT*, **42**, No. 550, 579–587, November (1970).
15. Bierwagen, G. P., "CPVC Calculations," *JPT*, **44**, No. 574, 47–55, November (1972).

6 *Oil Absorption Values*

The oil absorption value refers to the weight of linseed oil taken up by a given weight of dry pigment to form a paste. Two different ASTM procedures covering two different incorporation methods are presently specified. In the first (ASTM D281-31; reapproved 1974), a rub-out with a spatula, made on a glass plate or marble slab, thoroughly incorporates just enough linseed oil (drop-by-drop addition) with a weighed amount of pigment to produce a stiff, puttylike paste which does not break or separate. In the second method (ASTM D1483-60; reapproved 1974), which is referred to as the Gardner-Coleman method, a mild stirring and folding action (without rubbing or grinding) within a glass beaker adds just enough linseed oil to a weighed amount of pigment to produce a lump of soft paste.

It is obvious from the different methods of incorporation and the different stipulated end points (stiff versus soft paste) that these two test methods must yield quite dissimilar results. As expected, the spatula rub-out procedure with its more vigorous shearing action and earlier end point gives a much lower oil absorption value. Beyond this observation no simple expression has been found that relates the two oil absorption values.

Of the two the spatula rub-out test is the more important. The Gardner-Coleman test is used for estimating mill base consistencies for mixing operations or for characterizing the nature of a pigment. However, it is now little used in research studies and, except possibly for extender pigments, is infrequently reported in the literature. Hereafter in this book an oil absorption value, unless otherwise noted refers only to a value obtained by the spatula rub-out method.

SPATULA RUB-OUT OIL ABSORPTION TEST

An oil absorption value $\overline{OA}$ as obtained by the spatula rub-out method is defined as the weight of linseed oil per hundred weight of pigment (as lb/100 lb or g/100 g) that is required to reach the rub-out end point. Although reported quantitatively, the $\overline{OA}$ value has qualitative overtones in that the rub-out procedure is conducted manually and the end point is judged by visual and tactile observation. These human factors tend to introduce a certain amount of uncertainty into a reported $\overline{OA}$ value. Despite continuous efforts to reduce the human factor in the conduct of this test, operator bias can still influence the determination of an oil absorption value. Any one operator can usually reproduce his or her own oil absorption values with fair to good precision. Deviations of ±5% or so can be expected when following the loosely defined ASTM test. However, by following the more rigorously defined British Standards method for obtaining a rub-out oil absorption value or the NL Industries* improvement on the ASTM test, deviations for a single operator are normally reduced to about ±2 to 3%.

Unfortunately, duplication of $\overline{OA}$ values among operators from different companies has generally been only fair (on the order of ±15%). To rectify this situation, two approaches have been taken. In the first place, operator retraining, using standard samples, has been undertaken. The idea is to promote a stricter adherence to one of the more rigorously defined test methods and also to make sure that the critical end-point condition is properly recognized. In the second place, an operator bias factor may be established based on the individual's past experience in running the test. This bias factor is then applied as a correction to bring his or her reported oil absorption values in line with the general average for all operators. Both approaches may be necessary to arrive at an industrywide concensus on oil absorption values.

Some of the more important physical factors affecting the determination of an oil absorption value are energy input, pigment conditioning, acid number of linseed oil, type of equipment, and end point. These will be discussed in turn.

Energy Input

As far back as 1946, the problem of reproducing oil absorption values was critically examined by the New England Club (1). Its members concluded that the longer the rub-out time per unit weight of pigment (up to 30 min), the lower is the oil absorption value, the less the discrepancy among operators, and the better the agreement between duplicate runs. Some 5 years later, Bessey and Lammiman corroborated this conclusion on the basis of their study of oil ab-

*Method No. TP-P-OA-2, NL Industries, Inc. (Titanium Pigment Division).

sorption, including the use of mechanical compaction (a ramming device) (2). These two investigators even went so far as to state that, provided that sufficient mixing time is scheduled to ensure complete wetting and dispersion, the oil absorption value obtained is independent of both operator influence and the nature of the rub-out medium. From such work it is now conceded that a valid oil absorption test must involve a vigorous rub-out over a prolonged period. The idea behind this requirement is that the $\overline{OA}$ end point should represent the densest packing that can be obtained and not some intermediate stage on the way to this maximum dispersed condition.

Pigment Conditioning

The humidity conditions prevailing during the storage of a pigment before an oil absorption test controls to a major extent the moisture content of the pigment at the time of test. In turn the pigment moisture content may affect the oil absorption value. Experimental studies show that the $\overline{OA}$ value tends to increase rapidly with the first traces of pigment moisture content and then tapers off to some maximum level when the pigment moisture content reaches a value of about 2% (3). Hydrophilic pigments that are water sensitive (such as certain titanium dioxides) are most subject to an increased oil absorption due to the presence of moisture. On the other hand, hydrophobic pigments (such as talc) may be little or not at all affected by moisture conditions.

Acid Number of Linseed Oil

The acidity of linseed oil exerts a strong dispersing action on most pigments. Hence high-acid linseed oils facilitate pigment dispersion and reduce the time required to attain an oil absorption end point. High acidity may even promote a somewhat denser packing arrangement.

The original selection of linseed oil as the rub-out medium was entirely natural, since the ASTM test was first given tentative official recognition in 1928 at a time when linseed oil was the dominant paint binder. Although the rubber industry presently favors dibutyl phthalate as the dispersing liquid for rating carbon blacks, the paint industry has stayed with linseed oil as the best choice for evaluating coating pigments.

Equipment Type

The spatula used in the rub-out operation has been closely specified in a British Standards method for measuring oil absorption. This has also been done by NL

Industries in an attempt to improve the reproducibility of the ASTM D281 test method. Both represent efforts to standardize the equipment used in the $\overline{OA}$ test procedure. The ASTM test is indefinite on this point in that it merely calls for a sharp-edged steel spatula (no dimensions).

One proposed method for completely mechanizing the oil absorption test makes use of a miniature Banbury-type mixer with a high-shear cam-type blade (4). The torque produced during the mixing of pigment and linseed oil is registered on an indicating scale as linseed oil is slowly added at a standard rate over a 20-min period to a weighed amount of pigment (total volume of pigment and oil about 35 to 40 cm^3). The end point corresponding to the $\overline{OA}$ value is indicated by a sharp peak (reflecting a maximum consistency) that is registered on a torque scale. Oil absorption values obtained by this mechanical method are on the order of 6% higher than ones obtained by manual incorporation.

Oil Absorption End Point

The oil absorption end point corresponds to a condition of maximum consistency (puttylike, homogeneous mass). To prevent overstepping this end point, it is necessary to sense when it is being closely approached so that smaller increments of oil can be more slowly added and more strongly worked into the mixture. When this precaution is observed, reproducibility is improved both for a single operator and among different operators.

An unpublished study by the present author shows that pigments tend to exhibit one of two types of behavior, either dilatancy or pseudoplasticity, during the oil absorption test. The dilatant pigment is relatively easy to rub out because it offers resistance to the rub-out action. Furthermore, the end point of the dilatant pigment is easy to recognize, since the working in of an additional drop of oil at the end point can dramatically change the dull-appearing, coherent, and relatively rigid putty into a shiny, slumping paste (an overstep of the oil absorption end point). On the other hand, the pseudoplastic type of pigment is much more difficult to incorporate because the pigment/linseed oil mix tends to slip out from under the spatula. Also, the oil absorption end point is difficult to judge, since no striking change in appearance or behavior occurs when it is reached. Instead there is a gradual approach to maximum consistency (which must be recognized as the end point) and an equally gradual departure from this stiff paste condition as more oil is added.

It is suggested that, when a new pigment is to be tested, a preliminary rub-out be made to approximately locate the region of maximum consistency. Once this is established, a follow-up run can be made, with a slower addition of oil and a more vigorous rub-out through this critical region, thus permitting a more precise assessment of the oil absorption end point. Operator training is especially important for pseudoplastic pigments.

SIGNIFICANCE AND INTERPRETATION OF AN OIL ABSORPTION VALUE

The significance of an oil absorption value, considered from a theoretical standpoint, involves such concepts as adsorption, wettability, capillarity, and the seven S factors affecting pigment packing discussed in Chapter 5. However, from a practical standpoint, an $\overline{\text{OA}}$ value is simply a measure of the linseed oil required to substantially fill the voids among the particles of pigment that are compacted under the peculiar set of dispersion conditions specified by the test procedure. Here the term "peculiar" implies distinctiveness or uniqueness. In the literature the expression "degree of packing of the pigment particles commensurate with the degree of dispersion" is frequently used in discussing $\overline{\text{OA}}$ values. The important point is that it is possible to disperse and compact pigment particles in a variety of ways. With this in mind, it can be realized that any given oil absorption test is actually an attempt to specify some one standard and reproducible method for dispersing and compacting pigment particles into a coherent, puttylike mass. It follows that the more detailed the test procedure, the better is the chance that reproducibility will be achieved.

Since an oil absorption test is arbitrarily determined, it might be reasoned that it has little practical value. However, experience has shown that the dispersion and packing value obtained at a linseed oil absorption end point can be closely related to the CPVC value for the same pigment in paint and ink films. In the case of the improved NL Industries procedure, it has been demonstrated that, in general, the two packing conditions are substantially equivalent based on solvent-type coatings (5).

CRITICAL PIGMENT VOLUME CONCENTRATION (CPVC) RELATED TO OIL ABSORPTION VALUE ($\overline{\text{OA}}$)

An oil absorption value is conventionally expressed on a weight-for-weight basis as g/100 g. However, from both a practical and a theoretical standpoint, it is more meaningful to convert an $\overline{\text{OA}}$ value to a fractional volume content (equivalent to the pigment CPVC in linseed oil). This provides a volume relationship that is of prime importance in formulating paint compositions and in interpreting paint behavior. The conversion of an $\overline{\text{OA}}$ to a CPVC value is expressed by Eq. 1, where ρ is the pigment density. A graphical solution to this equation is given by the nomograph of Fig. 6-1.

$$\text{CPVC} = \frac{100/\rho}{(\overline{\text{OA}}/0.935) + (100/\rho)}$$

$$\text{CPVC} = \frac{1}{1 + \overline{\text{OA}}\rho/93.5} \qquad (1)$$

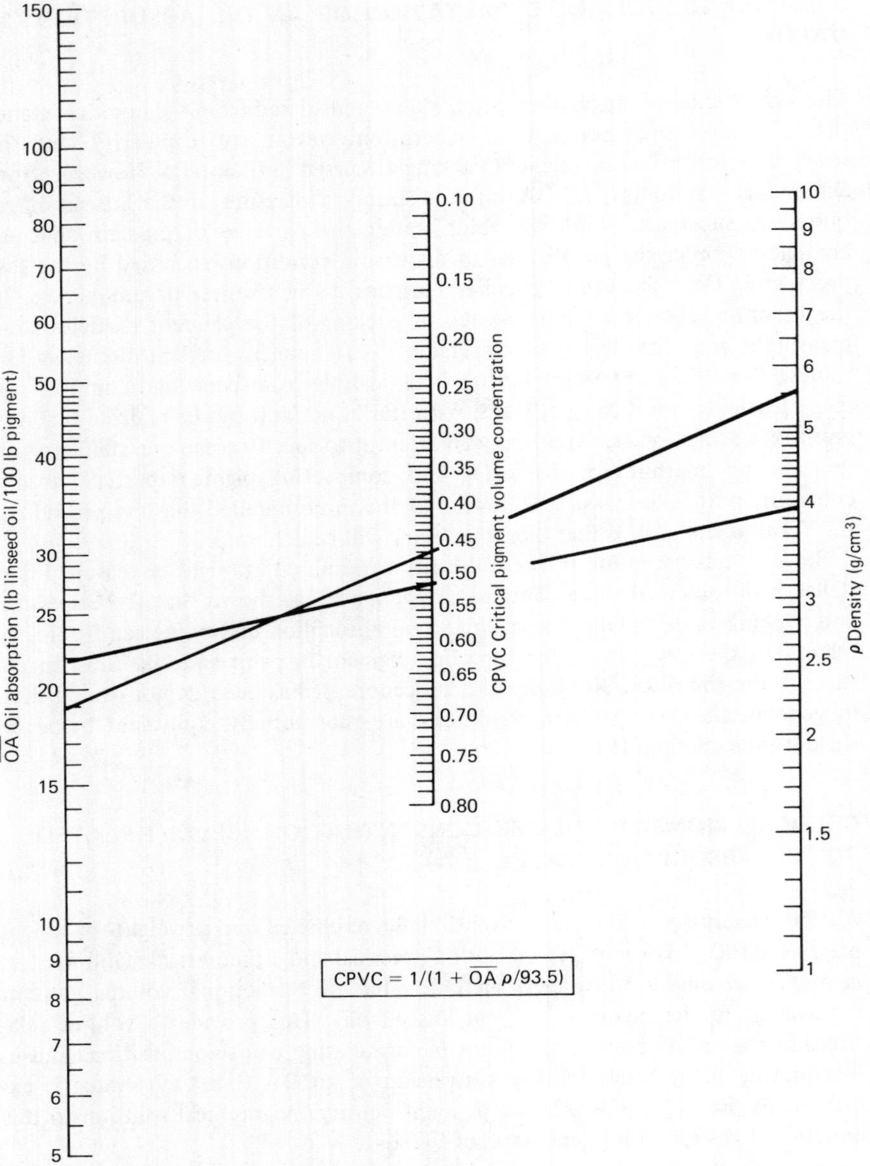

Fig. 6-1 Nomograph for determining the critical pigment volume concentration from the pigmentation density and the oil absorption value.

PROBLEM 6-1

Acicular zinc oxide ($\rho = 5.6$ g/cm^3) has a rub-out $\overline{\text{OA}}$ value of 19. Calculate its CPVC value.

Solution. Substitute the given data in Eq. 1, or read off the answer from the nomograph in Fig. 6-1.

$$\text{CPVC} = \frac{1}{1 + (19 \cdot 5.6)/93.5} = 0.47\ (47\%)$$

An oil absorption value actually involves both adsorption of oil on the exposed surfaces of the pigment particles and absorption of oil into the voids of the oil-wet particles. The layer of adsorbed oil acts as a barrier that effectively separates the pigment particles. Hence the oil-wet particles are less densely packed than they would be if no oil were present.

An expression relating the average particle diameter d, the thickness of the adsorbed layer a, and the two packing factors for the pigment (ϕ for the bare particles and ϕ_c' or CPVC for the oil-wet particles) is given by Eq. 2 (see Chapter 5 for derivation).

$$\phi \cdot d^3 = \text{CPVC} \cdot (d + 2a)^3 \tag{2}$$

The thickness of linseed oil adsorbed on carbon black particles has been calculated as about 0.0025 μ (roughly equivalent to two layers of coiled linoleic triglyceride molecules) (6). By using this as a reasonable value for pigments in general, Eq. 2 can be simplified to Eq. 3.

$$\phi = \text{CPVC}\left(1 + \frac{0.0050}{d}\right)^3 \tag{3}$$

PROBLEM 6-2

Calculate an approximate packing factor for bare (uncoated) titanium dioxide pigment ($\rho = 3.95$ g/cm^3; $d = 0.30\ \mu$) having a rub-out $\overline{\text{OA}}$ value of 22. Assume similar packing patterns for the bare and the oil-coated particles.

Solution. First calculate a CPVC value for the TiO_2 pigment from the given density and oil absorption data, using Eq. 1 or the nomograph of Fig. 6-1.

$$\text{CPVC} = \frac{1}{1 + (22 \cdot 3.95)/93.5} = 0.52$$

Substitute this CPVC value and the given particle diameter in Eq. 3 to obtain the pigment packing factor for the bare TiO_2 pigment particles.

$$\phi = 0.52 \left(1 + \frac{0.0050}{0.30}\right)^3 = 0.55\ (55\%)$$

An alternative liquid to linseed oil for the oil absorption test would have to be evaluated in terms of (*a*) its ability to disperse pigments and (*b*) its ability to tightly adsorb onto the surface of the pigment particles. Since liquids vary significantly in these two respects, it is not surprising that $\overline{\text{OA}}$ values obtained with other liquids fail to match linseed oil absorption values. For example, dibutyl phthalate (a liquid plasticizer for polyvinyl chloride and other resins) gives $\overline{\text{OA}}$ values that are anywhere from 36% (4 pigments) to 70% (21 pigments) higher than the values given by linseed oil (4, 7). A similar discrepancy has been demonstrated for tricresyl phosphate (another common plasticizer). Water has also been utilized as a rub-out liquid, and Ensminger (8) gives an expression relating water demand (a water absorption value) to an oil absorption value (Eq. 4).

$$\overline{\text{OA}}\ (\text{water}) = 16.7 + 0.67\ \overline{\text{OA}}\ (\text{oil}) \qquad (4)$$

According to this equation, the two oil absorption values are equal at an $\overline{\text{OA}}$ value of 50.6. Hence linseed values above 50.6 are higher than those for water; below 50.6 they are lower than the water values. At an $\overline{\text{OA}}$ value of 41.7, the two liquids give the same CPVC value. Scattered data from another source indicate that oil absorption values run about 10% less for water than for linseed oil (9). These data serve to emphasize the wide variations in $\overline{\text{OA}}$ values that are obtained in practice with alternative rub-out liquids.

As of now, a properly specified linseed oil still represents the ideal rub-out liquid for the coatings industry. It provides a reproducible and well-recognized $\overline{\text{OA}}$ value that serves both as a control parameter for monitoring pigment manufacture and, when converted to a CPVC value, a basic parameter for formulating and evaluating coating compositions.

REFERENCES

1. New England Club, "A Critical Examination of the A.S.T.M. Standard Method of Test for Oil Absorption of Pigments," *Off. Dig.*, **18,** 644–653 (1946).
2. Bessey, G. E. and K. A. Lammiman, "The Measurement and Interpretation of Oil Absorption," *JOCCA*, **34,** 519–546, November (1951).
3. Asbeck, W. K., D. D. Laiderman, and M. Van Loo, "Oil Absorption and Critical Pigment Volume Concentration (CPVC)," *Off. Dig.*, **24,** No. 326, 156–171, March (1952).
4. Hay, T. K., "Reaching an Objective in Oil Absorption Measurements," *JPT*, **46,** No. 591, 44–50, April (1974).
5. Stieg, F. B. and D. F. Burns, "The Effect of Pigmentation on Modern Flat Wall Paints," *Off. Dig.*, **26,** No. 349, 81–93, February (1954).
6. Patton, T. C., "Reflections of a Paint Engineer on Paint Flow, Interface Physics, and Pigment Dispersion," *JPT*, **42,** No. 551, 666–694, December (1970).

7. Sward, G. G., "Oil Absorption of Pigments," Chapter 3.5 in *Paint Testing Manual* (G. G. Sward, Ed.), American Society for Testing and Materials, Philadelphia, Pa., 1972.
8. Ensminger, R. I., "Efficient Operation for Pigment Dispersions," *Mod. Paint Coatings*, **65,** No. 5, 35, May (1975).
9. Patton, T. C., *Paint Flow and Pigment Dispersion* 1st ed., Wiley-Interscience, New York, 1964.

BIBLIOGRAPHY

Bainbridge, J. R., "Oil Absorption Tests for Pigments," *JOCCA*, **35,** 459–471, September (1952).

Marsden, E., "Oil-Absorption: A New Assessment: Part II," *JOCCA*, **42,** No. 2, 119–135, February (1959).

Mill, C. C. and W. H. Banks, "An Interpretation of the Oil Absorption of Pigments," *JOCCA*, **32,** 599–609, December (1949).

Newton, D. S., "Pigment Volume Concentration," *JOCCA*, **45,** No. 3, 180–199, March (1962).

7 Critical Pigment Volume Concentration (CPVC)

The physical aspects of a critical volume concentration were discussed and defined in Chapter 5 on pigment/binder geometry, where it was emphasized that at the CPVC a dramatic change in the geometry of the mixture occurs. Above the CPVC, voids (air spaces) are present; below the CPVC, the pigment particles are separated. These striking changes in the geometric configuration at the CPVC point are reflected by the equally striking changes that are observed in the properties and behavior of paint films. It is with these behavioral changes that this chapter is concerned.

Theoretically, a geometric CPVC represents an incisive intermediate value on the PVC scale. Practically, a CPVC is a narrow and more or less blurred transition region where properties to either side intrude on each other to exhibit a mixed condition. In turn this confused CPVC may result in less than abrupt changes and/or changes that anticipate or lag behind the geometric CPVC point. Paint and ink engineers must recognize, understand, and if necessary compensate for this departure from theoretical perfection in making use of the CPVC concept in formulating and evaluating coating compositions.

The dramatic changes in paint behavior that occur in passing through the CPVC can be used to determine the CPVC value. Conversely, by knowing the CPVC value, behavior at and to either side of the CPVC point can be anticipated and quantitatively predicted. Some of the physical, permeability, and optical properties that change more or less abruptly at the CPVC point are listed below under the appropriate headings.

Physical Properties

1. Density (1, 2)
2. Tensile strength (3-6)
3. Adhesion (6, 7)

Permeability Properties

4. Porosity (1, 2, 8-10, 15-18, 21)
5. Rusting (11, 12)
6. Blistering (11, 12)
7. Wet abrasion resistance (3, 4, 8, 13, 14)
8. Staining (3, 13, 14)
9. Enamel holdout (3, 4, 8, 9, 15)

Optical Properties

10. Light scattering (14)
11. Contrast ratio (4, 17, 20, 21)
12. Tinting strength (5)
13. Hiding power efficiency (10, 12)
14. Gloss (8, 11, 13)
15. Sheen (5, 19)

The nature of the changes in each of these 15 major properties at the CPVC point is shown in Fig. 7-1, where qualitative or quantitative changes have been graphed to illustrate the marked modification in property behavior that occurs through the CPVC transition stage. Other properties that display similar drastic changes in value are electrical conductivity, dielectric constant, block resistance, cold crack resistance, durability, and the like (12).

EXPERIMENTAL DETERMINATION OF THE CRITICAL PIGMENT VOLUME CONCENTRATION (CPVC)

CPVC from Density Measurements

The density of a dry paint film is suitable for determining a CPVC value, since the film density first undergoes an increase as the pigment volume concentration changes from an all-binder composition (PVC = 0) to the critical pigment volume concentration (PVC = CPVC), as shown in Fig. 7-1*a*. Beyond the CPVC the density decreases until the PVC reaches an all-pigment composition (PVC = 1.00).

One method for determining the point of reversal in density at the CPVC calls for the construction of two intersecting lines representing the reciprocal values

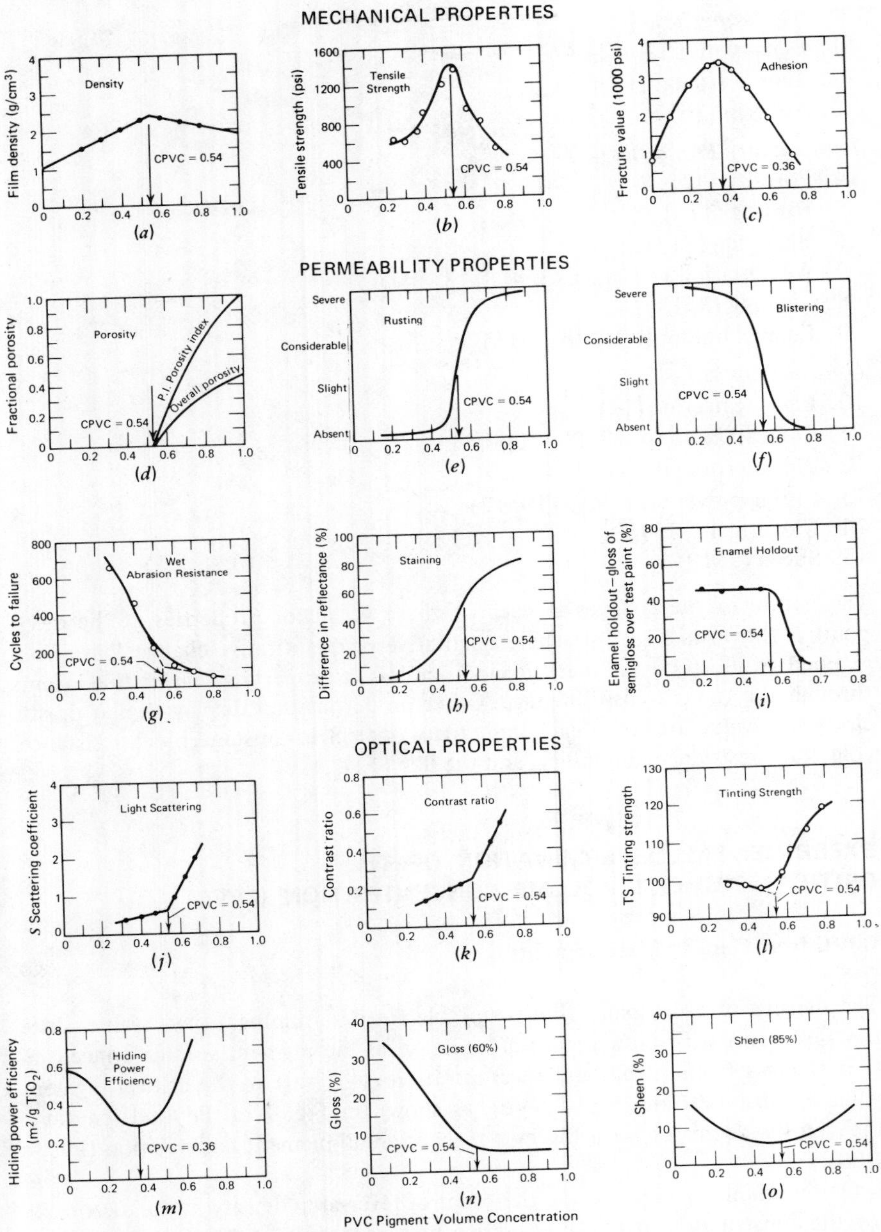

Fig. 7-1 Graphs showing the more or less abrupt changes in 15 major paint properties that occur at the critical pigment volume concentration.

of the paint film density through the regions above and below the CPVC as plotted against the fractional weight of pigment in the paint film (1, 2).

Let the subscripts p, b, and f identify quantities pertaining to the pigment, binder, and film, respectively, and consider a unit weight of film where w is the fractional weight of pigment and $(1 - w)$, by difference, is the fractional weight of binder. For the condition PVC < CPVC, voids are absent and the film density is given by Eq. 1.

$$\rho_f = \frac{1}{w/\rho_p + (1 - w)/\rho_b} \tag{1}$$

The reciprocal of the film density is given by Eq. 2.

$$\frac{1}{\rho_f} = \frac{1}{\rho_b} - w(1/\rho_b - 1/\rho_p) \tag{2}$$

For the condition PVC > CPVC, voids are present and the density of the paint film is given by Eq. 3.

$$\rho_f = \frac{1}{w/(\rho_p \cdot \text{CPVC})} \tag{3}$$

The reciprocal of the film density through this region is then expressed by Eq. 4.

$$\frac{1}{\rho_f} = \frac{w}{\rho_p \cdot \text{CPVC}} \tag{4}$$

Inspection of Eqs. 2 and 4 shows that for either one a plot of reciprocal film density $(1/\rho_f)$ against fractional pigment weight w gives a straight line. A theoretical graph is shown in Fig. 7-2, where Eq. 2 gives a straight line with a negative slope running from a value of $1/\rho_b$ at $w = 0$ to a value of $1/\rho_p$ at $w = 1.00$. Conversely, Eq. 4 gives a straight line with a positive slope running from zero at $w = 0$ to a value of $1/(\rho_p \cdot \text{CPVC})$ at $w = 1.00$. Presumably a CPVC value could be calculated directly from this last all-pigment point, since for this terminal condition Eq. 4 simplifies to Eq. 5.

$$\text{CPVC}(\text{at } w = 1.00) = \frac{\rho_f}{\rho_p} \tag{5}$$

However, a CPVC is more correctly-obtained from values of w and ρ_f that are read off at the intersection point Ⓢ for the two plotted lines. From these values the CPVC is computed using Eq. 6.

$$\text{CPVC} = \frac{w\rho_f}{\rho_p} \tag{6}$$

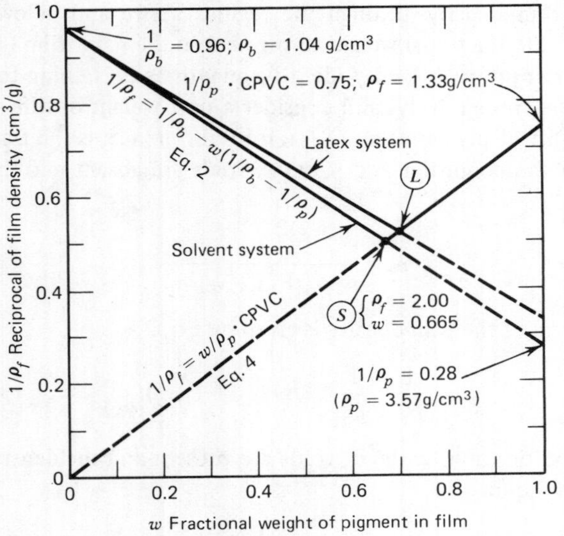

Fig. 7-2 Graph illustrating the manner of plotting film density data to determine a critical pigment volume concentration.

A CPVC value can be calculated solely from the value of w at the intersection point by using Eq. 7.

$$\text{CPVC} = \frac{w/\rho_p}{w/\rho_p + (1 - w)/\rho_b} \tag{7}$$

Good agreement with CPVCs obtained by alternative procedures is claimed for the density method when based on solvent coating compositions. Latex systems are far less satisfactory, since latex PVCs corresponding to solvent-derived CPVCs are less than perfect solids. This is due to the fact that the spherical latex particles of latex paints fail to completely wet the pigment particles and also fail to adequately deform into the interstices and clumps of pigment particles. As a result, holes (minute voids) are created in latex coatings before a solvent-derived CPVC is reached, leading to reduced film densities. This is reflected graphically in Fig. 7-2 by the gradual departure of the plotted line for the latex paint from the solvent paint line. It is true that a CPVC can be calculated from a value of w as read off the plot at the intersection point Ⓛ of the latex line, but this is not a valid CPVC. Although pinpointing a reversal in direction, this reversal is due to the change that occurs in progressing from a paint with extremely minute voids (due to incomplete latex particle compaction) to a paint with relatively larger voids (due to the transition to the fully porous stage).

The actual CPVC for a latex paint is significantly less and corresponds to an earlier stage where more latex particles must be present to yield an essentially solid film composition as pigment is added to binder.

PROBLEM 7-1

Calculate the CPVC for the solvent-based paint film as shown in Fig. 7-2.

Solution. Substitute appropriate data as read from Fig. 7-2 in any of Eqs. 5, 6, and 7.

$$\text{Eq. 5:} \quad \text{CPVC} = \frac{1.33}{3.57} = 0.37\ (37\%)$$

$$\text{Eq. 6:} \quad \text{CPVC} = \frac{2.00 \cdot 0.665}{3.57} = 0.37\ (37\%)$$

$$\text{Eq. 7:} \quad \text{CPVC} = \frac{0.665/3.57}{(0.665/3.57) + (0.335/1.04)} = 0.37\ (37\%)$$

Comment. Note that a single experimental point on the straight line given by Eq. 4 is sufficient to calculate the CPVC by Eq. 4. However, this observation becomes rather academic in practice, since normally a PVC ladder of compositions must be scheduled to bracket the CPVC region and provide several points to determine the intersection breakpoint. Furthermore, the CPVC calculated from the intersection point is a true CPVC (equal to ϕ_c' as discussed in a preceding chapter) and is preferred to the questionable CPVCs computed from values further up on this plotted line (which approach ϕ_d as an end value. Note also that the CPVC value calculated for the latex paint intersection using Eq. 7 is larger than the solvent-based CPVC value (0.39 versus 0.37). This result is unreasonable and indicates the dubious validity of a CPVC value calculated from a latex point intersection using Eq. 7.

A second graphical method for establishing a CPVC from density data utilizes a plot of a function of film thickness against PVC (1). Let x_m equal the paint film thickness as measured experimentally, and x_c be a theoretical solid thickness (no voids) as calculated from the densities of the paint components. For the region PVC $<$ CPVC, $x_m = x_c$ since the paint film is solid composition. On the other hand, for the region PVC $>$ CPVC, voids are present and $x_m > x_c$. Hence at the CPVC there is a departure of x_m from x_c that can be expressed as a dimensionless ratio independent of the thickness of the paint applied.

$$\text{Dimensionless ratio} = \frac{x_m - x_c}{x_c} \tag{8}$$

When this ratio parameter is plotted against PVC, the CPVC can be read off at the point where the plotted curve takes off from the base PVC scale (a point corresponding to a zero value for the dimensionless ratio).

A third graphical procedure calls for a plot of the pigment packing factor ϕ versus PVC (see Fig. 5-1). Here the positively sloped line for the region PVC < CPVC intersects the essentially horizontal line for the region PVC > CPVC at a point corresponding to the CPVC (1).

Experimental Procedures. The paint film volume may be calculated directly from area and thickness measurements or determined indirectly by a mercury displacement (immersion) procedure.

Volume measurement by area and thickness can be carried out by drawing down paint films on Teflon-coated metal sheet, drying, stripping, cutting to size (e.g., 1 by 3 in.), and measuring the thickness between microscopic slides, using a micrometer reading to 0.0001 in.

Volume measurement by the mercury immersion method (Archimedes' principle) is carried out by weighing the paint film in air, followed by weighing it in mercury using a metal denser than mercury (such as tungsten) for a sinker.

CPVC from Dry Pigment Compaction

The fact that a dry pigment powder is at one end of the PVC scale (PVC = 1.00) suggests that the pigment packing factor for the dry compacted power could serve as a base for calculating a valid CPVC(CPVC = ϕ_c'). The obvious problem here lies in determining just how tightly the powder should be compressed to simulate the packing of the pigment in a pigment/binder mix at the CPVC point. In work undertaken along this line with a number of pigments, the particles have been confined within a rugged cylindrical metal cell (22). By using vibration in conjunction with a ramming device that delivers multiple impacts and a final compression in a vice, pigment packing factors have been determined and reported.

PROBLEM 7-2

The powder densities for a number of pigments compacted first by multiple mechanical impacts and then by compression in a vice are listed below, together with their solid densities and oil absorption values. Compare ϕ and CPVC values as computed from these two data sources.

	Density (g/cm^3) Compacted Powder	Solid	Oil Absorption Value (g/100 g)
Calcium carbonate (whiting)	1.91	2.71	16.4
Zinc oxide (ZnO)	3.20	5.60	16.9
Barium carbonate			
Precipitated	2.62	4.10	23.7
Ground (barytes)	2.64	4.20	19.6
Lead carbonate, basic	4.52	7.1	11.9
Diatomaceous earth (SiO_2)	0.94	2.2	196.0

Solution. Compute packing factors for the dry compacted pigment from the expression $\phi = \rho_{pow}/\rho_{sol}$, where ρ_{pow} is the density of the compacted powder and ρ_{sol} is the density of the solid pigment. Calculate CPVC values using Eq. 1 of Chapter 6.

	Packing Factor for Compacted Powder ϕ	CPVC
Calcium carbonate (whiting)	0.70	0.68
Zinc oxide (ZnO)	0.57	0.50
Barium carbonate		
Precipitated	0.63	0.49
Ground (barytes)	0.63	0.53
Lead carbonate, basic	0.64	0.53
Diatomaceous earth (SiO_2)	0.43	0.18

In all cases the compacted powder gives the higher value, indicating denser packing.

Comment. Much of the difference between these values is due to the fact that particle-to-particle contact is being made in the case of the dry powders, whereas in the oil absorption test (CPVC values) the particles are separated by adsorbed layers of oil. In this respect, note the lower CPVC value for the fine precipitated barium carbonate than for the coarser grade ground barium carbonate. The much higher value for the mechanically compacted diatomaceous earth can be attributed in part to a fracturing of the fragile diatom structure under the impact of the mechanical ram.

Inspection of these data reemphasizes the unsolved problem of how to exactly adjust the powder compaction to give ϕ values that correlate with CPVCs derived from oil absorption data.

CPVC from Optical Properties

The dramatic changes in optical behavior that occur at the critical pigment volume concentration can be used to accurately establish a CPVC on the PVC scale. These changes are due primarily to the introduction of porosity into the paint film above the CPVC. This onset of porosity acts to significantly alter such optical properties as light scattering, contrast ratio, tinting strength, hiding power efficiency, gloss, sheen, and color uniformity.

Hiding Power Related to Porosity Index. The pigment/binder interfaces which are responsible for light scattering below the CPVC are partially replaced by pigment/air interfaces above the CPVC. Since air has a lower refractive index than any binder, this has the effect of reducing the average refractive index of the overall binder in which the pigment is dispersed.

The CPVC is probably most accurately defined as the PVC at which air interfaces are first introduced into the dry paint film (because of a deficiency of binder with respect to pigment). Hence the best method for determining the CPVC should be one that directly detects the first introduction of air into the film with no dependence on secondary factors such as film integrity, tensile strength, wet abrasion resistance, and enamel holdout.

The ability to scatter light depends, for white pigments, in major part on the difference between the refractive index n_p of the pigment and that of the binder n_b in which it is dispersed. An approximate expression relating hiding power to the refractive indices of pigment and binder involves a form of the Lorentz-Lorentz relationship given by Eq. 9 (23).

$$M = \frac{(n_p/n_b)^2 - 1}{(n_p/n_b)^2 + 2} \tag{9}$$

In turn, hiding power (HP) has been shown to be empirically proportional to M^2. Hence HP is approximately related to n_p and n_b by Eq. 10.

$$\mathrm{HP} = k \cdot \left[\frac{(n_p/n_b)^2 - 1}{(n_p/n_b)^2 + 2}\right]^2 \tag{10}$$

A reasonable refractive index for most solid organic binders can be taken as $n_b = 1.50$. When air with a refractive index of 1.00 is introduced into a paint film, the now porous binder exhibits an index of refraction that reflects the proportional contributions of the binder and the air to the total index of refraction as given by Eq. 11, where P.I. is the porosity index (10).

$$n_b = 1.00 \cdot \text{P.I.} + 1.50(1 - \text{P.I.})$$
$$n_b = 1.00 + 0.50(1 - \text{P.I.}) \tag{11}$$

Substitution of this value for n_b in Eq. 10 provides an expression giving hiding power in terms of porosity index and pigment refractive index. Since such an equation is cumbersome, it is best put into the form of a graph to give a clear visual portrayal of how these properties are related (see Fig. 7-3). This figure shows that the onset of porosity in a paint film produces only a nominal increase in hiding power for high-refractive-index pigments (e.g., TiO_2) but a dramatic increase for low-refractive-index pigments (e.g., calcium carbonate, and talc). For example, in the case of rutile titanium dioxide ($n_p = 2.76$) a change in porosity index from 0 to 0.10 increases the relative hiding power by about 15%, whereas for calcium carbonate or talc ($n_p = 1.60$) the increase is

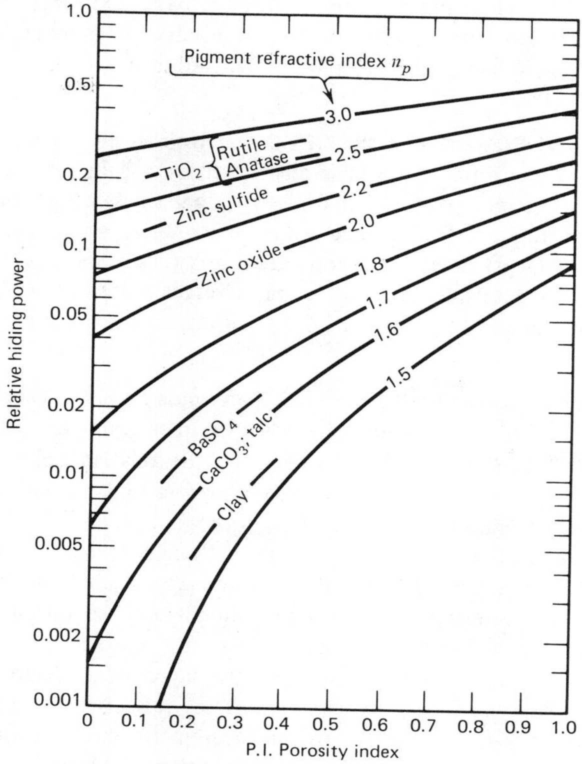

Fig. 7-3 Graph showing the relative increase in hiding power with increase in porosity for pigments with different refractive indices.

about 150%. Such widespread differences in HP increase with increased porosity explain why the determination of the CPVC by optical methods is excellent for extender pigments (low refractive indices) but much less so or even questionable for hiding pigments (high refractive indices) unless they are diluted with extenders (14, 20).

Several different optical properties have been used to determine the sudden change in visual appearance that signals the transition from an all-solid to a porous condition at the CPVC point.

Change in Optical Properties at the CPVC Determined by the Scattering Coefficient. The most fundamental procedure for establishing optical data over a range of PVCs is the direct measurement of the light reflectance and transmittance for a series of films, given by a PVC ladder. From such data, scattering coefficients (S values) can be determined by using Kubelka-Munk equations. A plot of scattering coefficients versus PVC yields two straight lines that intersect at the CPVC value. This spectroscopic method is basic and sensitive; the CPVC point can be located with certainty; and the results are independent of such factors as film integrity. A typical plot of scattering coefficient versus PVC is given in Fig. 7-1*j* (14).

Change in Optical Properties at the CPVC Determined by the Contrast Ratio. Measurement of contrast ratios generally specifies a substrate of black and white squares or stripes over which films of equal thickness prepared from a PVC ladder are applied. Reflectance measurements are made over the black (R_B) and white (R_W) areas, and contrast ratios (CRs) are calculated, where $CR = R_B/R_W$. A typical plot of contrast ratio versus PVC is shown in Fig. 7-1*k*. The intersection of the two plotted lines gives the required CPVC value (4, 17, 20, 21).

Change in Optical Properties at the CPVC Determined by the Tinting Strength. The ability of a colorant to change the color of an opaque paint is measured in terms of its tinting strength (TS). Conversely, the relative opacities of white pigmented coatings can be determined by comparing films tinted with the same fixed amount of colorant (23).

Determination of the CPVC by tinting strength calls for the preparation of a PVC ladder of opaque paint films that are tinted with a small but fixed amount of colorant. The scattering coefficient (S value) for these tinted paints arises primarily from the white hiding pigment, and the absorption coefficient (K value) from the colorant. By applying successive layers of paint, the paint thickness can be built up to a point where the coating reflectivity can be measured directly. Reflectivity R_∞ is a reflectance measurement on a film of such thickness that additional thickness produces no reflectance change. Reflectivity appears as the key parameter in the best known equation of the Kubelka-Munk theory (Eq. 12).

$$\frac{K}{S} = \frac{(1 - R_\infty)^2}{2R_\infty} \tag{12}$$

Since K for the PVC ladder is constant (fixed colorant additon), S values for films of varying PVC can be compared for reflectivity on the basis of Eq. 12. If the S value for some one film is known, it can be used as a standard. However, S values are ususally not known, and therefore advantage is taken of the fact that, for a controlled set of test conditions, tinting strengths are proportional to their S values. A TS value of 100 is arbitrarily assigned to one of the films. Based on this standard, TS values for the other films are calculated (using Eq. 13) and plotted against PVC values. From this graph the CPVC value is obtained as indicated in Fig. 7-1*l* (5).

$$\underset{\text{(Comparison sample)}}{\frac{\text{TS} \cdot (1 - R_\infty)^2}{2R_\infty}} = \underset{\text{(Standard, where TS = 100)}}{\frac{\text{TS} \cdot (1 - R_\infty)^2}{2R_\infty}} \tag{13}$$

PROBLEM 7-3

The reflectivity of a standard paint sample is 0.40, and that of a comparison sample is 10% less (= 0.36). Compare their tinting strengths.

Solution. Assume a TS value of 100 for the standard sample, and calculate the TS of the comparison sample by substitution into Eq. 13.

$$\frac{\text{TS}(1 - 0.36)^2}{0.72} = \frac{100(1 - 0.40)^2}{0.80}$$

$$\text{TS} = \frac{100 \cdot 0.45}{0.57} = 0.79$$

The tinting strength of the comparison sample is 21% less than that of the standard sample (0.21 = 1.00 - 0.79).

PHYSICAL PROPERTIES THAT EXHIBIT OPTIMUM PERFORMANCE AT THE CRITICAL PIGMENT VOLUME CONCENTRATION

Film density is one of the physical properties of a paint film that reaches a peak value at the CPVC, as discussed in a preceding section and as shown in Fig. 7-1*a* (1, 2).

Tensile Strength

The tensile strength of a paint film has also been shown to attain a maximum value at the CPVC, as shown in Fig. 7-1*b* (3–6). The reversal in tensile strength at the CPVC is now well recognized and is used to determine the CPVC on the PVC scale.

Adhesion also appears to improve with an increase in PVC up to the CPVC point, after which this beneficial trend is reversed and adhesion is rapidly lost.

Adhesion to Metal

The adhesion of alkyd paint to mild steel substrates, using such diverse pigmentations as anatase titanium dioxide, zinc oxide, and natural iron oxide, has been reported as optimal at the CPVC. Failure at this point of maximum bonding is mostly adhesive; failure below and above the CPVC, mostly cohesive (6). In this study the CPVC was determined by tensile strength measurements (maximum tensile strength), and the adhesion (bond) strength by a direct pull-off technique using a tensometer (fracture values ranged up to 3400 psi). The authors suggest that the point of maximum adhesion might well serve as a method for determining the CPVC.

Adhesion to Wood

In another study the maximum adhesion of latex films to primer and midcoated smooth pine panels, using such pigmentations as calcium carbonate (both ground and precipitated), TiO_2, barytes, clay, and talc, was found to occur at or 5 to 15 PVC points above the CPVC. In this case the CPVC was determined by gloss measurements (minimum gloss), and the adhesion by a procedure based on cross-hatching and a subsequent pull-off with pressure sensitive tape. It was noted in this work that increased adhesion was imparted by the finer particle size emulsions and also by the platy and acicular, as opposed to nodular, pigments (7).

PERMEABILITY-RELATED PROPERTIES THAT UNDERGO MORE OR LESS DRASTIC CHANGES AT THE CRITICAL PIGMENT VOLUME CONCENTRATION

The onset of porosity in a pigmented film at the CPVC, as shown in Fig. 7-1*d*, leads to a tremendous increase in the penetration of the paint by gases, vapors, and liquids.

Rusting

When a porous paint film is applied over a ferrous substrate, severe rusting can normally be anticipated because of the rapid intrusion of water and its contaminants. As indicated in Fig. 7-1*e*, even minimal porosity promotes corrosion attack on a metal substrate such as iron (12). The shift from no corrosion to severe corrosion is normally centered in a narrow PVC band around the CPVC point.

Latex paints applied to cold rolled steel panels and placed in a salt spray cabinet corrode very rapidly (in a matter of hours) when PVC > CPVC. A PVC ladder based on this effect can serve to locate the CPVC point with good accuracy.

Blistering

Blister resistance refers to the ability of a paint system to resist the action of moisture without raising blisters. If the film is porous, water behind the film can readily escape to the outside surface. When confined by an impervious paint film, any water present may build up pressure and blow a bubble of water-softened film outward from the substrate. Hence, to a major extent, blistering is a function of paint permeability; and, as shown in Fig. 7-1*f*, the change from severe to no blistering generally occurs within a narrow band centering on the CPVC point (11, 12). A standard method for evaluating the blistering of paints is ASTM D714-56 (1974).

Wet Abrasion Resistance

Wet abrasion resistance refers to the scrub resistance of paint films to aqueous cleaning solutions. Standard methods for determining the scrub resistance of interior flat latex paints are specified by ASTM D2486-74a and Federal Test Method Standard (FTMS) 141a (6141, 6142). As might be anticipated, the wet abrasion resistance of porous films is far inferior to that of solid films, since entry of the cleaning solution into the porous film structure undermines the film integrity and weakens the resistance of the paint film to the wet abrading action. This decline in wet abrasion resistance through the region centering on the CPVC point is clearly shown in Fig. 7-1*g* (3, 4, 8, 13, 14).

Stain Resistance

The correlation between staining and porosity is sufficiency good for it to have been made the basis of a standard test for measuring the porosity of paint film

(ASTM D3258-73). Degree of staining is conveniently measured in terms of the change in paint reflectance before and after a standard stain has been applied and removed. As might be expected, staining increases as the PVC increases. As shown in Fig. 7-1*h*, this increasing trend corresponds to an upsweeping curve that has its inflection point at the CPVC (3, 13, 14).

Enamel Holdout

Enamel holdout may refer to either (*a*) the ability of a substrate to prevent the entry of an enamel into its interior structure or (*b*) the inherent ability of an enamel to resist absorption into a porous substrate. In the present context, enamel holdout refers only to the substrate and its ability to hold out against the penetration of the enamel coating into its interstices (3, 4, 8, 9, 15). It is obvious that if the interstices are already filled, as is the case below the CPVC, the enamel holdout should be excellent. Conversely, in the region above the CPVC, with its associated porosity, the enamel holdout will be poor. This type of behavior is shown in Fig. 7-1*i*, where the abrupt decline in enamel holdout above the CPVC is quite striking. A subjective measurement of primer substrate holdout may be made according to FTMS Method 6261 ("Primer Absorption and Topcoat Holdout").

OPTICAL PROPERTIES THAT UNDERGO ABRUPT TO GRADUAL CHANGES AT THE CRITICAL PIGMENT VOLUME CONCENTRATION

Abrupt Changes in Optical Properties

Optical properties that are directly influenced by paint film porosity, such as light scattering, contrast ratio, tinting strength, and hiding power efficiency, are all subject to abrupt changes in value at the critical pigment volume concentration, as shown in Figs. 7-1*j*, *k*, *l*, and *m*. The first three optical properties have been made the bases for accurately determining the CPVC point, as discussed in a preceding section. Paints that are formulated through the CPVC region always demand a proper recognition of the CPVC point, since visual appearance is most sensitive to even minor shifts in PVC to either side of the CPVC (4, 5, 10, 12, 14, 20, 21).

Color uniformity is also optimum for deep-colored alkyd flat paints, which are formulated in a narrow band running from just below the CPVC (e.g., CPVC − 0.03) to the CPVC itself (15). Although color uniformity is not sufficiently precise for determining the CPVC, knowledge of this point permits the formulation of paints with excellent color uniformity (15, 17, 24).

The location of the PVC that ensures optimum color uniformity of tinted alkyd paints has been stated to be just below the CPVC point (25). Furthermore, it is generally acknowledged that paints that exhibit good color uniformity also provide the best combination of other desirable properties (25).

Gradual Changes in Optical Properties

Gloss is a surface property that causes paint to have a shiny or mirrorlike appearance. Sheen is specular gloss that corresponds to a near-grazing angle of incidence (e.g., 85°). Both optical properties are dependent on the surface rather than the bulk properties of a paint. Hence they are only casually responsive to the introduction of air into a paint film at the CPVC. For this reason they exhibit only gradual changes throughout the CPVC region, as shown in Figs. 7-1*n* and *o* (5, 8, 11, 13, 19). Gloss and sheen generally fail to provide the incisive data needed for a precise CPVC determination. Sheen difference in coatings over sealed and unsealed surfaces, however, is reported as minimal at the CPVC point (25).

REDUCED PIGMENT VOLUME CONCENTRATION Λ AS A KEY FORMULATING PARAMETER

The reduced pigment volume concentration is defined as the ratio of the pigment volume concentration to the critical pigment volume concentration, as given by Eq. 14 (12).

$$\Lambda = \frac{\text{PVC}}{\text{CPVC}} \qquad (14)$$

When $\Lambda > 1$, the CPVC has been exceeded and porosity (air pockets) is present in the paint film; when $\Lambda < 1$, the PVC is less than the CPVC and pigment separation is present in the solid pigment/binder film.

Typical pigmentation levels have been established in the past for different applications on the basis of pigment volume concentrations. For example, Fig. 7-4 gives the representative PVC pigmentation ranges for interior architectural finishes formulated in the 1960s (26). However, with the introduction of more modern pigments and vehicles these ranges have undergone revision. It is now recognized that the important parameter in formulating paints is not the PVC as such, but rather the relationship of the PVC to the CPVC (the reduced PVC or Λ). By setting up formulating guidelines based on Λ values, modifying factors such as the introduction of a new pigment or new vehicle are automatically taken into account. By using the reduced pigment volume concentration as a formulating tool, prediction of paint film behavior becomes more

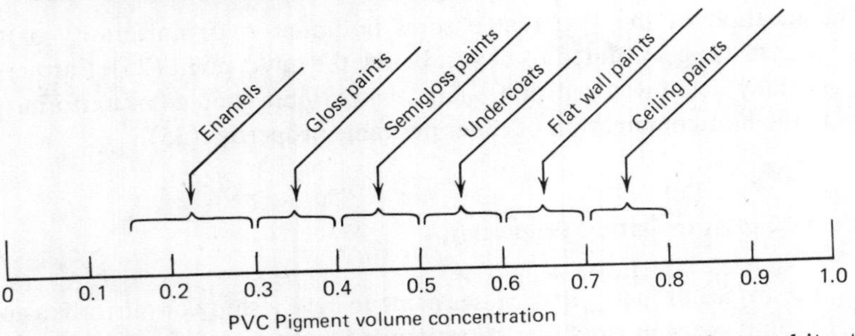

Fig. 7-4 Representative PVC pigmentation ranges for interior architectural finishes.

accurate and improvement in the film properties dependent on this quantity can be more precisely controlled. Optimal ranges for various types of paints based on the reduced pigment volume concentration are graphically shown in Fig. 7-5 (12).

The low Λ values for the gloss enamels ensure a generous excess of binder, which flows out to provide a highly reflective surface uninterrupted by an excessive number of surface particles that protrude and scatter incident light (reduce gloss). The semigloss enamels, intended for architectural coatings,

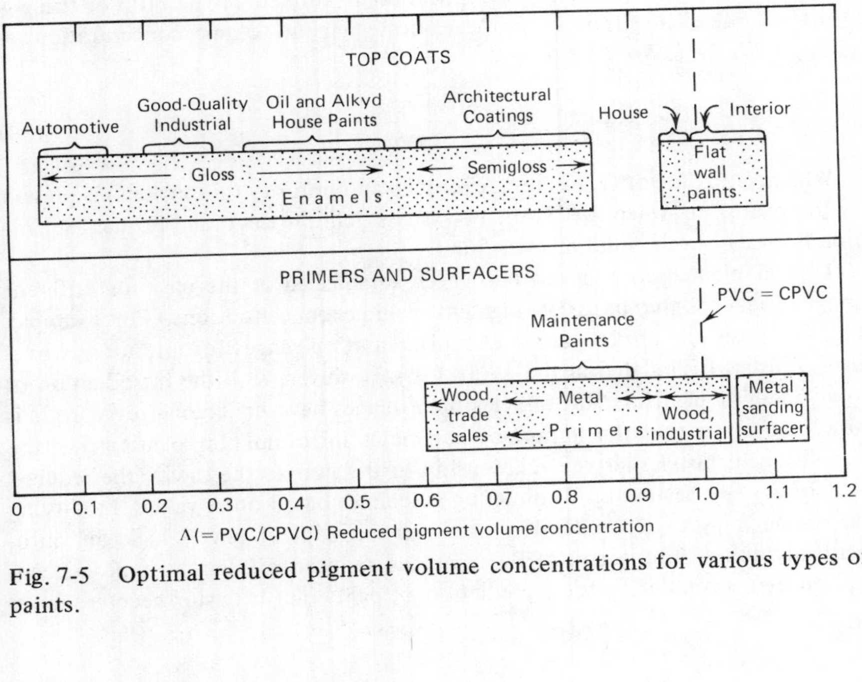

Fig. 7-5 Optimal reduced pigment volume concentrations for various types of paints.

have higher Λ values, since a high gloss over a vast expanse of space is esthetically disturbing. Actually, flatting agents or high-oil-absorption pigments may be necessary to reduce the gloss of architectural paints to an acceptable level. Flat architectural paints are formulated at or close to a Λ value of 1.00, a region where these paints exhibit optimum performance.

Maintenance primers are generally formulated in a Λ range of from 0.75 to 0.90 to secure optimal resistance to rusting and blistering. Since sanding surfacers must provide ease of sanding with minimal clogging of the sanding paper by binder, they are formulated considerably above the CPVC point (Λ values of 1.05 to 1.15). Experimental studies indicate that wood primers are best formulated through a narrow Λ band running from 0.95 to 1.05 to assure optimum overall performance.

Modifying Factors that Influence a Λ Value

Binder Shrinkage on Drying (Curing). A binder may be subject to more or less shrinkage during the conversion of the wet applied coating to a dry paint film. Hence calculations of Λ involving binder density should be based on the density of the binder in the dry film (12). The relation among the quantities PVC_{wet}, PVC_{dry}, and binder shrinkage s during drying is given by Eq. 15.

$$PVC_{dry} = \frac{PVC_{wet}}{1 - s + s \cdot PVC_{wet}} \tag{15}$$

As a rough approximation, the PVC can be considered to increase by 0.0025 point for each percent of shrinkage in the binder during the drying cycle.

PROBLEM 7-4

The PVC of a coating on a wet resin basis is 52%. The CPVC for the dry paint film is 54%. If the binder undergoes a shrinkage of 16% on drying (curing), is the CPVC exceeded by the dry paint film, and if so by how much?

Solution. Substitute the appropriate given data in Eq. 15, and solve for the dry film PVC.

$$PVC_{dry} = \frac{0.52}{1 - 0.16 + 0.16 \cdot 0.52} = 0.563$$

The CPVC is exceeded by 0.023 point (= 0.563 − 0.540).

This result could also have been approximately determined by multiplying 16% by the 0.0025 factor, giving an increase in the PVC of 0.04 on going from the wet to the dry condition (0.52 to 0.56). The CPVC is exceeded by approximately 0.02 point (= 0.56 − 0.54).

Comment. The change of a PVC from 0.520 to 0.563 is considerable and is especially significant because this shift takes place through the CPVC of 0.54, resulting in a change in the reduced pigment volume concentration from 0.963 to 1.043. Binder shrinkage on drying should always be considered as a modifying variable in interpreting experimental data or predicting performance based on Λ calculations.

Processing. Intensive and/or prolonged processing generally acts to promote denser packing. As a result a higher CPVC can normally be expected. This means, in turn, that for a given PVC the PVC/CPVC ratio or Λ is reduced.

Consider, for example, a metal primer that is scheduled for 8 hr of processing in a ball mill. Its PVC is fixed, but its CPVC increases with time of milling. If the processing step is not carried to completion and the primer is prematurely withdrawn, the reduced pigment volume concentration may be unduly high, resulting in impaired corrosion resistance.

Pigment Factors. Agglomeration of pigment leads to a less dense packing and a lower CPVC value. Conversely, an unusually efficient dispersant may provide a higher CPVC than normally expected. Such influences on paint performance can be conveniently assessed in terms of the reduced pigment volume concentration of the paint film.

ULTIMATE CRITICAL PIGMENT VOLUME CONCENTRATION (UPVC)

Since a more or less significant pigment/vehicle interaction occurs when pigment is dispersed in a vehicle, it is incorrect to state that the CPVC is dominated by either the pigment or the vehicle (27). Ramig offers evidence that the CPVC is a property of both the pigmentation and the binder in latex paints (20). The question of CPVC control even extends to the selection and level of the pigment dispersant that is present in the system.

The dramatic change in the CPVC that occurs with change in additive level in a titanium dioxide/mineral oil system has been reported by Asbeck and is shown in Table 7-1. At the highest CPVC the particles are presumably completely dispersed (monodisperse state), corresponding to what Asbeck calls the ultimate critical pigment volume concentration (UPVC) for that particular system. In the Asbeck paper a theoretical expression relating CPVC, UPVC, d (diameter of primary pigment particle), and D (diameter of pigment agglomerate) was developed as given by Eq. 16.

$$\text{CPVC} = \text{UPVC} - (\text{UPVC})^2 \left(\frac{D - d}{D}\right)^3 \tag{16}$$

Table 7-1 Packing of Titanium Dioxide Pigment in Mineral Oil as a Function of Dispersant Level (27)

Dispersant Level (%)	CPVC (%)	Relative Agglomerate Size (D/d)	Number of Primary Particles per Agglomerate
0.0	28	13.0	~2000
0.5	32	5.8	~200
1.0	36	3.4	40
2.0	41	1.3	2
3.0	43 (UPVC)	1.0	1

Note that two CPVC limits are set by this equation. For $D = d$ (monodisperse state), CPVC = UPVC; for $D >> d$ (very large agglomerates), CPVC $\approx$ UPVC(1 - UPVC). Based on the system shown in Table 7-1, the upper limiting value for the CPVC is 0.43; the lower limiting value, 0.25[= 0.43(1 - 0.43)]. These values serve to indicate the marked dependence of the CPVC value on the state of dispersion (relative aggolmeration) in a dispersion system. Asbeck makes the interesting observation that agglomerates tend toward a uniform size, with each unit of clustered particles having its fair share of surface dispersing molecules (the presence of smaller agglomerates would necessarily require a disproportionately large number of dispersing molecules for the surface/volume dimensions involved).

Equation 16 can be solved explicitly for the relative agglomerate diameter D as given by Eq. 17.

$$D = \frac{d}{1 - \sqrt[3]{(\text{UPVC} - \text{CPVC})/\text{UPVC}^2}} \tag{17}$$

PROBLEM 7-5

The CPVC of an incompletely dispersed system is 36% (0.36). If the UPVC for this particular system is 43% (0.43), calculate an approximate relative size for the undispersed particle agglomerates.

Solution. Substitute the problem data in Eq. 17, and solve for the ratio D/d.

$$\frac{D}{d} = \frac{1}{1 - \sqrt[3]{(0.43 - 0.36)/(0.43)^2}} = 3.6 \text{ (relative size)}$$

REFERENCES

1. Cole, R. J., "Determination of Critical Pigment Volume Concentration in Dry Surface Coating Films," *JOCCA*, **45**, No. 11, 776–780, November (1962).
2. Pierce, P. E. and R. M. Holsworth, "Determination of Critical Pigment Volume Concentration by Measurement of the Density of Dry Paint Films," *Off. Dig.*, **37**, No. 482, 272–283, March (1965).
3. Becker, John C., Jr., and D. D. Howell, "CPVC Concentration of Emulsion Binders," *Off. Dig.*, **28**, No. 380, 775–793, September (1956).
4. Schaller, E. J., "Critical Pigment Volume Concentration of Emulsion Based Paints," *JPT*, **40**, No. 525, 433–438, October (1968).
5. Wiita, R. E., "Vehicle CPVC," *JPT*, **45**, No. 578, 72–79, March (1973).
6. Reddy, J. N. et al., "Studies on Adhesion: Role of Pigments," *JPT*, **44**, No. 566, 70–75, March (1972).
7. Montreal Society for Coatings Technology, "Adhesion of Latex Paint: Part I," *JCT*, **48**, No. 617, 52–60, June (1976).
8. Armstrong, W. G. and W. H. Madson, "The Effect of Pigment Variation on the Properties of Flat and Semi-Gloss Finishes," *Off. Dig.*, **19**, No. 269, 321–335, June (1947).
9. Newton, D. S., "Pigment Volume Concentration," *JOCCA*, **45**, No. 3, 180–199, March (1962).
10. Stieg, F. B., Jr., "Particle Size as a Formulating Parameter," *JPT*, **39**, No. 515, 703–720, December (1967).
11. Asbeck, W. K. and M. Van Loo, "Critical Pigment Volume Relationships," *Ind. Eng. Chem.*, **41**, No. 7, 1470–1475, July (1949).
12. Bierwagen, G. P. and T. K. Hay, "The Reduced Pigment Volume Concentration as an Important Parameter in Interpreting and Predicting the Properties of Organic Coatings," *Prog. Org. Coatings*, **3**, 281–303 (1975).
13. Dobkowski, T. P., "Calcined Aluminum Silicate Pigments in Latex Paints," *JPT*, **41**, No. 535, 448–454, August (1969).
14. Rudolph, Julie P., "Determination of the CPVC of Latex Paints Using Hiding Power Data," *JCT*, **48**, No. 619, 45–50, August (1976).
15. Stieg, F. B., Jr., "Color and CPVC," *Off. Dig.*, **28**, No. 379, 695–706, August (1956).
16. Stieg, F. B., Jr., "Latex Paints and the CPVC," *JPT*, **41**, No. 531, 243–248, April (1969).
17. Stieg, F. B., Jr., "Particle Size Parameter and Latex Paints," *JPT*, **42**, No. 545, 329–334, June (1970).
18. Stieg, F. B., Jr., "Profiles of White Hiding Power in Flat Latex Paints," *JPT*, **45**, No. 576, 76–82, January (1973).
19. Philadelphia Society for Paint Technology, "Determination of CPVC by Calculation," *Off. Dig.*, **33**, 1437–1452, November (1961).
20. Ramig, A. R., Jr., "Latex Paints–CPVC, Formulation, and Optimization," *JPT*, **47**, No. 602, 60–67, March (1975).
21. Stieg, F. B., Jr., "Latex Paint Opacity–A Complex Effect," *JOCCA*, **53**, 469–486 (1970).
22. Bessey, W. K. and K. A. Lammiman, "The Measurement and Interpretation of Oil Absorption," *JOCCA*, **34**, 519–546, November (1951).
23. Mitton, P. B., "Opacity, Hiding Power, and Tinting Strength," in *Pigment Handbook* (T. Patton, Ed.), Vol. III, Wiley-Interscience, New York, 1973.

24. Stieg, F. B., Jr., and D. F. Burns, "The Effect of Pigmentation on Modern Flat Wall Paints," *Off. Dig.*, **26**, No. 349, 81–93, February (1954).
25. Garland, J. R. and S. Werthan, "A Graphical Method for Formulating Alkyd Flat Wall Paints," *Am. Paint J.*, p. 68, June 13 (1960).
26. Stieg, F. B., "Are Conventional Trade Sales Formulating Practices Wasteful?" *JCT*, **48**, No. 612, 51–59, January (1976).
27. Asbeck, W. K., "Dispersion and Agglomeration Effects on Coatings Performance," *JCT*, **49**, No. 635, 59–70, December (1977).

8 Latex Critical Pigment Volume Concentration (LCPVC)

In the case of solvent-type coatings, the CPVC value for the dry paint film can be closely calculated from rub-out oil absorption data (see Chapter 6). However, in the case of latex coatings, the mechanics of converting from a wet coating to a dry paint film is different, and a modified concept is necessary to reconcile a latex CPVC (LCPVC) with a solvent-type CPVC and an $\overline{\text{OA}}$ value.

The dispersion of pigments in latices and the manner of latex drying are unique. Latex vehicles are not solutions of binder in solvent, but rather are concentrated suspensions of discrete, spherical, and relatively tacky, resinous particles in water, with a particle size range that may compare closely with that of the pigments that they are called upon to disperse.

To visualize the drying of a latex/pigment dispersion (a latex coating), picture the latex particles as more or less tacky spheres that strongly adhere to each other on contact as water is lost from the latex composition (1). The loss of water may be due either to evaporation into the atmosphere or to absorption into a porous substrate. In turn the pigment particles in the latex/pigment mix become enmeshed in the resinous network that forms. As the wet latex paint shrinks to a dry coating, the latex particles coalesce and deform as best they can around the pigment particles to yield a more or less dense packing arrangement (a latex paint). The proportion of pigment to latex (the PVC) largely controls the gross physical aspects of the pigmented latex paint. This is shown in Fig. 8-1, where conditions corresponding to PVC $<$ LCPVC (Fig. 8-1*d*), PVC = LCPVC (Fig. 8-1*e*), and PVC $>$ LCPVC (Fig. 8-1*f*) are schematically illustrated in comparison with their solvent-type counterparts, Figs. 8-1*a*, *b*, and *c*.

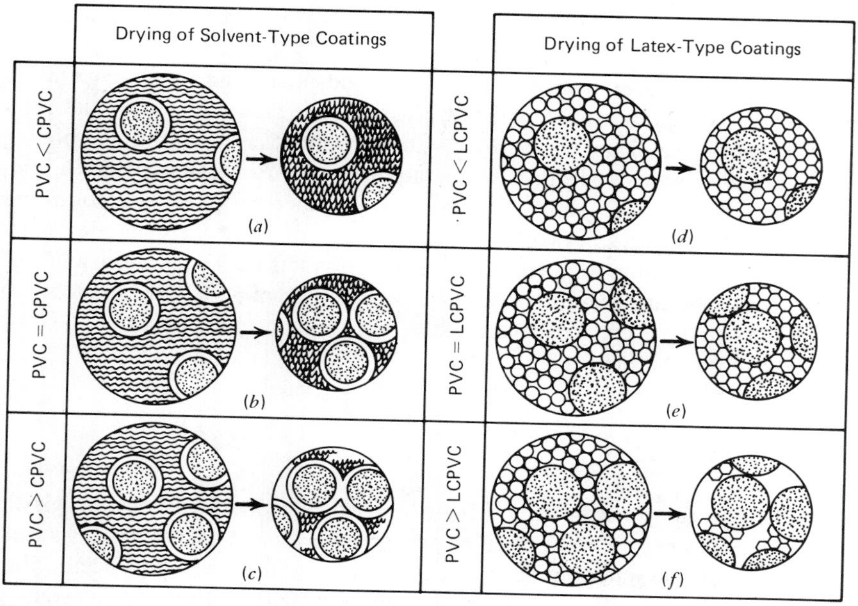

Fig. 8-1 Schematic illustration of the state of pigment dispersion for both solvent-type and latex paint films corresponding to conditions where the pigment volume concentration is less than, equal to, and more than the critical pigment volume concentration.

In the case of solvent-type coatings, the liquid vehicle flows rather than deforms around the pigment particles on drying. This flow assists the compaction process as the film shrinks because of loss of volatile solvent. Although the vehicle becomes more viscous as solvent evaporates, flow persists through most of the drying cycle. In the case of latex paints, there is a preliminary flow of the latex suspension. This takes place before the time when the latex particles are first forced to come into intimate contact because of initial water loss. However, after this relatively short but very important initial flow, pigment compaction to achieve a high CPVC is achieved mainly by plastic deformation and coalescence of the latex particles. From this it follows that the deformability of the latex particles, generally assisted by a coalescing agent, is an index of the ability of the latex to provide a tightly compacted (high-PVC) latex paint. The PVC that can be attained is also significantly affected by the size of the latex particles relative to the particles of the pigment (smaller particles give higher PVCs) (2, 3).

GLASS TRANSITION TEMPERATURE

The glass transition temperature T_g is a second-order transition temperature marking the change in a polymer condition from one characterized by glassiness (brittleness) to one characterized by flexibility or plasticity (melting and boiling points are first-order transition temperatures). The T_g value provides a convenient measure of the ability of the latex polymer particles to deform and coalesce into a resinous network (film) during the drying cycle of a latex coating. The glass transition temperatures of latex polymers may range from a low of -15 C, corresponding to a soft latex particle, to a high of 30 C, corresponding to a hard latex particle that requires a coalescing agent in order for it to manifest film-forming capabilities (2, 3).

COALESCING AGENT

A coalescing agent is essentially a fugitive plasticizer that facilitates the plastic flow and elastic deformation of latex particles. By the use of a coalescing agent, latex paints can be modified as necessary to provide improved coalescing (film-forming) properties over wide ranges of application temperatures, while avoiding an undue softness or tackiness of the final dry coating. Many coalescing agents fall in the glycol class (such as ethylene glycol, propylene glycol, and hexylene glycol) (4-6).

COMBINED EFFECT OF LATEX PARTICLE SIZE, LATEX GLASS TRANSITION TEMPERATURE, AND COALESCING AGENT ON LATEX CRITICAL PIGMENT VOLUME CONCENTRATION

Small latex particles move about in a latex coating with greater facility than larger ones and, when caught between pigment particles, allow a closer particle-to-particle approach. By acting more like a continuous medium (a solution coating) they also exhibit a correspondingly enhanced ability to penetrate into the interstices of pigment particles.

The ability to penetrate is a valuable asset when latex paints are applied to substrates covered with a chalky overlay such as might be encountered with weathered paints or porous prime coats. Thus finer size emulsion particles promote better adhesion to either chalky or prime coated surfaces. The better flow properties of the finer particle size latices also account for their use in the formulation of gloss latex paints. However, it should be noted that for critical situations latex coatings are still commonly fortified with solvent-type vehicles. Thus it is common practice to blend 10 to 30% of an alkyd vehicle into an outside latex house paint to ensure adequate penetration into the chalked surfaces of weathered houses, with consequent improved bonding to the underlying substrate.

The smaller particle size latices also give higher LCPVC values. The influence of latex particle size at a constant glass transition temperature (T_g = 20 C) is shown in Fig. 8-2, where the diameter of the latex particle is plotted versus LCPVC for three different levels of a typical coalescing agent. Close inspection of Fig. 8-2 reveals an unexpected finding, namely, that excessive coalescing agent is counterproductive (20% levels of hexylene glycol give lower LCPVC values than do 10% levels) (2, 5). Obviously there is an optimum level of coalescing agent that provides a maximum LCPVC value. This reversal in LCPVC with excess coalescing agent is presumably due to the premature coalescence that occurs during the film-forming process. It is postulated that the excessively softened latex particles coalesce too quickly, aggregates that prevent uniform particle distribution are immediately formed, and a porous structure rather than a solid resinous network of polymer particles results.

The effect of the latex glass transition temperature on the latex CPVC is shown in Fig. 8-3, where T_g is plotted versus LCPVC for two latex particle sizes (2).

The levels of a typical coalescing agent that provide maximum LCPVC values for acrylic latex paints pigmented with a medium size calcium carbonate pigment are plotted as a function of latex T_g and *d* values in Fig. 8-4. Although the

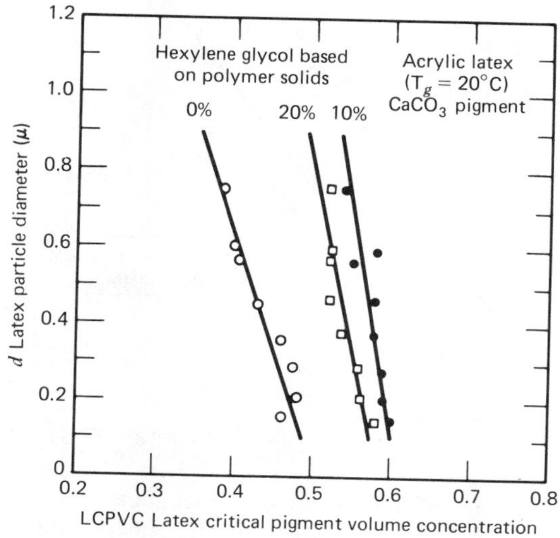

Fig. 8-2 Graph relating latex particle diameter to latex critical pigment volume concentration for three levels of coalescing agent.

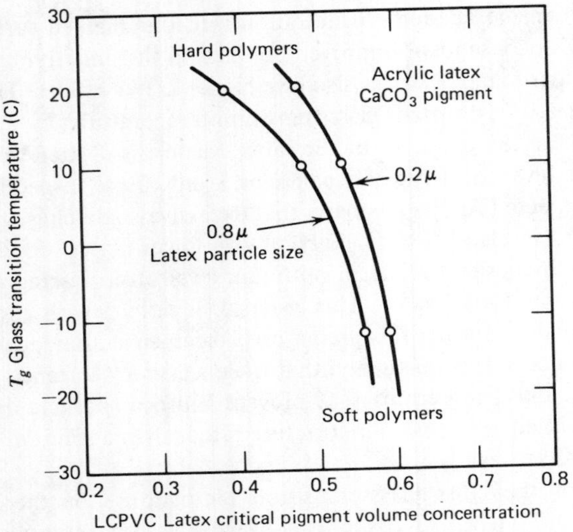

Fig. 8-3 Graph relating glass transition temperature to latex critical pigment volume concentration for two latex particle sizes.

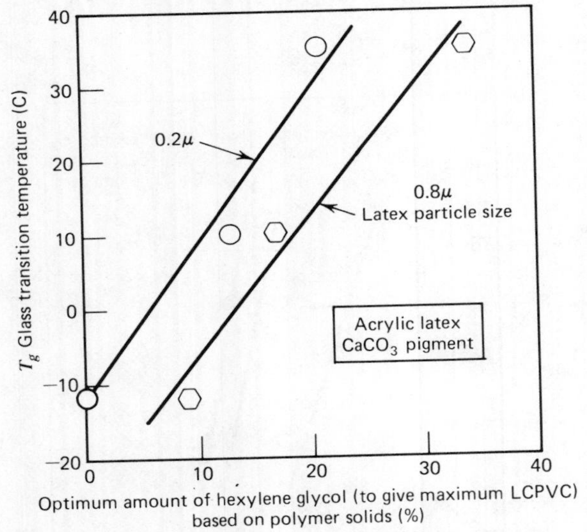

Fig. 8-4 Graph relating optimum percent of coalescing agent (hexylene glycol), in terms of obtaining the maximum latex critical pigment volume concentration, to latex particle size and glass transition temperature.

data are scattered, the trends are quite clear. This interplay among the several variables to produce optimum CPVC values merits study.

In making any comparison of the binding abilities of latices, it should always be remembered that the latex polymer contributes only two (T_g and d) of the three major factors affecting the CPVC. The third major factor is the selection and level of coalescing agent used in the test coating.

LATEX BINDER INDEX (BINDER EFFICIENCY)

The critical pigment volume concentration for a latex paint is invariably less than that for its solvent-type counterpart. In effect, this means that at their respective CPVC points a larger volume of latex polymer (V_{latex}) than of solution binder (V_{sol}) is required to bind a given solid volume of pigment. On observing this relationship, Berardi (7) was prompted to call the ratio of these volumes a binder index and suggested that this index be used to quantify the pigment binding ability of a latex polymer and, by implication, its usefulness as a latex paint binder. In this book the term "binder efficiency" (e) will also be used to represent the binder index.

$$e = \frac{V_{\text{sol}}}{V_{\text{latex}}} \tag{1}$$

A binder index or binder efficiency of 1.00 corresponds to a 100% resin binding ability. Latex polymers have efficiency values less than 1.00. The binder efficiency can be calculated from a knowledge of the measured LCPVC for a given pigmentation, together with the pigment density and oil absorption value.

The binder volume per unit pigment volume at the rub-out oil absorption end point is given by Eq. 2. This is also equal to the CPVC point.

$$\frac{V_b}{V_p} = \frac{\overline{\text{OA}}/0.935}{100/\rho_p} \tag{2}$$

The binder volume per unit pigment volume at the LCPVC is given by Eq. 3.

$$\frac{V_b}{V_p} = \frac{1 - \text{LCPVC}}{\text{LCPVC}} \tag{3}$$

By definition the binder efficiency (binder index) is the ratio of the binder volume of Eq. 2 to that of Eq. 3, since both expressions are based on unit weight of pigment.

$$e = \frac{\text{LCPVC} \cdot \overline{\text{OA}} \cdot \rho_p}{93.5 \cdot (1 - \text{LCPVC})} \tag{4}$$

Equation 4 can be solved explicitly for LCPVC to give Eq. 5.

$$\text{LCPVC} = \frac{1}{1 + \overline{\text{OA}} \cdot \rho_p / 93.5 \cdot e} \tag{5}$$

PROBLEM 8-1

Contrast ratios were used to establish the LCPVC values listed below for a vinyl acetate copolymer latex paint fortified with 10% of a coalescing agent. The LCPVC values correspond to four single pigments having the tabulated density and oil absorption values. Calculate and compare the binder efficiencies for this latex polymer.

Pigment	Density (g/cm^3)	$\overline{\text{OA}}$ Value (g/100 g)	LCPVC (%)
$CaCO_3$	2.71	13.3	67.1
Talc	2.85	38.8	40.0
TiO_2	4.16	22.1	44.5
Clay	2.27	59.6	35.2

Solution. Substitute in turn the tabulated values for ρ_p, $\overline{\text{OA}}$, and LCPVC in Eq. 4. Values of 0.786, 0.788, 0.788, and 0.786 for the binder efficiency e are obtained for the four pigments. These e values are essentially the same.

The constancy of the value of e for the four pigments in Problem 8-1 indicates that the binder efficiency is solely a function of the latex polymer plus modification by the coalescing agent. The binder index is independent of the pigmentation (5).

PROBLEM 8-2

Practical experience has shown that a PVC/LCPVC ratio of 1.08 provides good-quality latex flats (optimum opacity and enamel holdout) (5). It has also been demonstrated that a pigmentation consisting of 36%w of titanium dioxide, 49%w of calcium carbonate, and 15%w of clay gives a satisfactory balance of hiding and extender pigment. Using this information, calculate a proper PVC for a latex flat coating based on a latex emulsion system (including a coalescing agent) that has an overall binder efficiency of 0.70. Obtain the average density for the pigment mixture from the pigment data of Problem 8-1. However, use an $\overline{\text{OA}}$ value of 14.5 for the pigment mixture as determined from an actual rub-out, rather than calculating a value from the $\overline{\text{OA}}$ values of the individual pigments.

Solution. Calculate the volume composition of 100 g of pigment mixture, and from this the pigment mixture density.

Pigment	Weight (g)	Density (g/cm^3)	Volume (cm^3)
TiO_2	36	4.16	8.65
$CaCO_3$	49	2.71	18.08
Clay	15	2.27	6.61
	100		33.34

$$\rho_p \text{ (mixture)} = \frac{100}{33.34} = 3.00 \text{ g/cm}^3$$

Substitute this mixture density value and the given values for e and $\overline{\text{OA}}$ in Eq. 5, and solve for LCPVC.

$$\text{LCPVC} = \frac{1}{1 + 14.3 \cdot 3.00/93.5 \cdot 0.70} = 0.604$$

To provide a PVC/LCPVC ratio of 1.08, adjust the PVC relative to the LCPVC accordingly.

$$\text{PVC} = 0.604 \cdot 1.08 = 0.652 \ (65.2\%)$$

Comment. If an average $\overline{\text{OA}}$ value of 23.4 as calculated from the $\overline{\text{OA}}$ values of the individual pigments had been used in the computation, a LCPVC value of 0.473 would have been obtained rather than 0.604. This discrepancy in $\overline{\text{OA}}$ values is due to the packing of the finer pigments into the interstices of the larger particles, as discussed in Chapter 5 on pigment/binder geometry.

Problem 8-2 serves to emphasize the fact that oil absorption values for pigment mixtures must be obtained by an actual rub-out method, rather than from a calculated average $\overline{\text{OA}}$ value, in working out this type of problem.

LATEX POROSITY. INDEX AND BINDER EFFICIENCY

Since the binder efficiency e of a latex is always less than 1.00 ($e = 1.00$ for linseed oil), it follows that for the same pigmentation the LCPVC for a latex paint is always less than the CPVC obtained from an oil absorption test. The porosity indices for the two systems are given by Eqs. 6 and 7.

$$\text{P.I. (oil)} = 1 - \left[\frac{\text{CPVC}(1 - \text{PVC})}{\text{PVC}(1 - \text{CPVC})}\right] \tag{6}$$

$$\text{L.P.I. (latex)} = 1 - \left[\frac{\text{LCPVC}(1 - \text{PVC})}{\text{PVC}(1 - \text{LCPVC})}\right] \tag{7}$$

This general type of equation was developed and discussed in Chapter 5 on pigment/binder geometry. A graphical solution for this generalized equation is given by the nomograph in Fig. 5-13.

The expression within the brackets of either Eq. 6 or Eq. 7 represents the fraction of binder in the interstices of the pigment particles. The binder fraction in this void space for a given PVC is less for a latex than for linseed or oil solution binder because the latex paint is more porous. Let the ratio of the latex binder fraction to the solution binder fraction for a given PVC be e'.

$$e' = \frac{\text{LCPVC}(1 - \text{PVC})/\text{PVC}(1 - \text{LCPVC})}{\text{CPVC}(1 - \text{PVC})/\text{PVC}(1 - \text{CPVC})} = \frac{\text{LCPVC}(1 - \text{CPVC})}{\text{CPVC}(1 - \text{LCPVC})} \tag{8}$$

$$\frac{\text{LCPVC}}{1 - \text{LCPVC}} = e' \cdot \frac{\text{CPVC}}{1 - \text{CPVC}} \tag{9}$$

Substituting this value of LCPVC/(1 - LCPVC in Eq. 7 gives Eq. 10.

$$\text{L.P.I. (latex)} = 1 - e'\left[\frac{\text{CPVC}(1 - \text{PVC})}{\text{PVC}(1 - \text{CPVC})}\right] \tag{10}$$

At the critical pigment volume concentration, the CPVC for the solution paint is given by $V_p/(V_p + V_{\text{sol}})$, and the LCPVC for the latex paint by $V_p/(V_p + V_{\text{latex}})$. Substitution of these values for CPVC and LCPVC in Eq. 9 results in Eq. 11.

$$e' = \frac{V_{\text{sol}}}{V_{\text{latex}}} \tag{11}$$

But this value for e' is the same as the value for e given by Eq. 1. Therefore e and e' are one and the same. This equivalence might have been sensed intuitively but is derived here for the record.

Note also that the expression in the brackets of both Eq. 6 and Eq. 10 is the same. From this equality, Eq. 12 is immediately derived.

$$e = \frac{1 - \text{L.P.I.}}{1 - \text{P.I.}} \tag{12}$$

The same type of pigment/binder geometry applies to both latex and solution (solvent-type) paint films. Applicable equations are summarized below.

Solution Paint Films

$$\text{Overall porosity} \begin{cases} = 1 - \dfrac{\text{CPVC}}{\text{PVC}} & (13) \\ = \text{P.I.} \cdot (1 - \text{CPVC}) & (14) \end{cases}$$

$$\text{Porosity index (P.I.)} \begin{cases} = \dfrac{\text{PVC} - \text{CPVC}}{\text{PVC}(1 - \text{CPVC})} & (15) \\[2ex] = 1 - \dfrac{\text{CPVC}(1 - \text{PVC})}{\text{PVC}(1 - \text{CPVC})} & (16) \\[2ex] = \dfrac{\text{overall porosity}}{1 - \text{CPVC}} & (17) \end{cases}$$

Latex Paint Films

$$\text{Latex overall porosity} \begin{cases} = 1 - \dfrac{\text{LCPVC}}{\text{PVC}} & (18) \\[2ex] = \text{L.P.I.} \cdot (1 - \text{LCPVC}) & (19) \end{cases}$$

$$\begin{array}{c}\text{Latex porosity index}\\ \text{(L.P.I.)}\end{array} \begin{cases} = \dfrac{\text{PVC} - \text{LCPVC}}{\text{PVC}(1 - \text{LCPVC})} & (20) \\[2ex] = 1 - \dfrac{\text{LCPVC}(1 - \text{PVC})}{\text{PVC}(1 - \text{LCPVC})} & (21) \\[2ex] = \dfrac{\text{latex overall porosity}}{1 - \text{LCPVC}} & (22) \end{cases}$$

The two systems are related to each other by the binder efficiency e as given by Eqs. 23 and 24.

$$e \begin{cases} = \dfrac{\text{LCPVC}(1 - \text{CPVC})}{\text{CPVC}(1 - \text{LCPVC})} & (23) \\[2ex] = \dfrac{1 - \text{L.P.I.}}{1 - \text{P.I.}} & (24) \end{cases}$$

Other equations can be developed, but the ones given above suffice for most numerical calculations.

PROBLEM 8-3

Derive an expression for the latex porosity index at the solution CPVC point for a given pigmentation.

Solution. The porosity index for a solution-type paint film at the CPVC point is zero. Hence, from Eq. 24, L.P.I. = $1 - e$.

Comment. This simple equality clearly shows the porosity that rapidly develops with a decline in latex binder efficiency. For example, given a binder index of $e = 0.70$, a L.P.I. of 0.30 (= 1 − 0.70) or 30% can be expected at the CPVC point of the solution-type binder.

PROBLEM 8-4

A solution paint film made up to a 55% pigment volume concentration has an overall porosity of 8.0%. If the volume of solvent-type binder in this paint film is replaced by an equal volume of latex binder having a binding efficiency of 0.70, calculate the overall porosity of the resulting latex film.

Solution. First calculate the CPVC value for the solution paint film from Eq. 13.

$$0.08 = 1 - \frac{\text{CPVC}}{0.55}; \qquad \text{CPVC} = 0.506$$

From this CPVC value determine the paint film P.I. from Eq. 17.

$$\text{P.I.} = \frac{0.08}{1 - 0.506} = 0.162$$

Calculate the porosity index of the latex paint film from Eq. 24.

$$e = 0.70 = \frac{1 - \text{L.P.I.}}{1 - 0.162}; \qquad \text{L.P.I.} = 0.413$$

From Eq. 20 calculate the LCPVC for the latex film.

$$0.413 = \frac{0.55 - \text{LCPVC}}{0.55(1 - \text{LCPVC})}; \qquad \text{LCPVC} = 0.418$$

Calculate the overall latex porosity from Eq. 19.

$$\text{Overall porosity} = 0.413(1 - 0.418) = 0.240\ (24.0\%)$$

Comment. From this it is seen that the replacement of a given volume of solution binder by the same volume of latex binder markedly alters the overall porosity of a paint system. This effect is due, of course, to the added void space that the substitute latex binder introduces into the paint film. In the above problem the porosity was increased threefold by the volume-for-volume replacement!

WHITE HIDING POWER RELATED TO POROSITY AND OTHER VARIABLES

A practical procedure for manipulating paint compositions to achieve desired changes in white hiding power, while holding constant selected variables such as porosity and titanium dioxide content, has been outlined in depth by Stieg (9–

11). The system is based on the experimental observation that the dry hiding power (ft^2/gal) of any white paint is due to three major interacting factors: (*a*) the weight of TiO_2 per gallon, (*b*) a porosity function, and (*c*) a titanium dioxide dilution function. No reflectometer is required by the procedure, since hiding power is judged by visual matching of paint panels. The simplified laboratory technique calls for the cross-blending of two similarly shaded test batches to obtain a dry-film brightness that matches the standard, and the observance of the convention that formula modifications should contain the same solids content by volume as the original standard. This invaluable system was developed by Stieg through long experience and should be carefully studied by anyone responsible for the formulation of white paints.

ENAMEL HOLDOUT AND STAIN RESISTANCE RELATED TO EFFECTIVE BINDER FRACTIONAL VOLUME CONTENT

From enamel holdout and stain resistance testing it has been shown that equivalent results are apparently obtained for coatings of equivalent effective fractional binder volume content, rather than for coatings of equivalent fractional overall porosity (8). Since the effective fractional binder volume for a latex paint is equal to $e(\text{LCPVC}/\text{PVC} - \text{LCPVC})$, Eq. 25 is valid for two latex paints with similar holdout and stain resistant properties.

$$\left[e\left(\frac{\text{LCPVC}}{\text{PVC}} - \text{LCPVC}\right)\right]_1 = \left[e\left(\frac{\text{LCPVC}}{\text{PVC}} - \text{LCPVC}\right)\right]_2 \tag{25}$$

If one of the two paints is a solvent type, Eq. 26 applies.

$$\frac{\text{CPVC}}{\text{PVC}_1} - \text{CPVC} = e\left(\frac{\text{LCPVC}}{\text{PVC}_2} - \text{LCPVC}\right) \tag{26}$$

If both systems are solvent based, Eq. 27 applies.

$$\left[\frac{\text{CPVC}}{\text{PVC}} - \text{CPVC}\right]_1 = \left[\frac{\text{CPVC}}{\text{PVC}} - \text{CPVC}\right]_2 \tag{27}$$

By using these equations, a paint that has proved satisfactory from the standpoint of enamel holdout and stain resistance can be modified with an alternative pigmentation and/or an alternative binder to obtain a coating that provides the same satisfactory holdout properties.

PROBLEM 8-5

A latex flat wall paint with a LCPVC of 0.45 gives, when formulated to a PVC of 0.51, a washed stain brightness (brightness after gilsonite stain removal) that is

70% of its original brightness. The latex binder efficiency is 0.65. If a shift is made to an alternative titanium dioxide pigmentation that gives a LCPVC of 0.52, calculate the PVC required to provide equal stain resistance.

Solution. Substitute the given data in Eq. 25 and solve for the required PVC.

$$0.65\left(\frac{0.45}{0.51} - 0.45\right) = 0.65\left(\frac{0.52}{\text{PVC}} - 0.52\right)$$

$$\text{PVC} = 0.546\ (54.6\%)$$

REFERENCES

1. Stieg, F. B., "Latex Paints and the CPVC," *JPT*, **41,** No. 531, 243–248, April (1969).
2. Schaller, E. J., "Critical Pigment Volume Concentration of Emulsion Based Paints," *JPT*, **40,** No. 525, 433–438, October (1968).
3. Becker, J. C. and D. D. Howell, "CPVC Concentration of Emulsion Binders," *Off. Dig.*, **28,** No. 380, 775–793, September (1956).
4. Andrews, M. D., "Influence of Ethylene and Propylene Glycols on Drying Characteristics of Latex Paints," *JPT*, **45,** No. 598, 40–48, November (1974).
5. Ramig, A. R., Jr., "Latex Paint–CPVC, Formulation, and Optimization," *JPT*, **47,** No. 602, 60–67, March (1975).
6. Hoy, K. L., "Estimating the Effectiveness of Latex Coalescing Aids," *JPT*, **45,** No. 579, 51, April (1973).
7. Berardi, P., "Parameters Affecting the CPVC of Resins in Aqueous Dispersions," *Paint Technol.*, **27,** 24, July (1963).
8. Stieg, F. B., "Particle Size Parameter and Latex Paints," *JPT*, **42,** No. 545, 329–334, June (1970).
9. Stieg, F. B., "Are Conventional Trade Sales Formulating Practices Wasteful?" *JCT*, **48,** No. 612, 51–59, January (1976).
10. Stieg, F. B., "Formulating for Maximum Economy," *Am. Paint Coatings J.*, pp. 14–15, 18, 37–39, August 23 (1976).
11. Stieg, F. B., "The ABC's of White Hiding Power," *JCT*, **49,** No. 630, 54–58, July (1977).

9 *Surface Tension*

Pigment dispersion is not a simple process. Rather, it involves highly complex relationships that are still incompletely understood. Most important are the phenomena taking place at the interfaces (boundary areas) between pigment and vehicle. Therefore a proper understanding of pigment dispersion requires a study of interfacial forces.

The present interest is concerned only with the relatively weak influences that are associated with intermolecular or secondary bonding forces such as permanent dipole forces (nonadditive); induced forces (nonadditive); dispersion, London, or nonpolar van der Waals forces (additive, universal, always attractive); and hydrogen bonding forces. In this chapter the forces in this class will be treated as a single entity with no attempt to differentiate among them. The strong primary bonding forces at an interface due to a chemical reaction are excluded from the discussion. However, it should be noted that even a small degree of chemical reactivity can have a tremendous influence on wetting (bonding) behavior. Also, for the sake of simplicity, the influence of gravitational force is disregarded as a significant factor unless otherwise specified.

NATURE OF SURFACE TENSION

The forces at the interface between two separated phases (such as pigment and vehicle) differ from the forces within either phase for the reason that the interface forces are in a state of unbalance. There is an unsymmetrical force distribution at the interface that gives rise to so-called surface tension effects. In practice these effects are often greatly modified by arranging for a concentration of certain molecules (additives such as surfactants) at the interface, where the molecules become arranged in a uniform direction or are otherwise oriented to impart desirable interface properties.

At the outset it is instructive to consider a simple liquid in a container that is open to the atmosphere, as shown in Fig. 9-1. A molecule deep within this liquid (a bulk molecule) is attracted equally in all directions by similar neighboring molecules and hence is in a stage of balanced equilibrium with its surroundings. Its tendency to move in any one given direction is no more probable than that to move in any other direction. At the surface, however, this evenly distributed state of balance no longer holds. Instead, a surface molecule is subjected to an unsymmetrical force distribution, since the molecules of the parent liquid pull on the surface molecule to one side only. The net effect is an unbalanced force that tries to drag surface molecules into the body of the parent liquid. This net force is responsible for the surface tension effects exhibited by a liquid surface. When the surface boundary is between two liquids or between a liquid and a solid, the resulting force is called an interfacial tension.

One dramatic manifestation of this force unbalance is the drive of surface tension to reduce the surface area of a liquid, whereby surface molecules are brought from a state of unbalance at the surface to a state of balance within the liquid. Thermodynamically, this corresponds to decreasing the free energy of the system, which means that a reduction in surface area takes place spontaneously whenever a liquid is afforded this opportunity. Hence any volume of liquid strives to confine itself by a minimum of surface area. Since a sphere encloses a maximum volume for a minimum of surface, this is the ultimate geometric form toward which surface tension forces continually try to shape a liquid. Ample proof of this is given by the spherical shape assumed by rain drops, soap bubbles, latex particles, or droplets of one liquid suspended in another, incompatible liquid.

For the same reason an expanse of liquid surface tends toward planar smoothness rather than roughness, since rugosity (a wrinkled or contoured surface) represents an excess of surface as far as the containment of liquid is concerned. As a result of this urge to reduce surface area, the outside of a liquid assumes the

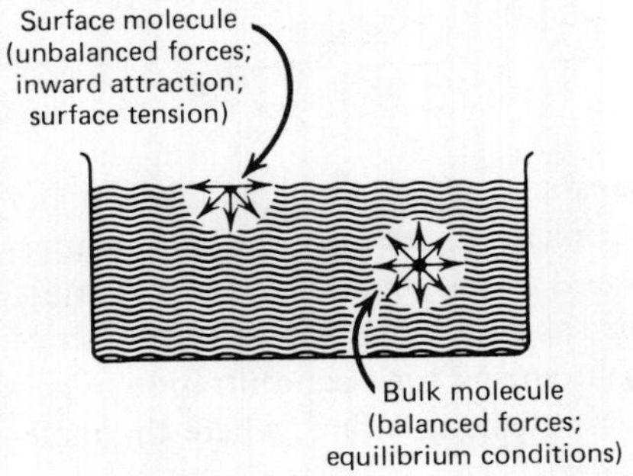

Fig. 9-1 Schematic diagram showing the balanced forces that act on a bulk molecule versus the unbalanced forces that act on a surface molecule.

characteristics of a stretched outer skin that is continually trying to contract. Surface tension derives its name from the surface forces that are responsible for this shrinking action.

Even though a solid cannot adjust to a surface tension drive, since it does not flow, it still is under a compulsion to bring its surface molecules within itself. On this basis it is assigned a so-called critical surface tension value.

DIMENSIONS OF SURFACE TENSION FORCES

Surface tension can be strikingly demonstrated experimentally by using the rectangular wire framework arrangement illustrated in Fig. 9-2. The single unattached length of straight wire L (placed loosely over the rectangular wire frame and free to move) is tied at its center with a thread so that it can be tugged back and forth on the framework. This assembly is then dipped into a soap solution to form an overall film. If after removal from the soap solution the soap film on the thread side is broken, the wire L can be pulled back and forth along the frame, the pull on the thread being exactly balanced by the film surface tension on the other side (like pulling a window shade up and down). All liquids exhibit this type of surface tension behavior.

By refining this simple aparatus, it can be shown that the force required to counterbalance the pull of the soap film is directly proportional to the width of the soap film (equivalent to the length of wire L in contact with the film). Furthermore, it can be demonstrated that this surface tension pull is independent of both the area and the thickness of the soap film (except for extreme thinness). Therefore a surface tension effect is measured solely in terms of the

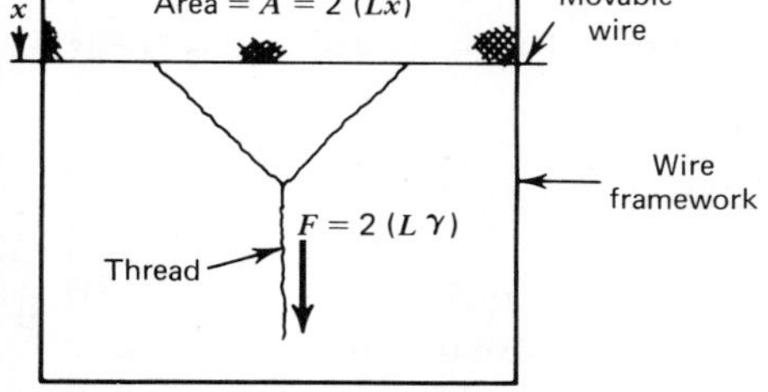

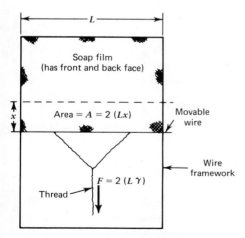

Fig. 9-2 Apparatus for demonstrating the nature of surface tension (a contracting force). Surface tension and specific free surface energy are shown to be numerically and dimensionally equivalent.

force acting along a unit length of surface. Surface tension is conventionally expressed in units of force (dynes) per unit length (cm).

Since the soap film attached to the wire has two surfaces (a front and a back face), the surface tension force exerted by the soap film on the wire in Fig. 9-2 equals $2L\,\gamma$, where γ is the soap film surface tension.

$$F = 2L\gamma \tag{1}$$

The behavior of the soap film on the wire assembly (under isothermal conditions) clearly indicates (*a*) that extension of the surface by pulling on the thread requires an input of mechanical work and (*b*) that the process is reversible (forces of extension and retraction and substantially equal). From this it follows that the work expended during stretching is stored up in the film during its extension. This stored energy, which resides in the front and back faces of the film, is termed free energy (thermodynamically speaking) in that it is recoverable. In magnitude, this stored free energy is equal to the work that is expended in the stretching process. It is equal to the reversible force applied $F(= 2L\gamma)$ multiplied by the distance x through which the force acts.

$$W(\text{work input}) = F \cdot x = 2L\gamma \cdot x \tag{2}$$

The new surface A created by stretching the film equals the effective wire length L multiplied by the distance x that the wire is moved under the application of force F.

$$A = 2Lx \tag{3}$$

It is this new surface that contains the newly created free surface energy. Hence the free surface energy stored per unit of new surface (the specific surface free energy E) is given by Eq. 4.

$$E = \frac{W}{A} = \frac{2L\gamma \cdot x}{2L \cdot x} = \gamma \tag{4}$$

Inspection of this equation reveals a most significant equivalence, namely, the liquid surface tension γ and the free energy residing in a unit area of liquid surface E are numerically and dimensionally equivalent quantities.

This unique equality permits the development of several important relationships that are presently discussed. Note that the free surface energy represents, not the total surface energy, but only the portion that is reversible in character.

SMOOTH VERSUS ROUGH SURFACES

It is important to distinguish between smooth and rough surfaces and to recognize that, whereas liquid surfaces are normally smooth or seek to aquire smoothness through flow, solid surfaces are normally rough. All surfaces strive to attain a minimal surface area, and, since the liquids flow, irregular or rough liquid surfaces that are openly exposed are soon reduced to smooth planar surfaces. Conversely, rough solid surfaces retain their roughness more or less indefinitely. The degree of roughness, rugosity, or irregularity i is conventionally measured by dividing the actual, true, or contour area A_i of the surface by its projected, grossly observed, outside geometric, or envelope area A_e. Surface roughness profiles have been measured and submitted by Hansen for a variety of metal substrates and coatings (1).

$$i = \frac{A_i}{A_e} \tag{5}$$

In general, for liquids, $i = 1.00$; for solids, $i > 1.00$.

EFFECT OF VAPOR PRESSURE ON SURFACE TENSION

A rigorous treatment of surface tension includes the influence of gases and vapor pressure on interfacial tension. However, except for liquids of high vapor pressure, the effect of the gaseous phase is generally negligible and will be excluded from the following analysis.

WETTING

In this book, the term "wetting" is used generically to embrace such specific wetting processes as adhesion, penetration, spreading, and cohesion. Each of these will be shown as a distinctively different type of wetting.

Free Surface Energy Can Involve Strong Wetting Forces

The normal manner of producing a new expanse of liquid is to extend a film that is already formed to a larger area by stretching (as by blowing a small soap bubble to a larger bubble, or by brushing out a paint over a substrate). However, it is possible to produce a new surface in the midst of a liquid by an outward pull (as by cavitation). In general, large forces are initially required to split apart at its center. Once it is disrupted, however, relatively small forces are usually sufficient to continue the surface extension.

Wetting Surface Energetics

The four different processes of adhesion, penetration, spreading, and cohesion will be defined and evaluated in terms of the surface energetics involved when smooth liquid surfaces contact rough solid surfaces.

Adhesion (Face-to-Face Contact)

Adhesion refers to the wetting situation where surfaces are brought into face-to-face contact. The conditions that apply to adhesive wetting are graphically illustrated in Fig. 9-3*a*, where an irregular solid surface of rugosity $i(= A_i/A_e)$ is brought into contact with a smooth liquid surface ($i = 1.00$). The model adhesion process considered here will be in terms of the free surface energies resident in the surfaces and interfaces before and after the liquid and solid surfaces are in contact. To simplify the analysis, the derivation will be based on a projected or smooth geometric area of 1.00 cm^2.

Before adhesive contact is made, the free energy resident in the smooth 1.00-cm^2 surface area of the liquid is γ_L (since the liquid surface tension and its specific free surface energy are equivalent). However, the free surface energy of the rough solid surface, corresponding to that resident in the contoured area under a geometric 1.00-cm^2 area, must be further modified by the roughness factor. Hence the free surface energy in the solid surface is $i\gamma_S$.

After adhesive contact is made, the situation resembles that depicted in the right section of Fig. 9-3*a*, where the final stage of adhesive wetting is illustrated. Note that the wetting may be incomplete. Let the area that is successfully contacted be a. If no contact is made, $a = 0$; if incomplete contact, $a = a$; if complete contact is established, $a = i$.

To wet the more extensive irregular, rough solid surface, the liquid surface (initially 1.00 cm^2) must be stretched. The extent of stretching will be in proportion to the solid surface area that the liquid succeeds in wetting in excess of 1.00 cm^2. This stretching factor is reflected in Eq. 7, which gives the expression for the final (postadhesion) stage of the adhesion process. Note that the area of liquid that fails to make contact (and remains unstretched) is equal to $(i - a)/i$.

Let the interfacial tension between the liquid and the solid be γ_I. Then the equations for the total free energies E_1 and E_2, applying to conditions before and after adhesion occurs, are given by Eqs. 6 and 7.

$$E_1 = \gamma_L + i\gamma_S \tag{6}$$

$$E_2 = \frac{\gamma_L(i - a)}{i} + \gamma_S(i - a) + \gamma_I a \tag{7}$$

Now, if $E_1 > E_2$, free energy is available, since more energy is present at the beginning than at the end. This results in spontaneous adhesion. If the reverse holds ($E_1 < E_2$), energy will have to be brought from the outside to promote the adhesion process.

Let ΔE be the difference between E_1 and E_2 for a wetting process.

$$\Delta E = E_1 - E_2 \tag{8}$$

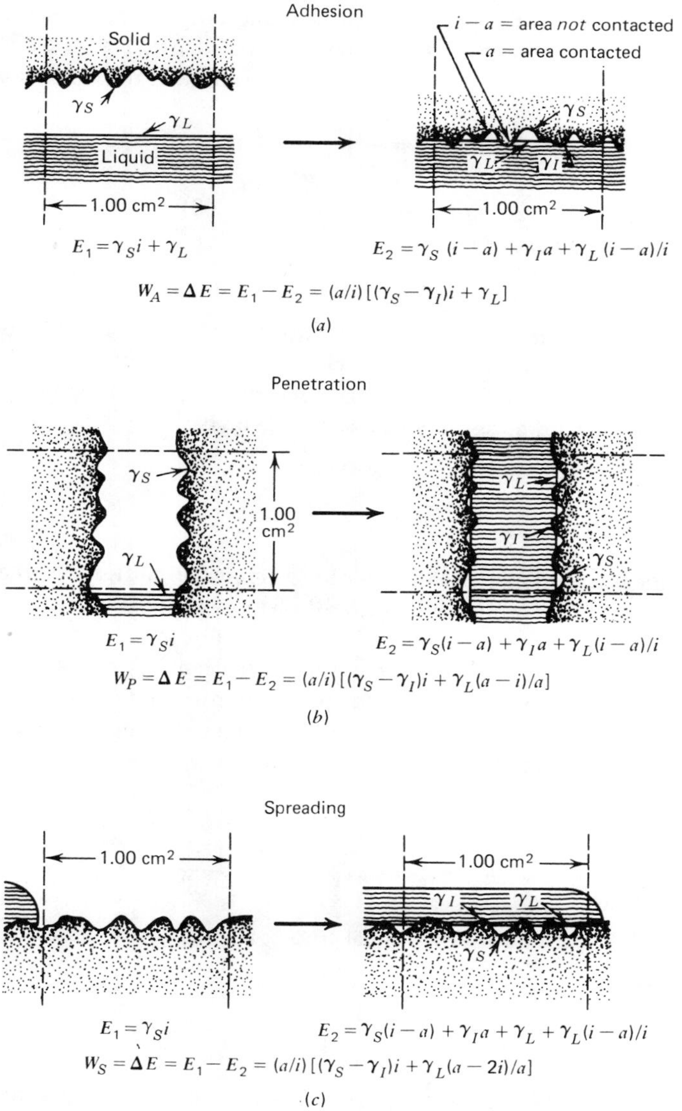

Fig. 9-3 Schematic diagrams illustrating the changes that occur during (*a*) the face-to-face adhesion of a liquid to a solid, (*b*) the penetration of a porous solid by a liquid, and (*c*) the spreading of a liquid over a solid surface. The paired left and right sections of each diagram represent conditions before and after the wetting process takes place.

As previously shown, if ΔE is positive ($E_1 > E_2$), the process is spontaneous; if negative ($E_1 < E_2$), the process is blocked and outside energy is necessary to force the wetting action along. Combining Eqs. 6 and 7 leads to the ΔE value of Eq. 9. This particular ΔE value is commonly referred to as the work of adhesion.

$$\Delta E = W_A = \frac{a}{i}\,[(\gamma_S - \gamma_I)\,i + \gamma_L] \tag{9}$$

If the solid surface is completely smooth ($i = 1.00$) and if complete adhesion is attained, Eq. 9 reduces to Eq. 10.

$$\Delta E = W_A = (\gamma_S - \gamma_I) + \gamma_L \tag{10}$$

Equation 10 is commonly given in the literature with no mention of the fact that it is applicable only to completely smooth solid surfaces. Without this qualification it can be misleading, since solid surfaces are rarely planar and adhesive contact cannot necessarily be assumed to be perfect.

Penetration

Penetration refers to the situation where a liquid permeates into the pores or open structure of a porous solid (such as a bed of pigment particles). The conditions that apply to penetration are graphically illustrated in Fig. 9-3*b*, where a liquid wets and works its way up the walls of the capillary of a porous solid. As in the case of adhesion, the model for penetration is considered in terms of the total free energies E_1 and E_2 that are resident in the surfaces and interfaces before and after penetration takes place.

$$E_1 = \gamma_S i \tag{11}$$

$$E_2 = \gamma_S(i - a) + \gamma_I a + \frac{\gamma_L (i - a)}{i} \tag{12}$$

The work of penetration W_P, equal to ΔE, is given by the difference between E_1 and E_2.

$$\Delta E = W_P = \frac{a}{i}\left[(\gamma_S - \gamma_I)\,i - \frac{\gamma_L (i - a)}{a}\right] \tag{13}$$

If the walls of the capillary are assumed to be completely smooth ($i = 1.00$) and if complete wetting penetration is attained, Eq. 13 reduces to Eq. 14.

$$\Delta E = W_P = \gamma_S - \gamma_I \tag{14}$$

Unless this qualification is given, Eq. 14 can be deceiving, since the internal capillaries of a porous solid can seldom be assumed to be completely smooth.

Spreading

Spreading refers to the situation where a liquid flows over a surface as a duplex film and not as a monomolecular layer. The conditions that apply to spreading are graphically illustrated in Fig. 9-3*c*, where a duplex film spreads over a rough solid surface. A duplex film is a film thick enough so that its two interfaces (liquid/solid and liquid/air) are independent and possess their own characteristic interfacial tensions. Duplex films can be only a few molecules thick. The model for spreading is considered in terms of the total free energies E_1 and E_2 that are resident in the surfaces and interfaces before and after spreading takes place.

$$E_1 = \gamma_S i \tag{15}$$

$$E_2 = \gamma_S(i - a) + \gamma_I a + \gamma_L + \frac{\gamma_L(i - a)}{i} \tag{16}$$

The work of spreading W_S, equal to ΔE, is given by the difference between E_1 and E_2.

$$\Delta E = W_S = \frac{a}{i}\left[(\gamma_S - \gamma_I)\, i - \frac{\gamma_L(2i - a)}{a}\right] \tag{17}$$

If the surface of solid over which spreading takes place is completely smooth (i = 1.00) and if complete wetting over the spread area is attained, Eq. 17 reduces to Eq. 18.

$$\Delta E = W_S = (\gamma_S - \gamma_I) - \gamma_L \tag{18}$$

More often than not this expression (Eq. 18) for the work of spreading is given in the literature with no explanation that it is applicable only to completely smooth solid surfaces. This can lead to confusion, since with the exception of some plastic and fused ceramic surfaces most solid surfaces are rough (wood, weathered metal, masonry, and the like).

Cohesion

The work of cohesion W_C refers to the special case of adhesion where two faces of the same liquid are brought into contact with each other. It corresponds to the decrease in free energy that occurs when the two faces of the liquid come into contact, adhere, and neutralize each other (their surface molecules become bulk molecules). Under these conditions the loss of free energy is complete, and the total expenditure of work for a geometric surface of 1.00 cm^2 (common to both) is equal to $2\gamma_L$.

$$W_C = 2\gamma_L \tag{19}$$

The several equations developed for the work of cohesion, adhesion, penetration, and spreading are given in Table 9-1 for ready reference. In all cases a plus W value indicates that the process is spontaneous. The listing is given in order of decreasing spontaneity (note the effect of the subtraction terms).

Two Regimes of Wetting Action Corresponding to $\gamma_S > \gamma_L$ and $\gamma_S < \gamma_L$

It is reasonable to assume that the free energies of liquid and solid surfaces must be mutually satisfied to some extent when the surfaces are brough into intimate contact. For example, if $\gamma_S > \gamma_L$, the interfacial free energy γ_I must also be less than γ_S. This means that for the wetting regime $\gamma_S > \gamma_L$ the expression in parentheses $(\gamma_S - \gamma_L)$ will always be positive. In view of the fact that a positive value for the work of wetting indicates a spontaneous process, it is seen by referring to Eqs. 20 and 21 that adhesion and penetration are assured when the solid surface tension exceeds the liquid surface tension, since only positive work

Table 9-1 Equations Relating Work of Cohesion, Work of Adhesion, Work of Penetration, and Work of Spreading to Surface and Interfacial Tensions and to Solid Surface Roughness and Fractional Amount of Interface Contact

Complete Equations

$$W_C = 2\gamma_L \qquad (19)$$

$$W_A = (a/i)\ [(\gamma_S - \gamma_I)\ i + \gamma_L] \qquad (9)$$

$$W_P = (a/i)\ [(\gamma_S - \gamma_I)\ i - \gamma_L (i - a)/a] \qquad (13)$$

$$W_S = (a/i)\ [(\gamma_S - \gamma_I)\ i - \gamma_L (2i - a)/a] \qquad (17)$$

Reduced Equations

Interface contact complete ($a = i$)

$$W_C = 2\gamma_L \qquad (19)$$

$$W_A = i(\gamma_S - \gamma_I) + \gamma_L \qquad (20)$$

$$W_P = i(\gamma_S - \gamma_I) \qquad (21)$$

$$W_S = i(\gamma_S - \gamma_I) - \gamma_L \qquad (22)$$

Interface contact complete ($a = i$) and solid surface completely smooth ($i = 1.00$)

$$W_A = (\gamma_S - \gamma_I) + \gamma_L \qquad (10)$$

$$W_P = (\gamma_S - \gamma_I) \qquad (14)$$

$$W_S = (\gamma_S - \gamma_I) - \gamma_L \qquad (18)$$

values can be obtained. Surface roughness acts only to promote these spontaneous wetting actions. Whether spreading occurs depends on whether $i(\gamma_S - \gamma_L)$ is larger than γ_L.

On the other hand, when $\gamma_S < \gamma_L$, no such generalizations can be made. From these statements it is obvious that it makes a tremendous difference whether $\gamma_S > \gamma_L$ or $\gamma_S < \gamma_L$. In the past much confusion has arisen because of a failure to differentiate between these two distinct regimes of wetting behavior. Since one regime merges into the other when $\gamma_S = \gamma_L$, this equality represents the critical point that separates the two. Different concepts and different procedures for handling wetting situations must be applied to either side of this critical point.

Spreading Coefficient

The term "coefficient" generally denotes a multiplying factor. However, in this instance, "spreading coefficient" is an alternative name given to the work of spreading over a completely planar surface with perfect wetting contact (see Eq. 18). Smooth solid surfaces are not often encountered in practice, but the validity of Eq. 18 can be checked experimentally by spreading a liquid of low surface tension (say a liquid hydrocarbon) over a liquid of high surface tension (say water). Here the two conditions of perfect contact ($a = i$) and smoothness ($i = 1.00$) are met. Let γ_W be the liquid with the higher surface tension in effect replacing the solid surface tension), and γ_L the liquid with the lower surface tension. Then Eq. 18 can be rewritten to express the tendency of a liquid of low surface tension (hydrocarbon) to spread over a liquid of high surface tension (water).

$$\text{Spreading coefficient} = W_S = (\gamma_W - \gamma_I) - \gamma_L \tag{23}$$

Since the three surface tensions of Eq. 23 are all susceptible to measurement, and since signs of spreading or retraction can be visibly observed when a drop of hydrocarbon is placed on a pool of water, experimental verification of Eq. 23 becomes possible. This type of testing has been carried out for a number of hydrocarbon liquids on water, and the results are plotted in Fig. 9-4. The correlation between predicted and observed behavior is excellent. This finding promotes confidence in the use of the other equations derived by theoretical reasoning.

Spreading coefficients have been further classified into initial, semi-initial, and final types. An initial spreading coefficient refers to the condition where the surfaces are fresh (uncontaminated liquids). A final spreading coefficient refers to a final condition where the liquids are mutually saturated with respect to each other. Initial spreading followed by retraction is possible as a result of the mu-

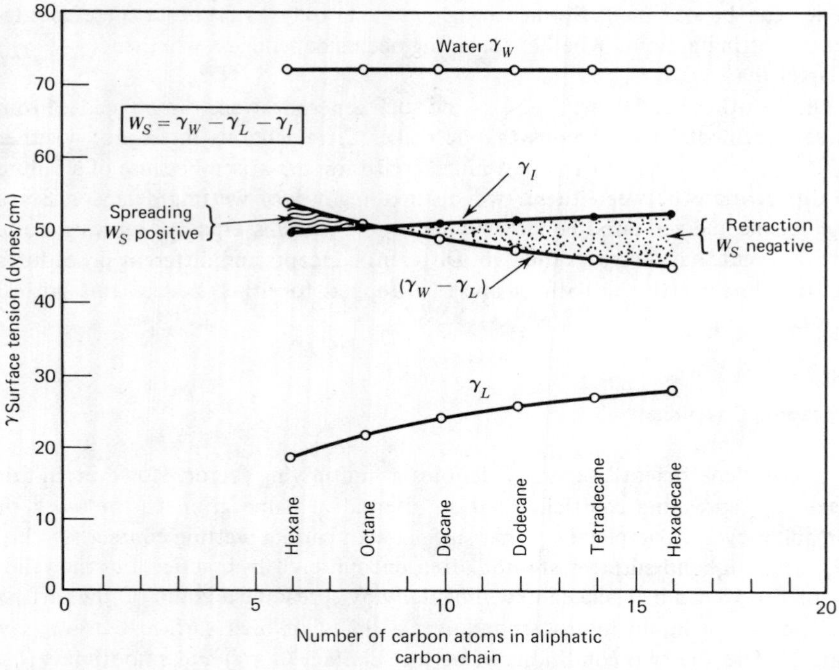

Fig. 9-4 Graph giving the surface and interfacial tensions for aliphatic hydrocarbons placed on water as a function of the number of carbon atoms in the aliphatic carbon chain. Regions of spreading and retraction are indicated by shaded areas.

tual saturation effect, although a mono-molecular layer may be left behind as part of the retraction process. A semi-initial spreading coefficient refers to some intermediate stage in the mutual saturation process.

Contact Angle and Work of Retraction

In preceding sections it was shown that for the regime $\gamma_S > \gamma_L$ spontaneous wetting is assured for both adhesion and penetration. Spreading may or may not occur.

For the regime $\gamma_S < \gamma_L$, wetting is less ubiquitous and the prediction of wetting behavior is more difficult. However, by taking advantage of the liquid/solid contact angle that forms when a liquid retracts over a solid, it is possible to calculate with reasonable accuracy the wetting behavior that can be expected when $\gamma_S < \gamma_L$.

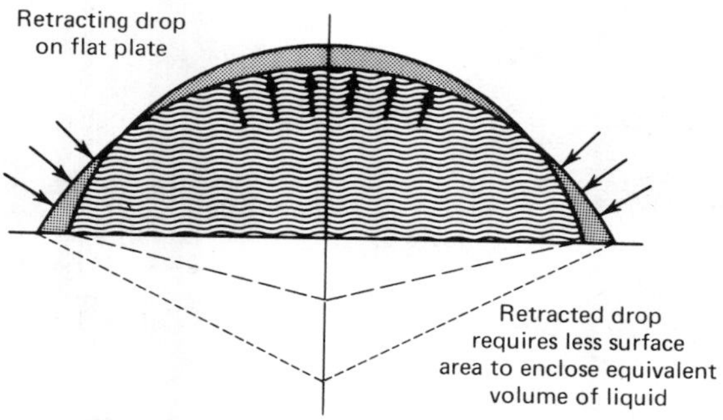

Fig. 9-5 Schematic diagram showing the geometry involved in the retraction of a liquid lens over a planar solid surface.

Inspection of Fig. 9-5 reveals that, when the liquid lens located on the solid surface retracts, its surface is diminished (the lens segment strives toward a condition of full sphericity). The reduction in liquid surface for a differential movement along the solid surface can be computed in terms of the contact angle that the liquid makes with the solid surface (contact angles are always measured through the liquid phase). Since the contact angle changes as retraction takes place, the process is not like the continuous wetting processes considered previously under the $\gamma_S > \gamma_L$ wetting regime. Instead, retraction of the lens proceeds to a point of equilibrium where the surface tension forces become exactly balanced.

The wetting conditions applying to liquid retraction are illustrated in Fig. 9-6, where the edge of a liquid lens is shown as retracting over an irregular solid surface of roughness i that has been completely wetted. The model process, however, will consider the more general situation where the solid surface is initially wetted over an area a. The retraction process is considered in terms of the free energies E_1 and E_2 that are resident in the faces and interfaces involved before and after retraction takes place. A geometric area of 1.00 cm^2, rather than a differential area, is taken for convenience of calculation.

$$E_1 = \gamma_L a + \gamma_S(i - a) + \frac{\gamma_L(i - a)}{i} + \gamma_L \cos \theta_i$$

$$E_2 = \gamma_S i$$

The work of retraction W_R is equal to ΔE, the difference between the initial and final energy expressions.

$$\Delta E = W_R = \frac{a}{i}\left[\gamma_I i + \frac{\gamma_L(i - a)}{a} + \gamma_L \cos \theta_i \left(\frac{i}{a}\right) - \gamma_S i\right] \qquad (24)$$

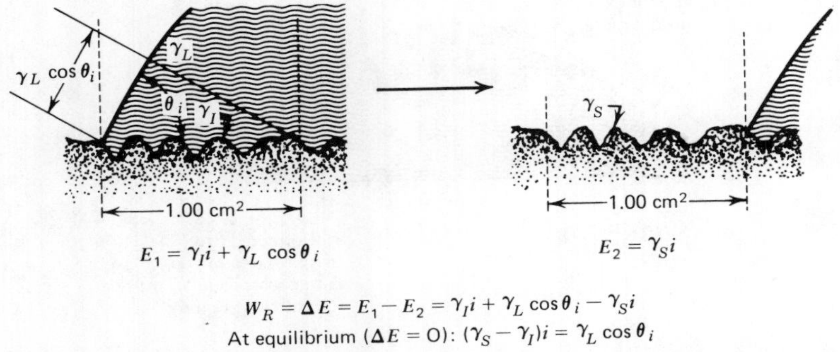

Fig. 9-6*a* Schematic diagram illustrating the change that occurs during the retraction of a liquid lens along an irregular solid surface representing conditions before and after retraction takes place.

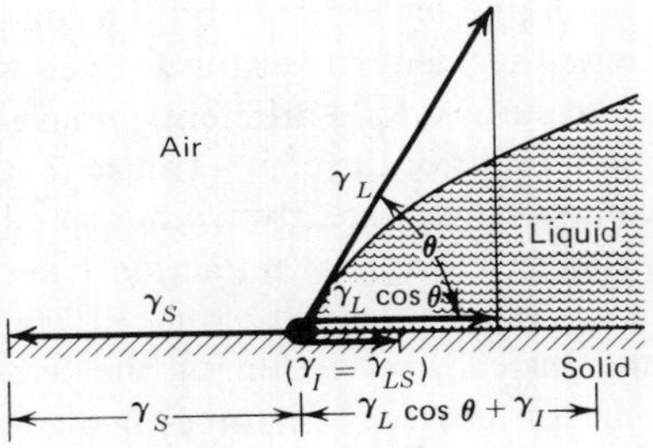

Fig. 9-6*b* Schematic vector diagram illustrating the surface tension forces acting at the vertex boundary line of a liquid contact angle on a planar solid surface.

For the equilibrium conditions corresponding to a stabilized contact angle θ_i over an irregular solid surface, there is neither a tendency for the liquid lens to retract further nor a tendency for it to spread. This occurs at the point where the work of retraction is zero ($W_R = 0$). Entering this condition into Eq. 24 and rearranging terms gives Eq. 25.

$$\gamma_L \left(\cos\theta_i + \frac{(i - a)}{i} \right) = (\gamma_S - \gamma_I)\, a \tag{25}$$

If complete wetting of the solid surface is assumed ($a = i$), Eq. 25 reduces to Eq. 26.

$$\gamma_L \cos\theta_i = (\gamma_S - \gamma_I)\, i \tag{26}$$

For a completely planar surface, Eq. 26 reduces still further to Eq. 27, where the absence of a subscript indicates a contact angle over a completely smooth surface.

$$\gamma_L \cos\theta = (\gamma_S - \gamma_I) \tag{27}$$

A classic diagram that is commonly shown for this planar surface condition in terms of vector forces is given by Fig. 9-6*b*. It is another way of visualizing the interplay of the surface tension forces that act at the vertex boundary line of a contact angle formed by a retracted liquid over a *smooth* solid surface (leads to same equality as given by Eq. 27).

Substituting the value of (γ_L cos θ) given by Eq. 27 for ($\gamma_S - \gamma_I$) in Eq. 26 gives a simple ratio for the rugosity factor in terms of the contact over the rough and smooth solid surfaces.

$$i = \frac{\cos\theta_i}{\cos\theta} \tag{28}$$

Wetting Behavior for the Regime $\gamma_S < \gamma_L$ in Terms of Contact Angle, Liquid Surface Tension, and Rugosity Factor

Although neither γ_S nor γ_I is easily susceptible to measurement, the difference between the two can be readily determined (Eqs. 26 and 27), and this is all that is necessary to set up work equations for adhesion, penetration, and spreading (or retraction) for the regime $\gamma_S < \gamma_L$. This is done by substituting the value of (γ_L cos θ_i) [or ($\gamma_L i$ cos θ)] for ($\gamma_S - \gamma_L$) i as given by Eq. 26 in the work equations of Table 9-1. This gives the set of work expressions listed in Table 9-2.

Table 9-2 Equations Relating Work of Adhesion, Work of Penetration, and Work of Spreading to Contact Angle, Liquid Surface Tension, and Rugosity Factor for the Regime $\gamma_S < \gamma_L$

Complete Equations

$$W_A = \gamma_L(\cos\theta_i + 1) = \gamma_L(i \cdot \cos\theta + 1) \tag{29}$$

$$W_P = \gamma_L(\cos\theta_i) = \gamma_L(i \cdot \cos\theta) \tag{30}$$

$$W_S = \gamma_L(\cos\theta_i - 1) = \gamma_L(i \cdot \cos\theta - 1) \tag{31}$$

Reduced Equations (Solid Surface Completely Planar, $i = 1.00$)

$$W_A = \gamma_L(\cos\theta + 1) \tag{32}$$

$$W_P = \gamma_L(\cos\theta) \tag{33}$$

$$W_S = \gamma_L(\cos\theta - 1) \tag{34}$$

The equations of Table 9-2 are most useful for predicting wetting behavior for the regime $\gamma_S < \gamma_L$. To facilitate visualization of the significance of these work equations there has been plotted in Fig. 9-7 a series of curves corresponding to contact angles from 0° to 180° and the rugosity factors ranging from a perfectly plane solid surface ($i = 1.00$) to solid surfaces with roughness factors of $i = 2.00$, $i = 3.00$, and beyond. The scale at the bottom gives the value of the functions of θ and i applying to adhesion, penetration, and spreading, respectively (see the expressions in parentheses in Eqs. 29, 30, and 31 in Table 9-2).

Since the maximum value for the cosine function is 1.00 (when the contact angle is 0°), it appears that a seemingly unreal situation arises when $\cos \theta_i$ as calculated from $i \cdot \cos \theta$ takes on a value greater than 1.00 ($\cos \theta_i = i \cdot \cos \theta > 1.00$). However, when this situation occurs, it means that the zero contact angle has already been reached and that the solid surface, by virtue of its roughness

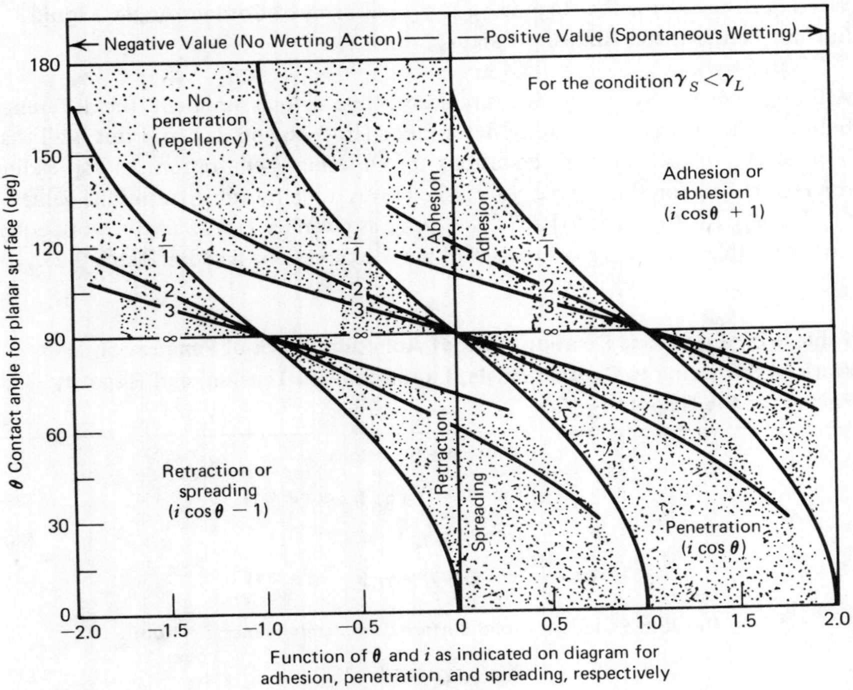

Fig. 9-7 Graph relating the contact angle over a smooth solid surface θ and the rugosity factor for a rough surface i to functions of θ and i that apply respectively, to adhesion ($i \cos \theta + 1$), penetration ($i \cos \theta$), and spreading ($i \cos \theta - 1$). Regions of adhesion versus abhesion, penetration versus repellency, and spreading versus retraction are shown by shaded areas.

has, in effect, gone beyond this point and shifted the wetting regime from $\gamma_S < \gamma_L$ to $\gamma_S > \gamma_L$. Angles corresponding to $i \cdot \cos\theta > 1.00$ may be imaginary, but the value of $i \cdot \cos\theta$ is still valid for estimating the work of adhesion, penetration, and spreading.

Note that in the work equations of Table 9-2 it is the sign of the expression in parentheses that determines whether the work is positive or negative (this is also noted at the top of Fig. 9-7). A positive value assures a spontaneous wetting action; a negative value, a blocked or reversed wetting action (abhesion rather than adhesion, repellency rather than penetration, and retraction rather than spreading).

Summary of Wetting Effects

From a study of the graph of Fig. 9-7 a number of important conclusions can be drawn that apply to the regime $\gamma_S < \gamma_L$. The contact angle referred to in the following summary is the contact angle over a perfectly smooth surface (θ).

Adhesion. Spontaneous adhesion always occurs for contact angles less than 90°. However, for a combination of a rough surface and a contact angle over 90°, adhesion may or may not occur. In fact, roughness becomes antagonistic to adhesion, and abhesion becomes more probable as roughness increases.

Penetration. Spontaneous penetration occurs for contact angles less than 90°, and does not occur for contact angles over 90°. The roughness of the solid surface accentuates either the penetration or the repellency action but has no influence on which type of wetting takes place.

Spreading. Retraction occurs for contact angles over 90° or over planar surfaces for any contact angle. However, spontaneous spreading for contact angles less than 90°, especially for small contact angles, may be induced by surface roughness.

Table 9-3 summarizes the conditions that lead to spontaneous wetting for both regimes of wetting behavior. Inspection of this information shows that adhesion is almost always obtained, and penetration generally obtained. Spontaneous spreading, however, is obtained for only a limited number of special situations.

PROBLEM 9-1

A 60° contact angle is given by a liquid over a smooth solid surface ($\theta = 60°$). What degree of roughness must be imparted to the solid surface to induce spontaneous spreading over the solid?

Table 9-3 Summary of Conditions That Lead to Spontaneous Wetting

	Surface Tension Regime					
	$\gamma_S > \gamma_L$		$\gamma_S < \gamma_L$			
Contact angle θ for smooth solid surface:			$<90°$		$>90°$	
Nature of surface:	Smooth	Rough	Smooth	Rough	Smooth	Rough
Type of Spontaneous Wetting						
Adhesion	Yes	Yes	Yes	Yes	Yes	[a]
Penetration	Yes	Yes	Yes	Yes	No	No
Spreading	[b]	[c]	No	[d]	No	No

[a] Yes for $i \cdot \cos\theta < -1.00$; no for $i \cdot \cos\theta > -1.00$.
[b] Yes for $(\gamma_S - \gamma_l) > \gamma_L$; no for $(\gamma_S - \gamma_l) < \gamma_L$.
[c] Yes for $i(\gamma_S - \gamma_l) > \gamma_L$; no for $i(\gamma_S - \gamma_l) < \gamma_L$.
[d] Yes for $i \cdot \cos\theta > 1.00$; no for $i \cdot \cos\theta < 1.00$.

Solution. Reference to Fig. 9-7 shows that rugosity values of $i > 2$ will induce spontaneous spreading (the intersection of the 60° contact angle line with roughness curves exceeding $i = 2$ all lie in the spontaneous spreading region).

The borderline roughness factor i separating the spreading and retraction regions could also have been obtained by setting W_S in the spreading equation (Eq. 31) equal to zero and solving for i.

$$W_S = 0 = \gamma_L (i \cdot \cos\theta - 1)$$

Set the expression in the parentheses equal to zero and solve for i.

$$i \cos 60° - 1 = 0; \qquad i = \frac{1}{0.5} = 2$$

MEASUREMENT OF LIQUID SURFACE TENSION

Three of the several methods for measuring surface tension (capillary tube, ring detachment, and drop weight) are routinely employed in industry. The physical principles applying to these procedures are based on relatively simple concepts. However, complications arise in reducing these principles to practice, so that the simple equations that are initially developed must be modified to provide accurate surface tension measurements.

Capillary Tube Method

The classical capillary rise (or depression) method for measuring liquid surface tension is probably the most familiar (2). It is capable of yielding very accurate results. The elementary theory considers that the weight of liquid pulled up to fill the cylindrical capillary tube is entirely supported by the surface tension force exerted by the liquid around the circumference of the meniscus contact line with the capillary wall (see Fig. 9-8*a*). The mass of the raised liquid (to a first approximation) equals $\pi r^2 L\rho$, where r is the inside capillary radius, L the length of the supported liquid column, and ρ the density of the liquid. Therefore the gravitation pull on this column of liquid is $\pi r^2 L\rho g$, where g is the gravitational acceleration (980 cm/sec^2). Opposing this downward force is the upward pull of surface tension, equal to value to $2\pi r\gamma_L \cos\theta$. The insertion of the cosine factor is necessary to give the vertical component of the surface tension force that acts directly upward to effectively support the liquid column weight.

Equating these two opposing forces and solving for the liquid surface tension yields Eq. 35 for equilibrium conditions.

$$\gamma_L = \frac{Lr\rho g}{2\cos\theta} \tag{35}$$

For the not uncommon case where the angle of contact θ is at or close to $0°$ ($\cos\theta = 1.00$) and where the bore of the capillary is sufficiently fine to give a hemispherical meniscus, it has been demonstrated that Eq. 35 can be written as Eq. 36, where the term $r/3$ corrects for the small weight of liquid in the meniscus region (2).

$$\gamma_L = \frac{(L + r/3)\, r\rho g}{2} \tag{36}$$

For approximate work the correction factor $r/3$ can be disregarded, leading to Eq. 37.

$$\gamma_L = \frac{Lr\rho g}{2} \tag{37}$$

PROBLEM 9-2

When one end of a clean capillary tube is partially immersed in water ($\gamma_W = 72.6$ dynes/cm), a liquid rise of 5.4 cm is observed. After cleaning, the same tube is similarly immersed in toluene ($\rho = 0.87$ g/cm^3), and a 2.1-cm rise is recorded. Assuming contact angles at or close to $0°$ for both liquids, calculate the surface tension of the toluene.

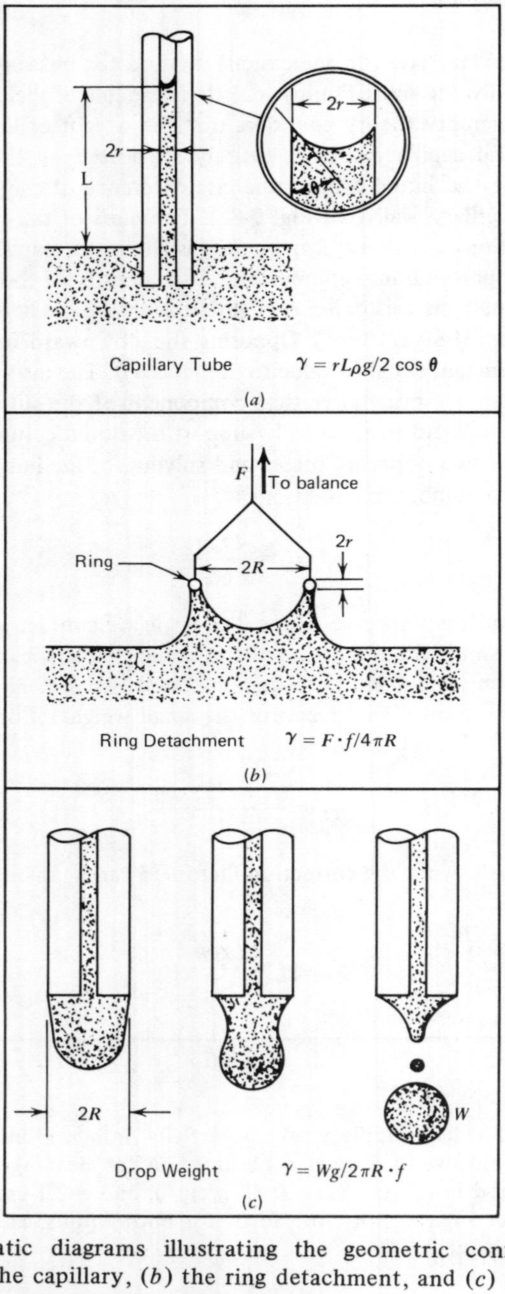

Fig. 9-8 Schematic diagrams illustrating the geometric configurations that characterize (*a*) the capillary, (*b*) the ring detachment, and (*c*) the drop weight methods for determining liquid surface tension.

Solution. Equation 37 can be rewritten as $rg/2 = \gamma_L/L\rho$. Since $rg/2$ is a constant for the capillary tube, the following equality holds.

$$\left(\frac{\gamma_W}{L\rho}\right)_{\text{water}} = \left(\frac{\gamma_L}{L\rho}\right)_{\text{toluene}}$$

Substitute the given data in the above equality to obtain the toluene surface tension.

$$\frac{72.6}{5.4 \cdot 1.00} = \frac{\gamma_L}{2.4 \cdot 0.87}; \qquad \gamma_L \text{ (toulene)} = 28 \text{ dynes/cm}$$

Comment. It is evident from this problem that a capillary tube can be assigned a calibration constant $K = rg/2$. This constant can be established by immersion of the tube in a liquid of known density and surface tension. The constant can then be used to determine the surface tensions of other liquids. In the above problem the capillary tube constant K is 13.44 (= 72.6/5.4). Note that the capillary inside radius must be 0.027 cm, since $r = 2K/g$.

RING DETACHMENT METHOD

In the ring detachment method a ring of thin metal wire (usually platinum) is horizontally positioned and then dipped below the surface of the test liquid. Thereafter it is gradually raised, as illustrated in Fig. 9-8*b*, until it becomes detached from the hanging liquid surface. In rising from the surface, the wire ring carries with it a film of liquid which is shaped like a hollow cylinder with an inner and an outer face. Liquid surface tension in the liquid film opposes the ring detachment by pulling down on the ring via both the inner and outer faces of the cylindrical liquid film. The total surface tension force opposing the ring detachment is equal to $4\pi R\gamma_L$, where R is the ring radius. The upward mechanical force required to detach the ring is equal to the difference between the total upward pull and the weight of the ring with its supporting framework (the ring assembly). Setting the net (effective) upward force F equal to the downward force of surface tension and solving for the surface tension gives Eq. 38.

$$\gamma_L = \frac{F}{4\pi R} \tag{38}$$

The derivation of Eq. 38 implicitly assumes that the cylindrical film lifted above the liquid surface is bounded by inner and outer walls that are vertically positioned. This assumption is actually quite erroneous, and hence a correction must be introduced to compensate for wall curvature complications. From both empirical experimentation and an elegant mathematical treatment (3, 4) it has been shown that a correction factor can be derived as a function of two dimensionless ratios, R/r and R^3/V, where R/r is the ratio of the ring to the wire radius and R^3/V is the ratio of the cube of the ring radius to the volume of the liquid lifted above the bulk liquid surface just before its detachment. Use of this correction technique permits accurate measurement of liquid surface tension by the ring detachment method, as with the du Nouy tensiometer (see ASTM D1331-56; reapproved 1975).

Drop Weight Method

Although proposed by Tate over 100 years ago (in 1864) (5), the drop weight method for measuring surface tension remains an amazingly simple and inexpensive procedure that is capable of yielding highly accurate liquid surface tension values. As its name implies, the method involves the gradual formation of a drop of test liquid at the flattened end of a small-bore tube, as illustrated in Fig. 9-8*c*. When the drop reaches a size where the surface tension can no longer support the weight of the drop, the drop breaks away and falls from the flat face at the end of the tube, at which point it is caught and weighed. A number of counted drops are allowed to collect in a tall, narrow vial until a sufficient quantity of liquid has been accumulated to permit accurate calculation of the weight of a single droplet.

Equating the gravitational and surface tension forces, which are exactly in balance when drop detachment occurs, leads to Eq. 39.

$$Wg = 2\pi R \gamma_L \tag{39}$$

The weight of the drop (in g) in Eq. 39 has been multiplied by the gravitational factor g to give the downward pull on the drop in dynes. By reference to Fig. 9-8*c* it is seen that this downward gravitational pull is opposed by the liquid surface tension acting around the outside circumference of the flat end of the tube ($= 2\pi R$).

In practice the actual drop weight is less than for the ideal drop conceived of in the model situation. This becomes immediately evident on observing the stages in the formation of the drop as schematically shown in Fig. 9-8*c*. Thus the actual detachment arises from the instability of the cylindrical neck that develops during drop formation. Hence the radius of this narrow section is the correct one to use in computing the liquid surface tension of the drop. In addition, a portion of the drop retracts at the time of breakaway and is retained at the end of the tube.

A mathematical treatment of the geometrics of drop formation was worked out by Lohnstein (6), and appropriate correction factors based on this derivation have been calculated and reported. Later, Harkins and Brown (7) developed a set of empirically derived correction factors based on extensive experimental data, and these are commonly employed today for correcting the drop weight type of surface tension measurement.

The correction factor f, as read from a table or graph, is then entered into the Tate equation (Eq. 39) to give the modified expression in Eq. 40.

$$Wg = f \cdot 2\pi R \gamma_L \tag{40}$$

The experimentally derived correction factor f is tabulated or graphed as a function of the dimensionless ratio $R/V^{1/3}$, where V is the volume of a single drop. Hence the use of this procedure means that, in addition to knowing the outside tube radius R and the drop weight W, the liquid density ρ must also be determined in order to calculated the drop volume ($V = W/\rho$).

Simplified Drop Weight Tensiometer Design. To circumvent these several steps at some small sacrifice in accuracy, although better than ±1.0% accuracy is retained, this author has proposed a much simpler technique for the engineer working with ink, paint, and coating materials (8). The key to this simplified method lies in the selection of a *critical* radius for the flat end of the capillary tube, which permits the assignment of a constant value of 0.606 for the correction factor. This factor is applicable to solvents, coatings, inks, and liquid materials in general. As developed by this author, this critical outside radius for the flat bottom end of the detachment tube is 0.274 cm. When these data are inserted into Eq. 40, along with a gravity acceleration of 980 cm/sec^2, there results the deceptively simple Eq. 41. Note that liquid density has been eliminated as a factor in calculating the liquid surface tension.

$$\gamma_L = 939W \tag{41}$$

A tensiometer design conforming to the above specifications is schematically diagramed in Fig. 9-9. It is simple and can be constructed at small cost. The test liquid is sucked up into the bulb above the capillary tube and then allowed to flow back dropwise into the vial container, the rate of drop detachment being regulated by the pinchcock clamp.

Three precautions must be observed:

Cleanliness. It takes only a minor amount of contaminant to vitiate a surface tension measurement. Make certain you are working with clean glass.

Slow Drop Formation. Arrange for slow drop formation that is free from vibration (a drop every 10 to 20 sec or more, but no less). If surfactants are present in minute amounts, still longer detachment times may be necessary.

Volatility. If the liquid is volatile, schedule an immediate weighing after the last drop is collected (and possibly make a correction for calculated volatile loss).

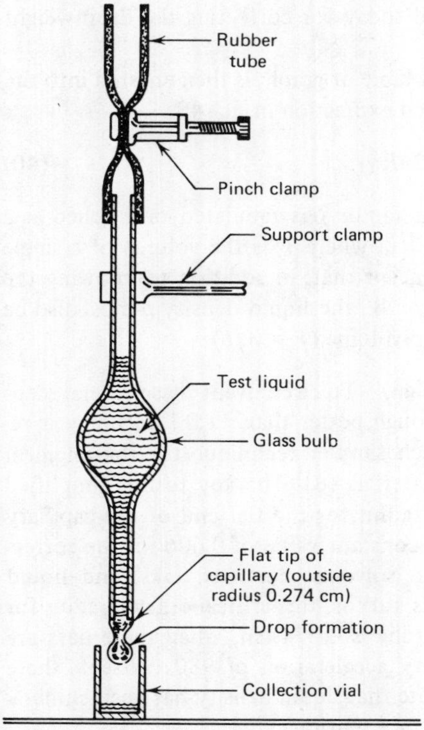

Fig. 9-9 Schematic diagram of a simple "drop weight" tensiometer designed for the routine measurement of liquid surface tensions (critical outside radius of capillary tip 0.274 cm).

For further details regarding the derivation and use of this unique modification of the drop weight method for determining liquid surface tension refer to the original paper (8).

PROBLEM 9-3

Twelve drops of paint solvent, collected over a 5-min period from a drop weight tensiometer (critical outside bottom radius 0.274 cm) weigh 0.362 g. Calculate the solvent surface tension.

Solution. The weight of a single drop is 0.0302 g (= 0.362/12). Substitute this single droplet weight in Eq. 41 to obtain the solvent surface tension.

$$\gamma_L = 0.0302 \cdot 939 = 28.5 \text{ dynes/cm}$$

Comment. The surface tension of this test solvent (toluene) as given in the literature is 28.4 dynes/cm.

If the bottom outside radius of the capillary tube is nearly but not exactly 0.274 cm, the tube can still be used by calibrating it with a liquid or liquids of known surface tension (e.g., water).

PROBLEM 9-4

It is desired to calibrate a drop weight tensiometer with an outside radius that is just off the critical value of 0.274 cm. Ten drops of water collected from the tensiometer over an 8-min period are found to weigh 0.777 g. Compute the calibration constant for the instrument.

Solution. The weight of a single drop of water (surface tension 72.6 dynes/cm) is 0.0777 g. Substitute this value in Eq. 41, and solve for the calibration constant.

$$72.6 = K \cdot 0.0777; \qquad K = 934$$

Comment. As a general rule it is wise to check any new tensiometer with liquids of known surface tension to confirm the 0.274-cm critical radius or, alternatively, to develop a calibration constant that will compensate for any slight noncomformity.

SOLID SURFACE TENSION

In 1805 Young postulated that, when two surfaces are brought into intimate contact (e.g., two immiscible liquids or a liquid and a solid), the surface with the lower free energy content neutralizes (to the extent of its free energy content) the free energy content of the surface with the higher free energy (9). This is equivalent to saying that the interfacial tension developed at an interface during a wetting process is approximately equal to the difference between the two initial surface tensions involved (in effect, the interfacial tension represents the left-over portion of the higher surface tension that is not neutralized).

Expressing this concept in algebraic notation for a liquid/solid interface for the regime $\gamma_S < \gamma_L$ gives Eq. 42, where γ_I is the interfacial tension.

$$\gamma_I = \gamma_L - \gamma_S \tag{42}$$

Young's neutralization principle is a simple idea and hence is appealing to the coatings engineer. However, it has been the source of much controversy over the years. To answer critics of the rule, Antonoff demonstrated in 1942 that the principle holds exactly for immiscible liquids that are mutually saturated with respect to each other (10). The specified condition of mutual saturation apparently had been disregarded by many investigators who had challenged the validity of the principle. Graphs illustrating the nonconformity and the conformity to the rule for fresh and mutually saturated solutions, respectively, are given for water/organic solvents in the literature (8).

In the case of a liquid/solid interface, mutual saturation is improbable. However, by assuming that neutralization is still approximately attained, it is possible to derive some useful results.

If the value of γ_I as given in Eq. 42 is substituted in Eq. 26, Eq. 43 results.

$$\gamma_L \cos \theta_i \approx (\gamma_S - \gamma_L + \gamma_S)\, i$$

$$\gamma_S \approx \frac{\gamma_L(\cos \theta_i + i)}{2i} \tag{43}$$

For a completely planar surface Eq. 43 reduces to Eq. 44.

$$\gamma_S \approx \frac{\gamma_L\,(\cos \theta + 1)}{2} \tag{44}$$

The approximate relationships of Eqs. 43 and 44 afford a means of obtaining reasonable values for the surface tension of solid surfaces. In the case of Eq. 44 both the contact angle and the liquid surface tension are readily susceptible to measurement.

If the contact angle is 90° (cos 90° = 0), the solid surface tension is approximately half of the liquid surface tension for both smooth and rough solid surfaces.

$$\gamma_S \approx \tfrac{1}{2}\,\gamma_L \text{ (90° contact angle)}$$

It was noted earlier that spontaneous penetration does not occur for contact angles over 90°. Hence the above expression indicates that, when $\gamma_S < \frac{1}{2}\gamma_L$, spontaneous penetration is blocked. This same conclusion was also reached by Shafrin and Zisman, who state that in designing liquid/solid systems in which capillary penetration is not possible the solid must be chosen so that $\gamma_S < \frac{1}{2}\gamma_L$ (11).

Critical Solid Surface Tension

If the contact angle is 0° (cos 0° = 1.00), the solid surface tension becomes substantially equal to the liquid surface tension for smooth solid surfaces (but not rough surfaces).

$$\gamma_S \approx \gamma_L \text{ (0° contact angle over smooth surface)}$$

On the basis of this observation, Zisman developed a method for ascertaining the surface tension of a smooth solid surface (12). The procedure consists in applying a series of liquids (preferable homologous) of known surface tension to the smooth test surface and measuring the contact angles that are obtained. The cosines of these contact angles are then plotted against the surface tensions of the respective liquids. Extrapolation of these data to a zero contact angle (cos 0° = 1.00) gives the solid surface tension, since at this point $\gamma_S \approx \gamma_L$. Zisman calls the surface tension obtained by this procedure a *critical* solid surface tension. In effect this procedure assumes that Young's neutralization principle is operating at the zero contact angle, a point where the solid and liquid surface tensions exactly match.

An involved empirical expression relating γ_L, γ_S, and θ was developed and reported in 1965 by a group of German investigators (13). It is stated to be valid when liquids that do not react chemically with the substrate are brought into contact with low-energy solid surfaces such as painted areas and plastic substrates. In the literature, diagrams of γ_S as a function of the liquid contact angle have been published (14).

In analyzing the reported material, this author found that essentially the same value for a solid surface tension could be obtained much more simply from the contact angle and liquid surface tension by using Young's equation with only slight modification.

$$\gamma_S = \frac{(\gamma_L - \theta/8)(\cos\theta + 1)}{2} \tag{45}$$

For solid surface tensions in the range from about 20 to 60 dynes/cm (those of most concern to the coatings engineer) Eq. 45 gives γ_S values that deviate only slightly (±1 dyne/cm) from the ones calculated by using the much more complex empirical equation developed by the German group.

PROBLEM 9-5

The contact angles tabulated below were determined for three liquids on a paint (two-component isocyanate/acrylic type), on a plastic (polyethylene), and on electrolytic tinplate (with fair lacquer wetting properties). Calculate the average solid surface tensions for the three solid surfaces.

Solid Surface	Water (W)	Formamide (F)	Ethylene Glycol (EG)
Liquid:	Water (W)	Formamide (F)	Ethylene Glycol (EG)
Surface tension (dynes/cm):	72	58	48
Paint	80°	60°	53°
Plastic	91	71	61
Tinplate	72	51	44

Solution: Substitute the appropriate data in turn in Eq. 45, and solve for the solid surface tensions.

Sample calculation (water/paint):

$$\gamma_S = \frac{(72 - 80/8)(\cos 80^\circ + 1)}{2}$$

$$= \frac{(72 - 10)(0.174 + 1)}{2} = 36 \text{ dynes/cm}$$

Calculated Solid Surface Tension (dynes/cm)

Solid Surface	Individual Values W	F	EG	Average
Paint	36	38	33	36
Plastic	29	32	30	30
Tinplate	41	42	36	40

SURFACE DEWETTING (CRAWLING OR BEADING UP)

Dewetting refers to the crawling or beading up that may occur when a liquid or coating is applied as a *film* to a solid substrate. This is different from the application of a *droplet* of the liquid or coating to the substrate. In the first case (dewetting) any contact angle formed is a receding one over a previously wetted surface; in the second case any contact angle formed is an advancing one over a dry surface. In practice, receding contact angles are usually found to be significantly less than advancing angles. At borderline conditions a receding angle may be absent (the applied coating remains intact as a spread film), yet an advancing angle is observed (an applied droplet fails to spread spontaneously).

Spontaneous spreading (or absence of an advancing contact angle) is the safest condition to ensure the bonding of a commercial coating to a solid substrate. However, absence of a receding contact angle (no crawling) may still be acceptable as an application condition even though an advancing contact angle exists.

Characterization of Solid Surface Tensions in Terms of Spreading Liquids

At least two procedures have been proposed for characterizing solid surface tensions in terms of the speading or nonspreading (crawling) behavior of selected liquids when applied as *films* to solid substrates. These methods determine the point at which dewetting (film breakup) first occurs, rather than determining a zero contact angle (at which point spontaneous spreading first occurs). Although characterization of a solid surface by this technique is quite approximate compared with the much more theoretical contact angle approach, it has the advantage of practicality (being a direct, fast, simple, and inexpensive procedure that provides immediate and useful information for coating and ink work).

A standard test for measuring the wetting tension (solid surface tension) of polyethylene and polypropylene films is specified by ASTM 2578-67 (reapproved 1972). Although designed primarily for rapidly evaluating the ability of these plastic films to retain inks, coatings, and adhesives, the method has much wider applicability. As with the Zisman method for obtaining a solid surface tension, the assumption is made that when the liquid/solid contact angle is 0°, the wetting (surface) tension of the solid surface equals that of the liquid surface. However, in this very practical method the technique used to arrive at a zero contact angle is somewhat unique.

Two reagent-grade liquids, formamide ($HCONH_2$) with a surface tension of 58 dynes/cm and ethylene glycol monoethyl ether ($CH_3CH_2OCH_2CH_2OH$) with a surface tension of 30 dynes/cm, are mixed in controlled proportions to yield a series of liquid mixtures with surface tensions ranging from 30 to 58 dynes/cm in 2-dyne/cm steps (a table showing the required proportions is given with the specification).

The very tip of a cotton applicator is wet with an amount of one of the mixtures just sufficient to permit spreading over 1 in.^{2} of plastic surface with no excess present. If a continuous liquid film is formed that holds for 2 sec or more, the liquid with the next higher surface tension is applied to a new surface of the film; if the film breaks into droplets in less than 2 sec, the next lower surface tension liquid is applied. Proceeding in the direction indicated by these initial tests (and using a new applicator each time), a mixture is selected that comes closest to wetting the solid plastic surface for 2 sec. The surface tension of this liquid mixture is taken as the solid wetting (surface) tension.

Since formamide and ethylene glycol monoethyl ether are strong solvents for many organic materials, their use may be contraindicated for application to some surfaces. However, the outlined technique represents a relatively simple procedure for determining a solid surface tension.

In the case of polyethylene, a solid wetting tension of 35 dynes/cm is normally regarded as acceptable for flexographic printing. Since smooth untreated polyethylene has a solid surface tension of about 30 dynes/cm, it follows that some type of surface treatment such as roughening is required to render the polyethylene plastic suitable for this type of printing.

A somewhat similar scheme for characterizing the solid surface tensions of paint substrates has been outlined by Hansen, based on the use of eight selected liquids of varying surface tension (15):

Water	72.8	Dimethyl sulfoxide	43.0
Formamide	56.0	Dimethyl formamide	35.2
Ethylene cyanohydrin	44.4	*n*-Methyl-2-pyrrolidone	39.0
2-Pyrrolidone	37.6	Acetone	23.7

As in the preceding test, the test liquid is applied as a film to the candidate substrate with a cotton-tipped applicator. A notation is made as to whether the liquid remains intact, crawls, or beads up completely. Experience has shown that results from this dewetting test correlate with major differences in the bonding performance of coatings in practice.

Incidentally, it has been found that the dried films of most trade sales products, including water-based coatings, are dewet by water. This is of interest from the standpoint of dirt collection, since approximately six times more water may be collected on a non-wetting vertical surface (by beading up) than on a vertical surface that the water wets and then runs off. As a result hydrophobic surfaces tend to be consistently dirtier than hydrophilic surfaces (15). Even roughness (which tends to encourage spreading) promotes a better appearing surface. This runs counter to a common conception that roughness promotes dirt retention.

CALCULATION OF A LIQUID SURFACE TENSION FROM ITS CHEMICAL STRUCTURE

The surface tension of an organic liquid is related to its chemical structure by a so-called liquid parachor. A liquid parachor $[P]$ is defined by Eq. 46.

$$[P] = \frac{M\gamma_L{}^{1/4}}{\rho - \rho_v} \tag{46}$$

In Eq. 46 M is the liquid molecular weight, and ρ and ρ_v are the densities of the liquid vapor, respectively. For approximate calculation work, the liquid vapor density can be excluded from consideration, since it is generally negligible in comparison with the liquid density. Introducing this simplification and solving for γ_L gives Eq. 47.

$$\gamma_L = \left(\frac{[P]\,\rho}{M}\right)^4 \tag{47}$$

A liquid parachor is an additive function in that it is obtained by summing the individual parachor values contributed by the several parts of the liquid chemical

Table 9-4 Representative Parachor Values

Elements (atomic weight in parentheses)					
C	Carbon (12.0)	4.8	S	Sulfur (32.1)	48.2
H	Hydrogen (1.0)	17.1	F	Fluorine (19.0)	25.7
N	Nitrogen (14.0)	12.5	Cl	Chlorine (35.5)	54.3
O	Oxygen (16.0)	20.0	Br	Bromine (80.0)	68.0
P	Phosphorus (31.0)	37.7	I	Iodine (127)	91.0

Organic Structures			
Bond		Ring structure	
Double	23.2	Three-member	16.7
Triple	46.6	Four-member	11.6
Ester linkage	-3.2	Five-member	8.5
		Six-member	6.1

structure. Typical parachor values for selected elements and structures are listed in Table 9-4. The procedure for calculating a liquid surface tension using parachor values is illustrated by the following problem.

PROBLEM 9-6

Calculate in turn the surface tensions of ethyl acetate, ethylamine, and toluene.

Solution. List the structural parts of each organic liquid in tabular form, with the associated parachor value of each part, and add the individual values to obtain a total parachor value. Use this total value to calculate the liquid surface tension, applying Eq. 47.

Ethyl Acetate ($CH_3CO_2C_2H_5$): $M = 88.1$; $\rho = 0.901$ g/cm^3

Structural Part	Number	Parachor Value	Total Contribution
C	4	4.8	19.2
H	8	17.1	136.8
O	2	20.0	40.0
Double bond	1	23.2	23.2
Ester linkage	1	-3.2	-3.2
			[P] = 216.0

$$\gamma_L \text{ (ethyl acetate)} = \left(\frac{216 \cdot 0.901}{88.1}\right)^4 = 23.8 \text{ dynes/cm}$$

Ethyl amine ($C_2H_5NH_2$): $M = 45.1; \rho = 0.69$ g/cm^3

Structural Part	Number	Parachor Value	Total Contribution
C	2	4.8	9.6
H	7	17.1	119.7
N	1	12.5	12.5
			$[P] = 141.8$

$$\gamma_L \text{ (ethyl amine)} = \left(\frac{141.8 \cdot 0.69}{45.1}\right)^4 = 22.2 \text{ dynes/cm}$$

Toluene (C_7H_8): $M = 92.1; \rho = 0.87$ g/cm^3

Structural Part	Number	Parachor Value	Total Contribution
C	7	4.8	33.6
H	8	17.1	136.8
Ring (6)	1	6.1	6.1
Double bond	3	23.2	69.6
			$[P] = 246.1$

$$\gamma_L \text{ (toluene)} = \left(\frac{246.1 \cdot 0.87}{92.1}\right)^4 = 29.2 \text{ dynes/cm}$$

Comment. These calculated values for the liquid surface tension vary by less than 1.0 dyne/cm from experimentally determined values.

MODIFYING SURFACE TENSION INFLUENCES

Up to this point, surface tension has been treated from a more or less simplified and ideal standpoint. In actual practice, however, there are many possible modifying factors. In particular, the tremendous influence of surfactants and their adsorption at the interface are the subject of a separate chapter (Chapter 13). Among other factors that modify surface tension phenomena are equilibrium versus nonequilibrium conditions, the vapor phase, advancing versus receding contact angles, and spreading by monomolecular films.

Equilibrium Versus Nonequlibrium Conditions

A differentiation should be made between the interfacial tension between surfaces on initial contact (nonequilibrium conditions) and that present after standing for some time (equilibrium conditions). As already pointed out, the interfacial tension between two uncontaminated (or neat) liquids on initial contact and the tension existing some time later when the two liquids are mutually saturated can be quite different.

Vapor Phase

Liquid vapor is a third phase that is normally present over exposed liquid and solid surfaces but has been neglected here for the sake of simplification. For liquids of low vapor pressure, the effect of vapor pressure can be safely ignored. For liquids of high vapor pressure, γ_S and γ_L can be considered as surface tension values (actually interfacial tension values) corresponding to the solid and liquid surfaces in contact with the liquid vapor. Except for research studies the influence of vapor pressure is usually not significant.

Advancing Versus Receding Contact Angles

The contact angle displayed by liquid advancing over a previously unwet solid surface may be significantly different from the contact angle displayed by a liquid receding over a solid surface that has previously been in intimate contact with the liquid. This discrepancy is referred to as contact angle hysteresis. An advancing contact angle is usually greater than a receding angle. Very clean, smooth surfaces show no hysteresis, suggesting that contact angle hysteresis is induced by surface contamination and/or surface roughness (12).

Spreading by Monomolecular Film

Spreading may occur by a monomolecular film, and yet not by a duplex film. This advance spreading by the monofilm may in turn induce a later spreading via a duplex film. On the other hand, liquids that are unable to spread by their own adsorbed oriented monolayers exist and are referred to as autophobic liquids (12). In certain instances the spreading monolayer may undergo chemical reaction (say hydrolysis of an ester), leading to the formation of products that block further spreading (12).

REFERENCES

1. Hansen, C. M., "Surface Roughness Profiles and Coatings Performance," *JPT*, **44**, No. 570, 61–66, July (1972).
2. Adamson, A. W., *Physical Chemistry of Surfaces*, Interscience, New York, 1960.

3. Harkins, W. D. and H. F. Jordan, *J. Am. Chem. Soc.*, **52,** 1751 (1930).
4. Freud, B. B. and H. Z., *J. Am. Chem. Soc.*, **52,** 1772 (1930).
5. Tate, T., *Phil. Mag.*, **27,** 176 (1864).
6. Lohnstein, Th., *Ann. Phys.*, **22,** No. 4, 767 (1907) *Z. Phys. Chem.*, **84,** 410 (1913).
7. Harkins, W. D. and F. E. Brown, *J. Am. Chem. Soc.*, **41,** 499 (1919).
8. Patton, T. C., "Reflections of a Paint Engineer on Paint Flow, Interface Physics, and Pigment Dispersion," *JPT,* **42,** no. 551, 666–694, December (1970).
9. Young, Thomas, "An Essay on the Cohesion of Fluids, III," *Phil. Trans. R. Soc. London,* Series A, **65,** 65 (1805).
10. Antonoff, G., "On the Validity of Antonoff's Rule," *J. Phys. Chem.*, **46,** 497 (1942).
11. Shafrin, Elaine G. and W. A. Zisman, "Upper Limits to the Contract Angles of Liquids on Solids," in *Advances in Chemistry* Series, No. 43: *Contact Angle, Wettability, and Adhesion,* American Chemical Society, Washington, D.C., 1964, p. 154.
12. *Advances in Chemistry* Series, No. 43: *Contact Angle, Wettability, and Adhesion,* American Chemical Society, Washington, D.C., 1964.
13. Driedger, A. W. et al., *Kolloid-Z Polym.,* **201,** No. 1, p. 52, and No. L/2, p. 101 (1965).
14. Lindberg, B., "Painting on Plastic Materials," *JOCCA,* **58,** 408–413, November (1975).
15. Hansen, C. M., "Surface Dewetting and Coatings Performance," *JPT,* **44,** No. 570, 57–60, July (1972); also "Characterization of Surfaces by Spreading Liquids," *JPT,* **42,** No. 550, 660–664, November (1970).

10 *Work of Dispersion, Work of Transfer (Flushing), and Work of Flocculation*

WORK OF DISPERSION

The expenditure of energy that attends the submergence of a solid in a liquid (a dispersion process) is described here in terms of the submergence of a 1-cm^3 cube in a liquid. Bringing the solid cube from a point outside the liquid to a point inside is visualized as a model three-stage process involving in turn the wetting processes of adhesion, penetration, and spreading, as schematically illustrated in Fig. 10-1. The sum of these separate inputs of energy equals the work of dispersion W_D.

Consider first the energy associated with the cube before it is brought into contact with the liquid. The cube has six faces, each with a surface area of 1 cm^2 and free energy content γ_S. When one face of the cube is brought into contact with the surface of the liquid, the work of adhesion is given by $W_A = i(\gamma_S - \gamma_I) + \gamma_L$. As the cube enters the liquid and is submerged to the stage where its top face is just level with the liquid surface, the work of penetration involved (four faces) is given by four times the unit work of penetration: $W_P = 4i(\gamma_S - \gamma_I)$. The last stage in the dispersion process involves the overrunning of the top face of the cube (one face), where the work of spreading is given by $W_S = i(\gamma_S - \gamma_I) - \gamma_L$. The total work or work of dispersion is given by the sum of the three separate work inputs.

$$W_D \text{ (1-cm}^3\text{ cube)} = 6i(\gamma_S - \gamma_I) \quad (1)$$

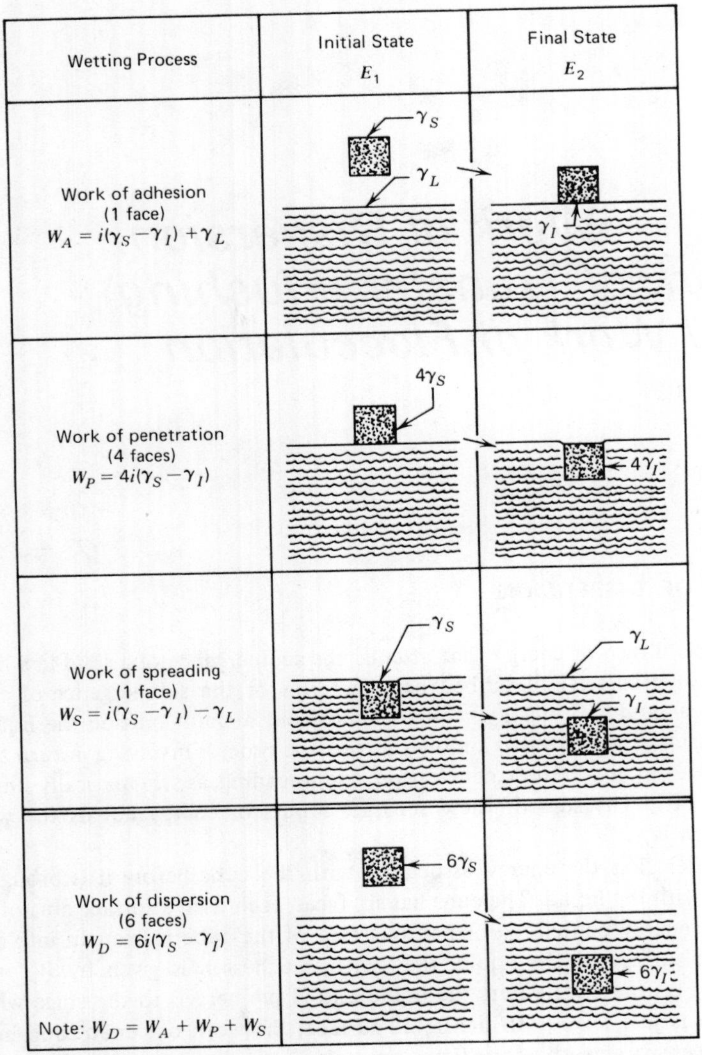

Fig. 10-1 Schematic illustration of the positioning of a solid cube during the three intermediate stages leading to its submergence in a liquid.

Although a 1-cm^3 cube was purposely selected to clearly illustrate the three-stage dispersion process, the results are universally applicable to the dispersion of a pigment particle of any shape in a vehicle. Although the end result could have been obtained directly from a consideration of the overall energy expenditure before and after submergence, this was purposely not done for the reason that some intermediate stage of the dispersion operation may possibly retard or block the attainment of a final dispersion. By focusing on the stepwise nature of the dispersion process, the reason and the cure for an intermediate dispersion difficulty become more readily apparent. Each consecutive stage in the sequence of adhesion, penetration, and spreading requires relatively more work input. Therefore it is possible that a pigment particle may penetrate into a liquid with ease but resist submergence.

WORK OF TRANSFER (FLUSHING)

Pigment flushing refers to the direct transfer of pigment particles that have been precipitated or prepared in an aqueous phase to a nonaqueous phase. In one common procedure, press-cake (aqueous or water-wet pigment phase) is mixed and agitated with a nonaqueous vehicle (oil, solvent, and/or resin phase). If the material compositions have been correctly formulated, the pigment particles preferentially transfer (flush) to the nonaqueous phase and the bulk of the water becomes essentially clear as it relinquishes its pigment content (1).

This transfer process is now considered in terms of surface energetics. Three substances are involved in pigment transfer (two immiscible liquids and a solid), leading to three interfacial tensions. These interfacial tensions will be denoted by subscripts, each consisting of two letters indicative of the two surfaces in contact. Since water is a common phase in transfer, the letter W is used to designate the liquid of higher surface tension (usually an aqueous solution), and L is reserved for the liquid of lower surface tension. The subscript S denotes the solid surface tension.

In using the dual subscript notation, the letter denoting the substance of higher surface tension is always placed first. Thus γ_{WL} is a meaningful designation for an interfacial tension, but γ_{LW} has no significance since by our convention γ_W is always larger than γ_L. Consider a system where the solid surface tension lies between the two liquid surface tensions ($\gamma_W > \gamma_S > \gamma_L$). Then the three interfacial tensions are $\gamma_{WS}, \gamma_{WL}, \gamma_{SL}$.

Let a smooth spherical particle with surface tension γ_S be lodged at the interface between two immiscible liquids having surface tensions γ_W and γ_L, as shown in Fig. 10-2. What relationship must hold to make this a stable state of equilibrium? From inspection of the diagram given in Fig. 10-2 it can be seen

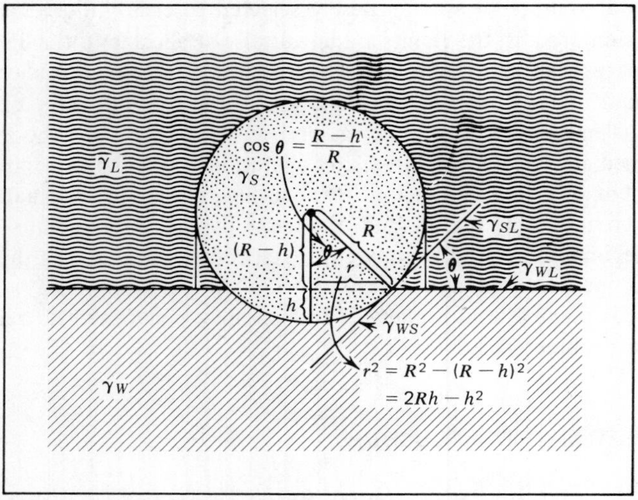

Fig. 10-2 Schematic diagram showing the three surface tension forces acting on a spherical particle positioned between two immiscible liquids of different surface tension. Also given is the geometry that applies to this configuration.

that the areas of interfacial contact are as follows:

$$W/S \text{ area} = 2\pi Rh$$

$$S/L \text{ area} = 4\pi R^2 - 2\pi Rh$$

$$W/L \text{ area} = \pi R^2 - \pi r^2 = \pi(R-h)^2$$

The last expression holds true because $r^2 + (R-h)^2 = R^2$ from a right triangle relationship. Note also that for purposes of calculation the outer radius limit of the W/L interface is arbitrarily taken as equivalent to the radius of the spherical particle.

The energy resident in the three interfaces is then given by Eq. 2.

$$E = \gamma_{WS} 2\pi Rh + \gamma_{LS}(4\pi R^2 - 2\pi Rh) + \gamma_{WL}\pi(R^2 - 2Rh + h^2) \tag{2}$$

For an equilibrium condition to exist, any tendency for the spherical particle either to penetrate further into or to withdraw from either liquid phase must be accompanied by zero work input. In terms of mathematics this is equivalent to saying that the differential of E with respect to differential h must equal zero. Carrying out the differentiation necessary to obtain dE/dh leads to Eq. 3.

$$\frac{dE}{dh} = 2\pi R\gamma_{WS} - 2\pi R\gamma_{SL} - \frac{2\pi R\gamma_{WL}(R-h)}{R} \tag{3}$$

By setting dE/dh equal to zero and noting that $\cos\theta = (R - h)/R$, Eq. 4, which is simply a modified form of Young's contact angle equation, is obtained.

$$\gamma_{WS} = \gamma_{SL} + \cos\theta\gamma_{WL} \qquad (4)$$

By taking advantage of Young's neutralization principle it is possible to derive an approximate expression for the contact angle from Eq. 4 in terms of the surface tensions of the solid and the two immiscible liquids.

$$\gamma_{WS} = \gamma_W - \gamma_S; \qquad \gamma_{SL} = \gamma_S - \gamma_L; \qquad \gamma_{WL} = \gamma_W - \gamma_L$$

Substitution of these values in Eq. 4 and solving for $\cos\theta$ gives Eq. 5.

$$\cos\theta \approx \frac{(\gamma_W - \gamma_S) - (\gamma_S - \gamma_L)}{\gamma_W - \gamma_L} \qquad (5)$$

Now, as γ_S approaches γ_W as a limiting value, the value of $\cos\theta$ approaches -1.0, corresponding to a contact angle of 180°. This means that the solid spherical particle is positioned essentially within the *W* phase when $\gamma_S = \gamma_W$. On the other hand, as γ_S approaches γ_L as a limiting value, the value of $\cos\theta$ approaches 1.0, corresponding to a contact angle of 0°. This means that the solid particle is positioned essentially within the *L* phase when $\gamma_S = \gamma_L$. It can be further demonstrated that, when $\gamma_S > \gamma_W$, corresponding to the overall condition $\gamma_S > \gamma_W > \gamma_L$, the solid particle migrates and enters the phase with the higher surface tension (the *W* phase). Conversely, when $\gamma_L > \gamma_S$, corresponding to the overall condition $\gamma_W > \gamma_L > \gamma_S$, the solid particle migrates and enters the phase with the lower surface tension (the *L* phase). These several sets of conditions are schematically illustrated in Fig. 10-3. From this it is concluded that particles with high solid surface tensions tend to migrate to or stay within liquids of high surface tension, whereas particles with low solid surface tensions tend to migrate to or stay within liquids of low surface tension.

Inspection of Eq. 5 shows that, by measuring the contact angle, the surface tension of the solid can be closely calculated from a knowledge of the two liquid surface tensions. Conceivably this could be made the basis for establishing the value of a solid surface tension.

Even though the foregoing concepts lead to approximate results (since Young's neutralization principle is not necessarily exact for liquid/solid interfaces), they still afford the paint engineer with valuable insight into migration behavior. Thus it is submitted that pigment flushing [migration from water (or *W*) phase to an organic (or *L*) phase] will have the best chance of success when the surface tension (or specific free surface energy) of the pigment particle, γ_S, is less than the surface tension (or specific free surface energy) of the liquid *L* phase, γ_L, into which the pigment is being flushed.

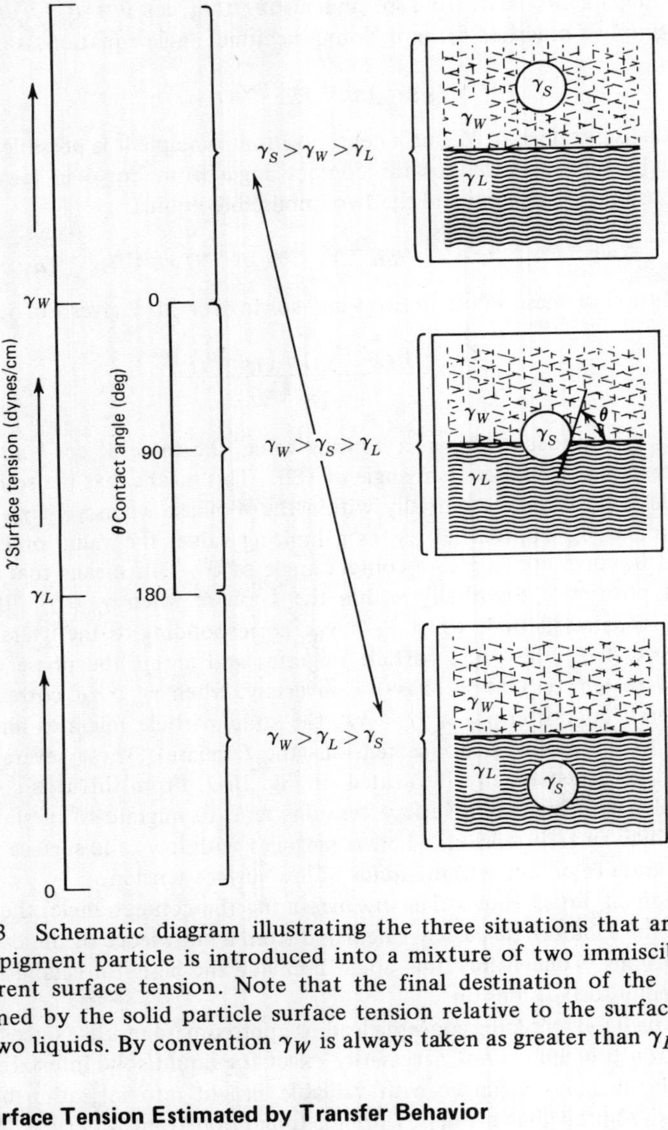

Fig. 10-3 Schematic diagram illustrating the three situations that are possible when a pigment particle is introduced into a mixture of two immiscible liquids of different surface tension. Note that the final destination of the particle is determined by the solid particle surface tension relative to the surface tensions of the two liquids. By convention γ_W is always taken as greater than γ_L.

Solid Surface Tension Estimated by Transfer Behavior

This author suggests that the specific free surface energy of pigment particles can be roughly estimated by shaking the candidate pigment particles in a bottle containing two immiscible liquids of known surface tensions. Depending on the migration behavior, a deduction can then be made as to the γ_S of the test particle in relation to the γ_W and γ_L values of the two immiscible liquids. One of three sets of conditions can occur for any given immiscible liquid pair, as shown in Fig. 10-3. By establishing sets of immiscible liquid pairs over a wide range of γ_W and γ_L values and by observing the migration behavior of a test pigment when shaken in these paired systems, it should be possible to roughly establish the relative solid surface tension of the pigment particles.

WORK OF FLOCCULATION

A simplified model for an ideal flocculation process is schematically illustrated in Fig. 10-4, where two dispersed solid cubical particles are brought into face-to-face contact. The specific free energy is either γ_{LS} or γ_{SL} for each face before contact and zero after contact. Hence the work of flocculation W_F is simply the difference between these two states as given by Eq. 6.

$$W_F = 2\gamma_{LS} \qquad \text{or } 2\gamma_{SL} \tag{6}$$

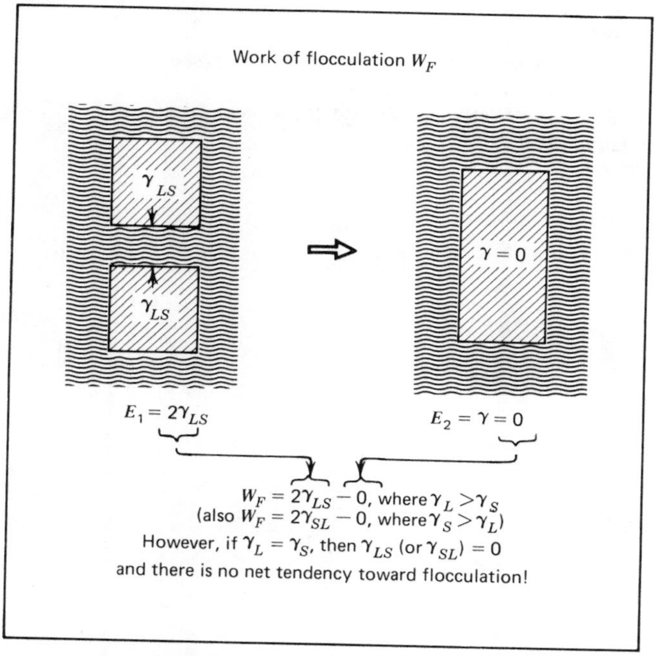

Fig. 10-4 Schematic diagram illustrating the initial and final states that apply to a pigment flocculation process.

According to Young's neutralization principle, γ_{LS} is approximately equal to $\gamma_L - \gamma_S$, and γ_{SL} is approximately equal to $\gamma_S - \gamma_L$. Substituting these expressions in Eq. 6 gives Eq. 7.

$$W_F = 2(\gamma_L - \gamma_S) \qquad \text{or } 2(\gamma_S - \gamma_L) \tag{7}$$

From these simple expressions a very important conclusion can be reached, namely, that the drive toward flocculation is directly related to the difference between the liquid and solid surface tensions in question. Hence, if the solid and liquid surface tensions are made equal, there is no tendency to flocculate.

A corollary to this statement is that pigment dispersion should also be facilitated by matching surface tensions. This has been demonstrated by Buechler et al., who note the importance of reducing the surface tension of the vehicle below that of the pigment particle to optimize pigment dispersion (2). In the case of the dispersion operation, lowering the vehicle surface tension just slightly below the solid surface tension is recommended as a safety factor to ensure spontaneous spreading of the vehicle over the pigment particles during the last stage of the dispersion process.

REFERENCES

1. Langstroth, T. A., "Pigment Flushing," in *Pigment Handbook*, Vol. III (T. Patton, Ed.), Wiley-Interscience, New York, 1973, pp. 447–455.
2. Buechler, P. R. et al., "Wetting as an Aid to Pigment Dispersion," *JPT*, **45**, No. 577, 60–64, February (1973).

11 *Capillarity*

Capillarity is defined as the observable motion or flow of a liquid, brought about by the liquid's own surface or interfacial tension forces. Capillary flow is caused by the pressure difference that is set up between two hydraulically connected surfaces of the same liquid with the flow direction being such as to decrease this pressure differential. When the pressure difference ceases to exist or when a further diminishment of the difference is not possible, flow ceases.

Differences in the curvature of the hydraulically connected liquid interfaces are responsible for capillary flow. However, in any real system it is necessary to consider, in the calculation work the possible contributions of other forces. Thus gravitational force must be introduced in computing the capillary rise in a small-bore tube or when calculating the equilibrium shapes of hanging drops. In many cases, however, gravitational force can be safely neglected.

The fundamental equation relating the capillary pressure p at an interface to the surface tension γ of a liquid and to the principal radii of curvature, r_1 and r_2, of the interface was derived by Laplace in 1806.

$$p = \gamma\left(\frac{1}{r_1} + \frac{1}{r_2}\right) \qquad (1)$$

It is convenient to assign a positive value to r when the radius is within the liquid of interest (liquid surface convex, i.e., rounded outward), and a negative value when the radius is outside the liquid (concave, i.e., rounded inward). With this convention, flow takes place from the region of higher capillary pressure to the region of lower capillary pressure.

For any given surface curvature the algebraic sum $(1/r_1 + 1/r_2)$ determines the net curvature. Note that r_1 and r_2 can have different signs, as in a saddle surface,

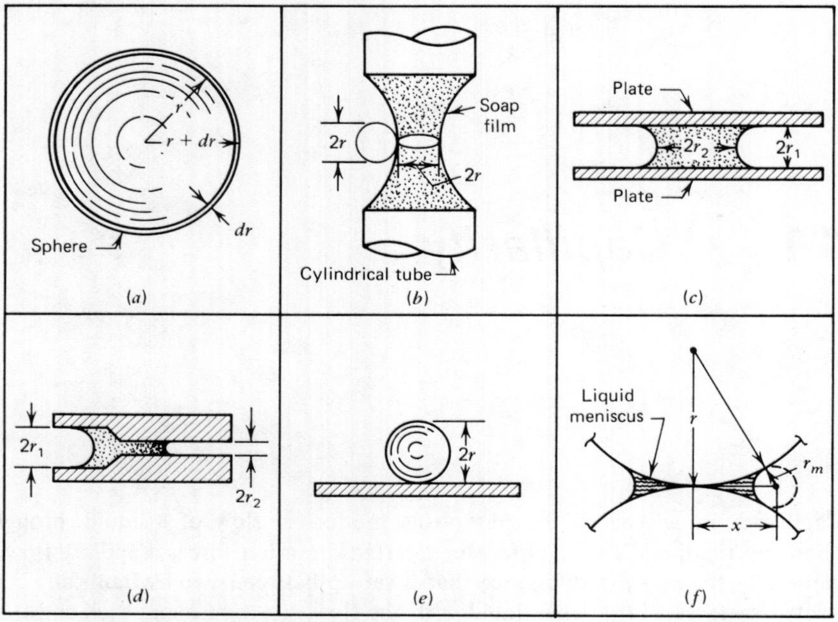

Fig. 11-1 Schematic illustration of (*a*) expansion of model liquid sphere, (*b*) soap film stretched between ends of two cylindrical tubes, (*c*) liquid positioned between two parallel plates, (*d*) liquid positioned in dual capillary (two bore sizes), (*e*) tiny liquid droplet located on horizontal plate, (*f*) liquid lodged in crevice space between two spherical particles.

as schematically illustrated in Figs. 11-1*b* and *c*. When the curved surface is spherical in contour, Eq. 1 reduces to Eq. 2 (since $r_1 = r_2$).

$$p = \frac{2\gamma}{r} \tag{2}$$

The validity of Eq. 2 is easily provable. Consider a spherical drop of liquid of radius r, and inject into the interior of this drop an increment of liquid such that the drop radius is increased by dr, as diagramed in Fig. 11-1*a*. The surface of the sphere is then correspondingly increased by $8\pi r \cdot dr$ (the second-order differential term is neglected).

$$\text{Sphere area (initially)} = 4\pi r^2$$

$$\text{Sphere area (after injection)} = 4\pi(r + dr)^2$$

$$\text{Sphere area increase} = 4\pi(r^2 + 2r \cdot dr + dr^2) - r^2$$

$$= 8\pi r \cdot dr$$

The work W done against the liquid tension force is $(8\pi r \cdot dr)\gamma$, an amount equal to the newly created free energy in the expanded spherical surface. Under equilibrium conditions this work is done by the injection pressure p pressing outwardly against the spherical area $4\pi r^2$. The work done by the total force exerted by this pressure $p(4\pi r^2)$ through the distance dr is equal to $p(4\pi r^2)\,dr$. Equating these two work quantities and simplifying gives the required proof.

$$(8\pi r \cdot dr)\gamma = p(4\pi r^2)\,dr$$

$$p = \frac{2\gamma}{r}$$

Note that, for a plane surface, $r = \infty$, and hence there is no capillary pressure. It is also possible to have a curved saddle surface with no net pressure. For example, consider the soap film stretched between the ends of two cylindrical tubes, as diagramed in Fig. 11-1*b*. At all points in the film the vertical curvature (with negative radius) is just balanced by the horizontal curvature (with positive radius). Since the two radii are numerically equal, p in Eq. 1 equals zero. Although equal radii are shown for the central waist location, the equality is applicable to any location on the film surface.

PROBLEM 11-1

A drop of toluene ($\gamma = 28.5$ dynes/cm) is inserted into the dual capillary tube diagramed in Fig. 11-1*d* to take up the indicated position. If the radii r_1 and r_2 are 0.60 mm and 0.20 mm, respectively, and if the toluene completely wets the capillary walls (0° contact angle), calculate the net capillary pressure and the direction of flow, if any.

Solution. The net capillary pressure is given by the difference in the pressures exerted by the menisci in the small and large bores of the dual capillary tube (radii in cm).

$$p = 2 \cdot 28.5\left(\frac{-1}{0.06} - \frac{-1}{0.02}\right) = 1900 \text{ dyne/cm}^2$$

The flow is from the larger to the smaller bore, corresponding to a flow from the higher capillary pressure (-950 dyne/cm^2) to the lower capillary pressure (-2850 dyne/cm^2).

PROBLEM 11-2

A drop of water ($\gamma = 72$ dynes/cm) placed between two parallel plates, as diagramed in Fig. 11-1*c*, completely wets the plate surfaces (0° contact angle). If the radii r_1 and r_2 are 0.50 and 10.0 mm, respectively, calculate the approximate capillary force that acts to hold the top plate against the bottom plate.

Solution. The capillary pressure exerted by the water can be computed from Eq. 2. Using consistent units, substitute appropriate data in this equation, taking into account positive and negative values.

$$p = 72\left(\frac{1}{1.0} - \frac{1}{0.050}\right) = -1370 \text{ dyne/cm}^2$$

The area A of the top plate being acted on by this negative pressure is 3.14 cm^2 ($A = \pi r^2 = 3.14 \cdot 1.0^2$). The total force F acting on the top plate is equal to the pressure times the top plate area ($F = pA$).

$$F = -1370 \cdot 3.14 = -4300 \text{ dyne (4.4 g)}.$$

PROBLEM 11-3

A spherical droplet of liquid of *unit* density and radius r is placed on a horizontal surface, as shown in Fig. 11-1*e*. Compare the hydrostatic and capillary pressures at the point of liquid/solid contact. Assuming a liquid surface tension of 50 dyne/cm and unit density, compute the capillary/hydrostatic pressure ratios for droplet diameters of 1.0 and 0.10 mm, respectively.

Solution. The hydrostatic pressure is given by multiplying the liquid height ($2r$) by the gravitational force (ρg).

$$p\text{(hydrostatic pressure)} = 2r \cdot 1.0 \cdot 980 = 1960r$$

The capillary pressure is given by Eq. 2.

$$p\text{ (capillary)} = \frac{2\gamma}{r}$$

The ratio of the two pressures is $\gamma/980r^2$.

$$\frac{p\text{ (capillary)}}{p\text{ (hydrostatic)}} = \frac{2\gamma/r}{1960r} = \frac{\gamma}{(980r^2)}$$

For a droplet diameter of 1.0 mm ($r = 0.05$ cm) the ratio is 20.4 [$20.4 = 50/(980 \cdot 0.05^2)$], and for a droplet diameter of 0.10 mm ($r = 0.0050$ cm) the ratio is 2040 [$2040 = 50/(980 \cdot 0.0050^2)$].

Comment. It is obvious that, as the radius of curvature decreases, the gravitational influence very rapidly becomes less significant. For purposes of reference, a droplet of this liquid, of 2.0-μ diameter (0.00010-cm radius), exerts a capillary pressure of about 1.0 atm.

$$p = \frac{2 \cdot 50}{0.00010} = 1{,}000{,}000 \text{ dynes/cm}^2 \text{(0.99 atm)}$$

PARTICLE ADHESION DUE TO CAPILLARITY

The particles of powders stick strongly to each other when dampened with a wetting liquid. Even in a humid environment, adhesion of hydrophilic powder particles is promoted by water vapor, which condenses to liquid water under saturation or near-saturation conditions and collects in the crevices between the powder particles. This type of sticking problem must always be considered in the storage of pigments.

The sticking of pigment particles due to capillarity can also produce serious mixing problems. It would probably be ideal if a pigment could be added to a vehicle by gradually sifting the pigment particles onto an agitated vehicle surface. However, it is more common to encounter a situation where clumps of pigment particles are abruptly immersed in a vehicle. The result may be that only the outer layer of the clumped pigment is wetted, so that a wetted or vehicle-impregnated dense outer layer envelops an inner core of dry pigment particles. This outer layer then effectively blocks the inner core from further vehicle contact. Capillary penetration of the vehicle into the dry core is also resisted by the trapped air inside, which becomes compressed by vehicle penetration from all sides.

The formation of such surface-wet/core-dry clumps (partially wetted pellets of pigment) can arise with any pigment, even such common ones as titanium dioxide, iron oxide, and chrome oxide, unless correct mixing procedures are scheduled. For example, in preparing a sand-mill premix, it is claimed that these pellets or partially wetted pigment clumps can usually be eliminated or minimized by cutting back on the initial charge of vehicle and by arranging for alternative pigment and vehicle addition, with the view of always maintaining a stiff mill base. Careful letdown with the remaining vehicle is scheduled last to provide a pellet-free mill base for charging to the sand mill.

MAGNITUDE OF CAPILLARY FORCES ACTING TO PROMOTE PARTICLE-TO-PARTICLE ADHESION

The capillary forces acting to promote the adhesion of particles to each other may be small on an absolute scale but tremendous relative to the particle weight.

Consider the case of two contacting spheres of radius r and density ρ that have a trace of wetting liquid (contact angle = 0°) collected around their point of contact, as schematically shown in Fig. 11-1*f*. Let r_m be the radius of the liquid meniscus, and x the distance from the point of particle-to-particle contact to the center of the radius of curvature of the liquid meniscus. Inspection of Fig. 11-1*f* shows that r, x, and r_m can be related by Eq. 3, using right triangle geometry.

$$r^2 + x^2 = (r + r_m)^2$$
$$x^2 = 2r \cdot r_m + r_m^2 \tag{3}$$

The pressure exerted by the capillary action is given by Eq. 4.

$$p = \gamma\left(\frac{1}{x - r_m} - \frac{1}{r_m}\right) \tag{4}$$

Consider the usual case where r_m is very small in relation to r and x. Equations 3 and 4 then reduce to the approximate expressions given by Eqs. 5 and 6, respectively.

$$x^2 \approx 2r \cdot r_m \tag{5}$$

$$p \approx \frac{-\gamma}{r_m} \tag{6}$$

The approximate capillary force acting to hold the two particles together is given by Eq. 7 ($F = pA$), where $A = \pi x^2$.

$$F = p\pi x^2 \approx \frac{-\gamma}{r_m}(\pi 2r \cdot r_m) \approx -2\pi r\gamma \tag{7}$$

Also, since the gravitational force acting on a sphere of radius r and density ρ is $\frac{4}{3}\pi r^3 \rho g$, the ratio of the capillary adhesion force to the gravitational force is given by Eq. 8.

$$\frac{F\,(\text{adhesion})}{F\,(\text{gravitational})} = \frac{2\pi r\gamma}{\frac{4}{3}\pi r^3 \rho \cdot 980} = \frac{0.00153\gamma}{\rho r^2} \tag{8}$$

PROBLEM 11-4

What is the largest radius that will just permit one spherical particle to support another spherical particle of the same radius when placed beneath it with only a trace of water between the two for adhesive support? Assume a water surface tension of 72 dynes/cm, a contact angle of 0°, and unit density for the hanging spherical particle.

Solution. For this equilibrium condition set the capillary force equal to the gravitational force. This is equivalent to setting Eq. 8 equal to 1.00. Insert the appropriate data in this expression, and solve for r.

$$1 = \frac{0.00153 \cdot 72}{1.00 \cdot r^2}; \qquad r = 0.33 \text{ cm}$$

From the above equations it can also be shown that the capillary adhesive force due to a trace of water between two spherical particles (diameter - 20 μ; $\rho = 1.0$ g/cm^3) is only about half a dyne ($0.45 \approx 2\pi 0.001 \cdot 72$). However, from Eq. 8 this is still 110,000 times the weight (expressed in dynes) of either particle alone ($110{,}000 \approx 0.00153 \cdot 72/1.00 \cdot 0.001^2$). It is this same type of capillary action that explains the tendency of latex particles to squeeze together as water is lost from a latex coating.

FILM FORMATION FROM LATEX SUSPENSIONS DUE TO CAPILLARITY

In the preceding section it was demonstrated that pigment particles are driven together by strong capillary forces. In a similar manner the spherical polymer particles of latex suspensions are also forced into intimate contact by capillarity as water is lost from the latex system. If the latex particles are sufficiently deformable, they will be strongly compelled to coalesce following particle-to-particle contact. To be assured of this compacting action it is necessary (*a*) that the liquid water be able to maintain its penetration of the latex particles, and (*b*) that the water surface remain unbroken. Fortunately, the water surface confined in the small capillary interstices is rendered highly resistant to rupture by its high curvature. Hence, as water evaporates, the adhering water that remains behind in the latex particle structure exerts a high negative pressure that drives the latex particles together.

It has been shown that, at the stage where the solid spherical latex particles just come into contact to form a continuous structure, the radius of the capillary water meniscus is equal to 0.155 times the radius of the contacting latex particles (1).

Let the radius r of the contacting latex particles in a latex structure be 0.10 μ (0.000010 cm). The corresponding radius of the water meniscus is then 0.15 times this value or $1.55 \cdot 10^{-6}$ cm. By assuming a zero contact angle and a value of 31 dynes/cm for the surface tension of the aqueous phase (typical of an aqueous surface tension with surfactant present), there is obtained from Eq. 2 a value of $4.0 \cdot 10^7$ dynes/cm^2 for the negative water pressure. Since this is equivalent to 580 psi, it represents a tremendous compacting action.

$$p = \frac{2 \cdot 31}{1.55 \cdot 10^{-6}} = 4.0 \cdot 10^7 \text{ dynes/cm}^2 \text{ (580 psi)}$$

For further discussion of film formation from latex suspensions due to capillarity and also the effect of the existence of a contact angle (θ not assumed as

0°), see references 1 and 2. Thus a liquid surface tension that is too high relative to the latex particle surface tension (say more than twice as great) has been shown to possibly negate any useful compacting action.

CAPILLARY FLOW

The expression for the rate of liquid flow (cm^3/sec) through a cylindrical capillary tube was derived in Chapter 2 and is repeated here for ready reference.

$$\frac{V}{t} = p \cdot \left(\frac{\pi r^4}{8 \eta L} \right) \tag{9}$$

where V = volume (cm^3)
t = time (sec)
p = pressure ($dynes/cm^2$)
r = capillary radius (cm)
η = viscosity (poises)
L = length of tube (cm)

Since V/t is equal to the product of the average liquid velocity u and the tube cross-sectional area $\pi r^2 (V/t = u \pi r^2)$, Eq. 9 can be rewritten as Eq. 10.

$$p = u \cdot \left(\frac{8 \eta L}{r^2} \right) \tag{10}$$

In a capillary tube the pressure exerted by the surface tension of a curved capillary meniscus is given by Eq. 11, where θ is the contact angle. The cosine factor is introduced into Eq. 2 in this derivation to give the portion of the surface tension that is effective in producing flow.

$$p = \frac{2 \gamma \cos \theta}{r} \tag{11}$$

Equating Eqs. 10 and 11 for pressure and solving explicitly for the average velocity u gives Eq. 12.

$$u = \left(\frac{\gamma \cos \theta}{\eta} \right) \left(\frac{r}{4L} \right) \tag{12}$$

Inserting dL/dt for u in Eq. 12 and rearranging gives Eq. 13.

$$L \cdot dL = \left(\frac{\gamma \cos \theta}{\eta} \right) \left(\frac{r}{4} \right) dt \tag{13}$$

If it is assumed that $L = 0$ at time $t = 0$, Eq. 13 integrates to Eq. 14.

$$L = 0.707\sqrt{(\gamma/\eta)(r \cdot \cos\theta)t} \tag{14}$$

Equation 14 can also be solved explicitly for t, giving Eq. 15.

$$t = 2L^2\left(\frac{\eta}{\gamma}\right)\left(\frac{1}{r \cdot \cos\theta}\right) \tag{15}$$

Two key equations have been developed. Equation 12 gives the instantaneous flow rate corresponding to a penetration distance L of liquid into the capillary. Equation 14 gives the depth of penetration achieved at the end of time t by capillary action, starting from zero penetration.

Equation 12 states that the rate of penetration of a vehicle into the interstices (capillaries) of, say, a bed of pigment particles varies directly with the size of the capillary openings, r; with the magnitude of the vehicle surface tension, γ; and with the cosine of the liquid/solid contact angle, $\cos\theta$; but indirectly with the vehicle viscosity, η, and the depth of penetration, L. Hence, for any given depth, penetration is facilitated by loosely packed, coarse pigment, high vehicle surface tension, a low contact angle (large $\cos\theta$), and low vehicle viscosity.

From the standpoint of obtaining rapid vehicle penetration into a bed of pigment particles by capillarity, it is unfortunate that the two factors in the desired combination of high vehicle surface tension and a low or zero contact angle are mutually antagonistic. Thus no penetration whatever is achieved if the liquid surface tension is so high as to give a contact angle of 90° or above. Of the two properties the more important requisite for effective penetration is a zero or near-zero contact angle.

Equation 14 states that the depth of vehicle penetration varies with the square root of the elapsed penetration time.

PROBLEM 11-5

A charge of packed pigment is dumped without agitation into a tank partially filled with bodied linseed oil ($\gamma = 25$ dynes/cm; $\eta = 35$ poises). Assuming that the linseed oil completely wets the pigment ($\gamma_S > \gamma_L$) and that the voids among the pigment particles correspond to capillary interstices of 2.0-μ effective diameter, calculate the depth of penetration at the end of 1.0 min, 1.0 hr, and 1.0 day. Neglect gravitational effects.

Solution. Substitute the problem information, expressed in consistent units, in Eq. 14 in turn, and solve for the depth of penetration for each of the three time periods. Note that $r = 10^{-4}$ cm and that $\cos 0° = 1.00$.

For the 1.0-min penetration period:

$$L = 0.707\sqrt{(25/35)(0.00010t)} = 0.0060\sqrt{t} = 0.0060\sqrt{60} = 0.046 \text{ cm}$$

For the 1.0-hr period: $L = 0.0060\sqrt{3600} = 0.36$ cm.
For the 1.0-day period: $L = 0.0060\sqrt{3600 \cdot 24} = 1.76$ cm.

The slow but persistent penetration of viscous liquids into the interstices of tightly packed pigment masses affords a logical reason for the common practice of "soaking" pigments in a vehicle overnight to obtain a significant wetting out of the pigment mass before continuing with the mechanical part of the dispersion process. It also renders more understandable the gradual "staining" of porous masonry by caulking compounds, wherein one of the caulking ingredients, retained as mobile liquid, is gradually sucked into the masonry capillaries by surface tension forces over a period of months or years. On the other hand, capillary penetration can be very rapid with low-viscosity fluids.

PROBLEM 11-6

One of the two open ends of a capillary tube (0.50-mm diameter) is brought horizontally into contact with a large drop of toluene ($\eta = 0.0059$ poise, $\gamma = 28.5$ dynes/cm). Assuming a zero contact angle and neglecting gravitational and kinetic influences, calculate the time required for the toluene to penetrate a distance of 2 in. (5.08 cm) into the capillary tube.

Solution. Using consistent units, substitute the given information in Eq. 15, and solve for the time in secs.

$$t = 2 \cdot 5.08^2 \left(\frac{0.0059}{28.5} \quad \frac{1}{0.025 \cdot 1.0}\right) = 0.43 \text{ sec}$$

Comment. The rapid entry of a low-viscosity liquid into a capillary tube with a radius on the order of that given in the problem is used to advantage in medical practice for securing a sample from the drop of blood that comes to the surface of a finger after being pricked by a needle.

Hydraulic Radius

Very few capillary systems of practical interest in the coatings industry are small-bore cylindrical tubes. However, for many noncylindrical capillary situations, reasonably valid results can be obtained by substituting the expression $2V/A$ for r in the capillary equations that have been developed, where V is the volume of liquid in the section of channel being considered and A is the total area of the solid/liquid interface in the same section. The ratio V/A is defined as the hydraulic radius r_h. For conduits of circular cross section the hydraulic radius is half the actual radius.

For the case of a bed of pigment particles, V can be considered as the pigment void volume and A the total surface area of the particles in the pigment bed.

Consider a 1.0-cm^3 bed of bare pigment particles. The fractional void volume of this pigment bed is $(1 - \phi)$, where ϕ is the pigment packing factor for the bare pigments. The total surface area of the particles in this 1.0-cm^3 volume is $\phi\rho S$, where ρ is the solid pigment density and S is the pigment specific surface area (area per unit weight of pigment). By definition the hydraulic radius of the capillary channels in this 1-cm^3 section of pigment particles is given by Eq. 16.

$$r_h = \frac{V}{A} = \frac{1 - \phi}{\phi\rho S} \tag{16}$$

An approximate average geometric radius for the capillary channels in the pigment bed is then given by Eq. 17, since $r \approx 2r_h$.

$$r \approx \frac{2V}{A} = \frac{2(1 - \phi)}{\phi\rho S} \tag{17}$$

PROBLEM 11-7

Calculate the approximate radii for the capillary channels in the four organic pigments for which data are listed below, and from the standpoint of this radius information list the organic pigments in decreasing order of vehicle penetrability.

Pigment	ϕ	S (m^2/g)	ρ (g/cm^3)
A	0.75	23	1.3
B	0.77	82	1.2
C	0.56	9	1.2
D	0.66	50	1.4

Solution. Substitute the given data in Eq. 17, and solve for each of the four capillary radii.

$$\text{A:}\quad r = \frac{2(1 - 0.75)}{0.75 \cdot 1.3 \cdot 23} = 0.022\ \mu$$

$$\text{B:}\quad r = \frac{2(1 - 0.77)}{0.77 \cdot 1.2 \cdot 82} = 0.0060\ \mu$$

$$\text{C:}\quad r = \frac{2(1 - 0.56)}{0.56 \cdot 1.2 \cdot 9} = 0.146\ \mu$$

$$\text{D: } r = \frac{2(1 - 0.66)}{0.66 \cdot 1.4 \cdot 50} = 0.0148\ \mu$$

Since rate of penetration varies directly with capillary radius (Eq. 12), vehicle penetration (other things being equal) is most rapid with pigment C, followed by pigments A, D, and B.

HOLDOUT (REVERSE OF PENETRATION)

The term "holdout" refers to nonpenetration. It may be used to describe either a substrate or a vehicle. A substrate that resists vehicle penetration (is weakly absorbent or nonabsorbent) has good holdout properties. A vehicle that has poor penetrating ability also has good holdout properties.

Strong penetration (lack of holdout) is necessary for the adhesion of a paint to a chalky or rusty surface. On the other hand, weak penetration (strong holdout) is desired if a glossy paint or enamel is to exhibit uniform gloss when applied over a primer or undercoating. Good holdout is also necessary for ensuring the color uniformity of an applied flat paint.

One subjective measurement of holdout is specified in FTMS 6261, "Primer Absorption and Topcoat Holdout," where actual applications of primers and topcoats are made over standard substrates. The uniformity of the dry paint is then visually assessed, usually in comparison with paint standards.

An approximate quantitative measure of vehicle penetration can also be obtained by placing the paint in contact with an absorbent surface. One such test is FTMS 4421, "Absorption Test," which specifies placement of a filter paper (Whatman No. 12, 12.5 cm) over the paint contained in the friction top cover of a half-pint paint can. After a period of 3 hr the average distance that the vehicle has migrated outward from the edge of the cover is recorded and used as a measure of the absorption (penetration) that has taken place.

Substrate holdout may also be determined indirectly by applying a standard stain to the subject substrate, followed by removal of the stain excess after a given time period. The depth or degree of absorbed stain is evaluated either by visual observation or by reflectance measurements. In turn these data are used to assess the relative holdout of the substrate.

Holdout and Color Uniformity of Flat Paints

As far back as 1947, Armstrong and Madson reported that uniformity in the appearance of flat paints over nonuniform porous surfaces was best (optimum holdout) when the paint saturation value was 1.00 (i.e., when PVC = CPVC) (3).

In 1956 Stieg demonstrated that the color variation exhibited by flat paints at laps, that is, at the boundaries between sealed and unsealed areas, and also the color variation between one- and two-coat applications of the same paint could be minimized or eliminated by formulating the paint to have a PVC at or just below the CPVC (4). In 1961 the Philadelphia Society for Paint Technology noted a general consensus that flat wall paints exhibit many optimum properties at or close to the CPVC and cited such properties as minimum color and sheen differences (as well as minimum change in hiding on drying) when flat paints were applied over substrates of different porosity (5). In 1967 Stieg proposed that the optimum performance of flat paints at the CPVC is linked to capillary flow, pointing out that above the CPVC the length of the capillary channels becomes shortened, whereas below this point the capillary channels become enlarged. Both departures from the CPVC condition result in increased capillary flow into a porous substrate, causing nonuniformity in appearance, especially with substrates of varying porosity (6).

In an effort to provide a quantitative background for the ideas advanced by Stieg, this author offers the following analysis of the relative rates of coating penetration into a porous substrate as a function of the coating PVC. The analysis is based on the assumption that the coating (say an alkyd flat paint) is pigmented with uniform spherical particles of radius r_p. The PVC values will be considered as varying to either side of a constant CPVC value.

Equation 10 can be rewritten as Eq. 18, where u is the velocity of vehicle penetration (absorption) into the porous substrate, p is the capillary pressure exerted on the entrance of the vehicle into the capillaries of the substrate, η is the vehicle viscosity, and r and L are the average radius and length, respectively, of the capillary channels in the applied coating.

$$u = \frac{pr^2}{8\eta L} \qquad (18)$$

Relative Penetration Rate for the Condition PVC < CPVC

Let n be the number of spherical particles of radius r_p in a unit volume of the coating for the condition where PVC < CPVC. The volume of solid particles in a unit volume is PVC, and the number of particles in this volume is given by Eq. 19.

$$n = \frac{\text{PVC}}{\frac{4}{3}\eta r_p^3} \qquad (19)$$

The surface area A of this number of particles is given by Eq. 20.

$$A = n \cdot 4\pi r_p^2 = \frac{3\text{PVC}}{r_p} \qquad (20)$$

The average capillary radius of the coating is then given by Eq. 21.

$$r = \frac{2V}{A} = \frac{2(1 - \text{PVC})}{3\text{PVC}/r_p} = \frac{0.666 r_p(1 - \text{PVC})}{\text{PVC}} \tag{21}$$

Based on Eq. 10, a comparison between the velocities of penetration for pigment volume concentrations of PVC and CPVC can now be made by assuming the same average channel length for the PVC and CPVC coatings (same applied thickness) and equal viscosities. Since p, L, and η are held constant, the ratio of the velocities of penetration is given by Eq. 22.

$$\frac{u_{\text{PVC}}}{u_{\text{CPVC}}} = \left(\frac{r_{\text{PVC}}}{r_{\text{CPVC}}}\right)^2 = \left(\frac{(1 - \text{PVC})\text{CPVC}}{(1 - \text{CPVC})\text{PVC}}\right)^2 \tag{22}$$

Relative Penetration Rate for the Condition PVC > CPVC

For PVC paints above the CPVC, the capillary radius stays the same, since all the particles are in contact at the CPVC condition and remain so for higher PVC values. However, the effective channel length L of the film becomes shortened with increasing PVC, since there is a decreased volume of vehicle. This volume is insufficient to fill the pigment capillaries.

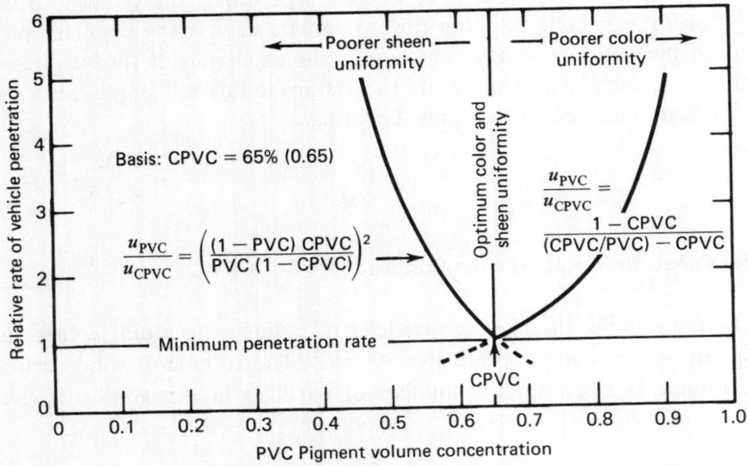

Fig. 11-2 Graph showing relative rate of vehicle penetration from an applied coating into a porous substrate versus pigment volume concentration. Relative rates apply to conditions both with solvent present (initially) and with solvent absent (only binder left after solvent escape). Graph data based on CPVC value of 65% (0.65).

A comparison between the velocities of penetration for pigment volume concentrations of PVC and CPVC can be made using Eq. 10 by assuming the average channel length to be proportional to the volume of vehicle present in the pigment capillaries. The quantities p, r, and η are assumed constant.

$$\frac{u_{\text{PVC}}}{u_{\text{CPVC}}} = \frac{L_{\text{CPVC}}}{L_{\text{PVC}}} = \frac{(1 - \text{CPVC})}{(\text{CPVC}/\text{PVC}) - \text{CPVC}} \tag{23}$$

It can be shown that these relative relationships (Eqs. 22 and 23) hold true whether or not solvent is initially present as part of the vehicle system.

To illustrate these equations in graphical form, Fig. 11-2 shows a plot of the relative penetration rates versus PVC values for a series of flat paints with a common CPVC value of 65% (0.65). Inspection of this graph clearly shows the minimum rate of coating penetration at the CPVC point. In actual practice there is a tendency to formulate flat coatings somewhat above the CPVC point to take advantage of the dry hiding that is obtained (7, 8). However, with colored flat paints it is probably best to stay as close to the CPVC point as possible.

REFERENCES

1. Brown, George L., "Basic Colloid Chemistry of Aqueous Coatings," *Off. Dig.*, **37**, No. 489, Part 2, 15 (1965).
2. Patton, T. C., "Reflections of a Paint Engineer on Paint Flow, Interface Physics, and Pigment Dispersion," *JPT*, **42**, No. 551, 666–694, December (1970).
3. Armstrong, W. G. and W. H. Madson, "The Effect of Pigment Variation on the Properties of Flat and Semi-Gloss Finishes," *Off. Dig.*, **19**, No. 269, 321–335, June (1947).
4. Stieg, F. B., "Color and CPVC," *Off. Dig.*, **28**, No. 379, 695–706, August (1956).
5. Philadelphia Socity For Paint Technology, "Determination of CPVC by Calculation," *Off. Dig.*, **33**, 1437–1452, November (1961).
6. Stieg, F. B., "Particle Size as a Formulating Parameter," *JPT*, **39**, No. 515, 703-721, December (1967).
7. Ramig, A., Jr., "Latex Paints–CPVC, Formulation, and Optimization," *JPT*, **47**, No. 602, 60–67, March (1975).
8. Bierwagen, G. P. and T. K. Hay, "The Reduced Pigment Volume Concentration," *Prog. Org. Coatings*, **3**, 281–302 (1975).

12 *Theoretical Aspects of Pigment Dispersion Stability and Pigment Flocculation*

A large body of literature dealing with the stabilization of dilute dispersions of particles has evolved over the past hundred years. Of special interest is the DLVO theory, which deals in a fundamental manner with the kinetics of flocculation and the stabilization of particle dispersions. The term "DLVO" was coined in honor of the scientists in two countries who worked out the theory independently (Derjaguin and Landau in Russia and Verwey and Overbeek in England) (1, 2).

Unfortunately the DLVO theory is beset with cumbersome equations; only uncertain data are available for the most part for entering into these equations; and their practical employment is limited chiefly to dilute concentrations. Nevertheless the DLVO theory provides useful insight into the technical aspects of dispersion stability that can be applied to concentrated dispersions and for this reason is briefly reviewed to afford a better understanding of dispersion technology in the coatings industry. Some of the equations in this chapter are rough approximations of the rigorously derived theoretical expressions, which tend to be quite complex. These simplified approximate expressions are submitted with the intent of providing a better picture of the interplay of the various factors that affect the stability of pigment dispersions systems.

BROWNIAN MOTION

The thermal activity (Brownian motion) of minute particles (from macromolecules to micron-sized pigments) is amply verified by theory and experimental evidence. Thermal motion is the driving force that causes a gas to expand spontaneously within a given volume of space, or a solution concentration to spontaneously equalize throughout the confines of a container. Brownian motion refers specifically to the activity of small particles and is named for the botanist Robert Brown, who in 1827 first observed and reported the ceaseless zigzag motion of pollen particles immersed in water as viewed through a microscope. Today it is acknowledged that very small particles never stand still.

The velocity v of a single particle due to its thermal motion decreases as its mass m increases in such a way that at any given temperature T all particles have the same kinetic energy, equal to kT.

$$mv^2 \text{ (particle kinetic energy)} = kT \tag{1}$$

The constant k (Boltzmann's constant) in Eq. 1 equals R/N, where R is the gas constant in the ideal gas law equation $PV = nRT$ ($R = 8.3 \cdot 10^7$ ergs/deg-mole) and N is Avogadro's number ($N = 6.06 \cdot 10^{23}$), the number of molecules in a gram molecular weight of any substance. Hence k has a value of $1.38 \cdot 10^{-16}$ erg/deg ($= 8.3 \cdot 10^7/6.06 \cdot 10^{23}$).

At a room temperature of, say, 24 C (75 F), equal to an absolute temperature of 297 K (= 24 + 273), kT has a value of $4.1 \cdot 10^{-14}$ erg ($= 297 \cdot 1.38 \cdot 10^{-16}$).

In discussing the technical aspects of dispersion stability the product kT is most important, since it represents the average energy associated with all small particles because of their thermal energy.

As previously pointed out, a dispersed particle is not a stationary particle. Rather, it is a violently active particle that is moving about on a microscopic scale in a most random fashion. This haphazard activity is caused by the continuous collisions of the particle with its fellow particles and with the walls of the container within which it is confined. Since, on the average, each minute moving particle is endowed with kT ergs of energy ($4.1 \cdot 10^{-4}$ erg at room temperature), it has been found convenient to relate other energy factors affecting pigment dispersion stability to this basic kT quantity.

FLOCCULATION RATE OF UNSTABILIZED PIGMENT DISPERSION

It is of interest to compute the theoretical rate at which pigment particles flocculate when each particle-to-particle collision results in such intimate contact that bonding occurs (a flocculate is formed).

The theory of such rapid flocculation, as developed by Smoluchowski (3), postulates that this type of collision occurs and also that the rate of decrease in the number of particles per cubic centimeter, n, with time t is proportional to n^2. This proportionality is shown in Eq. 2, where d is the collision radius of the particle (taken as equal to the particle diameter) and D is the diffusion constant.

$$\frac{-dN}{dt} = 4\pi d \mathrm{D} n^2 \tag{2}$$

The diffusion constant D measures the particle's ability to diffuse (spread or scatter) throughout a solution, as given by Eq. 3.

$$\mathrm{D} = \frac{kT}{3\pi\eta d} \tag{3}$$

It is apparent from Eq. 3 that the particle diffusion constant is directly related to its thermal energy kT. Also, as might be expected, D is inversely related to the particle diameter d and to the viscosity η of the solution.

Integrating Eq. 2 gives Eq. 4.

$$\frac{1}{n_t} - \frac{1}{n_0} = 4\pi d \mathrm{D} t \tag{4}$$

The time to halve the particle count $t_{1/2}$ as a result of flocculation is given by Eq. 5.

$$t_{1/2} = \frac{1}{4\pi d \mathrm{D} n_0} \tag{5}$$

Substituting D from Eq. 3 in Eq. 5 gives Eq. 6.

$$t_{1/2} = \frac{3\eta}{4kTn_0} \tag{6}$$

Equation 6 tells us that the rate of flocculation in terms of halving the particle count is directly related to the viscosity of the dispersion medium. Hence one way to reduce flocculation would be to arrange for a system that is highly viscous in the shear rate range that applies to particle thermal motion. However, viscosity alone is quite insufficient to completely stabilize dispersions of technical importance to the coatings industry. For example, consider a dilute 0.1%v concentration of TiO_2 particles of 0.4-μ diameter in pure water ($\eta = 0.01$ poise). The number n of TiO_2 particles per cubic centimeter at this concentration is $3 \cdot 10^{10}$. Substituting this value for n_0 in Eq. 6 gives 6 sec as the time for half the TiO_2 particles to be flocculated in the water solution. From this it follows that practical dispersion systems (concentrations $>$ 0.1%v) are inherently unstable unless some sort of energy barrier is erected that prevents particles from coming into close contact when they approach each other on a collision course.

DLVO THEORY: ATTRACTION AND REPULSION POTENTIALS

Three major interacting forces are generally involved when pigment particles approach each other: electromagnetic forces, which are always attractive in nature; electrostatic (Coulombic) forces, which may be either attractive or repulsive but are almost always repulsive when considering pigment dispersions; and steric hindrance forces (due to adsorbed layers), which are repulsive in nature (4). These several forces originate from quite different sources, may be evaluated separately, and can be added or subtracted to obtain a total effect (a net attraction or repulsion potential). The interplay of the electromagnetic and electrostatic forces forms the substance of the DLVO theory. The contribution of steric hindrance to the overall stabilization or flocculation effect is yet to be reduced to a well defined theoretical basis.

ATTRACTIVE FORCES (ELECTROMAGNETIC IN NATURE)

Attractive forces, which are electromagnetic in nature must be overcome to prevent the flocculation that arises from particle interaction. These ever-present forces are referred to as London-van der Waals forces and are due to the influence of the dipoles within the particles acting on each other. A further complicating factor arises when a normally nonpolar particle is induced to take on polar characteristics when in the vicinity of polar particles.

The overall theory of electromagnetic attraction is quite involved, leading to cumbersome theoretical expressions that often require computer programming for their numerical solution.

The net attractive force due to these electromagnetic forces is generally given in terms of an attractive potential V_A that is measured in energy units (ergs). However, the value of V_A is rendered more meaningful when it is compared with kT (the particle kinetic energy). Hence graphs illustrating particle dispersion situations often plot the ratio V_A/kT to show this comparative relationship.

A rough approximation that relates V_A/kT to particle diameter d and the distance of separation s between the outside surfaces of the particles is given by Eq. 7, where A is a constant (same unit for d and s).

$$\frac{V_A}{kT} \approx A\left(\frac{d}{s}\right) \tag{7}$$

A reasonably representative value for A when working with nonaqueous systems is 0.12. Hence for solvent-type dispersions, say for pigments dispersed in alkyd vehicles and the like, Eq. 7 can be written as Eq. 8.

$$\frac{V_A}{kT} \approx \frac{0.12d}{s} \tag{8}$$

Inspection of Eqs. 7 and 8 show that the attractive force between two similar particles increases with the particle diameter and decreases with the particle separation distance. A graph of V_A/kT versus s based on Eq. 8 is given in Fig. 12-1a for selected pigment particle diameters (lower half of graph). It is con-

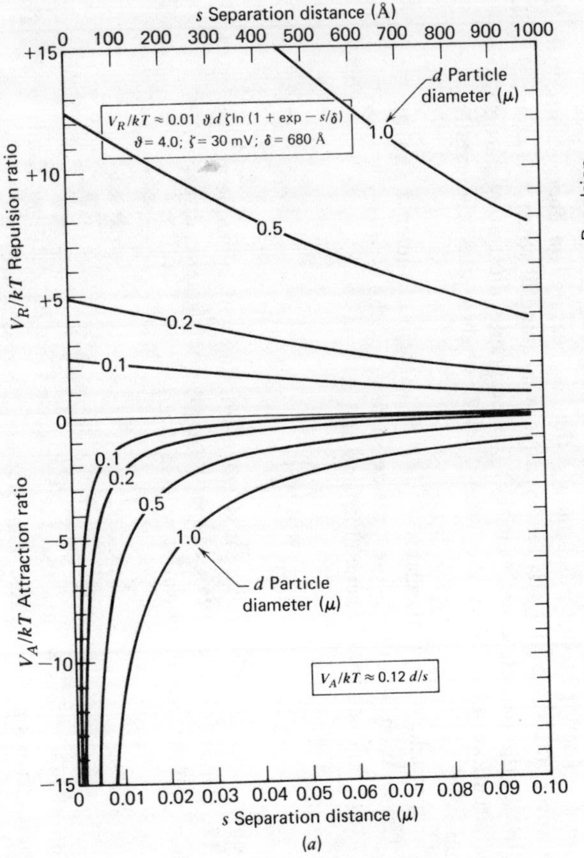

Fig. 12-1*a* Plot of the attraction and repulsion forces acting on two separated particles (for selected particle diameters) as a function of particle separation distance *s* (constant dielectric constant, zeta potential, and double layer thickness). Molarity is 10^{-6} mole/liter of univalent ions.

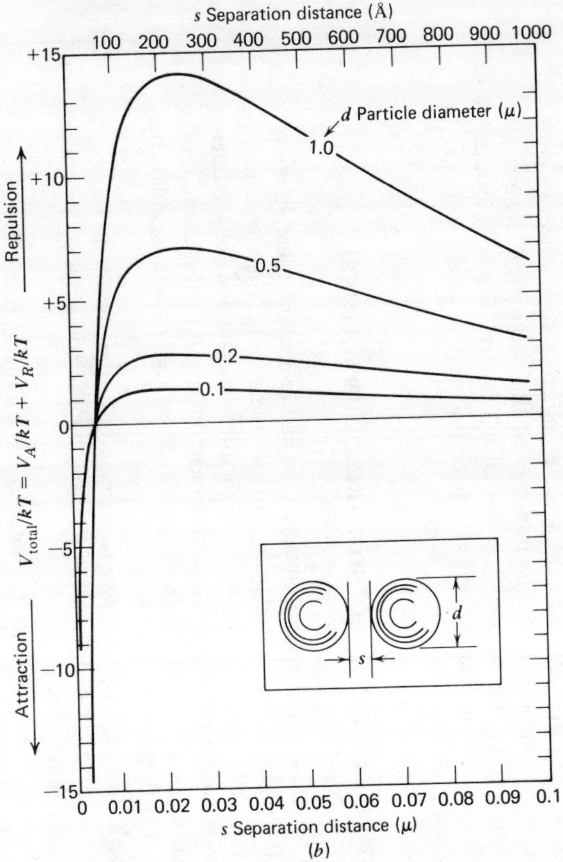

Fig. 12-1*b* Plot of the sum of the attraction and repulsion forces (net total) acting on two particles (for selected particle diameters) as a function of particle separation distance (constant dielectric constant, zeta potential, and double layer thickness).

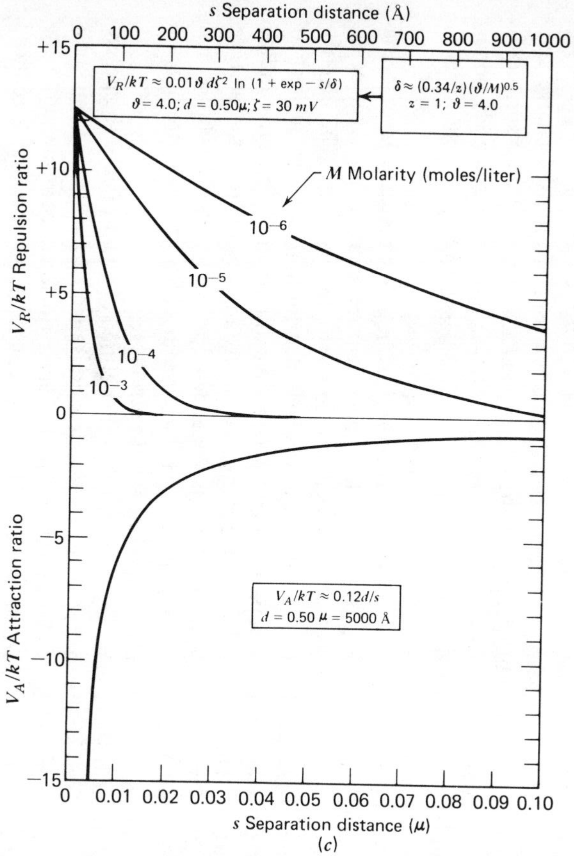

Fig. 12-1*c* Plot of the attraction and repulsion forces acting on two particles of diameter 0.50 μ (5000 Å) for selected solution molarities as a function of separation distance *s* (constant dielectric constant and zeta potential).

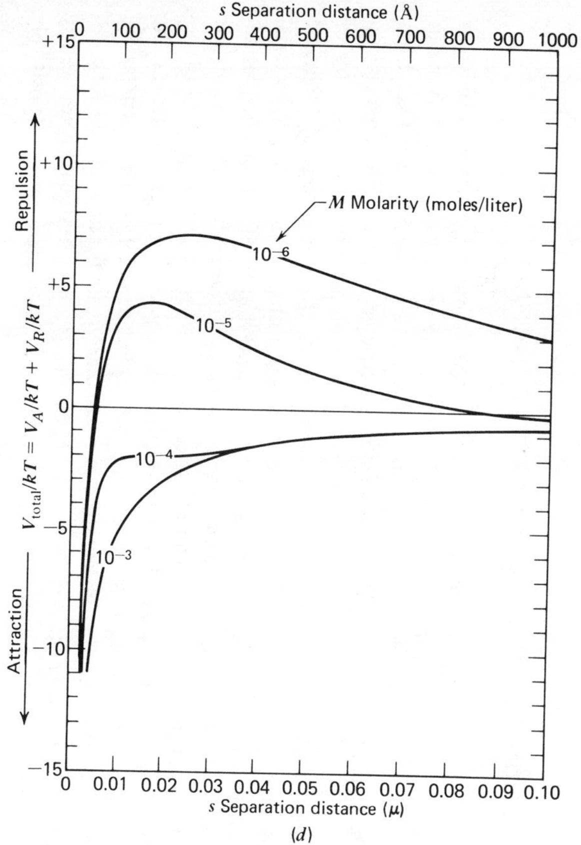

Fig. 12-1*d* Plot of the sum of the attraction and repulsion forces (net total) acting on two particles of diameter 0.50 μ (5000 Å) for selected molarities as a function of the separation distance *s* (constant dielectric constant and zeta potential).

ventional to assign a negative value to an attraction potential and a positive value to a repulsion potential.

REPULSION FORCE (ELECTROSTATIC IN NATURE)

One of the two major repulsive influences that stabilize a particle dispersion is the electrostatic force that arises from the unequal distribution of ions in solution around the particle and at its surface. These ions may originate from an ionogenic source associated with the particle surface (where dissociation occurs under certain conditions) or possibly from an unequal dissolving of particle ions of opposite charge. However, from the standpoint of technical importance, the most significant source of ions is probably the ionizable material (electrolyte) that is deliberately added to a dispersion to effect an unequal adsorption of one of the ion species on the particle surface. In all these cases the net result is the imparting of an electrical charge to the particle because of accumulation of an excess of either positive or negative charges lodged at the particle surface. More often than not, it is found in practice that particles are negatively charged. This observation is particularly true for aqueous dispersions, where the anions (negatively charged and normally less hydrated) are apparently more amenable to adsorption than are the more hydrated cations (positively charged).

Once given a charge due to the preferential adsorption of either negative or positive ions, the particle tends to attract solution ions of opposite charge and to bring them into the vicinity of the charged particle surface. Hence surrounding the charged particle is a layer of solution that is more or less densely populated with ions that are opposite in sign to the charge on the particle surface. These ions are referred to as counterions or gegenions.

The overall situation is that of a *double layer* of charge, one layer being localized on the surface of the particle and the other, neutralizing layer being present in a diffuse region extending outward into the solution. The electrostatic potential developed from this unequal distribution of ions, highest at the particle surface, drops off rapidly with penetration into the surrounding solution. The rate of decay of potential with distance away from the particle surface for a given liquid medium is dependent primarily on the solution ion concentration and the ion valence. An increase in either strongly promotes a rapid decline in electrostatic potential with distance.

Although the double layer has no sharply defined end point (since the potential decays exponentially), it is nevertheless convenient to assign it a thickness value designated by δ. This arbitrary double layer thickness corresponds to the distance from the particle surface where the electrostatic potential is reduced to 0.37 ($= 1/e$) of its original value. This double layer thickness δ in angstrom units (Å) can be calculated from Eq. 9, where ϑ is the dielectric constant of the liquid medium, z is the ion valence (the same for both anion and cation), and M is the ionic concentration (moles/liter).

$$\delta \text{ (in Å)} = \left(\frac{0.34}{z}\right)\left(\frac{\vartheta}{M}\right)^{0.5} \tag{9}$$

From this equation it is seen that the double layer thickness is reduced by an increase in either the ion valence or the ion concentration. In turn it will be shown that a reduced δ value leads to a reduced dispersion stability.

The repulsion potential V_R between two similarly charged particles can be calculated from Eq. 10, using the arbitrarily defined double layer thickness δ. Since the units to be entered into this particular approximate equation are not consistent, they will be briefly discussed.

$$\frac{V_R}{kT} \approx 0.01\vartheta d\zeta^2 \ln\left(1 + \exp - \frac{s}{\delta}\right) \tag{10}$$

where ϑ = dielectric constant: the dielectric constant for air is 1.00; for water, 80; and for alkyd resin solvent solutions, about 4.0
d = particle diameter (microns, μ)
ζ = zeta potential (millivolts, mV); the zeta potential is a measurable quantity that is used to approximate the particle surface potential
s = separation distance between surfaces of two similar particles (Å)
δ = double layer thickness (Å) as given by Eq. 9

Inspection of Eq. 10 reveals some pertinent relationships. Based on the dielectric constant, the repulsion between two particles in a water system should, other things being equal, be 20 times greater than that in an alkyd resin system, since the ratio of the dielectric constants is 20 (= 80/4). Actually the repulsion should be even greater, since in Eq. 9 the double layer thickness is increased 4.5 times (= $20^{1/2}$), and this increase acts favorably in Eq. 10 to increase the repulsion still more. However, things are not necessarily equal, since water strongly promotes ionization. This in turn acts to decrease the double layer thickness, and this reduction could more than offset the advantage of the high dielectric constant of the water.

For any given separation distance, repellency increases with particle size. Also, the particle charge, as measured by its zeta potential, exerts an especially strong repellent influence, since it enters Eq. 10 as the square of the zeta potential, ζ^2.

The influence of the ratio s/δ is more difficult to visualize, since it is locked up in the function $\ln(1 + \exp -s/\delta)$. However, it can be shown that more repulsion results from a smaller s/δ ratio. Hence large values of δ are desirable for dispersion stability.

To portray the effect of a change in particle diameter d on the repulsion potential V_R/kT as a function of the separation distance s there are plotted in Fig. 12-1*a* (top half) some representative data for dispersions of pigment particles of varying diameter in an alkyd resin solvent solution ($\vartheta = 4.0$). The zeta potential is taken as a nominal 30 mV, and the double layer thickness as 680 Å. Inspection of this figure shows that the larger particles are more forcefully repelled for any given separation distance than are the smaller particles.

To portray next the tremendous influence of ion concentration (which controls to a major degree the double layer thickness) there is plotted in Fig. 12-1*c* (top half) the repulsion potential ratio for a series of univalent ion concentrations (moles/liter), based on a constant particle diameter of 0.50 μ (same ϑ and ζ values). It is quite obvious that the repellency effect is drastically reduced as the ion concentration is increased. In fact progressing to higher ion concentrations essentially eliminates all repulsion reponse. The same type of repulsion collapse is also produced by an increase in ion valence. Dispersion stability, which depends on electrostatic repellency, evidently mandates that the ion concentration be held to a minimum and that ions other than univalent ones be avoided.

It is well established experimentally that dispersions stabilized by a double layer can be easily flocculated by the addition of only a trace amount of an indifferent electrolyte (salt), that is, an electrolyte that has no ions in common with the stabilizing (adsorbed) electrolyte. The flocculation power of an indifferent gegenion is proportional to the sixth power of its valence. Therefore the flocculating abilities of uni-, di-, and trivalent gegenions are in the ratio $1^6:2^6:3^6$ or 1:64:729. This points up the necessity of preventing the adventitious introduction of any gegenions, especially those that are di- or trivalent.

COMBINED EFFECT OF ATTRACTION (ELECTROMAGNETIC) AND REPULSION (ELECTROSTATIC) FORCES

In Figs. 12-1*b* and 12-1*d* are graphed the combined influences of the attractive and repulsive forces that are plotted separately in Figs. 12-1*a* and 12-1*c*. Inspection of these graphs is instructive in that it shows the distance of separation between two particles to be a controlling factor in determining whether the particles will be attracted or repelled.

In Fig. 12-1*b* an increasing repulsion between the particles is evident up to a certain point as they approach each other. Beyond this maximum the repellency effect rapidly diminishes, and on closer approach attraction takes over (flocculation occurs). Dispersion stability depends mainly on whether the particles have sufficient thermal energy to surmount this repulsion barrier and penetrate into the attraction region. It should be pointed out that the value kT for the energy assigned to a particle having Brownian motion is an average value and that, in any particle population, particles have energies above and below this mean. Hence to ensure stability against the more energetic particles the repulsion barrier should probably provide a V_{total}/kT value of +10 or more. A maximum repulsion value of +1 would probably be ineffective, since at this level a particle with slightly more than average energy kT could readily zigzag into the attraction zone with consequent flocculation (the particle energy slightly exceeds the maximum repulsion energy).

Inspection of Fig. 12-1*d* shows that, as the solution ion concentration increases, the repulsion barrier fades away. Even traces of salt addition can collapse a repulsion force with inevitable flocculation.

REPULSION FORCE (STERIC HINDRANCE)

An adsorbed layer of material on the surface of a particle can provide a steric (spatial) hindrance to close particle approach by interposing a mechanical barrier between approaching particles. Systems that are stabilized by a combination of steric hindrance and electrostatic repulsion tend to be quite stable. Colloidal materials such as proteins, gums, starches, and cellulosic derivatives are capable of adsorbing on a particle surface to provide a highly viscous and highly structured outside layer. The protective action is enhanced if the nonadsorbed portion of the colloid is especially compatible with the solution (is strongly hydrophilic in the case of aqueous dispersions). When acting in this capacity, these materials are referred to as protective colloids.

It is interesting to speculate on the mechanics of spatial barrier stabilization in the absence of electrostatic repulsion. Consider a situation where the pigment particles adsorb a layer of material (a protective colloid) that is 50 Å thick so that, when the outside surfaces of the adsorbed layers just come into contact, the particles are separated by an effective distance of 100 Å. In the lower half of Fig. 12-1*a* it is seen that for this separation distance the smaller particles are attracted to a lesser degree than are the larger particles. Now, even though the smaller particles are in an attraction zone, it should be remembered that the particle energy varies around a mean value of kT. Hence it is possible that some of the more energetic small particles may be able to break loose and disrupt and redisperse the flocculates as they are formed. On this basis particles with diameters of less than about 0.2 μ could conceivably, with the aid of a protective mechanical barrier, form a moderately stable dispersion even though under a net attractive influence. Particles much larger than about 0.2 μ would be too strongly attracted to break away or promote a redispersion.

In addition to inserting bulk material between two approaching particles, adsorbed layers may actually contribute a repulsive action (other than electrostatic). Theoretical and practical studies suggest that adsorbed layers exert a mutually repellent effect when forced into intimate contact.

COMMENT

From the above simplified approach to the theoretical aspects of pigment dispersion stability and pigment flocculation it is evident that many conflicting factors enter into the problem of ensuring dispersion stability. Many of these are only partially understood and lack quantification. Nevertheless the ideas discussed in this brief review provide the paint engineer with a reasonably sound basis for understanding the theoretical nature of a pigment dispersion.

REFERENCES

1. Derjaguin, B. V. and L. D. Landau, *Acta Physiocochim. URSS*, **14** (1941).
2. Verwey, E. J. W. and J. Th. G. Overbeek, *Theory of the Stability of Lyophobic Colloids*, Elsevier, Amsterdam, 1948.
3. Smoluchowski, M., *Z. Phys. Chem.*, **92,** 129 (1917).
4. Parfitt, G. D., *Dispersion of Powders in Liquids*, Wiley, New York, 1973.

BIBLIOGRAPHY

Adamson, A. W., *Physical Chemistry of Surfaces*, Interscience, New York, 1960.

Becher, P., "Colloid Stability," *JPT*, **41,** No. 536, September (1969).

Crowl, V. T. "Flocculation, Flotation, and Flooding in Phthalocyanine/Titanium Dioxide Pigmented Paints," *JOCCA*, **50,** 1023-1059 (1967).

Mysels, Karol J., *Introduction to Colloid Chemistry*, Interscience, New York, 1959.

13 *Interface Activity, Surfactants, and Dispersants*

Preceding chapters have dealt with the physical nature of surface tension and the theoretical aspects of dispersion stability and pigment flocculation. This chapter considers the reduction of some of these ideas to practice and takes up the profound and far-reaching effects exerted by the very minor amounts of surface active materials that become oriented in or at interface boundaries.

The dispersion of a pigment particle in a vehicle is a process primarily involving surfaces. The molecules residing below the surface merely accept and go along with the activity decided on by the surface molecules. In this regard it is instructive to estimate the ratio of surface to total molecules in a hypothetical spherical pigment particle. If the pigment particle is D units in diameter and each of the pigment molecules of which it is composed has an effective diameter of d units, it can be shown by simple geometry that the ratio of surface to volume molecules is given by Eq. 1.

$$\text{Ratio of surface to volume molecules} = \frac{6d}{D} \qquad (1)$$

For example, let a representative pigment particle be 1.0 μ (0.0001 cm) in diameter (D), and let it be composed of pigment molecules that are 2 Å (0.00000002 cm) in effective diameter (d). Then the ratio of surface molecules to total molecules in the pigment particle is 1 :833.

$$\frac{6 \cdot 0.00000002}{0.0001} = 0.0012 = \frac{1}{833}$$

This means that in this hypothetical particle each surface particle dictates the dispersion fate of some 832 interior molecules that are packed below the pigment surface. This is a rather simplified dramatization of the controlling influence of surface molecules, but it serves to emphasize that a relatively few layers of molecules situated at an interface boundary can completely dominate and determine surface activity such as pigment dispersion.

EFFECT OF SUBDIVISION ON DEVELOPMENT OF SURFACE AREA

For a given volume of solid material, the surface area that is opened up for activity becomes quite tremendous as the material is subdivided (pulverized, powdered, comminuted, disaggregated). For example, consider a cube of solid material 1 cm on edge which has an initial surface area of 6 cm^2 (six faces each 1 cm^3 in area). If the cube is sliced into smaller cubes, each having an edge dimension of L cm, the number of smaller cubes resulting will be $1/L^3$ in number, and each in turn will expose an area of $6L^2$ cm^2. The total area exposed by all the smaller cubes will then be $6/L$, as given by Eq. 2.

$$\frac{1}{L^3} \cdot 6L^2 = \frac{6}{L} \tag{2}$$

Since the initial area of the cube was 6 cm^2, the net result of the subdivision is a $1/L$ increase in overall surface area, as given by Eq. 3.

$$\frac{6/L}{6} = \frac{1}{L} \tag{3}$$

As a general rule, it can be stated that the development of new surface area due to particle subdivision is in inverse proportion to the reduction in some characteristic linear dimension. Thus, if roughly spherical particles are pulverized to one-tenth their original diameter, 10 times as much particle surface becomes exposed; if powdered to one-thousandth of their original diameter, 1000 times as much surface becomes exposed.

PROBLEM 13-1

Calculate the surface area of 1 g of solid carbon black in the form of a single sphere and in the form of tiny spherical particles, each 0.010 μ in diameter. Assume a value of 1.8 g/cm^3 for the carbon black density.

Solution. Calculate the volume for the single solid carbon sphere, and from this determine its area and diameter.

$$V = \frac{W}{\rho} = \frac{1.0}{1.8} = 0.556 \text{ cm}^3$$

$$D^3 = \frac{6V}{\pi} = \frac{6 \cdot 0.556}{3.14} = 1.06; \qquad D = 1.02 \text{ cm}$$

$$A = \pi D^2 = 3.14 \cdot 1.02^2 = 3.27 \text{ cm}^2$$

The ratio of the large particle diameter $D(= 1.02$ cm) to the tiny particle diameter $d(= 0.010\ \mu = 0.000001$ cm) is 1,020,000 : 1 (= 1.02/0.000001). The total surface area of the tiny particles is then 1,020,000 times the area of the single large sphere or 3,330,000 cm^2 (= 1,020,000 · 3.27). This area is equivalent to 333 m^2 or about one-twelfth of an acre.

From Problem 13-1 it can be seen that from a surface standpoint the very small dimensions of pigment particles are more than compensated for by their tremendously large numbers. Thus to surface-treat 12 g of the problem carbon black would require the spreading out of a much smaller quantity of a surface active chemical over *1 acre* of carbon black area.

Figure 13-1 presents a graphical comparison of the dimensions involved in the

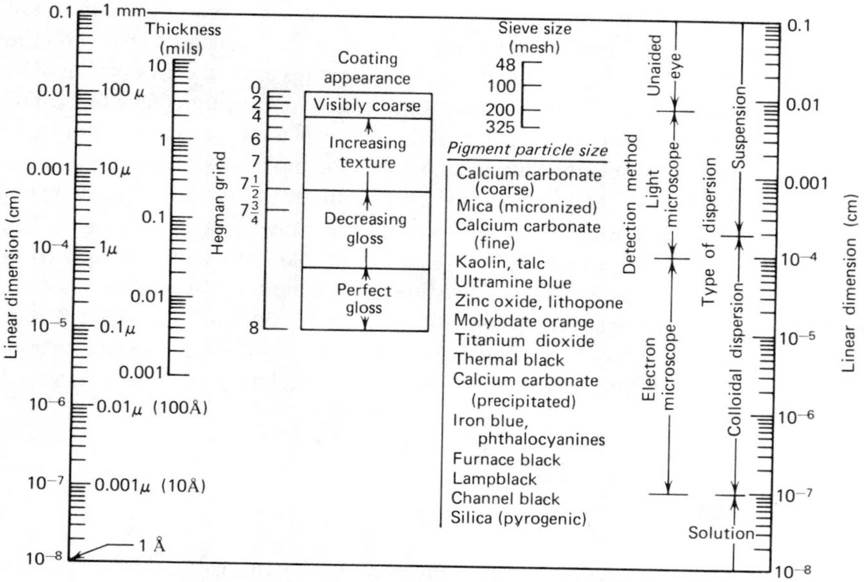

Fig. 13-1 Graphical presentation of the dimensions of pertinent quantities involved in the dispersion of pigments.

dispersion of pigments and serves to show the relation of pigment size to regions of interest. The locations of the boundaries are more or less arbitrary, as are the locations of the representative pigments, which in commercial production would normally be supplied to provide a choice of several average particle diameter grades.

As established earlier, surface tension derives from the unbalanced state of the forces acting on surface molecules. Actually any material exerts some degree of attraction for any and all molecules residing in or at its surface. For a pure material in an otherwise evacuated space the attractive force will obviously be restricted to its own molecular species. However, if the material is brought out of this protected environment and exposed to the atmosphere with its contents of O_2, N_2, CO_2, and H_2O (humidity), or if it is admixed with other materials, the inward attractive pull of the material becomes extended to embrace these foreign materials. From this viewpoint the situation must be highly confused at real interfaces, since the various molecules in a heterogeneous mixture will be competing for positions on the material surface. In the process, oriented layers of molecules often line up at the interface to arrive at an equilibrium condition.

For example, is a glass surface actually a glass surface? Probably not, since glass that has been aged at ordinary room conditions usually has an ultrathin layer of water on its surface that is several molecules thick, depending on the relative humidity. That such an adsorbed film of water is present is confirmed by the fact that methylchlorosilane vapor will react with the adsorbed water on the glass to make the glass water-repellent (methyl groups extend outward from the surface), whereas a completely dry glass (one baked in a vacuum) fails to exhibit significant water repellancy under the methylchlorosilane vapor treatment (1).

From this example it can be appreciated that the molecules at the surface of a pigment particle may, in large part, be quite different from the pigment particles themselves. In general, the surfaces of most pigments are more or less covered by molecules such as O_2, CO_2, and H_2O that have been adsorbed from the atmosphere and/or by salts persisting as adherent residues after washing operations during the pigment preparation. Quite commonly, a pigment is purposely surface treated with foreign molecules to endow it with a surface character and a dispersability performance that are completely alien to the actual pigment.

SURFACTANTS

If the electrically charged atoms of a molecule are so arranged that the plus and minus charges of the atoms are in perfect balance, the molecule is electrically neutral to the outside and is said to be nonpolar. Examples of nonpolar molecules are saturated hydrocarbons (octane, mineral spirits), ring hydrocarbons (toluene), and symmetrical chlorinated products (carbon tetrachloride). However, if the atoms of a molecule are so arranged that the plus and minus charges are not in balance, the molecule is no longer electrically neutral to the outside and is said to be polar. Examples of polar molecules are water, ethyl acetate, and butanol.

It is also convenient to designate certain groups as polar or nonpolar in nature. Examples of nonpolar groups are the alkyl radicals such as $-CH_3$ (methyl), $-C_2H_5$ (ethyl), and higher homolgues of this series. Examples of polar groups are $-OH$ (hydroxyl), $-COOH$ (carboxyl), $-SO_3H$ (sulfonate), $-OSO_3H$ (sulfate), $-NH_4^+$ (ammonium), $-NH_2$ (amino), and $(-OCH_2CH_2)_n$ (polyoxyethylene).

This duality immediately suggests the design of molecules with opposing characteristics, that is, molecules exhibiting polar behavior at one end and nonpolar behavior at the other end. This has been done, and such molecules with their antagonistic properties tend to gather at interfaces in an effort to satisfy their contrary natures. These dual molecules are termed surfactants, a contraction of the three words surface active agents.

Surfactants, then, are surface active agents that migrate to an interface such as the liquid/gas interface of a foam, the liquid/liquid interface of an emulsion, or the liquid/solid interface of a solid particle (pigment dispersion, slurry, paint). Here they alter the interface properties, that is, they exhibit surface active properties. Since they possess both polar and nonpolar characteristics, they are said to be amphipathic in nature.

According to some authorities, a true surfactant should not only strongly lower surface tension, but also (*a*) be soluble in at least one phase of a given disperse system and (*b*) form micelles (aggregates of surfactant molecules) when present in a system above its critical micelle concentration (CMC), that is, the lowest concentration at which micelles first form. However, the large-volume, low-cost ligno-sulfonates do not become oriented at interfaces, form micelles, or even lower surface tension very effectively at low concentrations, yet they are classed as surfactants in the compilation of U.S. statistics on surfactant products (2, 3).

Attempts have been made to relate the functional properties of single surfactants to their chemical structures. One significant finding is that strong wetting properties are apparently provided when the hydrophilic group is centrally located (as between two hydrophobic chains), whereas strong detergent properties result when the hydrophilic group is terminally located (as at the end of a hydrophobic chain) (2).

This is demonstrated by the two isomeric surfactants below, where the hydrophilic $-OSO_3Na$ group is located at a central and a terminal position, respectively.

$$
\begin{array}{l}
CH_3(CH_2)_5CH_2 \\
\qquad\qquad\quad | \\
\qquad\qquad HC{-}OSO_3Na \qquad \text{Strong wetter} \\
\qquad\qquad\quad | \\
CH_3(CH_2)_2CH_2
\end{array}
$$

$$CH_3(CH_2)_{10}CH_2{-}OSO_3Na \qquad \text{Strong detergent}$$

However, with mixed surfactant systems individual functional contributions are less clearly defined.

It is also a general rule that a surfactant becomes more efficient as it approaches the limits of its solubility in any given system.

In aqueous coating systems, where surfactants are of great concern to the coatings and ink engineer, amphipathic behavior is provided by surfactant molecules that contain both hydrophilic (polar, water-loving, solubilizing) and hydrophobic (nonpolar, water-hating) groups. At the aqueous/nonaqueous interface, the hydrophilic part seeks out the water phase, and the hydrophobic part the nonwater phase. By thus preferentially locating on either side of the interface, the contrary parts of the surfactant molecule satisfy their natural affinities. As a result the boundary between the two phases becomes, in effect, an oriented layer of surfactant molecules. Although this is an oversimplified picture of surfactant behavior, the general theoretical concept is valid.

Since there are dozens of both hydrophilic and hydrophobic groups to choose from in designing potential surfactants, it is understandable that literally thousands of surfactants have been synthesized. Of these, over 700 have been found of sufficient interest to be offered for sale in the United States by about 180 manufacturers (3).

One convenient way to break down this vast array of surface active agents is in terms of whether or not they ionize and, if so, the nature of the ionizing group. Thus surfactants are commonly classed as anionic, cationic, amphoteric (anionic or cationic, depending on the pH), and nonionic. Within each class a further subdivision can be made by chemical class, as shown by the pie-graph of Fig. 13-2. Surfactants that contribute multiple ions on a repeating basis (same molecule) are classed as polyelectrolytes. They are not included in the pie-graph statistics of Fig. 13-2, since conventional surfactants generally strongly reduce the surface tension of water, whereas polyelectrolytes seldom meet this requirement.

Since cost is usually a controlling consideration in selecting a surfactant, it should be noted that the average unit pricing for anionic, nonionic, cationic, and amphoteric surfactants is roughly in the ratio of 1:2:3:4, as shown in Table 13-1. However, as is evident from an inspection of the unit costs given in Fig. 13-2, there are wide variations in pricing within any given surfactant class. Because of the minimal cost of the anionics and nonionics, as well as their very favorable technical performance, these two classses constitute the "workhorse" surfactants of industry.

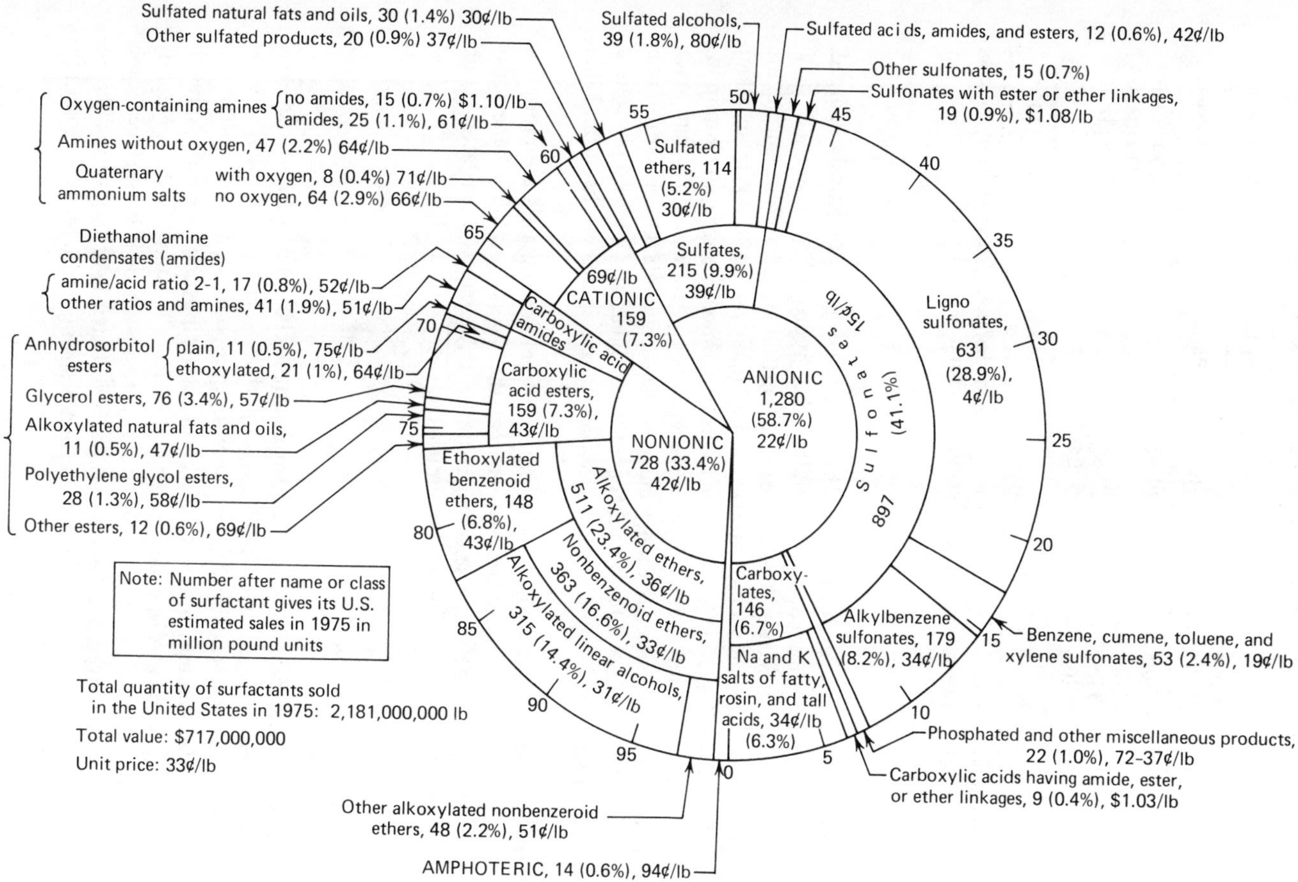

Fig. 13-2 Pie-graph giving the breakdown of surfactant sales in the United States for 1975. Production in 1975 was considerably higher for certain surfactant classes because of captive consumption.

Table 13-1 Surfactant Production[a] in the United States in 1975

Type of Surfactant	Production [million lb and (percent)]	Unit Value (¢/lb)	Unit Value Ratio	Dollar Value [million $ and (percent)]
Anionic	3063 (70.4)	22	1.0	674 (52.5)
Nonionic	1047 (24.1)	42	1.9	440 (34.3)
Cationic	226 (5.2)	69	3.1	156 (12.1)
Amphoteric	14 (0.3)	94	4.3	13 (1.1)
Total	4350 (100%)			1283 (100%)

[a]These production figures are significantly greater than the sales figures given in the pie-graph of Fig. 11-1 because of a large captive consumption and related factors.

Anionic Surfactants

An anionic surfactant is characterized by a relatively large, negatively charged hydrophobic group that consists generally of a hydrocarbon chain and/or a hydrocarbon ring structure. The hydrophilic portion of an anionic surfactant is commonly either an alkali sulfonate group (such as $NaSO_3^-$) or an alkali sulfate group (such as $NaOSO_3^-$). These hydrophilic moieties are quite common, since sulfuric acid and sodium hydroxide are cheap raw materials that react readily with organic materials to give inexpensive surfactants. Other hydrophilic groups found in commercial surfactants are the alkali carboxylate and phosphate structures. Examples of typical anionic surface active agents are listed in Table 13-2.

Surfactants based on sodium, potassium, and ammonium salts are more soluble in water than in hydrocarbons, whereas the reverse solubility behavior is exhibited by calcium, barium, and magnesium salts. Apparently the latter group is seldom used in the coatings and ink industries.

Cationic Surfactants

In contrast to the anionics, a cationic surfactant is characterized by a relatively large, positively charged hydrophobic group that is commonly built around pentavalent nitrogen (quaternary ammonium compounds). Of the surfactants available to the coatings industry, the cationics are possibly the most valuable for promoting pigment dispersion in *organic* vehicles. Presumably this capability is due to the marked ability of cationic agents to adsorb on solid surfaces (especially silicates) to give an attachment that is sufficiently strong to readily displace adsorbed gases and moisture and effect a strong bond. Many cationics for pigment dispersion have the general formula $RNH_3^+Cl^-$, with futher replacement of the hydrogen atoms by alkyl radicals quite common ($RNR_3^+Cl^-$). Examples of typical cationic surface active agents are listed in Table 13-2. Many of the quaternary cationics have germicidal, fungicidal, or algicidal activity. They also tend to be irritating and poisonous. With the exception of the ethoxylated cationics, the cationics are generally incompatible with the ionics, since their oppositely charged heavy ions react to form salts that are insoluble in water. Cationics are used mostly in either nonaqueous or acidic aqueous systems.

Amphoteric Surfactants

An amphoteric surfactant is characterized by chemical groups that permit it to function in either an anionic or a cationic capacity, depending on the pH of the system. The application of this group in the coatings or ink industry is extremely limited and will not be further discussed.

Nonionic Surfactants

As the name implies, a nonionic surfactant does not ionize in aqueous systems but rather hydrates in water, primarily by hydrogen bonding through its oxygen content. The hydrophilic portion of the nonionic molecule usually consists primarily of hydroxyl groups and/or ether linkages. Since both these groups are weakly hydrophilic, they must be present in relatively large numbers in the surfactant molecule to provide adequate water affinity. Also present in many nonionics are the even weaker ester and amide linkages. Hence the hydrophilic portion of the nonionic surfactant may be larger than its hydrophobic portion. One of the advantages of the nonionics is that they are compatible with ionic surfactants. Examples of typical nonionic surfactants are listed in Table 13-2.

Solubilization by polyoxyethylene is the key factor in the design of many nonionics, since the polyoxyethylene chain can be introduced into almost any organic compound with a reactive hydrogen. At least 60%w of the polyethylene oxide is usually necessary to obtain water solubility at room temperature; about 75%w, at 100 C. The higher content for the higher temperature is necessary, since the water solubility of polyoxyethylene surfactants *decreases* with rise in temperature. One test method for checking the properties of a polyoxyethylene surfactant is to determine the cloud point of the surfactant in a water solution by raising its temperature until a characteristic phase separation occurs.

Hard water or low electrolyte concentrations do not affect the surface activity

Table 13-2 Representative Surfactants Listed by Class

Anionics	
Sodium stearate (soap)	$C_{17}H_{35}COO^-Na^+$
Potassium lauryl sulfate (Conco Sulfate P)	$C_{12}H_{25}OSO_3^-K^+$
Sodium alkyl polyphosphate (Victawet 58B)	$R_5(P_3O_{10})_2Na_5$
Sodium dodecyl benzene sulfonate (LAS)	$CH_3(CH_2)_{11}(C_6H_4)SO_3^-Na^+$
Sodium diisopropylnaphthalene sulfonate (Aerosol OS)	$(C_3H_7)_2C_{10}H_5SO_3^-Na^+$
Sodium dioctylsulfosuccinate (Aerosol OT)	$C_8H_{17}COOCH_2$ — $C_8H_{17}COOCH{-}SO_3^-Na^+$
Sodium ethoxylated and sulfated lauryl alcohol (Avirol 100-E)	$CH_3(CH_2)_{11}(OCH_2CH_2)_nOSO_3^-Na^+$
Sodium ethoxylated and sulfated alkyl phenol (Triton X-200)	$RC_6H_4(OCH_2CH_2)_nOSO_3^-Na^+$
Cationics	
Ethoxylated fatty (soya) amine (Varonic L205)	$R(soya){-}N[(CH_2CH_2O)_5H]_2$
Stearylbenzyldimethylammonium chloride (Triton X-400)	$[C_{18}H_{37}{-}N(CH_3)_2{-}CH_2C_6H_5]^+Cl^-$

Lauryl pyridinium chloride	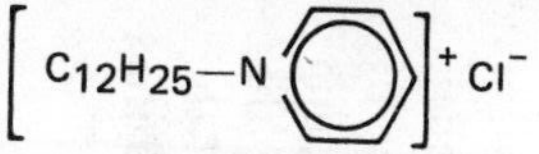
Dodecyltrimethylammonium chloride	$[C_{12}H_{25}—N—(CH_3)_3]^+ Cl^-$
Lauryl ether primary amine (Arosurf MG-70)	$C_{12}H_{25}O(CH_2)_3NH_2$
	Nonionics
Sorbitan monolaurate (Span 20)	$\underline{CH_2(CHOH)_2 \cdot CH}(CHOH)CH_2O(O)CC_{11}H_{23}$
Ethoxylated tridecyl alcohol (Renex 30)	$C_{13}H_{27}O(CH_2CH_2O)_{12}H$
Ethoxylated nonylphenol (Triton N—101)	$C_9H_{19}(C_4H_4)O(CH_2CH_2O)_{10}H$
Polyoxyethylene polyoxypropylene polyoxyethylene (Pluronic series)	$HO(CH_2CH_2O)_x \cdot (CH(CH_3)CH_2O)_y \cdot (CH_2CH_2O)_xH$
Alkylpolyoxyalkylene phosphate (Victawet 12)	$C_8H_{17}OP(O)\ (OR)_2$
Diethanolamide of lauric acid 2/1 (Emid 6540)	$C_{11}H_{23}CON(CH_2CH_2OH)_2 \cdot HN(CH_2CH_2OH)_2$
Tertiary acetylenic glycol (Surfynol 104)	$CH_3CH(CH_3)CH_2C(OH)(CH_3)—C \equiv C—C(OH)(CH_3) \cdot CH_2CH(CH_3)CH_3$

of polyoxyethylene nonionics. By adjusting the length of the polyoxyethylene chain almost any hydrophilic/hydrophobic balance can be obtained.

LOWERING OF SURFACE TENSION DUE TO ADDITION OF SURFACTANT

The following simple experiment serves to dramatize the immediate drop in surface tension that occurs when a minute amount of surfactant (soap) is introduced into a pure liquid (water) surface. Fill a clean dish with tap water, and sprinkle the surface lightly with talcum or similar powder. If the water is free from contamination, the talcum powder will appear as a continuous, slightly whitish, top layer. Take a toothpick, and touch one end to a bar of soap; then touch this same end to the water at the center of the dish. The resulting sudden drawback of the powder to the sides of the dish is a startling demonstration of the surface tension pull of the uncontaminated water (at the outside) tugging against the greatly lowered surface tension of the contaminated water at the center. The purpose of the powder is simply to ride along on the surface of the pure water as it draws away from the center and in this manner delineate the clear inside area, covered with a mono- or multimolecular layer of soap. This layer is oriented, in that the alkali portion of the soap molecule is anchored in the water phase with the hydrocarbon portion of the molecule extended upward and away from the water surface. This means that the surface tension of the contaminated surface is no longer a function of the water molecules but rather a function mainly of the hydrophobic moieties that override the water surface.

Only a fraction of a percent of a surfactant is required to drastically alter the surface properties of a liquid such as water. This is clearly shown by reference to Fig. 13-3, where the surface tensions for the water solutions of four representative surfactants have been plotted as a function of their weight percent concentrations. Inspection of this graph indicates that at about 0.1% concentration (1 part surfactant per 1000 parts of water by weight) the water surface is more or less fully covered by oriented surfactant molecules. Little or nothing is gained by adding further surfactant. As a matter of fact, for three out of the four surfactants it is seen that most of the possible surface tension reduction is already attained at about a 0.02% concentration.

Comparison of the surface tensions of the 0.1% water solutions of the four surfactants with the critical surface tensions of their hydrophobic groups shows that a reasonable match results. This is not surprising, since it is the hydrophobic groups on the outside surface that primarily determine the solution surface tension. Obviously the fluorocarbon surfactants have a tremendous built-in advantage over the hydrocarbon surfactants by virtue of their strongly hydrophobic fluorocarbon structure. In general, fluorocarbon and silicone surfactants can

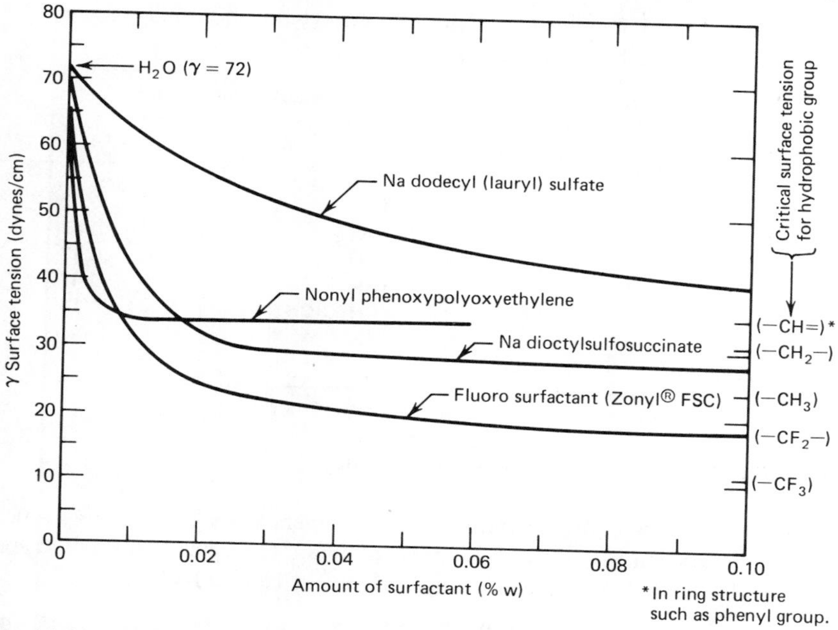

Fig. 13-3 Plot of surface tension for aqueous solutions of four surfactants versus their percent weight concentration.

reduce the surface tension of water (72 dynes/cm) to less than 20 dynes/cm; hydrocarbon surfactants are restricted to a lower limit of about 28 dynes/cm.

HLB NUMBERING SYSTEM FOR RATING THE RELATIVE HYDROPHILICITY OF A SURFACTANT

Of the systems that have been proposed for quantifying the relative strengths of the hydrophilic and hydrophobic (lipophilic) portions of a surfactant, one in particular has met with considerable success. This so-called HLB system is based on an arbitrary numerical scale where 0 is assigned to a surfactant that is overwhelmingly hyrdophobic (lipophilic) and 20 to a surfactant that is overwhelmingly hydrophilic (4). Any number that is then assigned to a surfactant represents a measure of the balance between its hydrophilic and hydrophobic strengths in terms of this HLB scale.

Expressions have been developed for computing approximate HLB numbers from analytic and compositional data. Extensive tables are also available that list the HLB values of many common surfactants (5).

Table 13-3 H and L Numbers for Representative Hydrophilic and Lipophilic (Hydrophobic) Chemical Groups

Hydrophilic Groups		Lipophilic Groups	
Group	H Number	Group	L Number
$NaSO_4$—	39.0	$-\overset{\mid}{C}H-$	0.47
KOOC—	21.0	$-CH_2-$	0.47
NaOOC—	19.0	$-CH_3$	0.47
HOOC—	2.1	=CH—	0.47
HO—	1.9	$-(CH_2C(CH_3)HO)-$	0.11
—O—	1.3		
$-(CH_2CH_2O)-$	0.36		

Calculation Methods for Determining HLB Value

One method proposed for computing HLB numbers is based on Eq. 4, which calls for the assignment of H or L *group* numbers to typical hydrophilic and lipophilic (hydrophobic) chemical groups as listed in Table 13-3 (6).

$$\text{HLB} = 7 + \Sigma\text{H} - \Sigma\text{L} \tag{4}$$

The larger the H or L group number, the greater is its contribution to the hydrophilic or lipophilic strength of the surfactant molecule.

PROBLEM 13-2

Calculate the HLB numbers for oleic acid, sorbitol monooleate, sodium oleate, sodium lauryl sulfate, and propylene glycol monolaurate.

Solution. Substitute the appropriate H and L group numbers in Eq. 4, and solve for the surfactant HLB number.

$$\text{HLB (oleic acid)} = 7 + 2.1 - (18 \cdot 0.47) = 0.6$$

$$\text{HLB (sodium oleate)} = 7 + 19. - (18 \cdot 0.47) = 17.6$$

$$\text{HLB (sorbitol monooleate)} = 7 + (5 \cdot 1.9) - (18 \cdot 0.47) = 8.0$$

$$\text{HLB (sodium lauryl sulfate)} = 7 + 39 - (12 \cdot 0.47) = >40$$

$$\text{HLB (propylene glycol monolaurate)} = 7 + 1.9 - (12 \cdot 0.47) = 3.3$$

Equation 4 is only moderately successful in obtaining the HLB number for a surfactant. However, expressions that are more limited in their surfactant coverage are more reliable. Thus the HLB values of most polyol fatty esters can be calculated from Eq. 5, where S is the saponification number of the ester and A is the acid number of the recovered acid (4).

$$\text{HLB} = 20\left(1 - \frac{S}{A}\right) \tag{5}$$

PROBLEM 13-3

Calculate the HLB number for polyoxyethylene sorbitan monolaurate, which has a saponification number S of 46 and an acid number A of 276 (lauric acid).

Solution. Substitute the given data in Eq. 5.

$$\text{HLB} = 20\left(1 - \frac{46}{276}\right) = 16.7$$

Comment. This is the same value as for a commercial offering of this surfactant (Tween 20; ICI Americas, Inc.).

Where the hydrophilic portion of a surfactant consists of ethylene oxide only, the HLB value can be calculated by Eq. 6, where w is the weight fraction of polyoxyethylene (4).

$$\text{HLB} = 20w \tag{6}$$

PROBLEM 13-4

Calculate the HLB numbers for two polyoxyethylene stearates, where stearic acid (MW = 285) is condensed with 8 and 40 moles, respectively, of ethylene oxide (MW = 44).

Solution. Calculate the weight fractions of the polyoxyethylene contents for the two surfactants, and substitute in Eq. 6 in turn.

8 moles EO (similar to Myrj 45 with listed HLB value of 11.1):

$$w \approx \frac{8 \cdot 44}{8 \cdot 44 + 285} = 0.553; \qquad \text{HLB} = 20 \cdot 0.553 = 11.1$$

40 moles EO (similar to Myrj 52S with listed HLB value of 16.9):

$$w \approx \frac{40 \cdot 44}{40 \cdot 44 + 285} = 0.860; \qquad \text{HLB} = 20 \cdot 0.86 = 17.2$$

Comment. Comparison of the calculated and listed HLB values (5) for these surfactants indicates a good reliability for Eq. 6.

Although many nonionic surfactants lend themselves to HLB calculation (Eqs. 5 and 6), others do not, such as those containing propylene oxide, butylene oxide, and nitrogen and sulfur atoms. Ionic surfactants are even less tractable. Therefore, for such surfactants, HLB values must be determined experimentally by a comparison procedure based on surfactants with established HLB numbers (4). Incidentally, a calculated HLB number greater than 20 simply represents an apparent HLB value that is to be used in combination with other surfactant numbers when calculating HLB values for surfactant blends.

Water Dispersibility Method for Estimating the HLB Number of a Surfactant

A surfactant HLB number can be roughly estimated by simply mixing a small portion of the surfactant with water, observing the nature of the resulting dispersion, and on the basis of this observation assigning a number to the surfactant, using the HLB ranges in Table 13-4 as a guide.

Use of the HLB System in Practice

Although the HLB system was developed originally to save time in the selection of emulsifiers (surfactants) for preparing emulsions, it has related applicability to the selection of surfactants for preparing pigment dispersions (7-9). Some of the guidelines set up for selecting surfactants for emulsions are briefly reviewed. Presumably by analogy and some technical imagination, they can be translated into the selection of test surfactants for pigment dispersions.

The HLB system postulates that each of the oils, waxes, solvents, and other materials (pigments) that are to be incorporated into a disperse aqueous system (emulsion or dispersion) has a characteristic HLB number. Selection of the surfactant system to achieve a satisfactory dispersed composition is then made on the basis of matching the surfactant HLB to that of the material being dispersed

Table 13-4 Estimation of Surfactant HLB Number from Its Dispersibility Behavior in Water

HLB Range	Nature of Dispersion
1–4	Immiscible (no dispersion)
5–6	Unstable or poor dispersion
7–8	Milky dispersion after vigorous shaking
9–10	Stable milky dispersion
11–13	Translucent or grayish dispersion
>13	Clear solution

or emulsified. Characteristic HLB numbers for the common waxes, oils, solvents, and plasticizers are well established and widely reported (4). Unfortunately relatively little information is available regarding pigment HLB numbers, although one listing for pigment colorants has been published and is reproduced in Table 13-5 (7). Also noted are HLB numbers taken from another source (10). The discrepancy in HLB values for some pigments in this table is attributed to the difference in the surface treatment of the pigment particles.

Certain principles have gradually evolved as a result of the continuing use of the HLB system. In the first place, the HLB of a surfactant blend can be simply calculated by computing the arithmetic average of the individual HLB contributions in proportion to their concentrations. In the second place, surfactant blends have been found to provide better stability than single surfactants alone. Presumably this is due partly to the fact that, with blends, proportions can be

Table 13-5 HLB Values for Pigment Colorants (7, 10)

Colorant	HLB Value	
	Ref. 7	Ref. 10
Inorganics		
Lampblack	10–12	
Iron oxide, red	13–15	10–12
Carbon black		14–15
Molybdate orange	16–18	
Titanium dioxide	17–20	
Chrome yellow, medium	18–20	
Iron oxide yellow	20+	
Organics		
BON red, dark	6–8	
Toluidine red, medium	8–10	14–15
Toluidine yellow	9–11	
Phthalocyanine green (blue shade)	10–12	
Phthalocyanine blue (red shade)	11–13	14
Quinacridone violet	11–13	
Nickel azo yellow	11–13	
Phthalocyanine green (yellow shade)	12–14	
Quinacridone red	12–14	
Azo yellow, high strength	13–15	
Hansa yellow		14
Phthalocyanine blue (green shade)	14–16	

readily adjusted to secure an exact HLB match. In the third place, a matching HLB number does not necessarily assure the best possible stability. It only provides the maximum stability that can be obtained with the surfactant system in question. This means that in practice a number of diverse surfactants must be checked out to determine the most efficient system among the many that can be formulated to the same HLB number. The HLB system is a tool that narrows the surfactant selection immeasurably. However, experimentation is still necessary to establish the best surfactant system out of the several different chemical combinations that can be formulated to match a target HLB number.

PROBLEM 13-5

Calculate the HLB value for a mixture of 72% of polyoxyethylene (20) sorbitan monostearate (HLB = 14.9) with 28% of sorbitan monostearate (HLB = 4.7).

Solution. HLB of mixture $= 0.72 \cdot 14.9 + 0.28 \cdot 4.7 = 12.0$.

RELATIVE HYDROPHILICITY OF PIGMENT PARTICLES

A rough qualitative indication of the hydrophilicity (water-loving nature) of a pigment is given by a simple flushing procedure. A small amount of the test pigment is dispersed in a test tube half filled with water, to which is then added about an equal volume of a hydrocarbon solvent (e.g., mineral spirits) and/or a drying oil (linseed oil). The test tube is then shaken, and an observation made as to whether the pigment particles tend to remain in the water phase, collect at the interface, or pass (flush) into the nonaqueous phase. The procedure is then reversed with the pigment particles first being dispersed in the nonaqueous portion, after which water is added and the mixture shaken. Depending on the overall pigment preference for either phase, an estimate can be made as to its relatively hydrophilic or hydrophobic nature.

It is an interesting observation that hydrophilic pigments act to stabilize oil-in-water dispersions, whereas lipophilic pigments tend to stabilize water-in-oil dispersions. Apparently in both cases the pigment reinforces the stability of the continuous phase.

PIGMENT DISPERSANTS

The primary function of a pigment dispersant is to surround a suspended pigment particle with a barrier envelope that by either ionic repulsion or steric hindrance prevents random contact with other particles. This in turn avoids adhesion and flocculation. Pigment dispersants tend to differ somewhat from emulsifiers in that a pigment dispersant attaches itself to the outside solid surface of the pigment, whereas in the case of an emulsifier a portion of the emulsifying agent tends to be submerged and becomes part of the liquid surface of the dispersed phase. The difference between attachment and submergence requires only a slightly different approach in discussing pigment dispersion as opposed to emulsification.

Dispersion of Inorganic Pigments

Most inorganic pigments are oxides (TiO_2, ZnO, Fe_2O_3, SiO_2) or insoluble salts ($CaCO_3$, $BaSO_4$, $PbCrO_4$). Both chemical types possess surfaces that attract and tend to retain an outside layer of water molecules. To a small degree they are self-dispersive. However, in practical pigment dispersions the separation barrier must be greatly improved by addition of an oriented surfactant that is commonly an ionizable molecule. Either the anion or the cation of the ionic surfactant must preferentially adsorb on the pigment surface with the counterion being remotely removed from the surface (a charged double layer). The potential due to the charge separation when the suspended particles are at rest is called the zeta potential (static conditions). When the suspended particles are in rapid motion (mixing, stirring), the repulsion potential increases and the charge difference is called a streaming potential (dynamic conditions).

Inorganic oxide surfaces by themselves exhibit a surface charge in water systems that is characteristic of the oxide and the pH of the water system. In this regard, each pigment has an isoelectric pH value corresponding to the pH where the negative and positive charges on the oxide pigment surface just neutralize each other. Representative isoelectric pH values for common coating pigments are listed in Table 13-6 (2). As the pH is adjusted away from this isoelectric pH, a charge imbalance starts to develop that leads to pigment particle repulsion. Since the introduction of an ionic dispersant into the oxide surface alters the relative balance of positive and negative charges on the surface, it also acts to alter the isoelectric pH value, as shown in Fig. 13-4. For any given pH, adding an ionic dispersant may improve or detract from dispersion stability, depending on whether the charge on the added dispersant acts to reinforce or to neutralize the charge already on the oxide surface.

Separation of pigment particles due to repulsion of like ionic charges on the particle surface is greatly facilitated by increasing the thickness of the steric barrier that separates them. Inorganic pigments by themselves possess a thin (about one molecule deep) adherent water layer. However, the introduction of an adsorbing surfactant acts to increase this water thickness many times to more

Table 13-6 Isoelectric pH Values for Oxide Pigments (Zeta Potential = 0)

Oxide Pigment	pH
Sb_2O_5 (antimony pentoxide)	0.3
SiO_2 (silica gel)	1.8
SiO_2 (quartz)	2.2
TiO_2 (rutile)	4.7
$Al_2Si_2O_5(OH)_4$ (kaolin)	4.8
$Al(OH)_3$ (hydrated aluminum oxide or gibbsite)	5.0
Fe_2O_3 (hematite)	5.2
TiO_2 (anatase)	6.2
Fe_2O_3 (maghemite)	6.7
Cr_2O_3 (chrome green)	7.0
ZnO (zinc oxide)	9.0
PbO (litharge)	10.3

effectively sterically hinder close interparticle contact. The enlistment of water molecules by the part of the surfactant that extends into the aqueous solution to assist in pigment dispersion is commonly referred to as hydrogen bonding. The weak surfactant water bonds that are formed act to (*a*) stabilize the buffer barrier between pigment particles and (*b*) contribute to the low-shear-rate viscosity of the system (thus reducing the need for thickeners in commercial products).

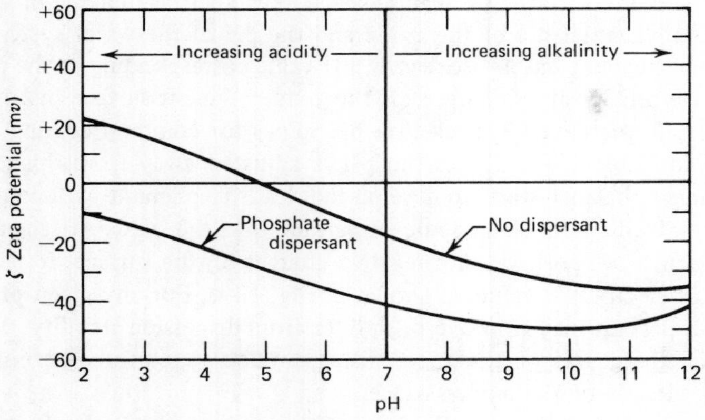

Fig. 13-4 Plot of the zeta potential exhibited by aqueous dispersions of kaolin pigment particles over a range of pH values with and without a phosphate dispersant present.

Dispersants for Inorganic Pigments

It is a general rule that (*a*) the more ionic the surfactant and (*b*) the greater the affinity of its adsorbed ion for the pigment particle surface, the more effective is the surfactant dispersive action. Among the materials ideally equipped to impart this type of ionic activity are the polyelectrolytes (polymeric salts) and materials closely related to them.

Polyelectrolytes. A polyelectrolyte is a macromolecule built up partly or entirely from monomeric units with ionizable groups. As opposed to a simple electrolyte (such as sodium chloride), in which the cation (Na^+) and anion (Cl^-) are relatively small and similar in size, a polyelectrolyte is characterized by a backbone macroion (a single large ion possessing a number of similarly charged groups connected by chemical bonds) and an equivalent number of small, independent, and oppositely charged balancing ions (counterions). Typical polyelectrolyte structures (repeating units) are shown in Table 13-7.

Because of their relatively high molecular weights, polyelectrolytes are also referred to as dispersing resins. They generally have only a moderate number of functional groups per unit weight and for this reason are used at fairly high concentrations (for a dispersant). They are commonly used with aqueous systems.

Since these polyelectrolytes are ionic in nature, they are generally water soluble. Although salts of the simple organic diacids (such as citric and succinic) have been used to provide pigment dispersion, they appear to be less effective than salts of the tri- and higher polyacids. However, this trend toward better

Table 13-7 Representative Polyelectrolyte Structures

Sodium polyacrylate:

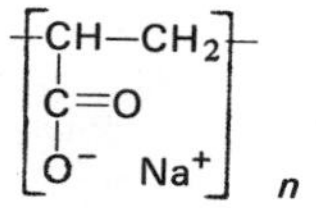

Ammonium poly(styrene/maleate):

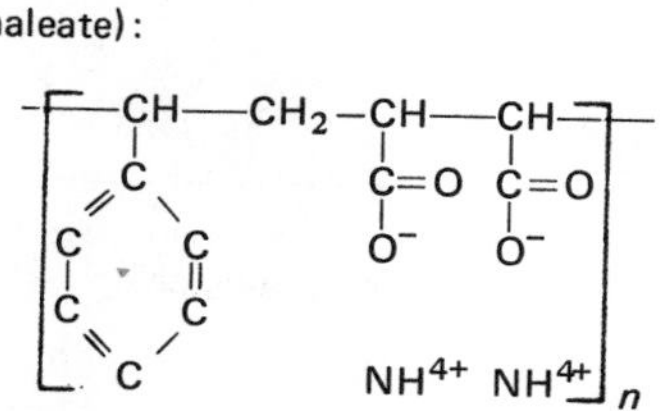

dispersion with increasing polymer size has an upper limitation. For example, polyacrylate salts usually impart optimum dispersing action when composed of between 12 and 18 monomer units.

Alkali Polyphosphates and Similar Alkali Inorganic Polymeric Salts. Polyphosphates are versatile ionic dispersants that at relatively low concentrations disperse a variety of inorganic pigments. Representative polyphosphates are the alkali metal salts of the phosphates listed in Table 13-8.

Unfortunately polyphosphates have a number of disadvantages such as (*a*) contributing to eutrophication (an acceleration of the metabolism and reproduction of assorted fungi on paint film and runoff areas), (*b*) blooming, although this occurs much less with potassium than with sodium ions and also less with longer phosphate chains, (*c*) inducing a viscosity increase in coating compositions on aging, and (*d*) being susceptible to "polyphosphate seeding," a chemical breakdown tendency (depolymerization) exhibited in the presence of silicates, certain transition metals, and pH values > 7 (2). However, polyphosphates are a very common ingredient in latex coatings.

Polyphosphates adsorb by two mechanisms, hydrogen bonding and chemisorption. Hydrogen bonding is usually easily effected at the surface of oxide/hydroxide pigments, since the distance between adjoining oxygen atoms is closely matched by the corresponding oxygen/oxygen distance of polyphosphate dispersants. Chemisorption (partial chemical reaction) also may occur, since polyphosphates tend to react with the surface molecules of many inorganic pigments (e.g., $CaCO_3$, $BaSO_4$, ZnO) to form adherent insoluble phosphates. In general, as the polyphosphate chain lengthens and more functional groups become involved, the more effective is the dispersant, at least up to a polyphosphate molecule with three potassium atoms.

Table 13-8 General Chemistry of Alkali Polyphosphates Where M Is a Monovalent Atom ($M_{n+2}P_nO_{3n+1}$)

Common Name	Ratio of Cationic to Anionic Oxide[a]	Number of Phosphorus Atoms	Representative Formula
Orthophosphate	3.00	1	K_3PO_4
Pyrophosphate	2.00	2	$K_4P_2O_7$ (TKPP)
Tripolyphosphate	1.667	3	$K_5P_3O_{10}$ (KTPP)
Metaphosphate	1.00	n[b]	$Na_n(PO_3)_n$ (Calgon)

[a] Cationic oxide: Na_2O, K_2O; anionic oxide: P_2O_5.

[b] Equivalent to a ring structure or an extremely long chain with the following approximate structure:

$$M(-O-\overset{\displaystyle MO}{\underset{\displaystyle \| \atop O}{\overset{|}{P}}}-)_nOM$$

Potassium polyphosphates are invariably selected over sodium polyphosphates, since the latter suffer from a snowing (blooming) defect. The bloom is a frosty crystalline deposit (sodium salt) that appears and collects on the surface of aged latex paints. Of the potassium polyphosphates, two in particular, potassium tripolyphosphate (KTPP) and tetrapotassium pyrophosphate (TKPP), possess a good balance of desirable properties. Although the two cost about the same, TKPP is claimed to provide lower initial viscosity and better long-term stability.

Inorganic surfactants, which resemble the phosphates by being alkaline in nature and possessing a polymeric structure, are also capable of providing dispersant activity. For example, the alkali polyborates, polyaluminates, and polysilicates function as useful dispersants for inorganic pigments, preferably those with acidic or hydroxylated surfaces (2).

Polyamines and Amino Alcohols. Polyamines impart excellent dispersion properties to many inorganic pigments. However, the polyamine selection must be such that the molecular chain length between the adsorbing polar amine groups is properly adjusted to adsorption sites on the solid particle surface. For example, propylene diamine is a stronger dispersant for clay particles than either ethylene diamine or butylene diamine.

Amino alcohols also provide excellent dispersing action because of their chemical structure, comprising an amino group that adsorbs on the inorganic particle surface and a hydroxyl tail that seeks out the water phase. Since the hydroxyl group of an amino alcohol engulfs more water than the amino group of a diamine, pigment systems dispersed with amino alcohols tend to be more viscous and thixotropic than those dispersed with diamines. The addition of another alcohol group to form an amine diol increases the viscosity of the dispersion still further, since more water is entrained by the dihydroxy tail. In general, the dispersion effectiveness of the amino alcohol increases with chain length.

Since polyamines and amino alcohols may cause dermatitis, they should be handled with care. A good representative amino alcohol is 2-amino-2-methyl-l-propanol, supplied under the trade name AMP-95 (11).

Dispersion of Organic Pigments

Organic pigments are invariably colorants that are used to impart decoration and identification to coating and ink compositions. Organic pigments are commonly crystalline in nature, are supplied in a diameter range from about 0.01 to 0.50 μ, have specific surface areas of around 10 to 100 m^2/g, and may be either hydrophobic or hydrophilic in character (see Table 13-5 for typical HLB values). They also vary greatly in their ease of dispersion, although the overall manufacturing trend is toward providing easier dispersing pigments since the surface treatment of pigments is becoming more sophisticated and more widely practiced.

The selection of a dispersant for a given organic pigment is generally highly specific, although it has been noted with some validity that if lecithin (a natural fatty phosphatide) is unable to improve a dispersion, no other dispersant is likely to do better. The surface treatment of organic pigment is quite diverse. Rosin has been used as a modifier since the 1920s, and since then rosin derivatives have been developed to improve the oxidative and dispersion stability of the surface treatment. In the 1930s the important amine-treated pigments made their appearance. More recently pigments coated with their own derivatives and with synthetic polymers have appeared on the market. In view of this proliferation of surface treatments it is not feasible to recommend specific dispersants for any given pigment type.

One suggested approach for determining a suitable dispersant system is to establish the HLB value of the pigment in question. Once this is known, the surfactant (preferably a blend) can be selected by matching the HLB of the surfactant system to that of the pigment. To this mix there may be added further fortifying dispersants, as discussed in the section entitled "Dispersants for Inorganic Pigments." On the basis of such practical tests as color strength development and freedom from flocculation, this procedure appears to work out well in practice (7-9).

In addition to the dispersants for the organic pigment, actual coating and ink systems generally contain several other assorted additives that are necessary to provide useful compositions. The final result is a confusion of interacting ingredients. In the selection of pigment dispersants, it is always hoped that the theoretical ideas advanced for attaining a stable dispersion will prove valid in practice and that serious incompatibilities in the final composition will be avoided.

ASSESSMENT OF DISPERSANT EFFICIENCY

The efficiency of a dispersant is commonly measured in terms of its effect on either (*a*) the settling behavior of a dispersed pigment suspension or (*b*) the rheological properties (viscosity) of the suspension.

Gravitational Settling of Pigment Particles

Pigment Sediment Volume. If pigment particles, which are initially separated, are prevented from flocculating (are well dispersed), settling to a compact sediment proceeds without interparticle interference (bridging is prevented). Conversely, if the separated particles are so poorly dispersed that they adhere on random contact, a weak network of interconnected pigment particles forms as settling proceeds, resulting in a loose, voluminous, fluffy sediment. Based on this diversity in settling behavior, the efficiency of a dispersant can be measured in terms of its relative ability to provide sediment compactness. Compactness of sediment is conveniently measured in terms of a so-called terminal sediment volume (TSV), which is defined as the ratio of the volume of the final compacted solids (produced by gravitational settling) to the theoretical dry solids volume. The TSV for a pigment suspension without a dispersant may be 10 to 30 times that for the suspension fortified with a very small concentration of a dispersing agent.

Turbidity of Supernatant Liquid. If pigment particles are initially poorly dispersed, then, on random contact, the larger particles hold onto the smaller particles with which they collide. On subsequent settling these smaller particles are swept down with the faster settling larger particles. The result is a clear supernatant liquid resting on an opaque pigment-settled phase with an abrupt interface separating them. Conversely, if the dispersion is prevented from flocculating, each particle settles in direct proportion to the square of its effective radius. As a result the larger particles rapidly settle out, leaving overhead the smaller particles, which drift downward at a much slower pace (or, if small enough, remain permanently in suspension because of Brownian motion). This results in a more or less sustained turbidity in the upper regions of the pigment suspension. Hence, on the basis of overhead appearance with long-term aging, dispersant efficiency can be rated in terms of the ability of the dispersant to impart a significant degree of continuing turbidity during the settling period. This method of assessing dispersant efficiency is less reliable than the pigment sediment volume method.

An initial screening for dispersant efficiency by gravitational settling can be quite simple. Fill a series of test tubes with very dilute solutions of the candidate dispersants (preferably at several dilute concentrations), and add a small, fixed quantity of the pigment to be dispersed. After vigorous shaking, set aside and observe for relative rate of settling, overhead turbidity, and final volume of settlement.

Viscosity of Pigment Dispersions

Measurement of viscosity is possibly the most common technique for assessing the effectiveness of pigment dispersants when formulating actual coating or ink compositions. This is so because the tremendous viscosity reduction provided by a dispersant affords a very practical indication of its dispersive powers.

In one procedure different and relatively small amounts of a test dispersant are added in turn to a fixed volume of the vehicle containing a fixed, high-solids level of the pigment(s) to be dispersed. This ladder series of mixtures is then subjected to intense mixing, and the mixture viscosities measured. From these data the concentration of dispersant that gives minimum viscosity is established.

Another method starts with a high-consistency mixture of the test pigment and vehicle. Into this are titrated, with agitation, successive small increments of the dispersant of interest. After each addition the viscosity is measured, and the concentration of dispersant giving minimum viscosity is determined.

Note that in both tests it cannot be assumed that a higher concentration of dispersant automatically provides a lower viscosity, since either a viscosity inversion (reversal of downward trend) or a plateau effect (region of little change) can be, and usually is, exhibited at relatively low dispersant concentrations. These viscosity inversion and plateau phenomena are illustrated in Fig. 13-5. Some of the reasons for these changes in viscosity trend with increasing surfactant concentration are discussed in Chapter 12 on dispersion theory. Obviously, an excess of dispersant is not only wasteful but also counterproductive.

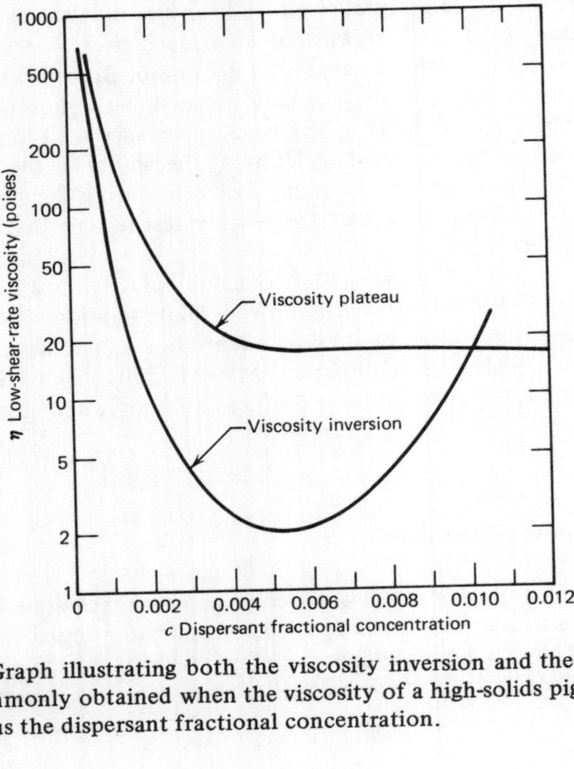

Fig. 13-5 Graph illustrating both the viscosity inversion and the plateau effect that are commonly obtained when the viscosity of a high-solids pigment slurry is plotted versus the dispersant fractional concentration.

It is also generally expedient to check the viscosity of the dispersed system at more than one shear rate in order to gain some idea as to the effect of the dispersant on the non-Newtonian nature of the system (dilatancy, pseudoplasticity, thixotropy).

Pigments vary greatly in the amount of dispersant necessary to effect their initial dispersion and/or to maintain this dispersion. They also vary in the amount of additional dispersant that may be and usually is introduced to maintain the pigment in a dispersed condition when mixed with the other ingredients (components that may tend to disrupt the dispersed state) in the finished formulation. The total amount of dispersant that is required to maintain the pigment in a dispersed condition in the face of such adverse influences is referred to as the *pigment dispersant demand*.

One technique that is useful for determining the pigment dispersant demand is the so-called concentration-aggregation method (12). Specifically, 50 g of the test pigment(s) is mixed with sufficient water to yield a slightly moist, stiff paste. A 10% water solution of the dispersant is then added in small increments, with agitation, until the mixture attains a fluid condition, characterized by the disappearance of surface ripples on gentle shaking. At this stage about 1.0 cm^3 of the fluid mixture is transferred to a watch glass, and 2 to 3 drops of a 6% solution of an ionic thickener such as sodium polyacrylate (Acrysol GS) are added. If insufficient dispersant is present, flocculation or even gelation results because of the increase in the ionic concentration contributed by the ionic thickener and also by a bridging effect among the pigment particles due to particle interconnections made by the long thickener molecule. If such flocculation occurs, the test procedure is repeated with additional dispersant until no evidence of flocculation is observed on the watch glass. Presumably, when the pigment dispersant demand is met, all or most of the adsorption sites on the pigment particles are occupied by dispersant molecules, so that little opportunity exists for thickener molecules to adsorb on the pigment particles and thus disrupt the dispersed condition.

SUMMARY

Dispersing agents include a wide and rather ill defined array of heterogeneous surface active chemicals. However, they are all alike in one respect: they adsorb at solid/solution interfaces to provide a boundary layer that prevents close particle approach. Stabilization of pigment dispersions occurs both by electrostatic repulsion between like charges and by steric hindrance (development of a solvation sheath enveloping the particles). Both stabilizing mechanisms are always present to some extent. Thus the ionic charges on molecules that are adsorbed at an interface not only repel a similarly charged surface but also enmesh water molecules to provide a buffer boundary between pigment particles (a hydration sheath). Nonionic surfactants stabilize completely or largely by the solvation mechanism, as do protective colloids.

REFERENCES

1. Rochow, E. G., *Chemistry of Silicones*, Wiley, New York, 1946.
2. Conley, R. F., "Design, Functionality, and Efficiency of Pigment Dispersants in Water-Base Systems," *JPT*, **46**, No. 594, 51–64, July (1974).
3. USITC Publication No. 804, "Synthetic Organic Chemicals, United States Production and Sales, 1975," U.S. Government Printing Office, Washington, D.C., 1977.
4. Chemmunique Reprint, "The HLB System," ICI United States, Inc., Chemical Specialties Division, Wilmington, Del., 1976.
5. Bulletin LG-60, "Atlas Surfactants," ICI Americas, Inc. Specialty Chemicals Division, Wilmington, Del., 1970.
6. Davies, J. T., *Proceedings of the International Congress on Surface Activity*, 2nd ed., Butterworths, London, 1957.
7. Pascal, R. H. and F. L. Reig, "Pigment Colors and Surfactant Selection," *Off. Dig.*, **36**, 839–852, August (1964).
8. Weidner, G. L., "Quantitative Aspects of Anionic-Nonionic Surfactant Interaction in Latex Paints," *Off. Dig.*, **37**, 1351–1369, November (1965).
9. Rapach, J., "The HLB Approach for Selecting Surfactants for the Coatings Industry," *Am. Paint J.*, pp. 92, 94, 96, 97, 100, May 15, (1967).
10. Bulletin 764-12, "Surfactants in Paints," ICI United States, Inc., Specialty Chemicals Division, Wilmington, Del., 1973.
11. NP Technical Bulletin No. 37, "AMP-95–the Versatile Ingredient for Water-Based Paint Systems," IMC Chemical Group, Hillside, Ill., 1976.
12. Staff article, "Surface Active Agents in Polymer Emulsion Coatings," *Resin Rev.*, **13**, No. 1, 18–32 (1963), published by Rohm & Haas Company.

14 *Solubility and Interaction Parameters*

The saying "Like dissolves like" has been used as a rough but fairly useful rule of thumb for predicting solubility ever since the earliest beginnings of the coatings industry (1, 2). Traditionally, "likeness" has been considered in terms of chemical composition. Thus, in general, hydrocarbon solvents (mineral spirits, toluene) are regarded as conventional solvents for hydrocarbon resins, and oxygenated solvents (esters, ketones) as conventional solvents for oxygenated resins (cellulosics, polyvinyl acetate). This is a convenient concept for estimating solubility behavior, and hence it is natural that the coatings engineer should cast about for more sophisticated techniques for specifying likeness. From an intensive study of the fundamental properties of matter there has now evolved a solubility parameter (interaction) system for evaluating likeness that is quite accurate, although refinements are still being sought to reconcile occasional differences and exceptions.

In this system likeness is specified in terms of a three-component parameter comprising dispersion (nonpolar), polar, and hydrogen-bonding forces. Every solvent (or solvent blend) and every polymer (or resin) is uniquely characterized by these three quantities. The more closely that materials (say solute and solvent) match each other in these three categories, the better is the solvency or compatibility. Such is the essential idea of the solubility parameter system. This chapter is devoted to the many important details that make the system workable.

SOLUBILITY PARAMETER

The concept that solubility is intimately related to the internal energy of solvents and solutes was first proposed in 1916 in a paper by Hildebrand, in which he postulated that molecules subject to the same internal pressure are most effective in attracting and interacting with each other (3). Internal pressure is defined specifically as the energy required to vaporize 1.0 cm^3 of a material (say a solvent). It is expressed by Eq. 1, where ΔE is the molar energy of vaporization, ΔH the latent heat of vaporization, R the gas constant, T the absolute temperature, and V the molar volume ($= M/\rho$). The internal energy/molar volume ratio $\Delta E/V$ is referred to as the cohesive energy density (CED).

$$\text{CED (cohesive energy density)} = \frac{\Delta E}{V} = \frac{\Delta H - RT}{V} \tag{1}$$

The cohesive energy density represents the energy that binds together the molecules in 1.0 cm^3 of liquid or solid material. If these molecules are to be separated into gaseous molecules (vaporized), this energy must be overcome. In the vapor state the energy of molecular interaction is considered as negligible.

Solubility theory postulates that, when a solute is surrounded by a solvent or solvent mixture of similar CED, dissolution occurs. Conversely, if the CEDs are significantly different, the molecules with the greater cohesive energy density cling together and tend to exclude the entry of molecules with the lesser CED value.

For solubility work it has been found convenient to resort to the square root of the cohesive energy density as a working quantity. It is designated a solubility parameter and is expressed by the symbol δ, as shown in Eq. 2.

$$\delta \text{ (solubility parameter)} = \sqrt{\text{CED}} = \sqrt{\Delta E/V} = \sqrt{(\Delta H - RT)/V} \tag{2}$$

If ΔE is expressed in cal and V in cm^3, δ has the unit dimensions of $(cal/cm^3)^{1/2}$. In recognition of Joel Hildebrand's outstanding contributions to solution theory this composite unit is designated a Hildebrand (h).

As with CEDs, solutes and solvents with like solubility parameters (similar hildebrand values) tend to be miscible or compatible; if significantly unlike, they tend to be immiscible or incompatible.

The total solubility parameter δ was the first and, for a time, the only quantity used for predicting solubility behavior. Later hydrogen bonding was established as a contributing factor, initially qualitatively (4) and later quantitatively (5). More recently polarity has been introduced as a third factor controlling solvency and compatibility (6). Today it is generally agreed that cohesive energy arises from the total contributions of the three components: (*a*) nonpolar interactions (dispersion forces) ΔE_d, (*b*) polar interactions ΔE_p, and (*c*) hydrogen-bonding forces ΔE_h, as given by Eq. 3.

$$\Delta E \text{ (total)} = \Delta E_d + \Delta E_p + \Delta E_h \tag{3}$$

Dividing Eq. 3 through by the solvent molar volume V and converting the resulting CED terms to total and partial solubility parameters yields Eq. 4.

$$\delta^2 = \delta_d^2 + \delta_p^2 + \delta_h^2 \tag{4}$$

$$\delta_d \text{ (dispersion component)} = \sqrt{\Delta E_d / V}$$

$$\delta_p \text{ (polar component)} = \sqrt{\Delta E_p / V}$$

$$\delta_h \text{ (hydrogen-bonding component)} = \sqrt{\Delta E_h / V}$$

It should be noted that a similarity between two total solubility parameters alone is an indication, but by no means an assurance, of solvency or compatibility. For example, the three solvents listed below are widely different in their solvency characteristics (compare the partial solubility parameter values), yet their total solubility parameters are quite close.

Solvent	δ	δ_d	δ_p	δ_h
Carbon disulfide	10.0	10.0	0.0	0.0
Nitrobenzene	10.7	9.0	5.0	3.0
2-Ethylhexanol	9.9	7.8	1.0	6.0

SOLUBILITY THEORY

The theoretical justification for matching solubility parameters to achieve solubility is given by Hildebrand's equation (Eq. 5) for the heat of mixing ΔH_M of two materials (say an amorphous polymer and solvent) whose internal energies arise from dispersion forces only. Incidentally, if there is no substantial change in the volume on mixing, as is generally the case, the energy of mixing ΔE_M and the heat of mixing ΔH_M are essentially equivalent. In Eq. 5 the subscripts b and s identify the volume fractions ϕ, the mole fractions x, the molar volumes V, and the solubility parameters δ of the polymer (binder) and solvent, respectively.

$$\Delta H_M \approx \Delta E_M \approx \phi_b \phi_s (x_b V_b + x_s V_s)(\delta_b - \delta_s)^2 \tag{5}$$

From a free energy standpoint, to attain dissolution or miscibility the heat of mixing should be as small as possible. Obviously, this is best achieved when the two solubility parameters have the same value. Such equality makes the factor $(\delta_b - \delta_s)$ in Eq. 5 equal to zero, which in turn makes the heat of mixing $\Delta H_M = 0$.

If the polymer and solvent have similar molar volumes, Eq. 5 reduces to Eq. 6, where V is the molar volume of the mixture.

$$\Delta H_M \approx \phi_b \phi_s \cdot V(\delta_b - \delta_s)^2 \tag{6}$$

Obviously, for this situation the most difficult solvation is experienced when the volume fractions of the polymer and solvent are equal ($\phi_b = \phi_s = 0.50$;

$\phi_b \phi_s = 0.25$). However, the chance that the molar volumes of polymer and solvent are essentially alike is highly unlikely. It is more reasonable to assume that the polymer has a relatively high molecular weight with a correspondingly high molecular volume. In fact, for many practical situations where the molecular weight of the solvent is negligibly small in comparison with the molecular weight of the polymer, it can be shown by algebraic manipulation that Eq. 5 reduces to Eq. 7.

$$\Delta H_M \approx \phi_b V_s (\delta_b - \delta_s)^2 \approx \phi_b \left(\frac{M}{\rho_s}\right) (\delta_b - \delta_s)^2 \tag{7}$$

Equation 7 holds for all but very high polymer volume fractions. Fortunately, the latter range is generally of minor interest to the coatings engineer.

The fractional relationships involved in Eqs. 5, 6, and 7 are shown graphically for a series of hypothetical polymer/solvent conditions in Fig. 14-1, where the expression $\phi_b \phi_s (x_b V_s + x_s V_s)$ from Eq. 5 is plotted versus the polymer fractional volume ϕ_b. The molar volume of the solvent V_s is held constant at 100 cm^3/

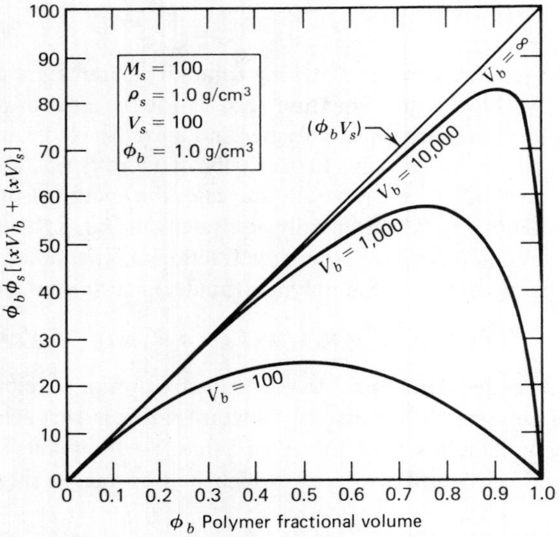

Fig. 14-1 Graph showing change in value of the expression $\phi_b \phi_s [(xV)_b + (xV)_s]$ in Eq. 5 with change in the polymer fractional volume ϕ_b for a series of hypothetical polymer/solvent mixtures. Polymer molar volume varied in steps from 100 to ∞ cm^3/mole; solvent molar volume fixed at 100 cm^3/mole; both densities kept equal to 1.0 g/cm^3.

mole, which the molar volume of the polymer V_b is varied in steps from 100 cm^3/mole to infinity. It is apparent that for equal molar volumes (lowest curve), dissolution is most difficult at equal fractional volumes. However, when the polymer molar volume is very high relative to the solvent molar volume, difficulty in dissolution increases in direct proportion to the increase in the volume fraction of the polymer (except at the very highest polymer concentration).

Inspection of Fig. 14-1 also reveals, as might be anticipated, that high-molecular-weight polymers (high-molar-volume polymers) are more difficult to dissolve. Thus, for any fixed polymer volume fraction ϕ_b, the readings on the $\phi_b \phi_s (x_b V_b + x_s V_s)$ scale corresponding to different V_b values increase as V_b increases. Note also that it is possible to have a concentrated solution of a polymer not too high in molecular weight that on dilution becomes insoluble ("kicks out"). This is indicated graphically on Fig. 14-1 by tracking along a V_b curve from a high to a lower polymer fractional volume. A rising trend indicates reduced solvation, and in certain instances this may be sufficiently severe to cause insolubility. This explains the "kick-out" of resin that is occasionally encountered in practice when cleaning up a manufacturing operation with excess solvent.

Equation 7 also informs us that solvation is less probable or at least more difficult as the molecular weight (molecular volume) of the solvent is increased.

In all the foregoing relationships it should be recognized that the key controlling determinant of solvency is the difference in value between the two solubility parameters $(\delta_b - \delta_s)$. Other equation factors are simply modifying influences. All the solvency phenomena discussed above are borne out by practical experience.

Equations 5, 6, and 7 are based on the assumption that only dispersion (nonpolar) forces are involved. However, the same types of relationships are presumably applicable to polar and hydrogen-bonding forces. As such they can be utilized with dispersion force equations to predict solvency behavior for any polymer/solvent system.

DETERMINATION OF THE TOTAL SOLUBILITY PARAMETER VALUE δ

At least three routes are available for obtaining the total solubility parameter value δ. It can be determined (*a*) by using some expression (such as Eq. 2) that relates δ to the other physical properties of the material, (*b*) by calculating δ from the chemical structure of the material, and (*c*) by systematically matching the solubility nature of the material against the solubilities of other materials of known δ values. These three procedures are considered in turn.

Calculation of δ from Physical Properties

The energy of vaporization per mole of material can be obtained by using Eq. 8, where ΔH is the molar latent heat of vaporization, R is the gas constant (1.986 cal/mole-°K), p is the material vapor pressure, and ΔV is the difference between the vapor and liquid molar volumes, all at absolute temperature $T(= \text{C} + 273)$.

$$\Delta E \approx \Delta H - RT \approx \Delta H - p\,\Delta V \tag{8}$$

At 25 C (77 F), Eq. 8 reduces to Eq. 9.

$$\Delta E_{25} \approx \Delta H_{25} - 592 \tag{9}$$

In the case of many solvents (mainly esters and hydrocarbon solvents), if the normal boiling point of the solvent is known, but not the latent heat of vaporization, the ΔH_{25} value can be conveniently calculated by using Eq. 10, where T_b is the normal boiling point of the solvent (7).

$$\Delta H_{25} = 23.7T_b + 0.020T_b^2 - 2950 \tag{10}$$

By substituting this value for ΔH_{25} in Eq. 9 and then the value thus obtained for ΔE in Eq. 2, there results Eq. 11.

$$\delta = \sqrt{(23.7T_b + 0.020T_b^2 - 3542)/V} \tag{11}$$

PROBLEM 14-1

Toluene ($M = 92.1$; density = 0.866 g/cm³) boils at 111 C under normal atmospheric pressure. Calculate its total solubility parameter at 25 C.

Solution. Calculate the molar volume of the toluene by dividing its molecular weight by its density.

$$V = \frac{M}{\rho} = \frac{92.1}{0.866} = 106.4 \text{ cm}^3$$

The normal boiling point of toluene (absolute temperature) is 384 K (= 111 C + 273). Substitute the above data in Eq. 11 to obtain the value of δ for toluene.

$$\delta \text{ (toluene)} = \sqrt{(23.7 \cdot 384 + 0.020 \cdot 384^2 - 3542)/106.4} = 8.9 \text{ h}$$

Comment. The computed δ value is the correct value for toluene as reported in the literature.

Calculation of δ from Vapor Pressure Data

The total solubility parameter can also be calculated from a knowledge of the change in the liquid vapor pressure with temperature (dp/dT), based on the well known Clapeyron equation, which involves the quantity ΔH (Eq. 12).

$$\frac{dp}{dT} = \frac{\Delta H}{T\Delta V} \tag{12}$$

If it is assumed that the vapor is ideal ($pV = RT$) and also that the liquid volume can be neglected relative to the vapor volume ($V = \Delta V$), Eq. 12 reduces to Eq. 13.

$$\frac{dp}{dT} = \frac{\Delta Hp}{RT^2} \tag{13}$$

Equation 13 can be integrated to yield the Clasius-Clapeyron equation (Eq. 14), which holds for restricted temperature ranges.

$$\ln \frac{p_1}{p_2} = \frac{\Delta H}{R}\left(\frac{1}{T_2} - \frac{1}{T_1}\right) \tag{14}$$

In Eq. 14, T_1 and T_2 are absolute temperatures (C + 273), ΔH is in cal when the gas constant is expressed in cal/mole ($R = 1.99$ cal/mole-°K), and p_1 and p_2 can be expressed either in mm Hg or in some other pressure unit as desired. Changing from natural to common logarithms and inserting for R the value of 1.99 yields Eq. 15.

$$2.3 \log \frac{p_1}{p_2} = \frac{\Delta H}{1.99}\left(\frac{1}{T_2} - \frac{1}{T_1}\right) \tag{15}$$

The vapor pressure of a liquid can normally be related to its absolute temperature through a limited range by Eq. 16, where p is the liquid vapor pressure in mm Hg and T the absolute temperature (= C + 273). A and B are empirical constants.

$$\log p = A - \frac{B}{T} \tag{16}$$

Substituting this equality in Eq. 15 gives Eq. 17.

$$\Delta H = 4.58B \tag{17}$$

PROBLEM 14-2

The vapor pressure of hexane ($M = 86.2$; $p = 0.66$ g/cm^3) in mm Hg through the range from about 10 to 40 C is given by the following expression:

$$\log p = 7.72 - \frac{1655}{T}$$

From the given data calculate the solubility parameter for hexane at 25 C.

Solution. Calculate the latent heat of vaporization ΔH for hexane from Eq. 17, recognizing that $B = 1655$ in the expression given in the problem.

$$\Delta H = 4.58 \cdot 1655 = 7580 \text{ cal/mole}$$

Substitute this value for ΔH in Eq. 8 to obtain a value for ΔE that in turn can be substituted in Eq. 2 to give the desired solubility parameter.

$$\delta \text{ (hexane)} = \sqrt{(7580 - 1.99 \cdot 298) \cdot 0.66/86.2} = 7.3 \text{ h}$$

Comment. The calculated value of 7.3 h for hexane checks with the literature value for this solvent.

Unfortunately the assumption of ideal gas behavior for the solvent vapor phase is often unrealistic, and for more accurate determinations of ΔH more complex expressions must be developed and used, such as Eqs. 18 and 19 (8).

$$p\,\Delta V = RT \cdot X \tag{18}$$

$$\frac{dp}{dT} = \frac{\Delta H p}{RT^2 X} \tag{19}$$

In Eqs. 18 and 19, X is computed from Eq. 20, where p_c and T_c are the critical pressure and critical temperature, respectively, for the test solvent.

$$X = \sqrt{1 - [(p/p_c) \cdot (T_c/T)^3]} \tag{20}$$

Eqs. 18, 19, and 20 have been used to generate an extensive tabulation of total solubility parameter values involving nearly 700 liquids, mostly solvents (8).

Calculation of δ from Surface Tension Data

The total solubility parameter δ has been related to surface tension γ because liquids with high δ values invariably yield high γ values. The expression of Eq. 21, which was first developed and used by Hildebrand and Scott (7), has since been employed to establish excellent correlation between these two quantities for many liquids (9).

Table 14-1 Recommended Values for the Constants K and a to Be Used in the Equation $\delta = K(\gamma/V^{1/3})^a$

Type of Liquid (Solvent)	K	a
Nonoxygenated		
Hydrocarbons		
Aliphatic, saturated	4.31	0.40
Aromatic	4.56	0.37
Halogenated	4.29	0.41
Amines		
Primary[a]	3.93	0.47
Secondary and tertiary	4.10	0.44
Overall (nonoxygenated)	4.21	0.43
Oxygenated		
Esters, ethers, and amides	3.58	0.56
Ketones[b]	5.96	0.25
Alcohols[b]	5.86	0.39
Carboxylic acids[b]	4.12	0.58

[a] Except ethylamine.
[b] Only fair correlation.

$$\delta = K\left(\frac{\gamma}{V^{1/3}}\right)^a \tag{21}$$

Table 14-1 lists recommended values for K and a that have been calculated for several liquid types (9). Except where noted, excellent δ values can be calculated through the use of Eq. 21.

PROBLEM 14-3

The surface tension of n-butyl acetate is 25.2 dynes/cm (20 C), and its molar volume is 132 cm^3/mole. Calculate its solubility parameter.

Solution. Substitute the given data for n-butyl acetate in Eq. 21, and solve for δ using the values for K and a given for esters in Table 14-1.

$$\delta\ (n\text{-butyl acetate}) = 3.58\left(\frac{25.2}{132^{1/3}}\right)^{0.56} = 8.8\ \text{h}$$

Comment. The calculated value of 8.8 h compares favorably with a literature value of 8.6 h for this solvent.

Equation 12 can be rewritten to give surface tension as a function of the total solubility parameter.

$$\gamma = \left(\frac{\delta}{K}\right)^{1/a} (V^{1/3}) \tag{22}$$

PROBLEM 14-4

Calculate the surface tension of *n*-octane, given its molar volume of 164 cm^3/mole and δ value of 7.55 h.

Solution. Using the constants from Table-1 that apply to aliphatic hydrocarbon solvents, substitute the given data in Eq. 22 and solve for the octane surface tension.

$$\gamma\ (n\text{-octane}) = \left(\frac{7.55}{4.31}\right)^{1/0.4} \cdot 164^{1/3} = 22.2 \text{ dynes/cm}$$

Comment. This value compares favorably with an observed surface tension for *n*-octane of 21.8 dynes/cm.

Calculation of δ from Chemical Structure

The total solubility parameter can be calculated from the chemical structure of a material by using in Eq. 23 Small's molar attraction constants as listed in Table 14-2 (10).

$$\delta = \left(\frac{\rho}{M}\right) \sum G = \left(\frac{1}{V}\right) \sum G \tag{23}$$

The individual molar attraction constants G are additive over the structural formula, and ΣG represents the sum of the attraction constants for the constituent atoms and chemical groupings in a unit molecule. This chemical structure method for establishing a δ value is reasonably accurate for many solvent classes, although less reliable for hydrogen-bonded materials (such as alcohols and carboxylic acids).

The unique merit of this evaluation system lies in its application to polymers that cannot be evaluated for δ from their other physical properties, as discussed in the preceding section. In working with polymers, the repeating unit of the polymer chain is made the basis for the computation.

PROBLEM 14-5

Calculate the solubility parameter for an epoxy resin (density 1.15 g/cm^3) in which the repeating mer unit has the following chemical structure:

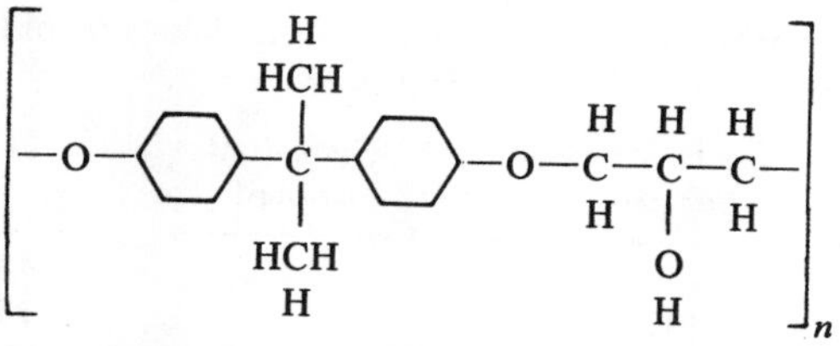

Solution. List the chemical groupings in the repeating mer unit for the epoxy resin; and, using the molar attraction constants listed in Table 14-2, calculate the sum of the molar attraction values (see Table 14-3). The molecular weight of the repeating unit is calculated from its chemical formula, $C_{18}H_{20}O_3$.

$$M = (18 \cdot 12) + (20 \cdot 1.0) + (3 \cdot 16) = 284$$

Calculate the solubility parameter by substituting the data developed in Eq. 23.

Table 14-2 Molar Attraction Constants at 25 C

Group	*G*	Group	*G*
Single-bonded carbon		H (variable)	80–100
$-CH_3$	214	O (ethers)	70
$-CH_2-$	133	Cl (mean)	260
$-\underset{\vert}{C}H-$	28	Cl (single)	270
		Cl (twinned as in $-CCl_2-$)	260
		Cl (tripled as in $-CCl_3$)	250
		Br (single)	340
$-\overset{\vert}{\underset{\vert}{C}}-$	−93	I (single)	425
		S (sulfides)	225
Double-bonded carbon		CO (ketones)	275
$=CH_2$	190	COO (esters)	310
$=CH-$	111	CN	410
$=\underset{\vert}{C}-$	19	CF_2 (*n*-fluorocarbon)	150
		CF_3 (*n*-fluorocarbon)	274
Triple-bonded carbon		SH (thiols)	315
$CH\equiv C-$	285	ONO_2 (nitrates)	ca. 440
$-C\equiv C-$	222	NO_2 (aliphatic nitro compounds)	ca. 440
Conjugation	20–30		
Cyclic structures		PO_4 (organic phosphates)	ca. 500
Phenyl	735	OH (hydroxyl)	ca. 320[a]
Phenylene (o, m, p)	658		
Naphthyl	1146		
Ring (5-membered)	110		
Ring (6-membered)	100		

[a] An empirically derived value, not originally presented in Small's listing of molar attraction constants.

Table 14-3 Summation of Molecular Attraction Values (Problem 14-5)

Group	Molecular Attraction Constant G	Number of Groups	ΣG (+)	ΣG (−)
$—CH_3$	214	2	428	
$—CH_2—$	133	2	266	
—CH—	28	1	28	
—C—	-93	1		-93
Phenylene	658	2	1316	
—OH	320	1	320	
—O—	70	2	140	
			ΣG = 2498	-93 = 2405

$$\delta \text{ (epoxy resin)} = \frac{1.15}{284} \cdot 2405 = 9.7 \text{ h}$$

Small's molar attraction constants were first developed and reported in 1953 (10). In 1970 these constants were reexamined over a wide spectrum of compounds, using multiple regression analysis, and a new set of values with expanded coverage was published (8). Improved correlation was obtained by assuming that carboxylic acids exist as dimers and that glycol ethers are intramolecularly hydrogen bonded. The concept was also advanced that certain molecules have the ability to assume the polar nature of their surrounding environment, in that in polar solvents the materials are capable of interacting as polar solvents whereas in nonpolar solvents the polar interactions are self-contained and the materials tend to behave in a nonpolar manner. This "chameleonic" principle has served to explain certain puzzling solvency anomalies such as those experienced when glycol ether solvents are used as coupling solvents for solubilizing relatively nonpolar resins in water or when they are used to prevent blushing and cratering of nonpolar resions during drying under humid conditions (8).

Determination of δ by Matching Solubility Performance

The solubility parameter of a given material can be determined experimentally by observing and comparing its solvency behavior with products of known δ values. This method assumes that the solubility parameter of a material of unknown δ value lies at the center of the range of δ values of materials with which it is miscible or compatible. If the material is a polymer of unknown δ value, it is conveniently evaluated against a spread of solvents of known δ values. Observations are made as to degree of solubility (complete, partial, borderline, insoluble) or degree of swelling. It has been suggested that, in general, visual observations of 10% solutions of solute in solvent are suitable for a solubility parameter study of this type, although lesser solute concentrations such as 3% or lower have also been used.

DETERMINATION OF PARTIAL SOLUBILITY PARAMETER VALUES

The dispersion (nonpolar), polar (dipole-dipole, dipole induced-dipole), and hydrogen-bonding forces that bind molecules together are referred to as secondary bonding forces. Together they are responsible for the cohesive energy density of a material. However, it is not an easy task to isolate these forces or to assign them quantitative values. Actually, hydrogen bonding is still not very clearly defined, nor is there any precise and generally accepted method for its measurement. Despite such theoretical and technical difficulties, procedures have been developed to closely approximate values for partial solubility parameters. Some of these techniques are considered.

Polar and hydrogen bonding forces are commonly grouped as a composite unit, and when this is done they are referred to collectively as association forces and assigned the subscript a, as indicated in Eq. 24.

$$\delta_a^2 = \delta_p^2 + \delta_h^2 \tag{24}$$

Determination of the Dispersion (Partial) Solubility Parameter Value

Since a saturated hydrocarbon material has no polar or hydrogen-bonding components, its dispersion energy equals its total internal energy. Because of this fact, saturated hydrocarbons can be made the basis for estimating the dispersion energy for association-type materials when there is a close steric resemblance between the two. Thus a solvent with association bonds can be systematically compared with a spectrum of saturated hydrocarbon solvents to determine the hydrocarbon solvent which most nearly matches in size (molar volume) and shape. Such a matching hydrocarbon is called a homomorph. An equivalency of ΔE (total energy of saturated hydrocarbon homomorph) and ΔE_d (dispersion energy of associated solvent) is then made at comparable reduced temperatures. For example, n-butanol has a critical temperature of 288 C (561 K) and a molar volume of 92 cm^3/mole. At a temperature of 25 C (298 K), its reduced temperature is $T_r = 0.531$ (= 298/561). At the same reduced temperature the energy of vaporization for the saturated hydrocarbon homomorph solvent (same molar volume as *n*-butanol) is 5600 cal/mole [homomorph energy value read off graph given in paper by Hansen (11)]. The dispersion solubility parameter for *n*-butanol is then calculated from Eq. 2.

$$\delta_d \text{ (n-butanol)} = \sqrt{5600/92} = 7.8 \text{ h}$$

Refinements in the homomorph concept have been made to account for occasional inconsistencies such as those arising from the presence of halogen atoms, double bonds, and the like.

Determination of the Polar (Partial) Solubility Parameter Value

A number of expressions have been derived relating the polar portion of a solubility parameter to the dipole moment. One of the simplest is given by Eq. 25, where μ is the dipole moment expressed in Debye units and V is the molar volume. Other equations are available that provide more reliable results, but they are more complex and involve additional variables.

$$\delta_p = \mu\sqrt{300/V} \tag{25}$$

PROBLEM 14-6

Calculate the polar (partial) solubility parameter values for *n*-butanol ($\mu = 1.7$; $V = 91.5\ \text{cm}^3/\text{mole}$) and nitroethane ($\mu = 3.6$; $V = 71.5\ \text{cm}^3/\text{mole}$).

Solution. Substitute appropriate data in Eq. 25, and solve for δ_p.

$$\delta_p \text{ (}n\text{-butanol)} = 1.7\sqrt{300/91.5} = 3.1 \text{ h}$$

$$\delta_p \text{ (nitroethane)} = 3.6\sqrt{300/71.5} = 7.4 \text{ h}$$

Comment. These values compare favorably with literature values of 2.8 and 7.6 h for *n*-butanol and nitroethane, respectively.

Determination of the Hydrogen-Bonding (Partial) Solubility Parameter Value

For the important case of hydrogen bonding by hydroxyl groups $(\text{—OH})_n$, approximate H-bonding δ values can be calculated by using Eq. 26, where n is the number of hydroxyl groups per molecule and V is the molar volume.

$$\delta_h = \sqrt{4530n/V} \tag{26}$$

PROBLEM 14-7

Calculate the δ_h values for *n*-butanol ($V = 91.5$ cm^3/mole), ethylene glycol ($V = 55.8$ cm^3/mole), and glycerol ($V = 73.3$ cm^3/mole).

Solution. Substitute appropriate data in Eq. 26, and solve for δ_h.

$$\delta_h \text{ (}n\text{-butanol)} = \sqrt{4650/91.5} = 7.1 \text{ h}$$

$$\delta_h \text{ (ethylene glycol)} = \sqrt{4650 \cdot 2/55.8} = 12.9 \text{ h}$$

$$\delta_h \text{ (glycerol)} = \sqrt{4650 \cdot 3/73.3} = 13.8 \text{ h}$$

Comment. These calculated values compare closely with literature δ_h values of 7.7, 12.7, and 14.3 for *n*-butanol, ethylene glycol, and glycerol, respectively.

Determination of Partial Solubility Parameter Values for Association Forces by Using Chemical Group Contributions

Partial solubility parameter values for polar and hydrogen-bonding forces can be calculated by resorting to tables listing the contributions of the attraction constants for different chemical groups to the two partial solubility parameters. The δ_p or δ_h value for a molecule as a whole is computed by summing up the individual contributions of the constituent groups to the molecule. Table 14-4 gives a list of some representative attraction constants for calculating δ_p and δ_h using Eqs. 27 and 28, respectively.

$$\delta_p = \left(\frac{1}{V}\right) \sum G_p \tag{27}$$

$$\delta_h = \sqrt{(1/V) \sum G_h} \tag{28}$$

PROBLEM 14-8

Calculate δ_p and δ_h for *n*-butyl acetate ($V = 132.5$ cm^3/mole) and ethoxyethanol ($V = 97.8$ cm^3/mole).

Solution. Substitute appropriate group attraction constants from Table 14-4 in Eqs. 27 and 28, and solve for the required partial solubility parameter values.

$$\delta_p \text{ (}n\text{-butyl acetate)} = \frac{250}{132.5} = 1.9 \text{ h}$$

Table 14-4 Attraction Constants for Chemical Groups Contributing to Association-Type Partial Solubility Parameter Values

Group	Polar Attraction Constant G_p (Eq. 27)	H-Bonding Attraction Constant G_h (Eq. 28)
$-CH_3$, $-CH_2$, $>CH-$	0	0
$-Cl$ (chloride)	300	100
$-O-$ (ether)	200	600[a]
$>C=O$ (ketone)	390	800
$-COO-$ (ester)	250	1250
$-OH$ (hydroxyl)	170	$4650/n$[b]
$-COOH$ (carboxylic acid)	220	2750

[a] Disregard contribution of ether group in calculating δ_h for ether-alcohols.
[b] Number of hydroxyl groups per molecule = n.

$$\delta_h\ (n\text{-butyl acetate}) = \sqrt{1250/132.5} = 3.1\ \text{h}$$

$$\delta_p\ (\text{ethoxyethanol}) = \frac{200 + 170}{97.8} = 3.8\ \text{h}$$

$$\delta_h\ (\text{ethoxyethanol}) = \sqrt{4650/97.8} = 6.9\ \text{h}$$

Comment. Note that in computing δ_h for the ether-alcohol the contribution of the ether group is disregarded.

PRACTICAL APPLICATION OF THE SOLUBILITY PARAMETER CONCEPT

A number of systems have been developed to make the solubility parameter concept useful to the practicing engineer. All are capable of yielding practical results. Resort is usually made to either two-dimensional (2-D) or three-dimensional (3-D) plotting to show graphically the interrelationships among the solubility parameter quantities. Four of these systems are listed in Table 14-5.

Presumably the Hansen system, based on the three partial parameters (δ_d, δ_p, and δ_h), will outsurvive the three competitive systems, since it is both fundamentally correct and amenable to practical utilization.

The system proposed by Gardon fails to quantitatively distinguish between the two association parameters δ_p and δ_h and is theoretically incorrect in using

Table 14-5 Some practical Solubility Parameter Systems

Main Author	Parameters	Comment
Gardon (6, 12, 13)	δ p (fractional polarity)	Interpretation modified by an H-bonding qualification factor; adapted to computer programming
Crowley (14–16)	δ μ (dipole moment) γ (H-bonding number)	A base 2-D chart of δ versus μ used for the contour mapping of H-bonding numbers; adapted to computer printout
Teas (17, 18)	$\delta_d/(\delta_d + \delta_p + \delta_h)$ $\delta_p/(\delta_d + \delta_p + \delta_h)$ $\delta_h/(\delta_d + \delta_p + \delta_h)$	Fractional quantities plotted on triangular graph; liberties taken with theory to simplify graphical computation work
Hansen (11, 19, 20)	δ_d δ_p δ_h	Plotting in 3-D space; the only theoretically correct system of the four listed

the total solubility parameter δ as one of the scale dimensions in mapping out solubility areas on two-dimensional plots.

The system proposed by Crowley also suffers from the fact that one of the scale dimensions is given in terms of δ. However, this system does distinguish between polar and H-bonding forces in a quantitative manner, although their measurement is made in arbitrary units (dipole moment and H-bonding number). The dipole moment is approximately related to δ_p by semiempirical expressions such as Eq. 25. The H-bonding number is defined in terms of the shift in wave number in the OD band of deuterated methanol when added in small amount to the test solvent in benzene or hexane (an H-bonding number unit equals a shift of 10 wave numbers). In general, an H-bonding number tends to be 1.5 to 4.0 times as large numerically as its corresponding δ_h value.

The system proposed by Teas is theoretically unsound because the ratios used in constructing the triangular charts have been reduced to expressions involving the first instead of the second powers of the partial solubility parameters. Also the quantitative aspects of the partial solubility parameters have been disregarded, since they are plotted only as fractional quantities.

The fact that all three of these questionable systems yield practical information despite their serious defects attests to the underlying validity and usefulness of the solubility parameter concept. However, the future would appear to favor the Hansen system, which is built on a theoretically sound basis.

Three-Dimensional Partial Solubility Parameter System

The visualization of three-dimensional plotting is difficult. It is even more difficult to routinely graph data in three-dimensional space. However, it is a fortunate circumstance that the three-dimensional space taken up by the three partial solubility parameters assumes the shape of a flat box (essentially 3 × 9 × 12 h). To further accentuate this flatness, the box can be sliced more thinly to provide two flat tiers representing low and high dispersion parameter values, respectively. This is shown schematically in Fig. 14-2, where the top and bottom tiers have the following dimensions:

	Range of Hildebrand Units (h)		
Tier	δ_d	δ_p	δ_h
Top (high δ_d)	1.6 (8.4–10.0)	9.0 (0–9)	12.0 (0–12)
Bottom (low δ_d)	1.3 (7.0–8.3)	9.0 (0–9)	12.0 (0–12)

For practical purposes each tier can be considered as substantially flat. With this as a premise, a spectrum of liquids (mostly solvents) has been graphed on two-dimensional charts (Figs. 14-3 and 14-4) for the low and high dispersion parameter ranges, respectively. Coding symbols have been introduced, as indicated on the plots, to show the upper and lower solvent positions within each tier. The data for these plots, as compiled by Hansen, are believed to be reasonably reliable.

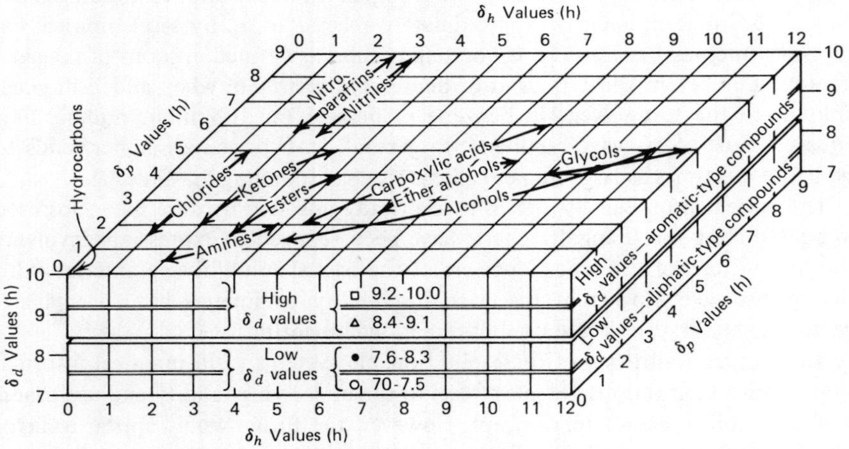

Fig. 14-2 Schematic diagram illustrating the arbitrary splitting of the δ_d-value scale of a 3-D partial solubility parameter space diagram into a lower (7.0–8.3 h) and a higher (8.4–10.0 h) tier to effectively flatten the 3-D space into two essentially plane (2-D) areas. Also indicated are specific regions occupied by the different solvent classes. The lower tier comprises mostly aliphatic-based solvents; the upper tier, mostly aromatic-based solvents.

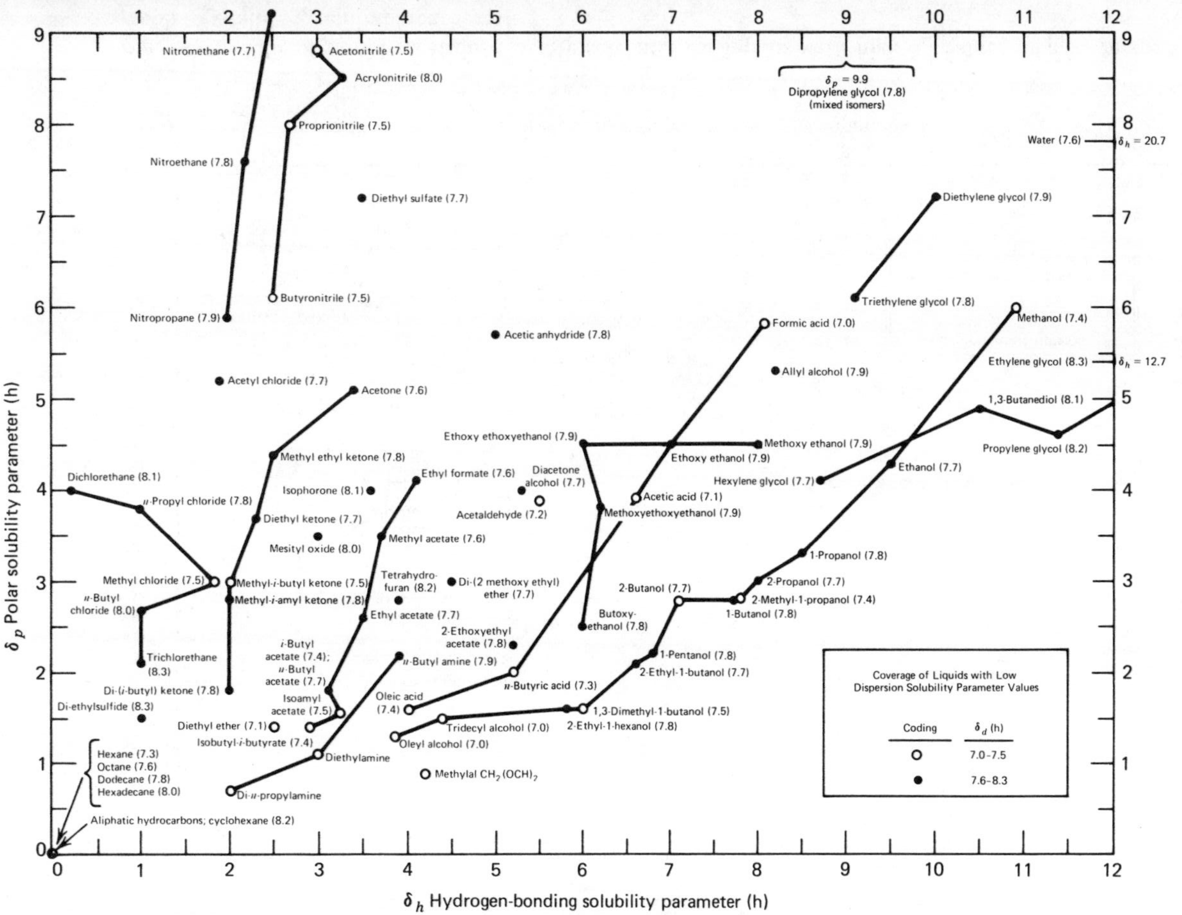

Fig. 14-3 Plot of solvent locations for solvents with low δ_d values on a δ_p versus δ_h graph (δ_d values given in parentheses).

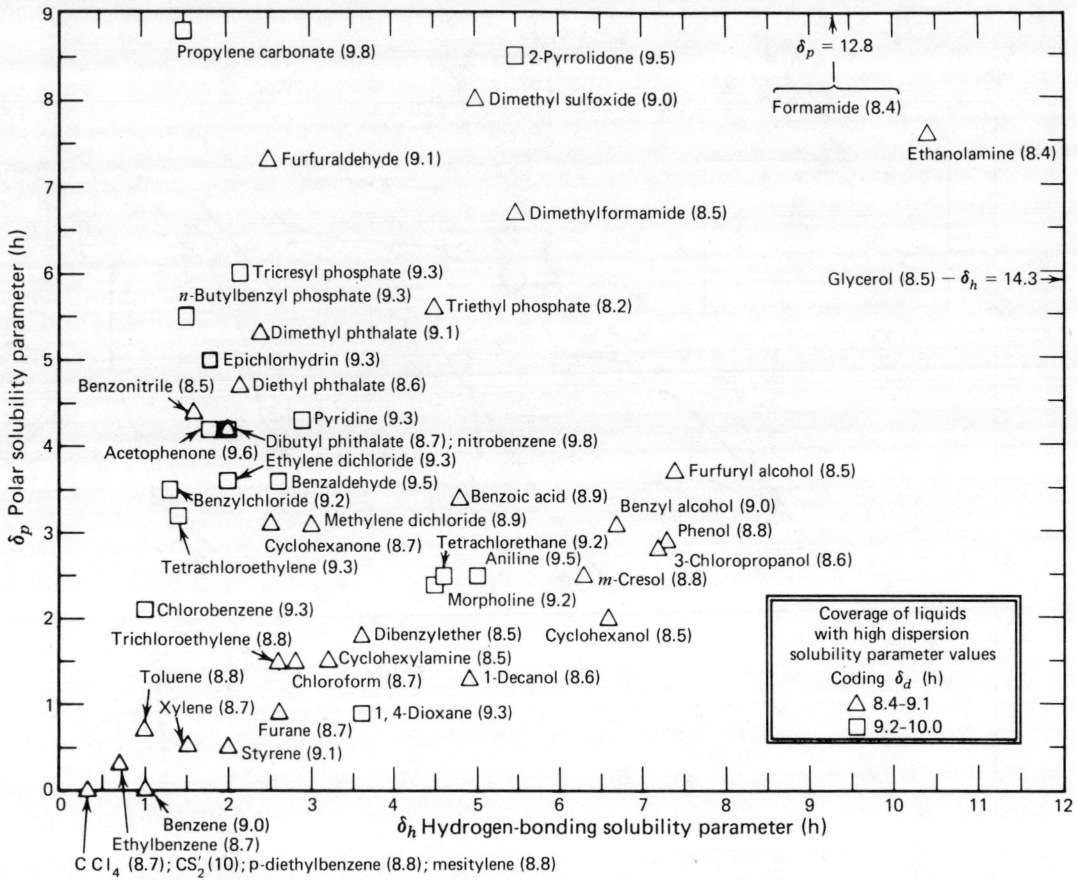

Fig. 14-4 Plot of liquid locations (mostly solvents) for liquids with high δ_d values on a δ_p versus δ_h graph (δ_d values given in parentheses).

Inspection of Figs. 14-2, 14-3, and 14-4 reveals five very important points.

1. All the hydrocarbon (HC) solvents are crowded into a tiny space at the lower left corner of either graph. In fact the aliphatic HC solvents are located at the graph origin.
2. Each class of organic solvent (ester, ketone, alcohol, nitroparaffin, and the like) is characterized by individual solvent points that stretch roughly over a specific linear region of the graph.
3. The hydroxyl-bearing solvents tend to fall in a region that sweeps upward to the right from near the bottom horizontal scale to a point for water that is way off the chart to the right ($\delta_p = 7.8$; $\delta_h = 20.7$).
4. The more preponderant the hydrocarbon portion of a solvent, the nearer is its position to the HC corner of the graph.
5. Aliphatic-based solvents are generally restricted to the lower tier; aromatic- and cyclic-based solvents, to the upper tier.

These very generalized observations serve to provide the coatings or ink formulator with a "feel" for expected solubility behavior. Except for the aromatic hydrocarbon solvents and a few others, it is notable that most of the mainstream solvents used in the coatings industry are located in the lower tier. This is quite convenient, for it permits simplified numerical and graphical solubility computations to be carried out with reasonable accuracy. Thus for numerical computations, it can be assumed that the δ_d values are substantially alike. Therefore the effect of the δ_d value can be neglected. For the same reason, graphical calculation can be reduced to plotting on two-dimensional graph paper. If a δ_d value happens to be significantly different, the end results can be correspondingly corrected to account for the difference on a qualitative or intuitive basis.

Polymer (Resin) Solubility. The relative solubilities or insolubilities (1.0 g solute/10 cm^3 solvent) of 34 polymers in 90 liquids (mostly solvents) involving over 3000 observations have been systematically tabulated (11). Based on these data, borderlines between regions of solubility (*S*) and insolubility (*I*) can be drawn on δ_p versus δ_h graphs for representative polymers. This has been done for eight polymers, as shown in Fig. 14-5. The *dotted S/I* parameter curves represent borderlines corresponding to low δ_d values; the *solid S/I* curves, to high δ_d values. Inspection of the eight graphs reveals that the solubility behavior of each polymer is quite different. This is clearly indicated by both the shape of the borderline *S/I* curves and the solubility areas that they encompass. With a little imagination it is possible to visualize the change in the locus of an *S/I* borderline curve in progressing from solvents of low to those of high δ_d values.

In general the borderline *S/I* curves are roughly elliptical or parabolic. Greatest solvency is exhibited within the center of the region, marginal solvency at the borderline, and insolubility outside, with the severity of insolvency depending on the distance from the borderline curve.

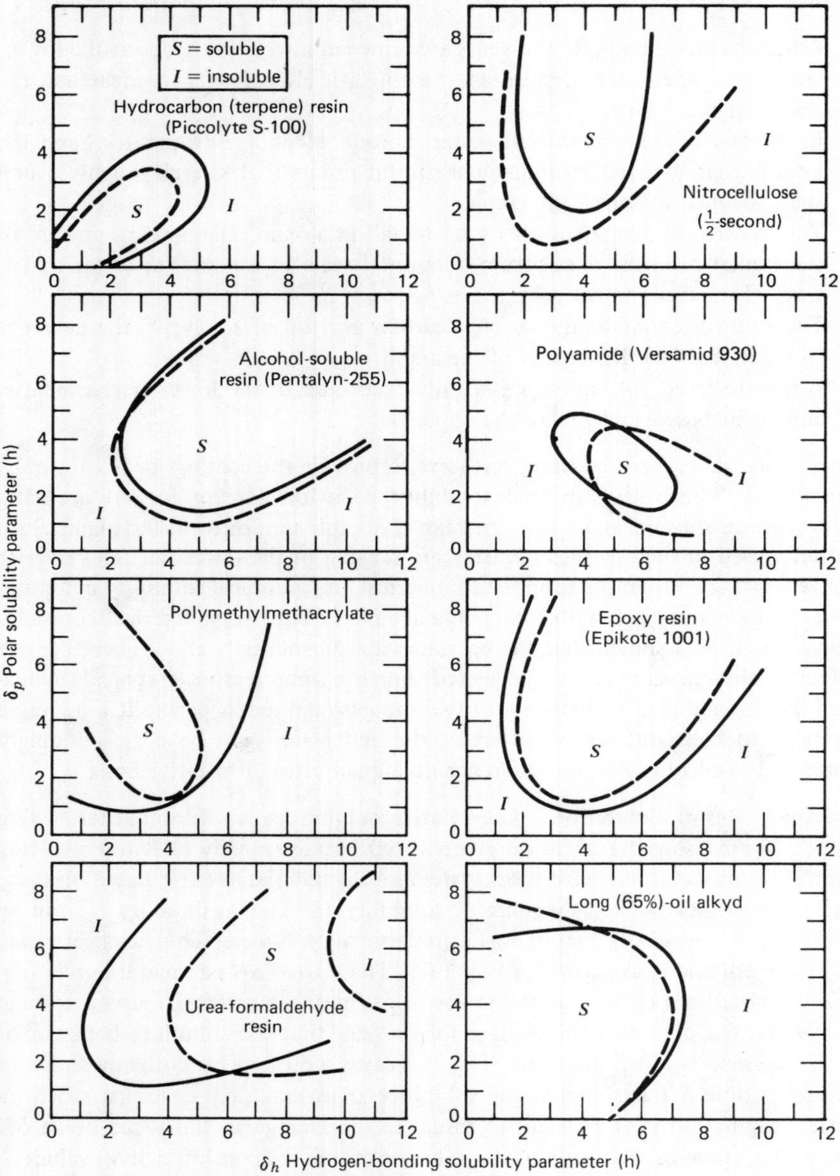

Fig. 14-5 Solubility parameter areas for eight polymers plotted on δ_p versus δ_h graphs. Dotted borderline S/I curves correspond to low δ_d values; solid S/I curves, to high δ_d values.

With most polymers, the formulator is afforded considerable leeway in selecting solvents on the basis of solvency alone. This is most helpful, since other solvent constraints (e.g., low solvent viscosity, proper evaporation rate, and, most important, low solvent cost) must be considered in establishing an acceptable final solventation.

Solvent Blends. Possibly the most compelling argument for accepting and using the solubility parameter system is that it affords an explanation of why polymers can tolerate large proportions of solvents that normally do not dissolve the polymers. It also makes understandable the extreme cases where a polymer is dissolved by a mixture of two solvents, either of which alone is incapable of dissolving the polymer. In both cases the explanation lies in the averaging out of the partial parameters for the solvent blend to match those of the polymer. The contributions of the three partial solubility parameters (δ_d, δ_p, and δ_h values) to the blend are made on the basis of the fractional volume contents of the blend components, as shown by Eqs. 29, 30, and 31.

$$\delta_d \text{ (blend)} = (\phi\delta_d)_1 + (\phi\delta_d)_2 + \cdots + (\phi\delta_d)_n \tag{29}$$

$$\delta_p \text{ (blend)} = (\phi\delta_p)_1 + (\phi\delta_p)_2 + \cdots + (\phi\delta_p)_n \tag{30}$$

$$\delta_h \text{ (blend)} = (\phi\delta_h)_1 + (\phi\delta_h)_2 + \cdots + (\phi\delta_h)_n \tag{31}$$

Two examples of a solvent and a nonsolvent acting in conjunction to produce polymer solvency are illustrated schematically in Fig. 14-6*a*. The polymer is $\frac{1}{2}$-sec nitrocellulose, and its *S*/*I* borderline curve for low δ_d values is indicated by the dotted line. The first solvent pair consists of the solvent methyl isobutyl ketone and the nonsolvent hexane; the second pair, the solvent acetone and the nonsolvent *n*-butanol. Solubility parameter values for these four substances are as follows:

Substance	(δ)	δ_d	δ_p	δ_h
Methyl isobutyl ketone	(8.3)	7.5	3.0	2.0
Hexane	(7.3)	7.3	0.0	0.0
Acetone	(9.8)	7.6	5.1	3.4
n-Butanol	(11.3)	7.8	2.8	7.7

Since the dispersion (partial) solubility parameters of the binary solvent blends and the polymer (low-δ_d range) are essentially alike, the contribution of the δ_d values can be neglected in the calculation. Hence plotting can be done on conventional two-dimensional graph paper. The solvent and nonsolvent points are plotted and connected by straight lines as shown on the graph of Fig. 14-6*a*. These lines are then converted to 100% scales (division into 10 equal parts has been shown on the graph). Inspection shows that these straight lines intersect

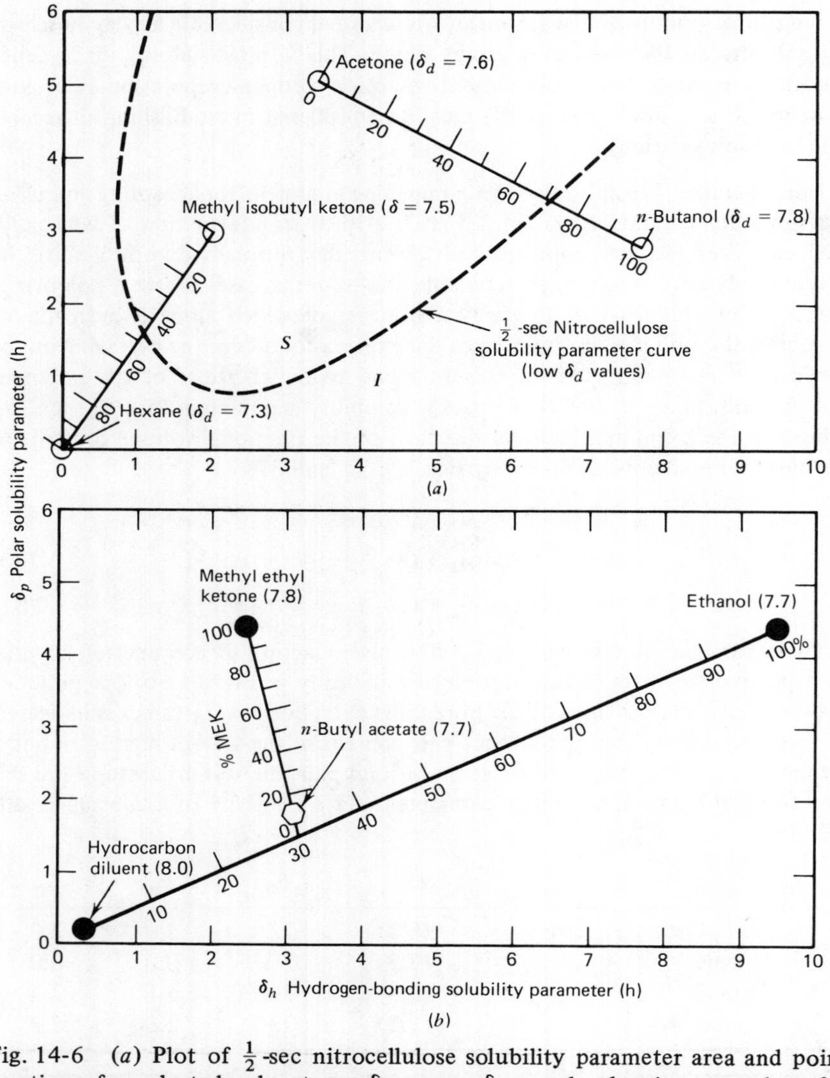

Fig. 14-6 (*a*) Plot of $\frac{1}{2}$-sec nitrocellulose solubility parameter area and point locations for selected solvents on δ_p versus δ_h graph, showing procedure for determining the maximum percentage of a nonsolvent in a nonsolvent/solvent blend that provides nitrocellulose solvency. (*b*) Illustration of graphical procedure to be followed on a δ_p versus δ_h graph in order to calculate the volume percentages of three solvents that will provide a three-component blend matching the partial δ values of a fourth target solvent.

the nitrocellulose *S/I* curves at points corresponding to 46% hexane/54% methyl isobutyl ketone and 71% *n*-butanol/29% acetone, respectively. From this it is concluded that any lesser percentages of hexane or *n*-butanol in the respective blends will result in nitrocellulose solvency, with solvency becoming more certain in progressing from the borderline region to the center of the solubility area.

The maximum ability of a solvent to tolerate a nonsolvent and yet produce polymer solvency is frequently expressed in terms of a dilution ratio. A dilution ratio is defined as the critical ratio of nonsolvent to solvent at which insolvency is first observed, as indicated by a persistent heterogeneity (hazing). In the case of cellulose nitrate solutions the observations are carried out at concentrations of 8.0 g of nitrocellulose in 100 cm^3 of the nonsolvent/solvent blend (ASTM D1720-62; reapproved 1973). For the case of the two binary blends shown in Fig. 14-6*a*, the dilution ratios are 0.85 (= 46/54) for the hexane/methyl isobutyl ketone combination and 2.4 (= 71/29) for the *n*-butanol/acetone combination.

In mapping out areas of solubility on δ_p versus δ_h graphs, resort is frequently made to the use of binary nonsolvent/solvent mixtures to more accurately locate the polymer *S/I* borderline curve. This is done because solvents are restricted to a single-point location, whereas blends traverse an extended line. At some point on this line the required borderline solvency condition will be exhibited and can be pinpointed on the graph.

Even more striking is the ability of two nonsolvents for a polymer to act in conjunction to dissolve the polymer. This occurs when the individual points for the two nonsolvents are so located that some portion of the straight line connecting them crosses the region of solubility for the polymer. The fractional volumes of the two nonsolvents bounded by the points of entry and exit represent mixtures that are capable of dissolving the polymer.

Many nonsolvent binary mixtures that are capable of dissolving a polymer have been reported in the literature; six of them are listed below.

Polymer	Binary Nonsolvent Mixture
Polymethylmethacrylate	Ethoxyethanol/aniline
	Nitromethane/carbon tetrachloride
Epoxy resin	Butanol/2-nitropropane
Polyamide	Ethanol (99%)/ethylene chloride
Chlorinated polypropylene	Cyclohexanol/acetone
Lignin	Methanol/nitromethane

Problem 14-9 serves to demonstrate further the general procedure that is employed in solving solubility problems.

PROBLEM 14-9

Calculate both algebraically and graphically a volume composition consisting of methyl ethyl ketone (MEK), ethanol, and a hydrocarbon solvent (diluent) that has the same solvency characteristics as *n*-butyl acetate. The total and partial solubility parameters for the individual solvents are as follows:

Solvent	(δ)	δ_d	δ_p	δ_h
n-Butyl acetate	(8.5)	7.7	1.8	3.1
Methyl ethyl ketone	(9.3)	7.8	4.3	2.5
Ethanol	(13.0)	7.7	4.4	9.5
Hydrocarbon diluent	(8.0)	8.0	0.2	0.3

Solution. Since the δ_d values for the several solvents are essentially alike, it is necessary only to match the δ_p and δ_h values for the solvent blend to those of the *n*-butyl acetate.

Algebraic Solution: Let ϕ_H and ϕ_K be the volume fractions for the hydrocarbon diluent and the MEK, respectively. Then the volume fraction for the remaining ethanol ϕ_A is equal to $(1 - \phi_H - \phi_K)$. Set up two equations for the polar and hydrogen-bonding equivalences, using appropriate δ_p and δ_h values.

Polar (partial) solubility parameter equivalency:

$$1.8 = 0.2\phi_D + 4.3\phi_K + 4.4(1 - \phi_D - \phi_K)$$

Hydrogen-bonding (partial) solubility parameter equivalency:

$$3.1 = 0.3\phi_D + 2.5\phi_K + 9.5(1 - \phi_D - \phi_K)$$

Solve these two equations simultaneously for the three volume fractions.

$$\phi_K \text{ (MEK)} = 0.103 \text{ (10.3\%v)}$$

$$\phi_A \text{ (ethanol)} = 0.280 \text{ (28.0\%v)}$$

$$\phi_H \text{ (HC diluent)} = 0.617 \text{ (61.7\%v)}$$

Comment. Note that, if a negative algebraic result had been obtained, no blend matching *n*-butyl acetate would have been possible.

Graphical Solution: Plot the several given δ_p and δ_h values on rectilinear graph paper, as shown in Fig. 14-6*b*. Draw a straight line between the points for the hydrocarbon diluent and ethanol, and divide this line into 10 equal parts to indicate 10%v divisions. Draw a straight line from the point for MEK through the point for *n*-butyl acetate, and extend to intersect the HC/ethanol line. Divide the portion of the line between the MEK point and this intersection into 10 equal parts to show 10%v divisions.

The point of intersection on the HC diluent/ethanol line corresponds to a blend of 31%v ethanol/69%v hydrocarbon diluent (read off the 100%v scale). To match the solubility behavior of *n*-butyl acetate, about 10%v of MEK must be mixed with 90%v of the blend (read off the other 100%v scale). The final composition is then as follows:

MEK = 10%v (0.10)

Ethanol = 28%v ($0.28 = 0.90 \cdot 0.31$)

HC diluent = 62%v ($0.62 = 0.90 \cdot 0.69$)

Comment. The graphical technique provides an answer that closely approximates the algebraic solution. Furthermore it gives a clear visual picture of the interrelationships among the component solvents. For example, the fact that the point for *n*-butyl acetate falls within the MEK/ethanol/HC diluent triangle means that a match using these three solvents is possible. The procedure of dividing the connecting lines between points and/or intersections into 10%v divisions can be carried out as many times as necessary to take care of any multicomponent system.

Although incompletely understood at the time, nonsolvents for nitrocellulose were used at an early stage in formulating cellulose nitrate solutions. These solvents that we now know to fall in the solubility area of the NC polymer were referred to as active solvents, whereas the hydrocarbon and oxygenated solvents that fell outside the solubility area were referred to as diluents and latent solvents, respectively. This terminology is still in use today. From a practical standpoint it is generally expedient to formulate solvent compositions with at least 10 to 20%v of active solvent to exert a "coupling" effect on the nonsolvents present and thus ensure the required solvency effect.

Kauri-Butanol (KB) Value. The kauri-butanol (KB) value of a solvent as specified by ASTM D1133-61 (1973), is defined as the volume (in cm^3) of the solvent (actually a nonsolvent under the test conditions) that is necessary to produce at 25 C a defined degree of turbidity when added to 20 g of a standard solution of kauri resin in *n*-butanol (3.33 g kauri gum/20.6 cm^3 *n*-butanol). This is a case of observing a series of nonsolvent/solvent mixtures for kauri gum borderline insolvency ("kick-out"), since kauri gum is soluble in *n*-butanol but not in hydrocarbon solvents. A KB value serves as an arbitrary measure of hydrocarbon solvent strength (KB hexane 31; KB mineral spirits 36; KB kerosene 34; KB gum turpentine 64; KB xylene 98; KB toluene 105).

Aniline Point (AP). The aniline point (AP) of a solvent, as specified by ASTM D611-64 (1973), is defined as the minimum solution (miscibility) temperature (F) for equal volumes of aniline and the test solvent. The test provides an arbitrary measure of hydrocarbon solvent power and takes advantage of the fact that aniline and HC solvents become more miscible both with increase in temperature and with increase in hydrocarbon solvent strength. Heptane (100%

aliphatic) has an aniline point of 158 F; toluene (100% aromatic), an aniline point of −56 F. The temperature spread between these two APs is 214 F° (= 158 + 56). Hence each 2.14 F° reduction in AP below 158 F corresponds roughly to a 1% gain in aromaticity and a corresponding gain in solvent strength. For example, a petroleum naphtha with an AP of 138 F can be considered as having an aromatic equivalent content of approximately 9.3% [= (158 - 138)/2.14].

For certain extreme conditions where the AP solution temperatures become so low that the aniline/HC solvent mix freezes or where carrying out the test becomes awkward, a so-called mixed aniline point has been established based on a mixture of 50%v aniline, 25%v *n*-heptane, and 25%v of the test hydrocarbon solvent.

Expressions giving the approximate relation between a KB value and an AP value are given by Eqs. 32 and 33.

$$KB = 84.3 - 0.37\ AP \tag{32}$$

$$AP = 228 - 2.7\ KB \tag{33}$$

The total solubility parameter for hydrocarbon solvents can be approximately computed from KB and AP values by using Eqs. 34 and 35, respectively.

$$\delta = 6.8 + 0.021\ KB \tag{34}$$

$$\delta = 8.57 - 0.0077\ AP \tag{35}$$

Plasticization. In a certain sense plasticizers may be regarded as permanent solvents. Thus, as opposed to regular solvents, which are volatile, plasticizers are substantially nonvolatile. In the former case, proper functioning depends on release of solvent; in the latter case, on retention of plasticizer. In both cases, however, solubility or compatibility is predicted by matching the partial solubility parameters of the materials involved.

Solubility parameter regions (low and high δ_d values) for polyvinyl chloride (PVC) have been plotted on a δ_p versus δ_h graph in Fig. 14-7. Also plotted are the solubility parameter points for several common plasticizers. With the exception of butyl stearate and triethyl phosphate, they fall within the high-δ_d-value region. Since the enclosed plasticisers also have high δ_d values, it is understandable that on the basis of matched parameter values they should account for the bulk of the commercial plasticizers used for plasticizing polyvinyl chloride.

Pigment Dispersibility. The hypothesis that solubility, miscibility, compatibility, dispersibility, and adhesion are manifestations of the same common phenomenon (physicochemical affinity) was advanced in 1963 by the Toronto Society for Paint Technology (21). In the introduction to the paper it was observed that natural phenomena are ordered, patterned, and supremely logical.

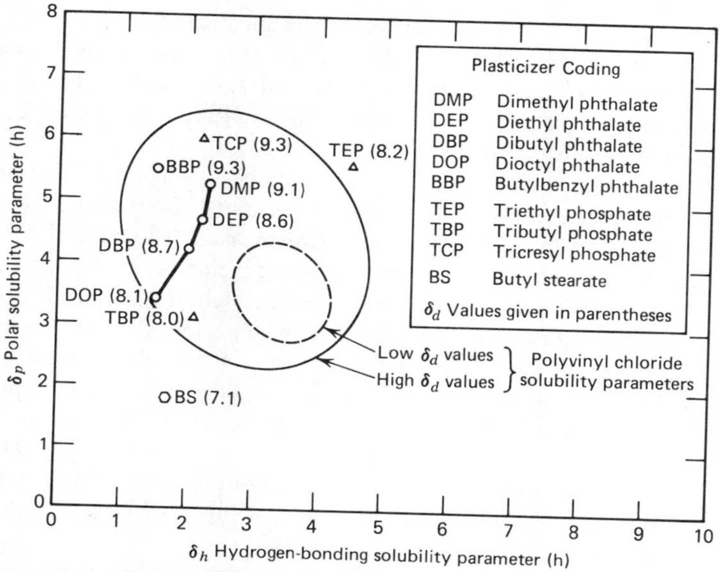

Fig. 14-7 Plot of the solubility parameter areas (low and high δ_d values) for polyvinyl chloride resin and point locations for selected common commercial plasticizers on a δ_p versus δ_h graph.

Hence, if some coating behavior appears capricious, it can be ascribed only to human difficulty in unscrambling the complex result of the physicochemical affinities.

The paper reported that titanium dioxide was effectively held in the dispersed (suspended) state by strongly H-bonded (hydroxyl-bearing) solvents such as propanol, butanol, and ethylene glycol, and that lampblack was effectively held in suspension by such moderately H-bonded solvents as butyl acetate and methyl ethyl ketone. On the basis of these observations and their overall experimental work, the authors concluded that hydrophilic pigments (titanium dioxide, yellow iron oxide, chrome yellow, molybdate orange) are well dispersed by strongly H-bonded solvents and resins, and that organophilic pigments (lampblack, phthalocyanines) are well dispersed by moderately H-bonded solvents and resins. In general, then, permanent pigment dispersions are best obtained by matching the pigment attractive forces to those of the solvent and resin. Other papers have since confirmed these conclusions (11, 16, 20, 22, 23).

It is not the bulk properties of a pigment that determine its dispersion behavior, but rather the nature of the pigment surface. Since pigment surfaces are usually modified during manufacture to a lesser or greater degree by surface treatment, it is somewhat unrealistic to expect pigments having the same bulk properties to necessarily have the same surface properties. Hence it is always wise, in discussing the specific dispersion properties of a pigment, to describe its nature precisely and completely, rather than identifying it loosely by some general chemical or blanket designation.

Characteristic suspension properties for a number of pigments have been reported, based on their suspension behavior in solvents of known partial solubility parameter values (11). A small amount of pigment (about 0.1 g for an organic pigment) is shaken in a test tube containing 5 cm^3 of solvent. This is done for 50 or more solvents of widely diverse partial solubility parameter behavior. The time and the degree of pigment settling are observed. Complete suspension over a significant time period is taken as indicative of complete solvent/pigment interaction (no flocculation); rapid settling, as indicative of little or no interaction (flocculation). Modifying influences such as particle size, solvent viscosity, and differences in solvent and pigment densities are taken into consideration, as well as possible chemical reactivity, in interpreting the final results. For example, if fine particles are absent from the makeup of a dense inorganic pigment, it is unreasonable to expect any appreciable suspension to be exhibited. On the basis of the observed settling behavior, regions of strong pigment/solvent interaction, as indicated by good to excellent suspension (dispersion), are then filled in on a partial solubility parameter graph. The center of this interaction region is then taken as characterizing the interaction center for the pigment in question. This center location is given in terms of a set of δ_d, δ_p, and δ_h values.

Results obtained for 25 test pigments in one such program revealed consistently high δ_d values (average of 10 h, with only three below 9.3 h) (11). Furthermore the values for δ_p and δ_h were more or less the same ($\delta_p = 1.00 \pm 0.40\delta_h$). The overall spread of these pigments on a δ_p versus δ_h graph is shown in Fig. 14-8.

General Comments. It is now generally recognized that the solubility parameter system or, more accurately, the interaction parameter system is capable of providing the coatings or ink engineer with a unifying concept linking pigments, binders (polymers and resins), and solvents into a reasonably orderly system. Further work is necessary to develop more accurately placed partial δ values, and the theory must be refined to explain yet unresolved inconsistencies. However, the basic idea is extremely useful, namely, that solubility, compatibility, and pigment dispersibility are best attained by matching interaction parameter values.

Most dispersion systems are composed of three components: pigment, binder, and solvent. In such systems the solvent and the binder compete for adsorption sites on the pigment surface. For optimum or acceptable pigment dispersion, the binder should be equally if not preferentially adsorbed (11, 22). In a study of

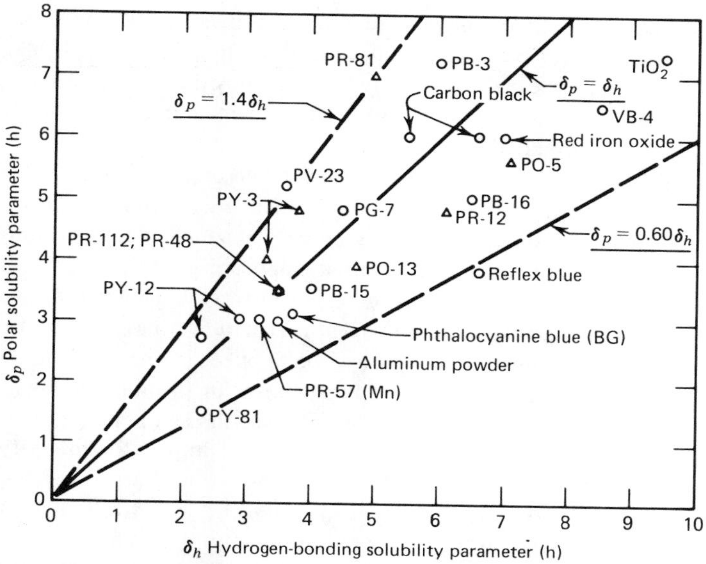

Fig. 14-8 Plot of pigment interaction centers on a δ_p versus δ_h graph as determined by settling behavior in a spectrum of solvents. All pigments characterized by relatively high δ_d values (average 10 h).

liquid inks it was demonstrated that, when the affinities among the pigment, binder, and solvent are strong, the pigment dispersion is satisfactory, with both binder and pigment being adsorbed. It is theorized that the presence of the binder/solvent mixture at the surface provides a diffuse layer that discourages flocculation and imparts the best flow (low viscosity) and print quality. However, the solvent adsorption must not be so strong as to interfere with an adequate pigment/binder interaction (22). Evidently the ideal situation is to have a reasonably good solvent for the binder, but with the binder interaction center still located between the centers of the solvent and pigment.

Certain general rules should be observed in using the parameter interaction system. In formulating a pigment mixture, pigments that are close together in interaction parameter value should be selected in order to avoid differing degrees of dispersion and consequent flocculation and floating. Resins in multiple-binder systems should also be chosen from these located in the same general interaction parameter region (maximum overlapping) to ensure optimum compatibility, satisfactory cure, and freedom from the formation of heterogeneous phases (hazing). Incidentally, in such multibinder systems, the pigment should be ground in the binder it most closely matches in partial δ value.

The technique of matching partial δ values is obviously applicable to many practical situations such as the formulation of primers or tie-coats for organic substrates, the correction of storage instability by adding a solvent of tolerably low volatility which is capable of moving the composition solventation nearer to the center of the binder solubility parameter region, and the formulation of tinting pastes with broad coverage (2).

Up to this point the discussion has been focused on matching partial δ values to achieve solubility, compatibility, or pigment dispersibility. However, an opposite goal is sometimes required, such as a planned incompatibility. Here again the interaction parameter system is applicable, but this time the components are selected from those with widely divergent parameter values. For example, nonmatching partial interaction parameter values can be used as a basis in designing a protective top coat to be applied, at the same time without smearing and reflow, to a slightly wet ink pattern, in formulating inks or coatings that do not interact with (swell) the rolls used in their application, or in the design of lacquers for plastics that do not craze or impair the integrity of the plastic to which they are applied (2, 16, 20).

Finally, troublesome formulations can often be effectively diagnosed by checking the components against their relative positions on a δ_p versus δ_h chart, with account being taken of their respective δ_d values as a possible modifying influence.

SOLUBILITY (INTERACTION) PARAMETER—THE UNIVERSAL CONSTANT

Although J. Hildebrand originated the solubility parameter theory, it was Harry Burrell who introduced the concept on a practical basis to the coatings industry (4, 24, 25). During the subsequent years Burrell has continued to publish highly useful papers that have contributed greatly to the education of the paint engineer in this important area of coatings technology. For the student desiring to probe deeper into the origin and development of the solubility parameter concept a review paper by Burrell published in 1968 is required reading (2).

In one paper Burrell refers to the solubility parameter as the universal constant, since it is a fundamental measure of intermolecular forces and hence serves to unify many diverse properties of matter. The relationships of three of these properties to the solubility parameter have already been considered in this chapter (Eqs. 36, 37, 38).

$$\Delta E = \frac{\delta^2 M}{\rho} \tag{36}$$

$$\Delta H = \frac{\delta^2 M}{\rho} + RT \tag{37}$$

$$\gamma = \left(\frac{\delta}{K}\right)^{1/a} \cdot \left(\frac{M}{\rho}\right)^{1/3} \tag{38}$$

Other related properties are shown in Eqs. 39, 40, and 41 (in Eq. 39 B is the material compressibility) (2).

$$\text{Coefficient of thermal expansion} = \frac{\delta^2 B}{T} \tag{39}$$

$$\text{Critical pressure} = \left(\frac{\delta}{1.25}\right)^2 \tag{40}$$

$$\text{Gas law correction factor} = \frac{\delta M^2}{1.2\rho} \tag{41}$$

Graphs of the solubility parameter versus the degree of crystallinity have been constructed and show that different polymer types tend to segregate in specific localized areas (26). In turn, the locations thus obtained have been related to other relevant properties (solubility, mechanical properties, electrical properties, and the like).

The present author suggests that the partial hydrogen-bonding parameter may also be casually related to the arbitrary HLB system. It is an interesting coincidence that the HLB system, which runs from 0 for a completely lipid (fatty) material to 20 for an overwhelming hydrophilic material, is comparable to the range for δ_h: from 0 for a completely hydrocarbon material to 20.7 for water itself.

The solubility parameter (interaction parameter) concept is still undergoing quantitative refinements to improve its accuracy and reconcile certain inconsistencies. Possibly the use of an acid/base modification would prove helpful (27). However, the concept is sufficiently developed to provide, first, a theoretically sound basis for understanding the interaction of pigments, binders, and solvents and, second, a logical system for either formulating practical coating compositions or correcting compositions that have proved unsatisfactory in performance. It is estimated that the interaction parameter system can be relied on to be correct about 95% of the time.

REFERENCES

1. Patton, Temple C., *Paint Flow and Pigment Dispersion*, 1st ed., Wiley-Interscience, New York, 1964.
2. Burrell, Harry, "The Challange of the Solubility Parameter Concept," *JPT*, **40**, No. 520, 197–208, May (1968).
3. Hildebrand, J., *J. Am. Chem. Soc.*, 38, 1452 (1916).
4. Burrell, H., "Solubility Parameters for Film Formers," *Off. Dig.*, **27**, No. 369, 726 (1955).
5. Lieberman, E. P., "Quantification of the Hydrogen Bonding Parameter," *Off. Dig.*, **34**, No. 444, 30 (1962).
6. Gardon, John L., "The Influence of Polarity upon the Solubility Parameter Concept," *JPT*, **38**, No. 492, 43–57, January (1966).
7. Hildebrand, J. and R. L. Scott, *The Solubility of Non-Electrolytes*, 2nd ed., 1936, and 3rd ed., 1949, Reinhold, New York.

8. Hoy, K. L., "New Values of the Solubility Parameters from Vapor Pressure Data," *JPT,* **42,** No. 541, 76–118, February (1970).
9. Lee Lieng-Huang, "Relationships between Solubility Parameters and Surface Tensions of Liquids," *JPT,* **42,** No. 545, 365–370, June (1970).
10. Small, P. S., "Some Factors Affecting the Solubility of Polymers," *J. Appl. Chem.,* **3,** 71 (1953).
11. Hansen, Charles M., "The Three Dimensional Solubility Parameters–Key to Paint Component Affinities: I. Solvents, Plasticizers, and Resins," *JPT,* **39,** No. 505, 104–117, February (1967): "II. Dyes, Emulsifiers, Mutual Solubility and Compatibility, and Pigments," *JPT,* **39,** No. 511, 505–510, August (1967).
12. Solvent Notes IC: 67-64SN, "Solvent Selection Based on Solution Theory," and IC: 68-24SN1, "Solvent Selection Based on Solution Theory: Part II. Solubility Maps of Various Resins," Shell Chemical Company Industrial Chemicals Division, New York, N.Y. 10020.
13. Nelson, R. C. et al., "Solution Theory and the Computer–Effective Tools for the Coatings Chemist," *JPT,* **42,** No. 550, 644–652, November (1970).
14. Crowley, James D., G. S. Teague, Jr., and Jack W. Lowe, Jr., "A Three-Dimensional Approach to Solubility," *JPT,* **38,** No. 496, 269–280, May (1966).
15. Crowley, James D., G. S. Teague, Jr., and Jack W. Lowe, Jr., "A Three Dimensional Approach to Solubility: II," *JPT,* **39,** No. 504, 19–27, January (1967).
16. Crowley, James D., "Fundamentals and Recent Developments in 3-Dimensional Solution Parameter Theory," paper presented at University of Utah Polymer Conference Series, June 23, 1970.
17. Teas, J. P., "Graphical Analysis of Resin Solubilities," *JPT*, **40,** No. 516, 19–25, January (1968).
18. Technical Bulletin 1206, "Predicting Resin Solubilities," Ashland Chemical Company, Division of Ashland Oil, Columbus, Ohio 43216, 1971.
19. Hansen, C. M. and K. Skaarup, "The Three Dimensional Solubility Parameter–Key to Paint Component Affinities: III. Independent Calculation of the Parameter Components," *JPT,* **39,** No. 51, 511–514, August (1967).
20. Hansen, Charles and Alan Beerbower, "Solubility Parameters," in *Encyclopedia of Chemical Technology,* 2nd ed., Supplement V, Wiley, New York, 1971.
21. Toronto Society for Paint Technology, "The Pattern of Solvent-Resin-Pigment Affinities," *Off. Dig.,* **35,** No. 466 1211–1231, November (1963).
22. Sorensen, P., "The Solubility Parameter Concept in the Formulation of Liquid Inks," *JOCCA,* **50,** No. 3, 226–243, March (1967).
23. Eissler, R. L., R. Zgol, and J. A. Stolp, "Solubility Parameters Applied to Sedimentation Volumes of Pigments," *JPT,* **42,** No. 548, 483–489, September (1970).
24. Burrell, H., "A Solvent Forumulating Chart," *Off. Dig.,* **29,** No. 394, 1159 (1957).
25. CDIC Production Club, "Solubility Parameters of Resins," *Off. Dig.,* **29,** No. 394, 1069 (1957).
26. Skeist, I., *Mod. Plastics,* May (1956).
27. Sorensen, Palle, "Application of the Acid/Base Concept Describing the Interaction between Pigments, Binders, and Solvents," *JPT,* 47, No. 602, 31–39, March (1975).

BIBLIOGRAPHY

Schreiber, H. P., "Solvent Balance, Dispersion, and Rheological Properties of Pigmented Polymer Compositions," *JPT,* **46,** No. 598, 35–39, November (1974).

15 *Volatility: Solvent and Water Evaporation*

Volatility is a key factor in the selection of a solvent. A balance between too fast a rate and too slow a rate of evaporation must be achieved to formulate a desirable solventation. In this regard there must be considered not only the initial rate, but also the intermediate and final evaporation rates, each of which has an influence on such properties as dry time, wet-edge time, leveling, sagging, streaking, pinholing, and cratering. The solvent also contributes to the wetting or nonwetting of the substrate to which the coating is applied. Finally, a solvent may strongly influence the orientation and properties of the solid structure it leaves behind on evaporation to the atmosphere.

VOLATILITY OF A NEAT SOLVENT (RELATIVE EVAPORATION RATE)

Undoubtedly the best clue to the volatility of a neat solvent (solvent only) is its vapor pressure. Although the boiling point of a solvent has also been used to predict solvent volatility, this approach has met with little success. Thus, although butanol has a boiling point (244 F) that is 17 F° below that of butyl acetate (261 F), its evaporation rate is roughly half as high. This type of discrepancy has discouraged the use of such terms as low-, medium-, and high-boiling to describe solvent volatility.

Molecular weight and density have also been used in combination with vapor pressure in attempts to more accurately predict the volatility behavior of a solvent.

Design of Solvent Evaporometer

Offhand it might seem that little difficulty should be experienced in measuring the volatility of a neat solvent: place it in a flat-bottomed dish that is open to the atmosphere, and monitor the loss due to evaporation by simply making periodic weighings over a reasonable time span. However, the task is not that simple, since volatile loss is strongly influenced by such factors as temperature, air movement, and the makeup of air in contact with the solvent surface. Therefore it is essential in any volatility measurement to strictly control the flow, temperature, and composition of the air over the evaporating surface.

The Shell Automatic Thin Film Evaporometer, originally reported in the literature in 1950 and since modified several times to improve its operation, probably represents the most sophisticated design for an evaporometer to date (1). It consists of an inner chamber, where the evaporation and weighing take place, and an outer chamber that provides a constant temperature environment. Standard conditions are a temperature of 25 C and a flow of air or nitrogen (0 to 5% relative humidity) at the rate of 21 liters/min through the apparatus. An accurately calibrated hypodermic syringe dispenses 0.70 cm^3 of test solvent evenly onto the surface of a suspended 9-cm diameter filter paper disk within a 10-sec period. Subsequent weight loss (0 to 100% evaporation) as sensed by an optical electrobalance is then continuously displayed on a strip chart recorder as a function of time.

More recently it has been demonstrated that the relative rates of neat solvents evaporating from a smooth nonporous surface do not parallel the corresponding relative evaporation rates obtained with filter paper (2). Thus water and alcohols evaporate more slowly from filter paper than from a metal surface, presumably because of hydrogen bonding. This adds one more factor that must be controlled to ensure consistent and reproducible results.

The Shell evaporometer provides precise and reproducible volatility information corresponding to an arbitrary set of conditions. But another evaporometer of different design might provide equally valid data for another set of arbitrary conditions where the weight losses were different. To reconcile such differences, it is convenient to rate the volatility of a test solvent relative to the volatility of some standard solvent. By convention, *n*-butyl acetate has been selected to serve as the standard solvent. It is assigned an evaporation rate of 1.00 (or sometimes 100). In the case of the Shell evaporometer the instrument is adjusted (calibrated) so that 90% of *n*-butyl acetate (99% ester) is evaporated in 470 ± 10 sec from filter paper. However, when using a smooth aluminum metal surface (1.5-cm diameter) and a 0.13-cm^3 test sample, the Shell evaporometer is adjusted so that 90% of *n*-butyl acetate (99% ester) is evaporated in 2902 ± 25 sec.

In the following discussion relative evaporation rate refers to the evaporation rate of the test solvent compared with that of *n*-butyl acetate under the same set of conditions.

Expressions for Calculating the Relative Evaporation Rate of a Neat Solvent

The relative evaporation rate E is defined by Eq. 1, where t_{90} is the time for 90% of the solvent to be evaporated under a set of strictly controlled conditions.

$$E = \frac{t_{90}\ (n\text{-butyl acetate})}{t_{90}\ (\text{test solvent})} \tag{1}$$

The 90% evaporation time is normally selected for the computation, since it reflects the true volatility behavior of the solvent more accurately than the 100% evaporation time (the loss of the last 10% of solvent tends to be affected by extraneous factors). The relative evaporation rate can be expressed on either an equivalent weight (E_W) or an equivalent volume (E_V) basis. On the assumption that n-butyl acetate ($\rho = 0.878$ g/cm^3) is the standard solvent, conversion from a volume to weight basis (or vice versa) can be carried out using Eqs. 2 and 3, where ρ is the solvent density.

$$E_W = \left(\frac{\rho}{0.878}\right) E_V = 1.14\rho E_V \tag{2}$$

$$E_V = \left(\frac{0.878}{\rho}\right) E_W \tag{3}$$

All too often literature values for evaporation rates fail to distinguish between E_W and E_V. With the advent of smooth surface evaporation data it will become necessary to add this as another distinguishing feature.

A number of equations have been proposed relating relative evaporation rate to vapor pressure p, either alone or in conjunction with other solvent properties such as density ρ and molecular weight M. Typical expressions of this type are given by Eqs. 4 to 7, which may be resorted to when direct evaporation data are not available. In all four expressions, n-butyl acetate is the reference standard, with an arbitrarily assigned evaporation rate of 1.00 (25 C).

$$E_W\ (\text{hydrocarbon, esters}) = 0.10p^{0.9} \tag{4}$$

$$E_W\ (\text{ketones, alcohols}) = 0.08p^{0.9} \tag{5}$$

Equations 4 and 5 were derived by the present author from some of the generalized curves for smooth surface evaporation given in a paper by Rocklin (2).

$$E_V = 0.0082p\sqrt{M} \tag{6}$$

This theoretical relative evaporation rate (Eq. 6) was submitted as a footnote in a paper by Dillon (3).

$$E_V\ (\text{oxygenated solvents}) = \frac{0.00064pM}{\rho} \tag{7}$$

Equation 7 appeared in a company educational bulletin (4). It is presumed that Eqs. 6 and 7 apply to relative evaporation rates from a filter paper substrate.

PROBLEM 15-1

Calculate the volume relative evaporation rates (E_V) for *p*-xylene ($p = 8.75$ mm Hg; $M = 106.3$; $\rho = 0.856$ g/cm^3) over a smooth metal surface, and for methyl ethyl ketone ($p = 90.2$ mm Hg; $M = 72.1$; $\rho = 0.800$ g/cm^3) over a porous cellulosic substrate.

Solution. *p*-Xylene: First calculate a relative weight evaporation rate for the *p*-xylene over a smooth surface, using Eq. 4.

$$E_W = 0.1 \cdot 8.75^{0.9} = 0.704$$

Convert to a volume relative evaporation rate, using Eq. 3.

$$E_V = \frac{0.878}{0856} \cdot 0.704 = 0.72$$

Methyl ethyl ketone (MEK): Calculate the volume relative evaporation rate for MEK over a cellulosic substrate, using either Eq. 6 or 7.

$$(6){:}\quad E_V = 0.0082 \cdot 90.2\sqrt{72.1} = 6.3$$

$$(7){:}\quad E_V = \frac{0.00064 \cdot 90.2 \cdot 72.1}{0.800} = 5.2$$

Comment. The value of 0.72 calculated for the E_V of *p*-xylene compares favorably with an experimentally determined value of 0.75 (smooth surface). However, in the case of MEK the E_V values of 5.2 and 6.3 are only rough approximations to an experimentally determined value of 4.1 (filter paper substrate).

Effect of Temperature Change on Relative Evaporation Rate of a Neat Solvent

The fact that the relative evaporation rate of a solvent is so intimately related to its vapor pressure suggests that any expression relating vapor pressure to temperature can also be adapted in modified form to relative evaporation rate to temperature. For example, the classic Clausius-Clapeyron equation (Eq. 8) is a fundamental expression relating vapor pressure to absolute temperature (C + 273) over limited temperature ranges, where ΔH is molar latent heat of vaporization (cal/mole) and R is the gas constant (= 1.986 cal/C°-mole).

$$\ln \frac{p_1}{p_2} = \frac{\Delta H}{R}\left(\frac{1}{T_1} - \frac{1}{T_2}\right) \tag{8}$$

Examination of Eqs. 4 and 5 suggests that relative evaporation rate (weight basis) from a smooth surface is proportional to the 0.9 power of solvent vapor pressure, as shown by Eq. 9.

$$E_W = kp^{0.9} \tag{9}$$

This can be solved explicitly for p (Eq. 10).

$$p = \left(\frac{E_W}{k}\right)^{1/0.9} \tag{10}$$

Assigning subscripts 1 and 2 to conditions applying to two different temperatures and substituting in Eq. 8 equivalent values for p as given by Eq. 10 yields Eq. 11.

$$\frac{1}{0.9}\ln\frac{E_{W_1}}{E_{W_2}} = \frac{\Delta H}{R}\left(\frac{1}{T_2} - \frac{1}{T_1}\right) \tag{11}$$

Changing from natural logarithms to the base 10, inserting the numerical value for R, and simplifying gives Eq. 12.

$$\log\frac{E_{W_1}}{E_{W_2}} = 0.197\Delta H\left(\frac{1}{T_2} - \frac{1}{T_1}\right) \approx 0.2\Delta H\left(\frac{1}{T_2} - \frac{1}{T_1}\right) \tag{12}$$

PROBLEM 15-2

The latent heat of vaporization for n-butyl acetate is reported as 10,600 cal/mole at 25 C. Assuming this to be an essentially constant quantity for the temperature changes involved, calculate the relative evaporation rates for n-butyl acetate at 15 C and 35 C.

Solution. The relative evaporation rate of n-butyl acetate at 25 C (298 K) is by definition 1.00. Substitute these values in turn in Eq. 12 for the two required temperatures of 15 C (288 K) and 35 C (308 K).

$$\log\frac{1}{E_{W_2}} = 0.2 \cdot 10{,}600\left(\frac{1}{288} - \frac{1}{298}\right); \qquad E_{W_2} = 0.57 \text{ (15 C)}$$

$$\log\frac{1}{E_{W_2}} = 0.2 \cdot 10{,}600\left(\frac{1}{308} - \frac{1}{298}\right); \qquad E_{W_2} = 1.70 \text{ (35 C)}$$

As shown by Problem 15-2, significant changes in relative evaporation rate result from comparatively minor changes in temperature. Thus the relative evaporation rate of n-butyl acetate is shown to increase on the average by about 6%/C° in going from 25 to 35 C.

Evaporation Cooling Effect. Some portion of the latent heat of vaporization required for solvent volatilization will be extracted from the solvent itself. If the volatile loss is rapid, the heat loss from the solvent will be correspondingly rapid, resulting in a significant self-cooling of the solvent. In some instances this evaporative cooling can be sufficiently pronounced to actually condense moisture vapor at the solvent surface.

The cooling of the solvent by evaporation also reduces its relative evaporation rate. This effect introduces another problem in running an evaporometer, namely, maintaining a constant solvent temperature despite the cooling effect, especially in the case of fast-evaporating solvents. In this regard, a metal substrate with its relatively large mass and excellent heat conduction is more effective in maintaining a constant temperature than a filter paper substrate.

EVAPORATION FROM MIXED SOLVENTS (SOLVENT BLENDS)

The relative evaporation rate for a solvent from a solvent blend is dependent on (*a*) its own concentration in the blend, (*b*) its individual (neat) relative evaporation rate, and (*c*) its escaping coefficient for the blend in question. An escaping (activity) coefficient is simply a compensating factor that adjusts for the interactions existing among the several solvent components making up the blend. The more dissimilar the solvent chemical types, the greater is the value of the escaping coefficient (4).

The relative evaporation rate for the total solvent blend is given by the sum of the individual evaporation rates as expressed by Eq. 13, in which the relative evaporation rates E, the concentrations c, and the escaping coefficients e are expressed in a set of consistent units.

$$E_T\,(\text{total}) = (ceE)_1 + (ceE)_2 + \cdots + (ceE)_n \tag{13}$$

Equation 13 provides a good index of the initial total evaporation rate from a known solvent blend. However, in real (not ideal) blends the relative losses of the different solvents to the atmosphere by volatilization do not match the original solvent composition. Hence, as the evaporation process proceeds, the composition of the solvent that is left behind changes. The direction of the change in the composition of residual liquid solvent can be qualitatively determined by comparing its initial fractional composition with the fractional composition of the escaping vapors. From such a comparison the solvent fractions that become depleted or enriched (percentagewise) with time can be identified, and from this trend the direction of the change in solvent blend can be qualitatively established. This can be quite critical, for once a trend is under way it tends to intensify. The fractional composition of the escaping vapors can be computed by reducing Eq. 13 to fractional form by dividing through by E_T, as given by Eq. 14.

$$1.00 = \frac{(ceE)_1}{E_T} + \frac{(ceE)_2}{E_T} + \cdots + \frac{(ceE)_n}{E_T} \quad (14)$$

Comparisons of c_n with $(ceE)_n/E_T$, in turn, provides the insight required to characterize the qualitative nature of the change in the liquid solvent composition as the evaporation process proceeds. This type of calculation depends, of course, on some access to data for escaping coefficient values. Such information can be quite accurately developed with computer assistance from involved theoretical data (5), or it can be obtained approximately by estimating the necessary escaping coefficient values from generalized graphs that have been reported in the literature (4). Such graphs are based on the observation that escaping coefficients are dependent mainly on chemical groups and, more generally, on the class of solvent. With this background three major solvent classes have been specified, with the solvents in any one class acting very much the same in volatility behavior. Escaping coefficients applying to any one class of solvent are then applied to any solvent within that class. Hence an escaping coefficient for any given solvent can be determined in terms of its class relative to the other solvent classes in the blend and its concentration.

Three graphs from which escaping coefficients can be estimated based on volume fraction concentrations v are given in Fig. 15-1*a* for hydrocarbon solvents, Fig. 15-1*b* for ester/ketone solvents, and Fig. 15-1*c* for alcohols/ether-alcohols (4). The two curves on each graph apply, respectively, to two binary systems (bottom scale and appropriate curve) as indicated. When tertiary systems are involved, interpolation between the curves is made, corresponding to the proportions of the two cosolvent classes involved. Problems 15-3 and 15-4 illustrate the practical use of these graphs in formulating coatings.

PROBLEM 15-3

A proposed polyamide ink solventation consists of equal volumes of isopropyl alcohol ($E_V = 1.6$) and VM&P naphtha ($E_V = 1.2$). Calculate the relative evaporation rate for the blend, and comment on the change in the nature of the liquid blend as evaporation proceeds.

Solution. Read off the escaping coefficient for isopropyl alcohol (1.55) on Fig. 15-1*c* (opposite the intersection of the $v = 0.5$ line with the hydrocarbon curve) and the escaping coefficient for VM&P naphtha (1.78) on Fig. 15-1*a* (opposite the intersection of the $v = 0.5$ line with the alcohol/ether-alcohol curve). Substitute these data in Eq. 13 to obtain the evaporation rate and in Eq. 14 to obtain the fractional composition of the escaping vapors.

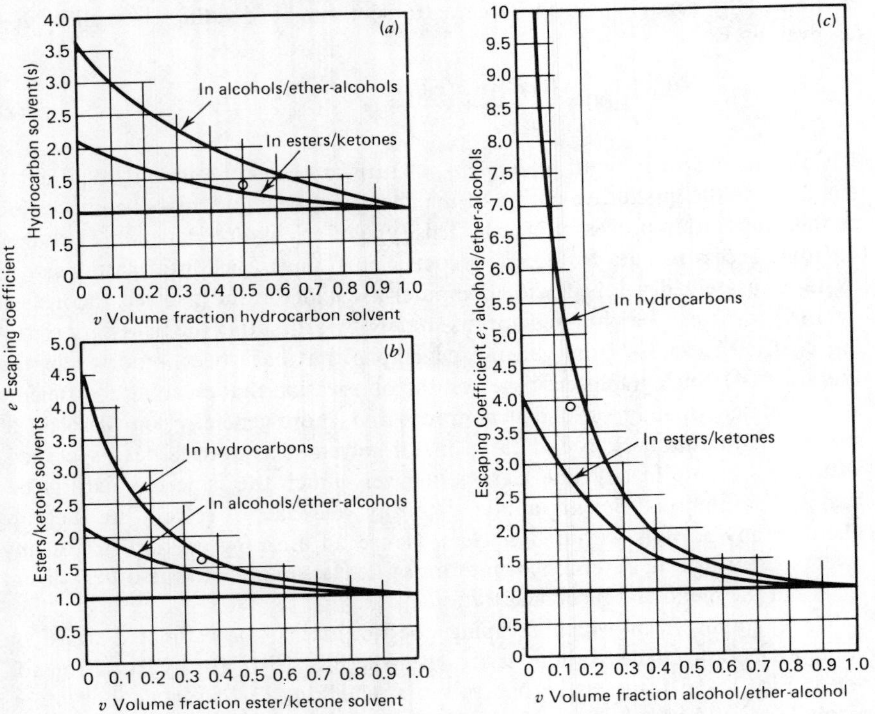

Fig. 15-1 Graphs giving escaping coefficients for solvents in terms of solvent class and solvent concentration for (*a*) hydrocarbon solvents, (*b*) ester and ketone solvents, and (*c*) alcohol and alcohol-ether solvents.

$$(13):\ E_T = (0.50 \cdot 1.55 \cdot 1.6) + (0.50 \cdot 1.78 \cdot 1.2)$$

$$= 1.24 \text{ (isopropanol)} + 1.07 \text{ (VM\&P naphtha)} = 2.31$$

$$(14):\ 1.00 = \frac{1.24}{2.31} + \frac{1.07}{2.31}$$

$$= 0.54 \text{ (isopropanol)} + 0.46 \text{ (VM\&P naphtha)}$$

The higher volume concentration of the alcohol in the vapor phase than in the original solvent blend (0.54 vs. 0.50) results in the systems becoming more alcohol-deficient as evaporation proceeds.

Comment. In the first place, it should be noted that the relative evaporation rate is greater for the blend (E_T = 2.31) than for either solvent alone. This is to be expected for mixed solvent classes. In the second place, the high-solvency component (active solvent) for the polyamide (isopropanol) may be fractionating off too rapidly relative to the VM&P naphtha (diluent). Such an out-of-balance situation could conceivably lead eventually to an inadequate solvency condition, with the possible development of imperfections such as cratering, pinholing, and blushing.

PROBLEM 15-4

A proposed solventation for nitrocellulose consists of 35%v of *n*-butyl acetate, 50%v of toluene ($E_V = 2.0$), 10%v of ethanol ($E_V = 1.7$), and 5%v of *n*-butanol ($E_V = 0.4$). Calculate the relative evaporation for the solvent blend, and comment on the change in the composition of the system as evaporation proceeds.

Solution. Read off the escaping coefficients for the four solvents from the graphs given in Fig. 15-1. These values are located on the graphs by circled points for easy reference. Substitute these data in Eq. 13 to obtain the evaporation rate for the blend and in Eq. 14 to obtain the fractional composition of the escaping vapors.

$$(13):\ E_T = (0.35 \cdot 1.6 \cdot 1.0) + (0.50 \cdot 1.4 \cdot 2.0) + (0.10 \cdot 3.9 \cdot 1.7) + (0.05 \cdot 3.9 \cdot 0.4)$$

$$= 0.57\ (n\text{-butyl acetate}) + 1.41\ (\text{toluene}) + 0.67\ (\text{ethanol}) + 0.08\ (n\text{-butanol}) = 2.73$$

$$(14):\ 1.00 = \frac{0.57}{2.73} + \frac{1.41}{2.73} + \frac{0.67}{2.73} + \frac{0.08}{2.73}$$

$$= 0.21\ (n\text{-butyl acetate}) + 0.52\ (\text{toluene}) + 0.24\ (\text{ethanol}) + 0.03\ (n\text{-butanol})$$

The lower concentration of *n*-butyl acetate in the vapor phase than in the original solvent blend (0.21 vs. 0.35) results in the system becoming enriched in this high-solvency component as the evaporation proceeds. This enrichment occurs partly at the expense of toluene diluent, which is being reduced in concentration (0.52 in the vapor phase vs. 0.50 in the solvent blend).

Comment. As in Problem 15-3, the evaporation rate of the blend is faster than that of any of the component solvents alone. The proposed solvent blend for nitrocellulose is well balanced in that with time there is a trend toward higher solvency, thus averting such adverse effects as pinholing, cratering, and blushing.

Analysis of Solvent Blend Compositions

Since the initial composition of a solvent blend is usually known, both the initial relative evaporation rate and the initial trend of the change in the starting composition with time can be determined. It can also be generally stated (*a*) that subsequent evaporation rates will be lower than the initial rate but equal to or higher than the relative evaporation rate of the slowest evaporating solvent, and (*b*) that the initial trend of the change in the compositional makeup of the solvent system is not likely to be reversed and, in fact, should be accelerated. These critical observations generally suffice to provide the coatings engineer with a reasonable base from which to make decisions. However, the researcher is forced to inquire more deeply into the nature of a changing solvent system.

Two procedures have been used to analyze solvent blends, radiotracer and gas chromatography techniques (6, 7). In the radiotracer technique the concentration of a given solvent is measured by introducing into the solvent structure during its manufacture, a small proportion of a labeled element (such as the isotope carbon-14, which is weak emitter of beta-rays and hence requires no shielding). The solvent content is then measured in terms of its relative radioactivity, which is detected either directly by a Geiger-Müller counter or indirectly by the light that is emitted by a scintillating agent introduced into the solvent system. However, the most convenient method of measuring the solvent composition is undoubtedly by gas chromatography, and presumably this is the technique most commonly employed by researchers in this field (6, 7).

EVAPORATION FROM WATER SYSTEMS

Water has a relative volume evaporation rate of 0.31 from filter paper and 0.56 from a smooth metal substrate at 25 C as measured by the Shell evaporometer, which uses circulating air at 0 to 5% relative humidity (RH). However, if the circulating air was adjusted to 100% RH, the relative evaporation rate for water would be reduced to zero, since the air overhead (at 100% RH) would already be completely saturated with water vapor. On the other hand, wide-ranging changes in relative humidity have little or no effect on the relative evaporation rates of organic solvents. This means that, since relative humidity selectively controls water evaporation rate, it also indirectly determines whether the water evaporation rate will be greater or less than the rates of the cosolvents with which it may be blended.

Critical Relative Humidity

The overriding importance of relative humidity in determining the drying properties of water borne coatings has led to the concept of a critical relative humidity (3). This concept is graphically illustrated in Fig. 15-2*a*, where a hypothetical binary system (constant temperature conditions) consisting of water and a miscible organic cosolvent is shown. The bottom scale gives the volume fraction of organic cosolvent, and the vertical scale the relative humidity. Two regions are delineated by the critical relative humidity curve. Above the curve the cosolvent evaporates faster than the water (the water evaporation is suppressed by the higher RH), and the system becomes depleted in the fraction of cosolvent that is present. Below the curve the water evaporates faster than the cosolvent (the drier air encourages more rapid water evaporation), and the system becomes enriched with a higher cosolvent fraction.

Consider a situation corresponding to point *A* on the 56% RH line of the graph on Fig. 15-2*a*. At this point the cosolvent is being lost at a faster rate than the water, resulting in a reduction in the volume fraction of the cosolvent (this corresponds to a movement to the left along the 56% RH line, as indicated). Conversely, at point *C* the water is evaporating faster than the cosolvent, leading to a reduction in its volume fraction (this corresponds to a movement to the right along the 56% RH line, as indicated). Point *B* represents a point of *unstable* equilibrium, since sooner or later some slight shift in conditions will precipitate a trend to either the left or the right that will then accelerate to the left or right, respectively. Conversely point *E* on the 56% RH line represents a point of *stable* equilibrium, since any movement away from *E* toward *D* or *F* will be immediately self-reversed and equilibrium will again be established (as indicated by arrows).

A study of the graph given in Fig. 15-2*a* explains the not uncommon situation where a cosolvent that is becoming enriched in a system during evaporation at one given RH reverses this trend and starts becoming depleted at a somewhat higher RH (see movement from point *G* to point *H* on Fig. 15-2a as illustrative of this situation).

Organic cosolvents with water at typical use concentrations of $<\sim 20\%$ tend to fall into three classes based on their critical relative humidities (CRHs). First there are the fast solvents, which have CRH values well below the RHs that are associated with air-dry flash-off conditions of coating lines (25 to 75% RH). As a result they are invariably depleted in volume fraction during the flash-off stage (see the CRH curve for propanol in Fig. 15-2*b* for a representative fast solvent). At the other extreme are the slow solvents, which have CRH values well above the RHs of coating lines and hence almost always become enriched (see the CRH curve for 2-hexoxyethanol in Fig. 15-2*b* for a representative slow solvent). Finally there are the intermediate solvents, where control of the RH relative to their CRHs is of great importance in attaining consistent quality coatings (see the CRH curve for 1-propoxy-2-propanol in Fig. 15-2*b* for a representative intermediate solvent).

The three curves of Fig. 15-2*b* apply to a temperature of 25 C and an air motion of 0 mph. However, as shown by Fig. 15-2*c*, a nominal rise in tempera-

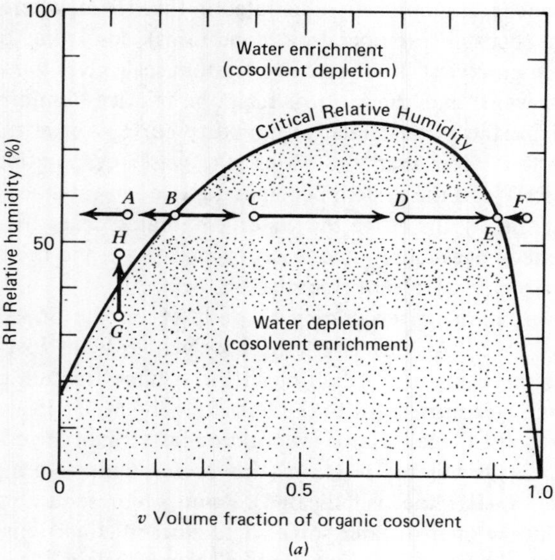

Fig. 15-2a

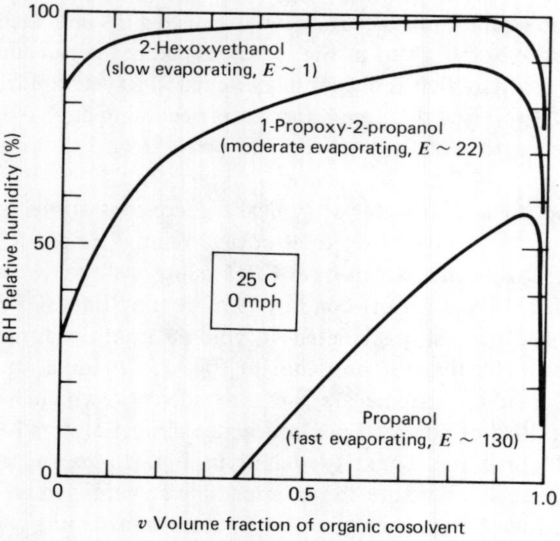

Fig. 15-2b

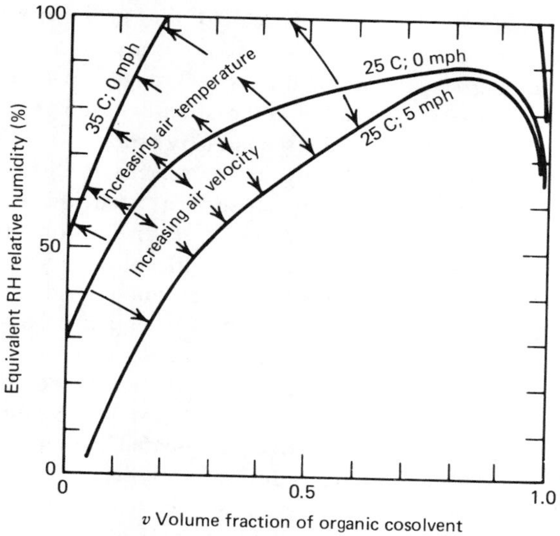

Fig. 15-2c

Fig. 15-2 Plots of relative humidity versus volume fraction of cosolvent, showing (*a*) a hypothetical critical relative humidity curve that separates the region of water enrichment (cosolvent depletion) from that of water depletion (cosolvent enrichment); (*b*) critical relative humidity curves for fast-, medium-, and slow-evaporating cosolvents, respectively; and (*c*) curves showing the effect of increased temperature and increased air velocity on the relocation of the critical relative humidity curve.

ture acts to greatly raise the effective CRH curve, whereas an increase in air motion generally lowers the CHR curve. These controlling factors of temperature and air motion furnish a means of manipulating the relative loss of water versus cosolvent during the air-dry flash-off stage. Fortunately, during subsequent baking conditions the higher temperatures involved generally lead to solvent enrichment, with the possible exception of very fast solvents.

EVAPORATION OF VOLATILES FROM APPLIED ORGANIC COATINGS

The evaporation of solvent from an applied coating introduces an added complexity in that the dissolved polymer exerts a strong effect on the evaporation process, especially during the final stage of drying, where a substantial retention of residual solvent may persist for weeks. For example, from 6 to 9% of solvent may be present in a nitrocellulose lacquer after 1 or 2 weeks of normal drying although it is completely dry to the touch (6). It is now generally recognized that the evaporation of solvent from a coating occurs in two consecutive but somewhat overlapping stages. In the first (wet) stage the pattern of solvent volatilization more or less follows the evaporation behavior that has been outlined for solvent blends alone. In the second (dry) stage, however, the volatile loss becomes controlled by the ability of the solvent to diffuse from within the relatively dry polymer to the polymer surface, from which it then escapes (6, 8). The second-stage diffusional process is relatively slow, and as a result coatings applied from solution often retain significant quantities of solvent for extended periods of time. In some instances retention of residual solvent creates no problems, but in other cases difficulties may be experienced such as delayed film hardening, impaired water and chemical resistance, and residual odor (6).

The drying curve given in Fig. 15-3 for a hypothetical evaporation pattern for an applied coating distinguishes between the initial wet stage, during which the evaporation is relatively rapid and surface dominated (solvent escapes from a liquid surface), and the final dry stage, during which the solvent loss is very slow and completely diffusion controlled (solvent that reaches the surface escapes almost at once from a substantially dry substrate). Note that a logarithmic time scale has been used to dramatize the complete difference in relative evaporation

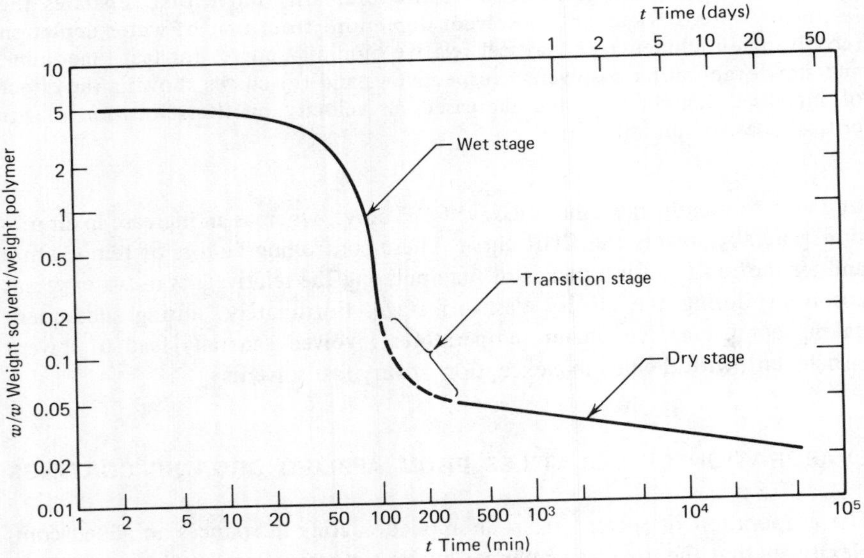

Fig. 15-3 Plot of solvent retention as a function of time, showing a hypothetical sequential drying pattern characterized by an initial wet stage and a final dry stage.

times during the initial and final stages of solvent volatilization. The following review is concerned only with the final dry stage of evaporation, since the earlier wet stage is covered essentially by a consideration of solvent blends alone.

The evaporation process during the diffusion-controlled dry stage is markedly dependent on the thickness of the applied polymer film. In two independent studies it has been shown that loss of residual solvent with time is given quite accurately by Eq. 15, where c is the concentration of the solvent in terms of weight of solvent per unit weight of polymer (w/w), x is the film thickness, and A and B are constants applying to the given coating composition (8).

$$\log c = A \log \frac{x^2}{t} + B \tag{15}$$

Except for the loss of the very last vestiges of solvent Eq. 15 holds very well for most of the dry stage period. The commanding role of thickness is shown by its entry into Eq. 15 as a square power. Thus for a given polymer/solvent composition the ratio x^2/t is a constant in reaching any specified dry stage concentration c. Hence it can be stated that, in general, retention time is inversely proportional to the square of the applied film thickness. For example, if the film thickness is doubled, the retention time is quadrupled.

PROBLEM 15-5

A thermoplastic vinyl chloride/vinyl acetate copolymer dissolved in methyl isobutyl ketone (MIBK) is applied to a substrate to form a 7.0-μ dry film. Retained solvent is found to be 12.2% (0.122) based on the polymer at the end of 1.0 hr and 8.6% (0.086) at the end of 1 day. What is the retained concentration (w/w) at the end of 2 weeks? Also, what would the retention have been if the film had been only 3.0 μ thick?

Solution. First substitute the given data in turn in Eq. 15 to obtain values for the constants A and B.

$$\log 0.122 = A \log \frac{49}{1} + B$$

$$\log 0.086 = A \log \frac{49}{24} + B$$

Subtract one equation from the other to eliminate B, and solve for A. Then substitute A in either equation to obtain B. Express Eq. 15 in terms of the calculated constants.

$$\log c = 0.11 \log \frac{x^2}{t} - 1.10 \tag{16}$$

Substitute the value of $t = 336$ hr in Eq. 16 to obtain the solvent retention at the end of 2 weeks.

$$\log c = 0.11 \log \frac{49}{336} - 1.10; \qquad c = 0.064\ (6.4\%)$$

The retention for the 3.0-μ thick film is calculated in a similar manner.

$$\log c = 0.11 \log \frac{9}{336} - 1.10; \qquad c = 0.053\ (5.3\%)$$

Comment. These results underwrite the persistent retention of solvent over long periods of time even for very thin films.

Factors Affecting Solvent Retention

The effect of film thickness on solvent retention with time is given by Eq. 15. However, other factors affecting solvent release are less amenable to mathematical treatment. Nevertheless, certain qualitative observations have been made that are helpful in predicting solvent retention behavior (6, 8).

It is somewhat unexpected that solvent release does not appear to closely parallel either solvent volatility or solvent power. This is shown in Table 15-1, which lists 22 common solvents in the approximate order of increasing reten-

Table 15-1 Selected Solvents Listed in Approximate Order of Increasing Retention by Polymers.
Relative evaporation rates E_v (*n*-butyl acetate = 1.0) noted for reference.

Solvent	E_v	Solvent	E_v
1 Methanol (*least retained*)	4.1	14 2-Nitropropane	1.5
2 Acetone	10.2	15 *m*-xylene	0.8
3 2-Methoxyethanol	0.5	16 Methyl isobutyl ketone	1.4
4 Methyl ethyl ketone	4.5	17 Isobutyl acetate	1.7
5 Ethyl acetate	4.8	18 2,4-Dimethyl pentane	5.6
6 2-Ethoxyethanol	0.4	19 Cyclohexane	5.9
7 *n*-Heptane	3.3	20 Diacetone	0.1
8 2-Butoxyethanol	0.1	21 Methylcyclohexane	0.3
9 *n*-Butyl acetate (*reference standard*)	1.0	22 Methylcyclohexanone (*most retained*)	0.2
10 Benzene	5.4		
11 2-Methoxyethyl acetate	0.4		
12 2-Ethoxyethyl acetate	0.2		
13 Toluene	2.3		

tion by polymers (6). On the other hand, solvent retention is apparently related to both the size and the shape of the solvent molecule. Thus larger and more irregularly shaped (bulkier) molecules evidently experience greater difficulty in diffusing through the molecular interstices (holes) of the bulk polymer to reach the surface and escape. For example, isobutyl acetate (branched) is retained more than *n*-butyl acetate (linear); 2,4-dimethylpentane (branched) is retained more than *n*-heptane (linear); and cylclohexane (nonplanar) is retained more than toluene (planar).

The nature of the polymer also exerts a positive effect on solvent retention. However, this is a general influence and is not specific for any solvent in particular (otherwise the generalized listing of Table 15-1 would not have been possible). Soft resins release solvent faster than hard resins. In this respect a plasticizer that serves to soften (plasticize) a resin is found to be most effective in facilitating the escape of solvent by diffusion. The influence of resin hardness is further confirmed by the finding that solvent loss becomes markedly increased just above the glass transition temperature T_g of a polymer. Any increase in temperature, of course, accelerates the release of solvent, but a rise in temperature from below to above the T_g generally results in a dramatic increase in solvent release by the diffusion process.

In the case of solvent blends, it often occurs that one of the more volatile solvents is essentially dissipated by the time the polymer/solvent composition reaches the dry stage. After reaching the dry stage, solvent retention more or less follows the general rules covering the release of single solvents.

EVAPORATION OF WATER AND COSOLVENTS FROM APPLIED LATEX SYSTEMS

Like the evaporation of organic solvents from a nonaqueous applied film, the evaporation of volatiles from an applied latex film can be considered as a two-stage process (9). In the initial wet stage solvent loss is substantially what might be expected for the volatiles alone; in the second stage volatile loss becomes largely controlled by the rate at which the volatile components can diffuse to the film surface and escape.

It has also been proposed that the latex evaporation process be broken down into three stages: (*a*) an initial stage, during which the latex particles move freely and the evaporation is at a constant rate; (*b*) an intermediate stage, where the particles come into irreversible contact and the water evaporation rate drops to 5 to 10% of its initial value; and (*c*) a final stage, in which the water escapes very slowly by internal diffusion to the film surface (10). The mechanisms that describe the evaporation trend of volatiles from a latex system are reasonably well characterized by the curve shown in Fig. 15-3.

A still more elaborate, four-stage film-forming process has been described for the physical process of evaporation, but the complex equations involved render this approach more of an exercise in mathematics than a practical tool for the coatings engineer, although it may be of interest to the researcher (11).

The presence of cosolvents with water in latex systems apparently has little effect on the initial evaporation rate of the water. As previously discussed, relative humidity, air motion, and temperature are the main parameters that control the wet stage of the evaporation process. The second, dry stage presents a much more complex situation. Thus the cosolvent tends to partition between the water and polymer phases, and the water evaporation is influenced by this type of distribution (9).

In certain instances the cosolvent may be completely lost by the time the dry stage is reached (this is typical of ethylene glycol monomethyl ether), or the cosolvent may distribute almost exclusively into the polymer phase (this is typical of ethylene glycol monobutyl ether acetate). In such cases the effect of the cosolvent on the water evaporation is correspondingly slight (9).

Cosolvent evaporation rates from latex films more or less follow their neat evaporation rates.

Ethylene glycol monomethyl ether (fastest).

Ethylene glycol monobutyl ether.

Ethylene glycol monobutyl ether acetate.

Ethylene glycol.

Diethylene glycol monobutyl ether (slowest).

It is notable that the presence of ethylene glycol facilitates coalescent (cosolvent) evaporation, since it acts to provide a swollen, continuous hydrophilic network through which the coalescent can more easily diffuse. Also, the more polar coalescents tend to evaporate faster during the dry stage, since they preferentially partition into the hydrophilic network, from which they can more readily diffuse and escape.

Since ethylene glycol is more hygroscopic than propylene glycol, ethylene glycol latex paints exhibit slower overall drying. Hence for exterior paint application the use of propylene glycol is preferable from the standpoint of providing less sensitivity to water washdown during the drying period (12).

From a long term standpoint (several weeks) relatively little water or cosolvent remains in the latex film. Any residual volatile component that is left is generally too minute in amount to exert a detrimental effect.

PRACTICAL CONSIDERATIONS

A very rapidly evaporating solvent tends to impart poor flow and may induce possible moisture condensation (film blushing). A very slowly evaporating solvent suffers from deficient dry, retarded hardening, questionable print resistance, and the like. At one time there was a trend to limit solvent selection to solvents that avoided evaporation extremes. More recently, however, the use of solvent systems with a minimum of solvent content in the intermediate range has been proposed to provide such advantages as little "dry spray," avoidance of "orange peel," improved color and luster, application of thicker coats, minimal floating, and/or clouding with iridescents (13).

The vapor pressure of most solvents increases by about 3 to 5% for each 1 F° rise in temperature. Hence a rise in temperature from 68 F (winter) to 84 F (summer) is sufficient to increase the vapor pressure of the solvent about 1.9 times ($1.87 = 1.04^{16}$). The increased volatility imparted by this nearly 90% increase in vapor pressure must generally be compensated for by replacing a portion (say 10 to 20%) of the more rapidly evaporating solvent with a slower type.

Since most solvent vapors are denser than air, they tend to remain in the vicinity of the film unless flushed from the region by air currents. When solvent vapor accumulates to saturate the space above a wet film, evaporation is seriously retarded. Hence the air movement past a wet paint film exerts a major influence on the evaporation process.

A marked cooling always accompanies the evaporation of solvent from a binder film, since heat must be abstracted from the binder system (in part) to furnish the heat of vaporization necessary to endow the solvent molecules with sufficient energy to escape from the binder surface into the atmosphere. If this reduction in temperature chills the binder below the dew point of the ambient air, the resultant moisture condensation frequently produces a so-called moisture blush. This undesirable deposition and subsequent retention of moisture by the binder is induced by the faster evaporating solvents. Corrective measures to overcome moisture blush consist in the substitution of either slower evaporating solvents or solvents that become miscible with water as water is condensed on the binder film (say isopropanol).

Organic solvent evaporation is always associated with the hazard of possible ignition and fire. Vapor pressure and lower explosive limit are the two major parameters controlling the flash-point behavior of organic solvents. This subject is very well covered by Ellis in two publications (7, 14).

REFERENCES

1. Apparatus Series No. 109, "Shell Automatic Thin Film Evaporometer," Roxana Machine Works, 300 Michigan Avenue, South Roxana, Ill. 62087.
2. Rocklin, A. L., "Evaporation Phenomena: Precise Comparison of Solvent Evaporation Rates from Different Substrates," *JCT*, **48**, No. 622, 45–57 November (1976).
3. Dillon, P. W., "Application of Critical Relative Humidity, an Evaporation Analog of Azeotropy, to the Drying of Water-Borne Coatings," *JCT*, **49**, No. 634, 38–49, November (1977).

4. Bulletin PC-69-4, "Predicting the Evaporation Properties of Mixed Solvent Blends," Shell Chemical Company, Petrochemicals Division, 1212 Avenue of the Americas, New York 10036.
5. Walsham, J. G. and G. D. Edwards, "A Model of Evaporation from Solvent Blends," *JPT*, **43,** No. 554, 64–70, March (1971).
6. Newman, D. J. and C. J. Munn, "Solvent Retention in Organic Coatings," *Prog. Org. Coatings*, **3,** 221–243 (1975).
7. Ellis, W. H., "Solvents," in *Paint Testing Manual* (G. G. Sward, Ed.), 13th ed., American Society for Testing and Materials, 1916 Race Street, Philadelphia, Pa. 19103, pp. 30–149.
8. Newman, D. J., C. J. Nunn, and J. K. Oliver, "Release of Individual Solvents and Binary Solvent Blends from Thermoplastic Coatings," *JPT*, **47**, No. 609, 70–78, October (1975).
9. Sullivan, D. A., "Water and Solvent Evaporation from Latex and Latex Paint Films," *JPT*, **47,** No. 610, 60–67, November (1975).
10. Vanderhoff, J. W., et al., *J. Polym. Sci.*, Symposium No. 41, 155 (1973).
11. Krzyzanowska, T., "A New Mechanism of the Physical Film-Forming Process," *Prog. Org. Coatings*, **3,** 349–360 (1975).
12. Andrews, M. D., "Influence of Ethylene and Propylene Glycols on Drying Characteristics of Latex Paints," *JPT*, **46,** No. 598, 40–48, November (1974).
13. McMaster, W. D., "A Fancy Formula for Finer Finishes," *Paint Varn. Prod.*, **46,** 27, February (1956).
14. Ellis, William H., "Solvent Flash Points—Expected and Unexpected," *JCT*, **48**, No. 614, 38–57, March (1976).

BIBLIOGRAPHY

Rocklin, Albert L., "Effect of Humidity and Other Ambient Conditions on Evaporation of Ternary Aqueous Solvent Blends," *JCT*, **50,** No. 646, 46–55, November (1978).

16 *Paint Flow Relationships (Coating Rheology)*

There is probably no true Newtonian coating or ink system. Instead, commercial or trade sales coatings and ink compositions are deliberately designed to exhibit non-Newtonian flow. It is with the latter type of rheology that this chapter is mainly concerned.

A conventional coating is subjected during manufacture, application, and dry (or cure) to rheological extremes. This is clearly indicated in Fig. 16-1, which shows some typical shear rate ranges that are encountered in the coatings industry. Thus during paint preparation in a high-speed impeller disperser the shear rate probably ranges from about 1000 to 10,000 sec^{-1} in the immediate vicinity of the disperser blade, whereas at the top of the container where the paint contacts the walls the motion may be substantially arrested (shear rate on the order of 1 to 10 sec^{-1}). Hence in one piece of equipment a range of shear rates is being encountered. After drumming off the paint for storage, pigment settling may occur. Here the shear rate tends to fall in the ultralow range of 0.001 to 0.5 sec^{-1}. On subsequent application by brushing, spraying, or roller coating, the shear rates will be at least 1000 sec^{-1} and can reach 100,000 sec^{-1} in certain roller coating applications. Following these vigorous types of application, there ensue entirely different conditions where such phenomena as sagging, leveling, and penetration occur. Here typical shear rates are usually less than 1.00 sec^{-1}. Paralleling the extreme variations in shear rate are corresponding extreme shifts in paint viscosity (e.g., from 1.0 to 10^6 poises). Typical shear rate ranges for representative coating operations are listed in Table 1-3. Faced with these extremes in rheological values, the engineer is almost forced to resort to logarithmic scales to adequately portray viscosity information in graphical form.

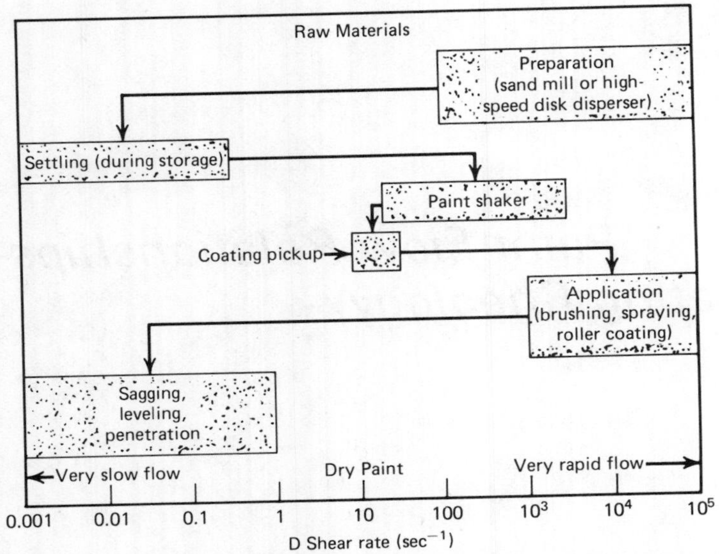

Fig. 16-1 Schematic illustration showing shear rate regions and approximate shear rate values applying to various aspects of paint flow.

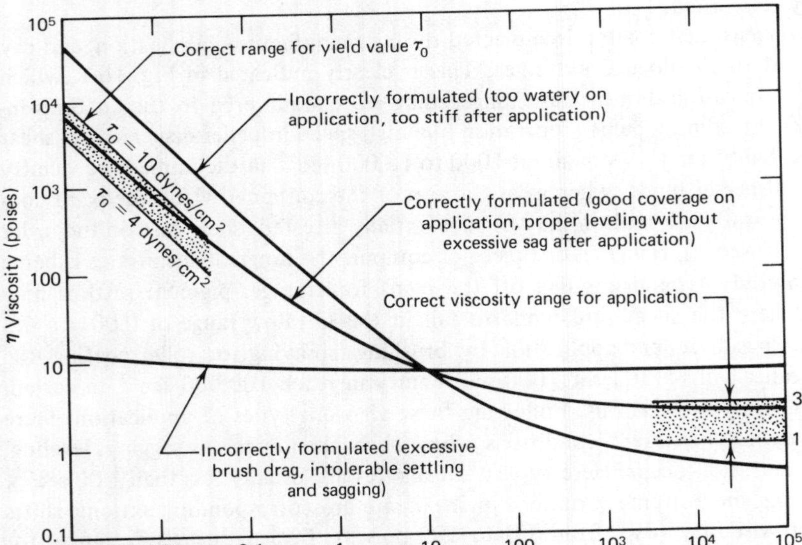

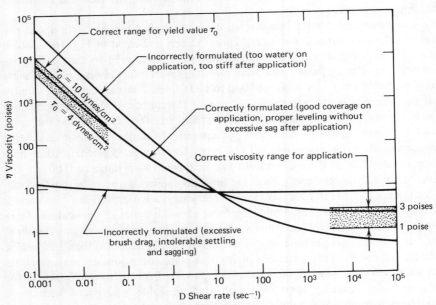

Fig. 16-2 Graph giving viscosity profiles for three coatings, one correctly formulated and two incorrectly formulated.

From experience this author is convinced that selection of the log shear rate for the abscissa (horizontal) scale and the log viscosity for the ordinate (vertical scale) is best for visualizing the interplay among the several variables.

To indicate how viscosity data can be routinely developed and plotted on such a graph, Fig. 16-2 shows viscosity curves for three paints, one correctly formulated (from a flow standpoint) and the other two incorrectly formulated. Note that near the center of the shear rate range (10 sec^{-1}) all three paints are similar in viscosity, whereas to either side they exhibit wide divergence. This spread in viscosity in the low- and high-shear-rate regions indicates the fallacy of attempting to extrapolate viscosity data obtained at some intermediate point to lower and higher shear rates. It also points up the fact that information derived from a single-point viscosity value is of limited use except for plant control purposes.

VISCOSITY PROFILE RELATIONSHIPS (CASSON EQUATION)

Each of the viscosity curves in Fig. 16-2 (log viscosity versus log shear rate) is referred to as a viscosity profile (1). Of the several equations that have been proposed to express a viscosity profile in mathematical form, the Casson equation (Eq. 1) appears best equipped to provide a valid relationship that can be made to fit a viscosity profile over its entire reach (2).

$$\eta_D^n = \eta_\infty^n + \left(\frac{\tau_0}{D}\right)^n \tag{1}$$

where τ_0 = yield value ($dynes/cm^2$)
D = shear rate (sec^{-1})
η_D = viscosity at shear rate D (poises)
η_∞ = viscosity at infinite shear rate (poises)
n = exponent (commonly 0.5)

Equation 1 is far superior to power law expressions that yield only straight lines on graphs of log viscosity versus log shear rate and for this reason are incapable of expressing the curved viscosity profile over any extended range (3, 4). Casson's equation is a mathematical expression that uniquely fills the need for an *overall* viscosity profile equation.

In Eq. 1 η_∞ represents the viscosity of the coating system at infinite shear rate, a condition where all structural viscosity due to pigment flocculation, colloidal aggregation, or any other source has been broken down and eliminated. Its value can be closely approximated by measuring the viscosity of the coating system at a very high shear rate (preferably above 10,000 sec^{-1}). The yield value of the coating τ_0 is more difficult to measure, but values can be approximated by the sophisticated viscometer instrumentation available today (5, 6). A yield value can also be approximately determined by properly interpreting the data afforded by the tail-end values of the viscosity profile curve in the ultralow-shear-rate range, or it can be closely calculated from three widely scattered points as read off the overall profile curve (see Problem 16-2).

The exponent n in Casson's expression (Eq. 1) controls the degree of bending through the intermediate portion of the viscosity profile curve, as shown in Fig. 16-3*a*. A typical value of n for coating compositions is 0.50. Larger values increase the bending; lower values decrease it.

The low to ultralow portion of the viscosity profile curve is controlled by the yield value τ_0, since the curve approaches the yield asymptotically as the shear rate decreases. This influence of yield value is shown in Fig. 16-3*b*, where the variable τ_0 takes on values of 0.1, 1.0, 10, and 100 dynes/cm^2. On the other hand, the high-shear-rate portion of the viscosity profile curve is controlled by η_∞, since the viscosity must asymptotically approach infinite-shear-rate viscosity as the shear rate increases indefinitely.

Study of the profile curves given in Fig. 16-3 shows that practically any viscosity curve for a coating can be expressed by Eq. 1 with good fidelity, the range running from vanishingly small shear rates to infinitely large ones.

The controlling factors as indicated in Fig. 16-3 can also be deduced from an analysis of Eq. 1. Thus, as the shear rate D increases, the term (τ_0/D) becomes correspondingly smaller in comparison to η_∞. Eventually the term becomes a negligible factor, making $\eta_\text{D} = \eta_\infty$. On the other hand, as the shear rate decreases, it is the quantity η_∞ that becomes vanishingly small compared with the term (τ_0/D), making $\eta_\text{D} = \tau_0/\text{D}$ or $\tau_0 = \eta_\text{D} \cdot \text{D} = \tau$. This is a mathematical way of saying that at an ultimate ultralow shear rate the viscosity η_D and the shear rate D must be such as to give a shear stress τ that falls on the yield value (τ_0) line.

Let the subscripts h and l denote high- and low-shear-rate conditions, respectively. Then by algebraic manipulation Eqs. 2 and 3 can be derived from Eq. 1.

$$\eta_\infty = \left[\frac{(\text{D}\eta)_h^n - (\text{D}\eta)_l^n}{\text{D}_h^n - \text{D}_l^n}\right]^{1/n} \tag{2}$$

$$\tau_0 = \left(\frac{\eta_l^n - \eta_h^n}{1/\text{D}_l^n - 1/\text{D}_h^n}\right)^{1/n} \tag{3}$$

Equations 2 and 3 are quite intractable because of the elusiveness of the exponent n. However, a reasonable first approximation for a value for n is 0.50, reducing Eqs. 1, 2, and 3 to Eqs. 4, 5, and 6.

$$\sqrt{\eta_\text{D}} = \sqrt{\eta_\infty} + \sqrt{\tau_0/\text{D}} \tag{4}$$

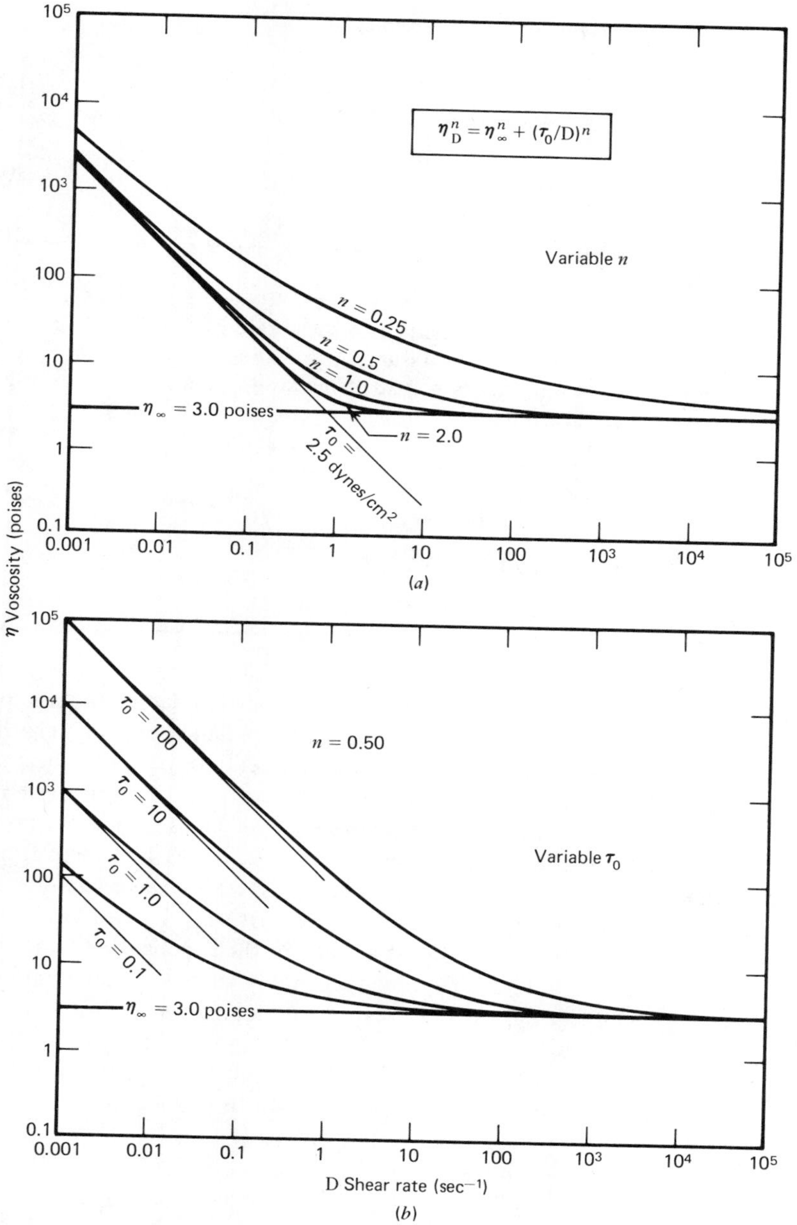

Fig. 16-3 (*a*) Curves of log shear rate versus log viscosity on graph developed from the Casson equation for various values of the exponent n (values of η_∞ and τ_0 held constant). (*b*) Curves of log shear rate versus log viscosity on graph developed from the Casson equation for various values of the yield value τ_0 (values for n and η_∞ held constant).

$$\eta_\infty = \left[\frac{\sqrt{(D\eta)_h} - \sqrt{(D\eta)_l}}{\sqrt{D_h} - \sqrt{D_l}}\right]^2 \tag{5}$$

$$\tau_0 = \left(\frac{\sqrt{\eta_l} - \sqrt{\eta_h}}{1/\sqrt{D_l} - 1/\sqrt{D_h}}\right)^2 \tag{6}$$

PROBLEM 16-1

The viscosity of a coating is 3.18 poises at a shear rate of 1000 sec^{-1} and 307 poises at a shear rate of 0.01 sec^{-1}. Assuming that the coating viscosity profile follows Casson's expression when $n = 0.50$, calculate the coating yield value and its viscosity at infinite shear rate.

Solution. Substitute the problem data in Eqs. 5 and 6.

$$\eta_\infty = \left(\frac{\sqrt{1000 \cdot 3.18} - \sqrt{0.01 \cdot 307}}{\sqrt{1000} - \sqrt{0.01}}\right)^2 = 3.0 \text{ poises}$$

$$\tau_0 = \left(\frac{\sqrt{307} - \sqrt{3.18}}{1/\sqrt{0.01} - 1/\sqrt{1000}}\right)^2 = 2.5 \text{ dynes/cm}^2$$

The more general expression (Eq. 1) is less responsive to numerical or graphical solution, since three constants are involved and they are tied up in a complex manner. However, an approximate solution can be obtained by assuming that, for practical coating systems, a viscosity measured or extrapolated to a shear rate of 100,000 sec^{-1} is substantially equal to the viscosity at infinite shear rate (η_∞). The inevitable, close, asymptotic approach of the viscosity η on the profile curve to the η_∞ line, with ever-increasing shear rate, is clearly evident from a study of the profile curves shown in Fig. 16-3. By taking this as an effective equality, it is possible by algebraic methods to derive Eq. 7 from Eq. 1. This derived equation is capable of developing a value for the exponent n that is quite accurate. Three sets of data are required, preferably in the ultralow, intermediate, and very-high-shear ranges. A numerical solution to the equation is probably obtained most rapidly by successively substituting values for n in Eq. 7 and adjusting after each substitution until a value is obtained for n that makes Eq. 7 essentially equal to zero. Problem 16-2 is given to illustrate the procedure.

$$\left[\eta_m\left(\frac{D_m}{D_l}\right)\right]^n + \eta_\infty^n - \left[\eta_\infty\left(\frac{D_m}{D_l}\right)\right]^n - \eta_l^n = 0 \tag{7}$$

The subscripts l, m, and h serve to identify quantities corresponding to low, medium, and high shear rates, respectively. The viscosity at infinite shear rate η_∞ is estimated by extrapolation from the η_h value.

PROBLEM 16-2

The viscosity of a house paint is 3.2 poises at a shear rate of 10,000 sec^{-1}, 20 poises at 1.0 sec^{-1}, and 800 poises at 0.01 sec^{-1}. By intuitive extrapolation it is estimated that the viscosity at infinite shear rate should be close to 3.0 poises (data plotted on log viscosity versus log shear rate graph). Develop an equation that accurately describes the viscosity profile of the house paint.

Solution. Substitute the problem data in Eq. 7 to provide the expression for the trial calculations based on successive n-value substitutions.

$$\left(20 \cdot \frac{1.0}{0.01}\right)^n + 3.0^n - \left(3.0 \cdot \frac{1.0}{0.01}\right)^n - 800^n = 0$$

$$2000^n + 3^n - 300^n - 800^n = 0 \tag{8}$$

Start with a value of 0.50 for n, and adjust as indicated to arrive at an n-value that gives a value of zero for the sum of the terms in Eq. 8.

n	Sum of terms
0.5	0.849
0.4	-1.822
0.48	-0.089
0.482	-0.007 (≈ 0)

The value of 0.482 for n gives the required zero value in Eq. 8. Next calculate τ_0 from Eq. 9 (a rearrangement of Eq. 1), using the low-shear-rate viscosity data (best).

$$\tau_0^n = \mathrm{D}^n \cdot (\eta_\mathrm{D}^n - \eta_\infty^n) \tag{9}$$

$$\tau_0^{0.482} = 0.01^{0.482}(800^{0.482} - 3^{0.482}); \qquad \tau_0 = 6.9$$

The expression describing the viscosity profile curve for the house paint is given by Eq. 10.

$$\eta^{0.482} = 3^{0.482} + \left(\frac{6.9}{\mathrm{D}}\right)^{0.482} \tag{10}$$

The viscosities given by this equation are identical with the measured viscosities except for the high shear rate of 10,000 sec^{-1} (3.2 poises measured; 3.1 poises calculated).

Comment. The calculations called for by Eq. 8 are readily carried out using an inexpensive hand computer. No more than four or five trial calculations are usually necessary to arrive at a value for n that gives the required zero (or close to zero) result.

A graph of log shear rate versus log viscosity is capable of spreading out rheological information uniformly over tremendous ranges of values. On such a graph a plot using the Casson equation (Eq. 1) results in a curve for the plotted data, as shown in Fig. 16-3. Occasionally, however, it is desirable to sacrifice a full range of values in order to obtain a straight-line relationship from the Casson expression. Examination of Eq. 1 shows that a plot of η^n versus $(1/\mathrm{D})^n$ yields a straight-line relationship. The intercept at the viscosity scale equals the viscosity at infinite shear rate (η_∞), and the slope of the straight line equals τ_0^n.

An example of this type of plotting, where $n = 0.5$, is given in Fig. 16-4, adapted from a graph in the 1977 Matiello lecture for a commercial one-coat house paint. It is clearly demonstrated that the plotting is limited to the lower range of viscosities and the higher range of shear rates.

If Eq. 4 is multiplied through by $\sqrt{\mathrm{D}}$, a second equation is generated that provides a straight line relationship as given by Eq. 4a.

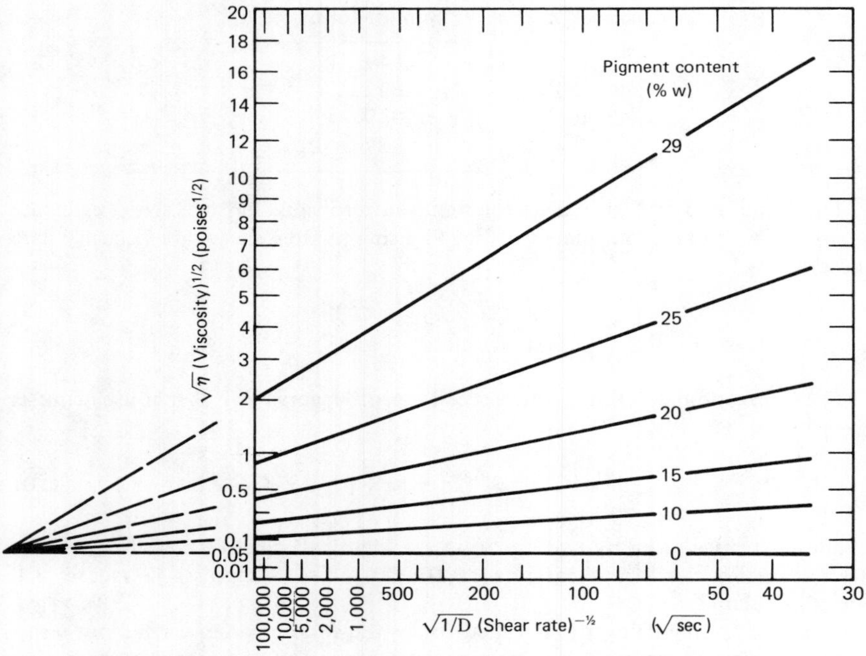

Fig. 16-4 Graph of the square root of the reciprocal shear rate versus the square root of the viscosity for selected pigment concentrations, illustrating the straight-line relationship given by the Casson equation for this type of plotting (commercial one-coat house paint).

$$\sqrt{\tau} = \sqrt{\eta_\infty} \cdot \sqrt{D} + \sqrt{\tau_0} \qquad (4a)$$

When $\sqrt{\tau}$ is plotted against $\sqrt{D}$, the intercept of the resulting straight line with the $\sqrt{\tau}$ scale gives the square root of the yield value $\sqrt{\tau_0}$ and the line slope gives $\sqrt{\eta_\infty}$. From this plotting technique, a pseudoplastic system with a yield value (a Casson system) can be reasonably characterized in terms of the constants (parameters) τ_0 and η_∞. This plotting method assumes that $n = 0.5$ applies in the more general Eq. 1.

An approximate but rapid experimental method for developing data that uses a modified form of Eq. 4a has been reported recently based on the single cylinder type of rotational viscometer (Brookfield Synchro-Lectric Viscometer) (7). Details of this practical semi-empirical approach for approximately characterizing Casson-type systems are well outlined including submitted specimens of the special graph paper required ($\sqrt{\text{dial reading}}$ vs $\sqrt{\text{rpm}}$) and a copy of the transparent overlay that provides a fast readout of critical numerical values.

PRACTICAL RANGES FOR HIGH-SHEAR-RATE VISCOSITY AND YIELD VALUE

On the basis of practical studies it is now possible to correlate individual coating properties with rheological measurement (1, 6). Two properties of key importance are the viscosity of the coating at high shear rate (~10,000 sec^{-1}) and the coating yield value.

In the first edition of this book (1964), considerable space was devoted to showing that brushability should be evaluated in terms of viscosity at high shear rates. Today this is an accepted fact, and attention is now directed toward the details of this correlation. A recommended range for trade sales coatings based on organic solvent systems is from 1.0 to 3.0 poises at a shear rate of 10,000 sec^{-1}. A suggested viscosity range for latex paints is from 2.5 to 5.0 poises at a shear rate of about 1370 sec^{-1}, although viscosities above 1.0 or 2.0 poises are not always easy to obtain (6). These ranges (indicated in Figs. 16-2 and 16-5) are also applicable to roller coating application (6). The viscosity of a coating for spray painting should be from about 0.25 to 2.5 poises at a spray shear rate of about 2500 sec^{-1} (3). In general, high application viscosities are desirable from a standpoint of high build (thick coating) with good leveling. On the other hand, excessively high viscosity can lead to application problems.

Coating yield value is not only more difficult to measure than viscosity at high shear rate but is also subject to variation (sheared and unsheared values). However, target yield values can be set up, based on undisturbed (unsheared) and sheared (after application) conditions.

An unsheared (undisturbed) yield value for a coating system based on organic

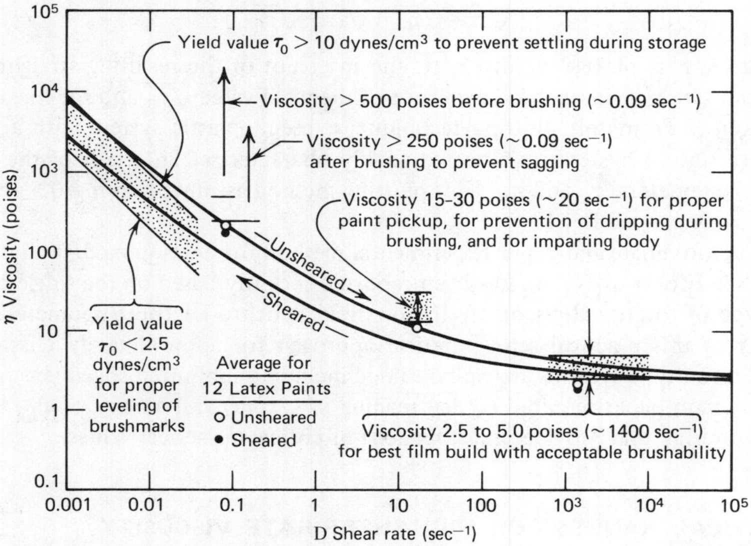

Fig. 16-5 Graph of log shear rate versus log viscosity, showing optimum regions for yield value, viscosity before brushing, viscosity for paint pickup, and brushing viscosity. Curves are for typical unsheared and sheared coating viscosity profiles.

solvents should at least exceed 4.0 dynes/cm². No limit above this is usually specified, although higher values are certainly desirable. After application the yield value for the sheared conditions may be found to be temporarily significantly reduced (thixotropic system) or only slightly lower (pseudoplastic system). In either case the sheared yield value (which initially must be brought below 4 dynes/cm² for proper flow performance) should soon recover to a range between 4 and 10 dynes/cm². The formulated system should be such that the recovery of the yield value will be paced so as to permit satisfactory leveling without undesirable sagging.

For latex coatings the yield value before application should be greater than 10 dynes/cm² to prevent pigment settling (during storage) but somewhat less than 2.5 dynes/cm² immediately after application to allow for proper leveling of brushmarks (6). A recovery of the yield value to 5 dynes/cm² or higher shortly after application is necessary to arrest sagging. Again there is a delicate balance between preventing undue sagging by prompt recovery of the yield value and allowing sufficient time for flow and leveling. Figures 16-2 and 16-5 illustrate the yield value ranges involved.

Inspection of Fig. 16-3*b* suggests that there should be some degree of correlation between yield value and viscosity as measured at some fixed ultralow shear rate. From data reported on 12 different latex paints (6) this author offers the expression given in Eq. 11 as a step in this direction.

$$\tau_0 = 0.0056\ \eta^{1.35}\ \text{(shear rate 0.085 sec}^{-1}\text{)} \tag{11}$$

PROBLEM 16-3

A latex coating with excellent leveling, sagging, and film buildup properties but moderate brush drag has an unsheared viscosity of 245 poises and a sheared viscosity of 97 poises, both at a shear rate of 0.085 sec^{-1}. Calculate the approximate latex yield values for these two conditions.

Solution. Substitute the given viscosities in turn in Eq. 11.

$$\tau_0\ \text{(unsheared)} = 0.0056 \cdot 245^{1.35} = 9.4\ \text{dynes/cm}^2$$

$$\tau_0\ \text{(sheared)} = 0.0056 \cdot 97^{1.35} = 2.7\ \text{dynes/cm}^2$$

Comment. These values compare roughly with measured yield values for the latex paint of ~6 and ~2, respectively.

PRACTICAL RANGE FOR PICKUP AND TRANSFER OF PAINT BY BRUSH

The shear rates operative during brush transfer of paint have been estimated to be in the range of 15 to 30 sec^{-1} (6). It has also been reported that the undisturbed viscosity profile can be directly related to the appearance of the undisturbed paint in the can and to the ability of the brush to pick up paint for application without dripping (1). Experience indicates that viscosities above 15 poises at about 15 to 30 sec^{-1} shear rate are fully adequate for good brush transfer performance (6).

CONTROLLING EFFECT OF COATING COMPONENTS ON VISCOSITY

The paint formulator is fortunate in that the three major components of a coating (binder, solvent, pigment) completely dominate the coating flow properties in the high-shear-rate range, whereas a minor amount of rheological flow additive, pigment flocculation, and/or the colloidal characteristics of the binder completely dictate the rheological properties of the coating in the low-shear-rate range. Hence, at the risk of over-simplifying the overall flow situation, it can be said that the formulator need only adjust the binder, solvent, and pigment composition to achieve proper application properties (high-shear-rate range). In this range the effect of any rheological additive present or the effect of any residual pigment flocculation can be safely ignored. A range from about 1.0 to 3.0 poises at 20,000 sec^{-1} shear rate can be set up as a reasonable target toward which to aim in formulating a coating system at such high shear rates.

On the other hand, the viscosity contributed by the binder/solvent/pigment composition in the low-shear-rate range is essentially negligible compared with the viscosity increase produced by a small amount of flow additive, pigment flocculation, and/or the aggregation of the molecules of a thixotropic binder (thixotropic alkyd). Thus most well-formulated coatings with a flow control additive present exhibit a yield value. This means that at the extreme end of the ultralow-shear-rate range the coating viscosity is essentially infinite. The contribution of the binder/solvent/pigment composition per se to this viscosity is insignificant. It is the level and nature of the flow additive that normally dictates the viscosity behavior through this low-shear-rate region.

In essence, then, the paint formulator can and should design a coating for two shear rate regions, each of which tends to be completely independent of the other and each of which is dominated by a different combination of coating components (see Fig. 16-6). This represents a logical approach to designing coating systems. It also means that viscometers must be available that can measure viscosities at both low and high shear rates. By resorting to a single viscosity

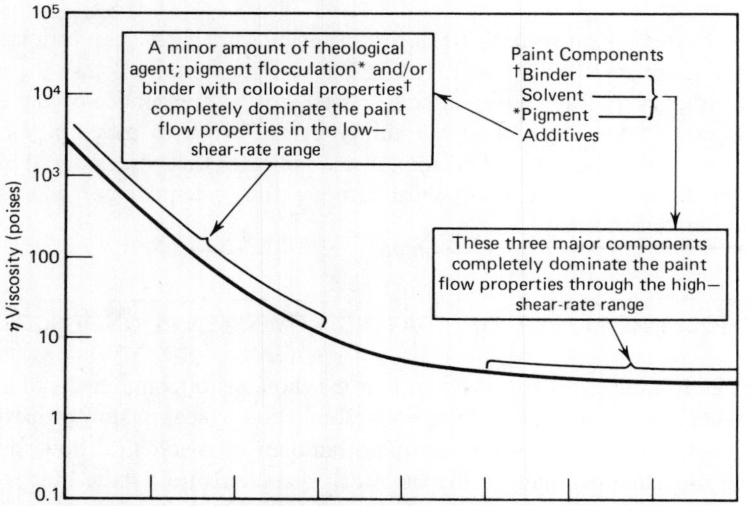

Fig. 16-6 Schematic graph illustrating the controlling influence of certain coating ingredients on coating flow in the ultralow- and high-shear-rate regions.

measurement at an intermediate shear rate, a formulator is attempting to work with insufficient information. Except for control purposes, a single-point viscosity is an invitation to frustration.

Nature of the Low-Shear-Rate Viscosity Imparted by Flow Control Additives

The viscosity at low shear rates that is imparted by a rheological additive is due to the loosely knit colloidal structure that the additive builds up within the coating system. A flow additive can be visualized as providing a tenuous network of swollen strands that permeate the coating. These serve to enmesh pigment and more or less enclose and immobilize pockets of vehicle.

A flow additive added to an organic solvent coating system is commonly referred to as a thixotrope or gellant; a flow additive designed for aqueous systems, as a thickener.

The networks provided by colloidal additives have little strength and are easily disrupted by even a moderate stress. However, when confronted by stresses of very low order such as those encountered with settling, sagging, paint pickup on the brush or roller coater, leveling, and penetration, such structures can offer considerable resistance to flow. In fact, if the stress is sufficiently low, the colloidal structure, weak as it is, can completely block low (the viscosity becomes infinite for the ultralow-shear condition being considered).

The function of the rheological additive, then, is to build up sufficient colloidal structure in the coating system to block or retard flow shear rate stresses. It is true that in moderate-to high-stress situations the additive structure is destroyed and structural influence becomes negligible. However, on cessation of the high shear stressing, the additive structure starts to rebuild at a rate that is a function of the flow additive in question. If the recovery is immediate, the system is purely pseudoplastic in nature with no thixotropic overtones present. Gellant additives tend to be characterized by this swift type of rebuild. On the other hand, if the recovery takes time, thixotropy is involved. This explains why an agent imparting this delayed type of recovery is commonly called a thixotrope.

The rate at which colloidal structure is rebuilt after being disrupted is critical in the selection of a flow control agent. For example, an excessively fast recovery prevents coating flow and leveling; too slow a recovery permits sagging.

Colloidal Structure Preferred to Pigment Flocculation

Two main routes are available to the coatings engineer for imparting structural viscosity to a coatings system at low shear rates, namely, colloidal structure provided by a flow control additive, and pigment flocculation.

In the older pigment flocculation route a surface active condition is arranged for where separated pigment particles are induced to lightly adhere to each other to produce flocs. As a result of this weak bonding, particle chains are formed that, taken together, provide a pigment network. It is this tenuous structure that provides the low-shear-rate viscosity exhibited by flocculated coating systems. A little flocculation can be tolerated, and with no alternative method of flow control available older paint systems were, perforce, often formulated along such lines. However, this method of control is a backward step, for it undoes, in part, the effort initially expended in dispersing the pigment particles. Furthermore, if the flocculation gets out of control, irreversible pigment seeding can (and all too often does) occur. Even a minor amount of pigment flocculation can degrade film integrity, and of course hiding and color development are automatically reduced in proportion to the degree of pigment flocculation that takes place.

The modern route to imparting structural viscosity at low shear rates is to develop a colloidal network with a gellant and/or thixotrope (solvent systems) or a thickener (water systems). This approach leaves the pigment dispersion unaffected, since now the flow control additive provides the extended chains and tentacles that touch each other and immobilize the coating composition. This route to controlling low-shear-rate viscosity is well recognized today, and pigment flocculation is rapidly being phased out as a method of flow control.

EFFECT OF PIGMENT VOLUME FRACTION ON MILL BASE VISCOSITY

Many expressions have been proposed for relating the pigment volume fraction V of a mill base to the mill base viscosity at infinite shear rate η_V. Although the first edition of this book considered three equations, the present edition takes up only one of these, since Eq. 12 is now considered far superior to the other two. It has been found applicable to both practical work (8) and theoretical studies (9).

$$\log \frac{\eta_V}{\eta_0} = \frac{K \cdot V}{U - V} \tag{12}$$

where η_V = viscosity of the mill base for a given V value at infinite shear rate
η_0 = viscosity of mill base vehicle
V = pigment volume fraction in mill base
U = ultimate pigment volume fraction
$K = k \cdot U$ = constant (commonly 0.50)

The ultimate pigment volume fraction is defined as the volume fraction where all the interstices of the packed pigment are just filled with vehicle. It differs from the UPVC in that in the latter case only binder is involved in filling the pigment interstices, whereas in the case of a U value it is the complete vehicle (binder and solvent) that just fills the spaces among the packed pigment particles. The two are the same, however, for an all-binder vehicle.

Inspection of Eq. 12 shows that when $V = 0$ the righthand term also equals zero, making $\log(\eta_V/\eta_0) = 0$ and in turn $\eta_V = \eta_0$. Also, since the viscosity of a mill base at its U value is so large as to be substantially infinite, both terms become infinite when V reaches the U value (note that here $U - V = 0$). Hence Eq. 12 is capable of representing conditions for a mill base through an extreme range running from an unpigmented vehicle to a completely loaded vehicle. Experience shows that Eq. 12 represents with good fidelity the viscosity/fractional pigment volume relationship in the intermediate range bounded by these two extremes.

A value of 0.50 for K has been shown to apply to a variety of titanium dioxide/extender mixtures in vehicles of different viscosities (8).

PROBLEM 16-4

An alkyd vehicle ($\eta_0 = 0.40$ poise) is pigmented with a TiO_2/extender pigment mixture to a pigment fractional content V of 25%. From an oil absorption test it is estimated that the U value for the pigment mixture is 45%. Compute the viscosity of the pigmented alkyd mill base when processed in a high-shear-rate dispersion mill.

Solution. Substitute the problem data in Eq. 12, using the value $K = 0.50$, which applies to this pigment mixture.

$$\log \frac{\eta_V}{0.40} = \frac{0.50 \cdot 0.25}{0.45 - 0.25}; \qquad \eta_V = 1.7 \text{ poises}$$

Equation 12 can be used to establish a starting formulation for a new mill base composition that is to be charged to a specific piece of grinding equipment by using prior plant performance data. This is done by determining from past experience the optimum mill base viscosity that has been found suitable for this particular mill. Once this is established, it is necessary only to know the new vehicle viscosity and the U value of the pigment mixture (obtained, say, from an oil absorption measurement) in order to calculate the optimum V value for the new mill base composition. This is done by using Eq. 13, a rearrangement of Eq. 12. The value of K can be taken as 0.50, or this also can be established from past experience with the particular equipment in question.

$$V(\text{mill base}) = \frac{\log(\eta_V/\eta_0) \cdot U}{K + \log(\eta_V/\eta_0)} \qquad (13)$$

where η_V is the optimum viscosity for any mill base to be charged to a particular mill in the plant.

PROBLEM 16-5

Past experience has demonstrated that a particular high-speed disperser in a plant provides optimum dispersion when the mill base has an η_V value of 60 poises. If a new pigment mixture to be dispersed has a U value of 52%v (= 0.52) and the grinding vehicle has a viscosity of 3.0 poises, calculate the optimum pigment fractional volume for the new mill base charge.

Solution. Substitute the problem data in Eq. 13, using a value of 0.50 for K.

$$V(\text{mill base}) = \frac{\log(60/3.0) \cdot 0.52}{0.50 + \log(60/3.0)} = 0.38\ (38\%\text{v})$$

Situations may arise where the ultimate pigment packing value U is not known or cannot conveniently be developed experimentally, yet data are available relating mill base viscosities to V values. In such cases the value for U may be established through either an algebraic or a graphical solution. A minimum of two sets of V versus η_V data is required for either type of solution, but more extensive data are preferred. The graphical solution resorts to a mathematical trick in order to provide a straight-line relationship connecting the plotted data. It consists in substituting $k \cdot U$ for K in Eq. 12, yielding Eq. 14.

$$\log \frac{\eta_V}{\eta_0} = \frac{k \cdot UV}{U - V} \tag{14}$$

This equation is then expressed in reciprocal form as given by Eq. 15.

$$\frac{1}{\log(\eta_V/\eta_0)} = \frac{1}{kV} - \frac{1}{kU} \tag{15}$$

Equation 15 is of such a form that, if $1/\log(\eta_V/\eta_0)$ is plotted against $1/V$, a straight-line relation is obtained in which the slope of the line is equal to $1/k$ and the intercept of this line with the $1/V$ scale is equal to $1/U$.

PROBLEM 16-6

Viscosity values of 1.95, 2.95, and 6.34 are measured for three alkyd mill bases where the alkyd vehicle (η_0 = 1.5 poises) is pigmented to fractional pigment volumes of 0.10, 0.20, and 0.30, respectively. Develop an expression reducing these data to equation form.

Solution. *Algebraic*: Substitute two sets of values in Eq. 12 in turn, and simultaneously solve the two equations for the required answer.

$$(10\%\text{v}):\ \log \frac{1.95}{1.5} = \frac{K \cdot 0.10}{U - 0.10}$$

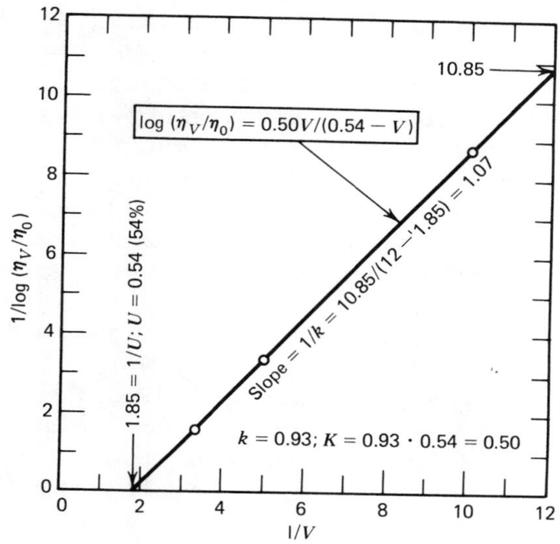

Fig. 16-7 Graph of $1/V$ versus $1/\log(\eta_V/\eta_0)$, illustrating procedure to be followed to obtain values for U and the constant k (see Eq. 15).

$$(30\%\text{v}):\ \log\frac{6.34}{1.5}=\frac{K\cdot 0.30}{U-0.30}$$

$$U = 0.54\ (54\%\text{v});\qquad K = 0.50$$

Graphical: Plot $1/\log(\eta_V/\eta_0)$ versus $1/V$ on a graph with uniform scales, and draw a straight line through the plotted points (see plotting as carried out in Fig. 16-7). The intercept on the $1/V$ scale of 1.85 gives the value for $1/U$. Hence $U = 0.54$ (= 1/1.85). The value for the slope $1/k$ is 1.07 (= 10.85/(12.00 – 1.85). Hence k is equal to 0.93 (= 1/0.93), and in turn K is equal to 0.50 (= 0.93 · 0.54). The relationship among the viscosity and V data is then given by Eq. 16.

$$\log\frac{\eta_V}{\eta_0}=\frac{0.50\cdot V}{0.54-V} \qquad (16)$$

MILL BASE DILATANCY

The quantity η_V in Eq. 12 is the viscosity of a mill base having a pigment volume fraction V at infinite shear rate. Although an infinite shear rate is an imaginary concept, it can be assumed to be essentially equivalent to the very high shear rates generated by most pigment dispersing equipment (~ 10,000 to 100,000 sec^{-1}).

Dilatancy is said to occur when there is an increase in viscosity with an increase in shear rate. It describes the rheological condition where a disproportionately high resistance is offered by a fluid system to forced flow. Dilatancy is most frequently encountered with highly pigmented systems where the pigment particles are tightly packed in a compact arrangement. When disturbed (sheared), these particles are tumbled about and forced to assume a less compact arrangement with an ensuing overall volume enlargement or dilation (hence the term "dilatancy"). In such systems, as exemplified by dilatant mill bases, the dilation effect leads to mutual particle interference. The closer the packing of the particles and the faster they are forced to move, the greater is their jamming action. This dependency of dilatancy on pigment volume fraction V and shear rate D is clearly evident in the model situation shown in Fig. 16-8.

Even though dilatancy is generally dependent on high pigment concentration and high shear rate, not all highly pigmented systems exhibit this type of behavior. It is also possible for a completely unpigmented vehicle such as a calcicoater (a highly polymerized limed drying oil) to be strongly dilatant. In this case the dilatancy is due to the jamming of the calcicoater macromolecules under this high stressing action. Low pigmentations of dilatant vehicles may result in dilatant mill bases, a condition that should not be confused with dilatancy due to high pigment loading.

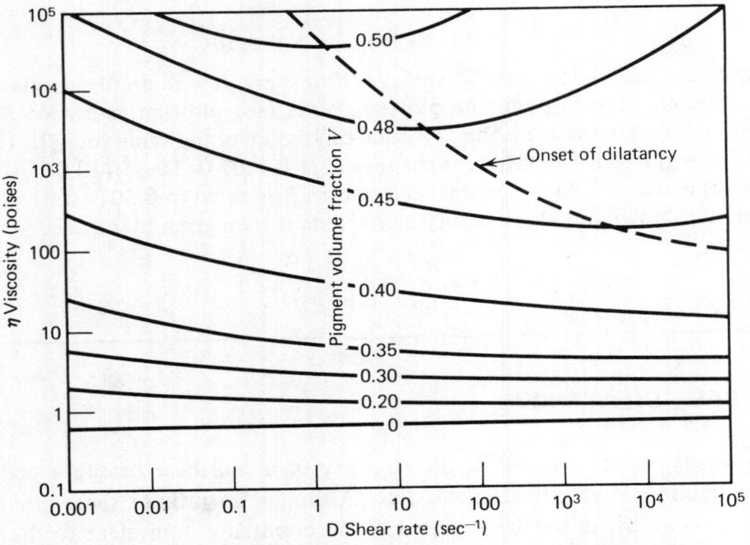

Fig. 16-8 Hypothetical plot of viscosity profile curves corresponding to increasing pigment volume fractions in a mill base with region of dilatancy indicated.

Mill bases intended for high-speed disk dispersers or roller mills are usually formulated to provide borderline dilatancy, since by so doing the stress applied to the surface of the mill base by the solid shearing surface is communicated throughout the bulk of the mill base with maximum effectiveness. Roller mills are capable of handling more dilatancy than disk dispersers. However, excessive dilatancy is counterproductive in either type of equipment.

Inspection of Fig. 16-8 shows that at the high pigment levels characteristic of dilatant systems only a small increase in the pigment volume fraction is required to change a nondilatant mill base into a strongly dilatant one.

One method for evaluating the relative dilatancy of pigment/vehicle mixtures calls for a determination of their wet and flow points (10). The wet point is determined by kneading a specified amount of the pigment mixture with just enough vehicle to form a soft coherent paste (end point identical with Gardner-Coleman oil absorption point). The flow point is determined by noting what further vehicle is required to provide a mixture that just drops from a vertically held spatula. Furthermore, the flow point mixture at this point must allow a rapid motion of the spatula across a layer of the mill base (about 1/16 in. thick) without strong drag, without permanent ridges being left behind, and with an immediate regain of a glossy appearance. The degree of dilatancy is gauged by the width of the gap separating the wet point (WP) from the flow point (FP). This observed flow behavior can be measured in terms of a dilatancy index as given by Eq. 17.

$$\text{Dilatancy index} = \frac{\text{FP} - \text{WP}}{\text{WP}} \tag{17}$$

A flow point condition requiring only 5 to 15% more vehicle than the wet point is considered strongly dilatant; from 15 to 30%, moderately to weakly dilatant; and >30%, substantially nondilatant. Obviously an abrupt change in consistency is indicative of strong dilatancy. A mill base formulation is commonly obtained by scaling up the composition of the pigment/vehicle mixture corresponding to the flow point.

The practical use of the dilatancy index in formulating mill bases is illustrated by a series of titanium dioxide/alkyd mill bases as shown in Fig. 16-9, where the volume fraction of the binder in a long-oil alkyd vehicle is plotted versus the dilatancy index (the fractional addition vehicle required to reach the flow point from the wet point). The volume fraction of the pigment at the wet point was essentially 0.50 for all five mill bases. Inspection of this graph shows that optimum mill base grinding is obtained with the moderately dilatant mill bases falling in the dilatancy index range from 15 to 30%.

However, these dilatancy index benchmarks are to be considered only as approximate guidelines, since some nondilatant systems show a narrower wet point/flow point gap than dilatant systems. For example, an alkyd (20%w NV)/

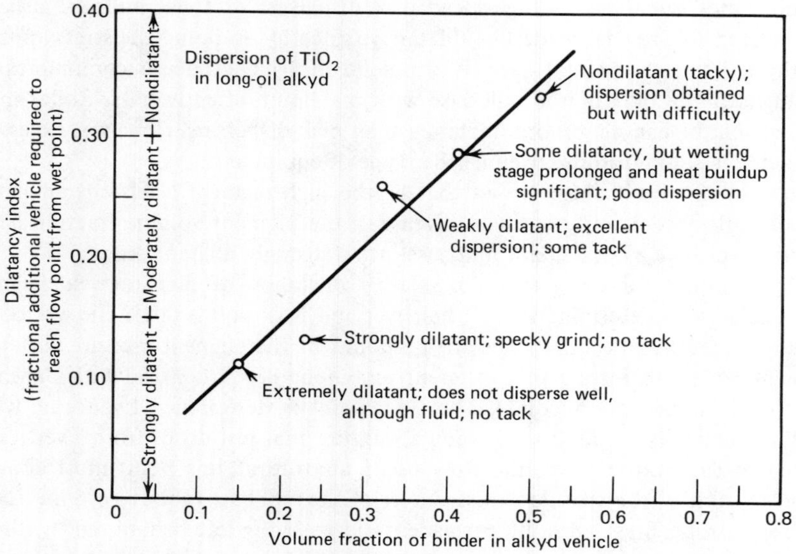

Fig. 16-9 Graph of volume fraction of binder in an alkyd vehicle versus the dilatancy index for a titanium dioxide/long-oil alkyd mill base system.

TiO_2 + $CaCO_3$ mill base without lecithin had a dilatancy index of 24% (0.24) but processed poorly (too dilatant). On addition of 4% of lecithin (based on the vehicle), the dilatancy index dropped to 13% (0.13), a value for which even more dilatancy was predicted, but actually the mill base became nondilatant and good processing was obtained. Oddly, the lecithin addition also induced denser packing, which should also have contributed to higher dilatancy. Apparently the powerful dispersing action of lecithin is such as to provide slippage among the pigment particles so that contact and jamming are enormously reduced.

Both strongly dilatant and nondilatant (strongly flocculated) mill bases are unsuitable for smearer-type dispersion equipment. On the other hand, the exact intermediate, weakly dilatant system that is best for a particular smearer mill is not always easy to pinpoint, since slight variations in equipment design or operating conditions can affect flow behavior. However, the wet point/flow point technique is a simple laboratory tool for determining with reasonable accuracy suitable initial mill base formulations for high-speed disk impellers and the like. If the degree of dilatancy proves to be too high in actual practice, it can usually be reduced by decreasing the pigment volume fraction. In the case of latex mill bases, dilatancy can also be reduced by altering the additives and/or their level (anionic surfactants tend to induce dilatancy; polyphosphates and hydroxyethylcellulose tend to suppress dilatancy) (10). With latex systems it is also possible to judiciously split the introduction of the additives between the mill base grind and the let-down vehicle to closely control the dilatancy during the dispersion step (10).

REFERENCES

1. Ehrlich, A., T. C. Patton, and A. Franco, "Viscosity Profiles of Solvent Based Paints: Their Measurement and Interpretation," *JPT*, **45**, No. 576, 58–67, January (1973).
2. Casson, N., in *Rheology of Disperse Systems* (C. C. Mill, Ed.), Pergamon Press, New York, 1959, pp. 84–104.
3. Wu, Souheng, "Rheology of High Solid Coatings: I. Analysis of Sagging and Slumping. II. Analysis of Combined Sagging and Leveling," ORPL Preprint, American Chemical Society Meeting, August 1977.
4. Camina, M. and D. M. Howell, "The Leveling of Paint Films," *JOCCA*, **55**, No. 10, 929–940, October (1972).
5. Van Wazer, J. R., J. W. Lyons, K. Y. Kim, and R. E. Colwell, *Viscosity and Flow Measurement: A Laboratory Handbook of Rheology*, Wiley Interscience, New York, 1963.
6. Sarkar, N. and Robert H. Lalk, "Rheological Correlation with the Application Properties of Latex Paints," *JPT*, **46**, No. 590, 29–34, March (1974).
7. Rosen, Meyer R. and Foster, William W., "Approximate Rheological Characterization of Casson Fluids," *JCT*, **50**, No. 643, 39–48, August (1978).
8. Ensminger, R. I., "Techniques for Efficient Pigment Dispersion Operations," *Mod. Paint Coatings*, pp. 29–35, May (1975).
9. Smith, Thor L., "Rheological Properties of Dispersions of Particulate Solids in Liquid Media," *JPT*, **44**, No. 575, 71–79, December (1972).
10. Daniel, F. K., "Determination of Mill Base Compositions for High Speed Dispersers," *JPT*, **38**, No. 500, 534–542, September (1966).

17 *Introduction to Pigment Grinding (Dispersion) into Liquid Vehicles*

The primary objective of pigment grinding is to incorporate pigment particles into a liquid vehicle to yield a fine particle dispersion. From the standpoint of the coatings and ink engineer, "grinding" does not have its usual connotation namely, the pulverizing or crushing of a solid material to fine bits of powder, although comminution may be a minor or incidental effect. Rather, the definition of pigment grinding is restricted to the process of incorporating pigment, already minutely divided, into a coating vehicle to yield a dispersion of primary particles. Pigment particles as received at the paint or ink plant may be agglomerated, that is, they may be and usually are collected into relatively soft packed clusters. Such aggregation may develop from a number of sources. Thus pigment particles may be stuck together by the interstitial deposits remaining from the evaporation of pigment wash liquors, by the incipient sintering of particles during a high-temperature preparation method, or by the pressure of compacting forces that develop when bags of pigment are piled on top of each other. However, the ultimate particle size of the pigment as supplied is normally sufficiently fine for the production of most coatings.

Pigment specifications frequently list the maximum residue that can be retained on a 325-mesh screen (equivalent to a diameter in excess of 44.5 μ or alternatively, to a $4\frac{1}{2}$ reading on a Hegman gage). For a high-grade enamel, it is necessary that this maximum size be reduced during the milling operation to at least a 7 Hegman reading ($<13\ \mu$).

Dispersing and mixing are not synonymous terms. In fact, processing conditions suitable for mixing may be quite unsuitable for dispersion and vice versa (1). In line with this distinction, a coating or ink should be properly defined as a dispersion of pigment particles in a vehicle rather than as a mixture of pigment and vehicle. Mixing implies stirring together, mingling, or blending, and the resulting mixture may be relatively gross; dispersing implies scattering, particle separation, and an ultimate distribution of primary particles with the dimensions involved being relatively minute.

DISPERSION PROCESS

The intimate incorporation of the pigment particles into the coating or ink vehicle (grinding or dispersion operation) can be visualized as occurring in three stages, although the stages overlap in any actual grind.

Wetting. Wetting refers to the displacement of gases (such as air) or other contaminants (such as water) that are adsorbed on the surface of the pigment particles, followed by a subsequent attachment of the wetting vehicle to the pigment surface.

Grinding. Grinding refers to the mechanical breakup and separation (deagglomeration) of the particle clusters to isolated primary particles.

Dispersion. Dispersion refers to the movement of the wetted particles into the body of the liquid vehicle to effect a permanent particle separation.

The displacement of air within the interstices of the pigment mass is initially facilitated by charging the pigment to the liquid vehicle. When this is done, the liquid advances into the channels and spaces of the pigment mass from the bottom, while the displaced air retreats and escapes from the top. The velocity u with which a vehicle penetrates a bed of pigment particles is given by Eq. 1, where r is the average capillary radius of the pigment bed and η is the vehicle viscosity. For the significance of k refer to Eq. 12 in Chapter 11 on capillarity.

$$u = k \cdot \frac{r}{\eta} \tag{1}$$

From this expression it follows that both tightly packed pigment aggolmerates (small pores) and high-viscosity vehicles oppose the penetration of vehicle into the pigment mass. Even under favorable conditions, if mechanical separation is not provided, wetting through capillarity proceeds but at a slow rate and to only a limited extent. To complete the wetting process, fresh surfaces must be exposed to the liquid vehicle.

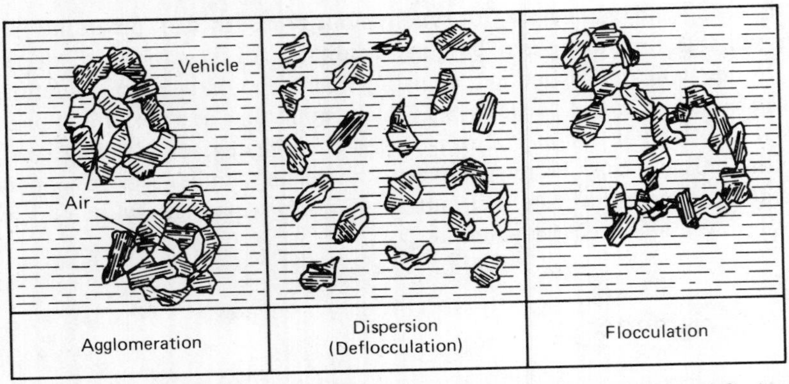

Fig. 17-1 Schematic illustration of pigment agglomeration (aggregates) pigment dispersion (deflocculation), and flocculation (flocculates).

The purpose of the grinding is to break up the particle agglomerates mechanically so that the entire surface of each particle is made available for wetting. Pigment particles that are clumped or packed together must be ripped apart and their internal faces exposed to the wetting action. The mechanical conditions necessary to ensure effective grinding will be considered in detail in Chapters 18 to 24 on various types of grinding equipment.

Once a particle is wetted with vehicle, the next stage consists in moving the particles into the body of the vehicle to effect a permanent separation among the wetted particles. Here the idea is to surround each particle with sufficient vehicle so that particle-to-particle contact is thereafter prevented. If and when the particles do tend to cluster again in the vehicle, the effect is called flocculation. Figure 17-1 schematically illustrates states of aggolmeration, dispersion (deagglomeration), and flocculation.

MICRONIZED OR JET-MILLED PIGMENTS

In recent years there has been a steady trend toward the expanded use of a fluid energy attrition process called micronizing or jet-milling (no moving parts) that is capable of significantly upgrading the dispersion qualities of many types of pigments (2, 3). The process consists essentially of impinging one particle against another at supersonic speed by means of a highly energized fluid (compressed air, superheated steam). The operation is carried out in a pulverizing chamber that is also capable of classifying the pigment particles by centrifugal force as they whirl around in the jet-mill chamber. The net effect of this disintegration process (a continuous operation) is threefold: a marked reduction in the number of larger or oversized aggregates that are present in the pigment mixture, a smaller and more uniform particle size distribution (as much as 98% of the jet-milled pigment may fall within a 5-μ diameter range), and a rounding off of the particle edges and surface irregularities to provide smoother and rounder particle shapes. By thus providing a narrower distribution of smaller and more evenly sized particles of rounded shape and by breaking up the more refractory particle aggregates before charging the pigment to the grinding equipment, subsequent pigment dispersion is greatly facilitated.

COMPOSITION OF GRINDING VEHICLE

The preparation of a coating is normally carried out in two stages. In the first or grinding stage, concern is centered exclusively on providing the conditions that will achieve the most satisfactory pigment dispersion. In the second or letdown stage, interest centers on properly reducing or letting down this grind dispersion with additional vehicle (generally of a far different solvent binder ratio than that for the grinding vehicle) to give a final coating composition having optimum application and performance properties. Figure 17-2 schematically illustrates the sequence of operations and the coating components involved. It is seen that the coating engineer must not only formulate a paint for ultimate service, but also formulate portions of the coating for these intermediate grinding and letdown operations.

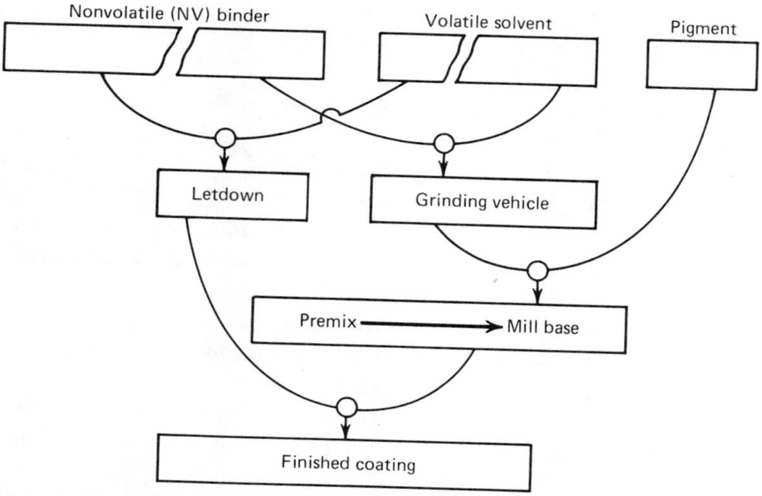

Fig. 17-2 Relationship of terms used to describe the initial, intermediate, and final products in the preparation of a coating composition.

GRINDING (DISPERSION) EQUIPMENT

The effectiveness of the dispersion operation is directly related to the ability of a given vehicle to wet out the pigment particle aggolmerates, to spread out over the particle surfaces, and finally to effect a lasting particle separation. The success of this operation is closely tied in with the ability of the grinding equipment to work in harmony with the grinding vehicle by expeditiously breaking up and exposing particle surfaces to the vehicle wetting and dispersing action. Since many types of pigment grinding equipment are in use in the paint and ink industries, it is not unexpected that a mill base that is quite satisfactory for one type of equipment (say a ball mill) is quite unsatisfactory for another type of equipment (say a three-roll mill). Hence a mill base must be formulated to fit the grinding equipment that is to carry out the dispersion process.

From a purely wetting standpoint an all-solvent vehicle should come close to being ideal, since solvents are extremely low in viscosity and low in surface tension and hence are able to penetrate pigment agglomerates quickly and to spread rapidly over exposed pigment surfaces. However, this same low viscosity and the inherent volatility of a solvent (among other factors) render any dispersion of the pigment particles a temporary affair. Furthermore, it is the binder that should preferentially be wetting the pigment surface, rather than the solvent. To effect a lasting mill base dispersion, sufficient binder must be present to provide a continuous and permanent coating around the pigment, thus ensuring stable separation of the pigment particles.

Smearing versus Smashing Dispersion Equipment

The breakup of particle aggolmerates can be accomplished by either smearing or smashing types of dispersion equipment, with smashing having two alternative approaches, namely, hammering and projectile impingement (1). These breakup operations are schematically illustrated in Figs. 17-3*a* and *b*. With a little imagination one can visualize the sorting out of paint and ink grinding equipment into smearer or smasher categories or into a mixture of these two operations (hybrid equipment).

Table 17-1 shows an approximate classification of pigment grinding equipment. The hybrid dispersion equipment partakes of both smashing and smearing effects in bringing about the pigment dispersion.

Although the above treatment is somewhat superficial, it permits some basic concepts to be developed in regard to the proper conditions for the mill base to be charged to a given type of grinding equipment. Thus, for the optimum breakup of particle clusters, smashers work most effectively when unhindered by viscous resitance, whereas smearers (conversely) must encounter strong viscous resistance to be effective at all.

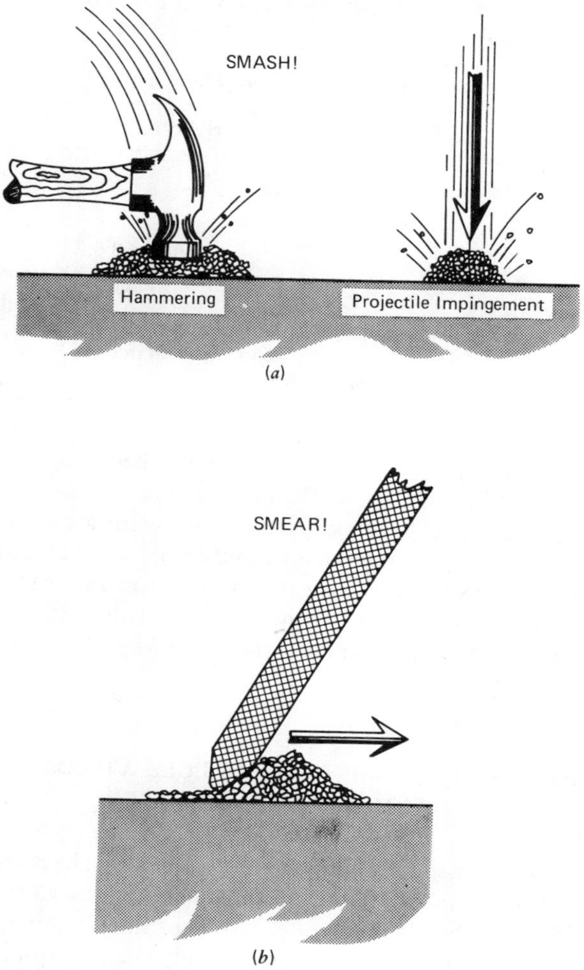

Fig. 17-3 Schematic drawings illustrating the breakup of pigment agglomerates by (*a*) a smasher type of processing, with the two subdivisions of hammering (as for an agglomerate caught between two colliding balls) and projectile impingement (as for an agglomerate being flung against a solid surface) and (*b*) a smearer type of processing, where a highly viscous mill base is required to achieve an effective shearing action.

Table 17-1 Classification of Dispersion Equipment Based on Type of Milling Action

Viscosity (high shear rate)	Type of Equipment: Smashers	Hybrids	Smearers
Very high			Three-roll mill
High			High-speed stone mill Colloid mill
Moderate		Shot mill High-speed disk impeller Attritor	
Low		Ball and pebble mills Sand and bead mills	
Very low	High-speed impingement mill		

In general, there are three distinct differences between a smasher and a smearer type of grind. First, the optimum direction of motion for a smasher dispersion is perpendicular to the substrate, whereas for a smearer it is parallel to the substrate. Second, the mill base viscosity for a smasher must be minimal, whereas it must be maximal for a smearer. Third, a smasher calls for maximum velocity (at the moment of impingement), whereas with a smearer velocities are nominal. Excellent pigment dispersions can be obtained by observing in practice these theoretical concepts.

GENERAL RANGES FOR GRINDING VEHICLE VISCOSITIES AND PIGMENT VOLUME FRACTIONS OF MILL BASES

To a major extent, the viscosity of the mill base grinding vehicle and the pigment volume fraction of the mill base control the high-shear-rate viscosity of the mill base composition (see Eq. 12 in Chapter 16). Thus smearer grinding equipment, which calls for high mill base viscosity, is generally charged with a mill base that is formulated with a high-viscosity vehicle and/or a high fractional pigment volume content. On the other hand, for smasher-type grinding equipment, the mill base is generally formulated with a low-viscosity vehicle and/or a low fractional pigment volume content. Representative ranges for these η_0 and V values (V expressed as a V/U ratio) are given in Fig. 17-4 (4). The quantity U is the volume fraction of pigment that is present when the vehicle just completely fills the interstitial space within the tightly packed pigment particles. A reasonable working value for U can be calculated from the oil absorption value for the pigment mixture by using Eq. 2.

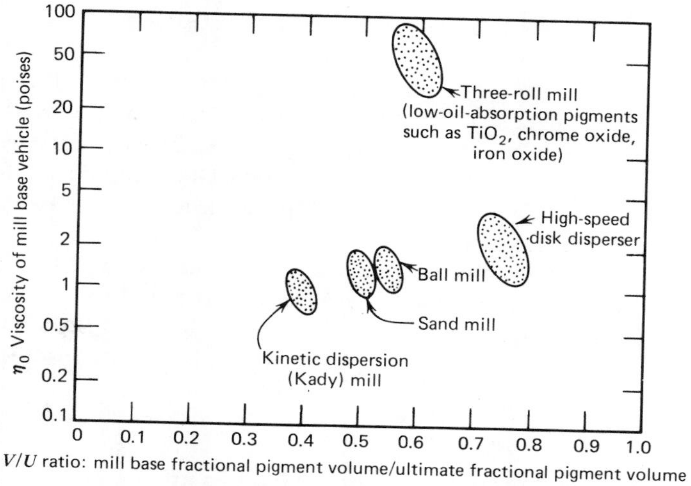

Fig. 17-4 Optimum mill base compositions plotted in terms of mill base viscosity versus V/U ratio. The V/U ratio is the ratio of the mill base pigment fractional volume to the mill base ultimate pigment fractional volume (the limiting fractional pigment volume where all the pigments are in contact, representing the ultimate or densest packing).

$$U = \frac{1}{1 + \overline{OA}\,\rho/93.5} \tag{2}$$

It is significant that viscosity is a controlling variable in the equations used to calculate the power demand for smearer grinding equipment based on laminar flow (three-roll mill, high-speed disk disperser), whereas viscosity is excluded from power demand equations applying to smasher or hybrid dispersion equipment characterized by turbulent flow (ball and pebble mills, sand and bead mills) (5).

DANIEL WET POINT AND FLOW POINT

A useful method for characterizing two consistency stages in the take-up of vehicle by a bed of pigment particles is the Daniel wet point and flow point technique (6). It is based on the amount of vehicle required to produce two critical and reproducible consistencies.

The first critical consistency (a wet point) is defined as the stage in the titration of a pigment mass with vehicle where just sufficient vehicle (as incorporated by vigorous kneading with a spatula) is present to provide a soft coherent mass. This condition corresponds to the "soft paste" end point of the Gardner-Coleman rub-out oil absorption procedure. For a given pigment mixture, wet points obtained with different vehicles may vary significantly.

The second critical consistency (a flow point) defines a later stage in the take-up of vehicle where just sufficient additional vehicle is incorporated to produce a flow or falling off from the vertical blade of a horizontally held spatula. Flow points also vary widely.

An index of the type of pigment dispersion that can be expected for a pigment/vehicle grind is afforded by the difference between the wet and flow point values. Good dispersive action can generally be anticipated for closely spaced wet and flow points; poor dispersive action, for widely spaced wet and flow points.

Behavior Pattern for Closely Spaced Wet and Flow Points (Indicative of Good Dispersibility)

As vehicle is incorporated with pigment by a kneading action with a spatula, the arrival at the wet point stage is characterized, in the case of good dispersibility, by a significant resistance to strong or sudden pressure, with the wetted mass turning dull in appearance (incipient dilatancy). When the wet point is exceeded, with little further vehicle addition, the pigment/vehicle mass abruptly coalesces to a smooth glossy paste. The additon of a small increment of vehicle at this stage converts the pasty mass to a mobile flowable dispersion corresponding to the flow point.

This type of dispersion is characterized by deflocculated particles, minimum vehicle demand, and a relatively close packed system. When the shear stress applied to this dispersion is of low order, sufficient time is allowed for the particles to slip and slide around each other without contact, and flow with minimum viscosity resistance results. However, when the shear stress is high, adjacent particles tend to ram against each other, and the viscosity resistance is greatly increased. Smearer-type grinding equipment requires such strong viscosity resistance for proper functioning.

Behavior Pattern for Widely Spaced Wet and Flow Points (Indicative of Poor Dispersibility)

As vehicle is incorporated with pigment by kneading with a spatula, the arrival at the wet point is characterized, in the case of poor dispersibility, by an adhesive, dull appearing, puttylike consistency. There is no resistance to sudden pressure,

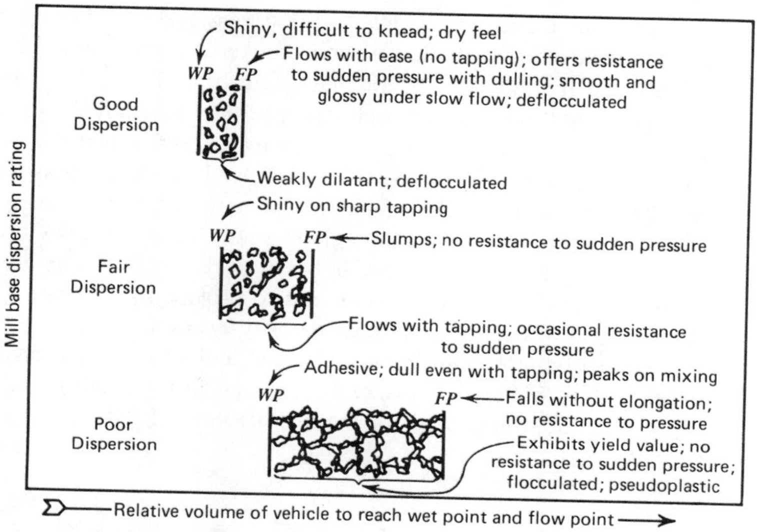

Fig. 17-5 Schematic plot of dispersion state in terms of mill base dispersion rating versus relative volume of vehicle required to reach characteristic wet point and flow point stages on a rub-out of the pigment with test vehicle.

and the wet end point tends to be elusive. Continued addition of vehicle results in a proportionately softer consistency, and considerable vehicle is usually required in order to reach the flow point stage where the pigment/vehicle mass (still dull appearing) drops under its own weight from a horizontally held spatula.

This type of dispersion is characterized by flocculated particles, as manifested by a demonstrable yield value (between the wet and flow points the pigment/vehicle mass hangs on the spatula with no sign of flow).

Some of the differences in dispersion behavior that are encountered in carrying out wet and flow point determinations are schematically illustrated in Fig. 17-5. For further description of these dispersion states and for explicit details of this novel evaluation scheme, reference should be made to the original article (6).

GENERAL CONSIDERATIONS

Heavy-Duty Mixers

A class of equipment not usually regarded as dispersion oriented in the coatings industry is the heavy-duty mixer. This category includes such processing units as double-blade mixers (W & P, Banbury), kneaders, pug mills, and dough mixers. As a class they are utilized specifically for processing systems of very high viscosity (pastes, putties, thermoplastics, uncured rubber compositions). They are efficient mixers and if properly used provide good dispersion. Today they are used only occasionally in the coatings industry, and then generally for preparing a premix such as a stiff well-mixed paste for charging to a three-roll mill.

The most common heavy-duty type is the double-blade mixer, consisting of a contoured interior chamber within which two counterrotating blades revolve either at different speeds (no overlapping) or at the same speed (with overlapping of the rotating blades). Clearance between the blades and chamber walls is restricted to give good overall circulation and to provide the smearer type of action that is required when processing high-viscosity systems.

With the advent of the easy-to-disperse organic and white hiding pigments, the use of heavy-duty mixers has markedly declined in the coatings industry. However, they are still widely used for the manufacture of putties and similar systems.

Dispersion (Grinding) Equipment

In the next seven chapters seven classes of conventional grinding (dispersion) equipment will be discussed. Two of the seven classes represent essentially smearer-type mills (three-roll mill, high-speed stone and colloid mills), and one is essentially a smasher-type mill (high-speed impingement mill). The remaining four classes are hybrid mills that provde both smashing and smearing activity in varying degrees to accomplish pigment dispersion.

Some of these dispersion mills, because of their inherent mechanical design, are automatically restricted to batch operations (ball and pebble mills, high-speed disk impeller, high-speed impingement mill, SWMill); other mills operate only on a continuous basis (three-roll mill, high-speed stone and colloid mills, sand and bead mills). Some mills require a homogeneous feed mixture (three-roll mill, high-speed stone and colloid mills, sand and bead mills); other mills accept unmixed raw materials as part of their overall capability (ball and pebble mills, batch attritor, high-speed disk disperser, high-speed impingement mill). All such factors must be considered in selecting milling equipment, as discussed in the next seven chapters.

The economics of paint manufacture, including the optimal use of plant dispersing equipment, has been discussed in two excellent papers by Daniel et al. (7, 8). The coatings engineer is advised to carefully study these two contributions covering the financial aspects of coatings manufacture (a topic not considered in this book) to round out his or her overall understanding of pigment dispersion technology.

REFERENCES

1. Patton, Temple C., "Theory of High-Speed Disk Impeller Dispersion," *JPT*, **42,** No. 550, 626–635, November (1970).
2. Moore, C. W., "Jet Milled Pigments," *Off. Dig.*, **22,** No. 304, 373 (1950).
3. Wade, W. G., and B. A. Taylor, "Dispersion of Micronized Pigments," *Paint Manuf.*, **30,** 355 (1960).
4. Ensminger, R. I., "Techniques for Efficient Pigment Dispersion Operations," *Mod. Paint Coatings*, May (1975).
5. Patton, T. C., "Paint Rheology and Pigment Dispersion," in *Pigment Handbook*, Vol. III, Wiley-Interscience, New York, 1973.
6. Daniel, Frederick K. and Pauline Goldman, "Evaluation of Dispersions by a Novel Rheological Method," *Ind. Eng. Chem.*, **18,** 26, January 15 (1946).
7. Daniel, F. K., "General Observations on Economics of Paint Production," *JPT*, **45,** No. 577, 65–68, February (1973).
8. Daniel, F. K. et al., "Economic and Technical Parameters of Pigment Dispersions," *JCT*, **49,** No. 631, 74–77, August (1977).

18 *Roller Mills (Three-Roll Mill)*

The three-roll mill diagramed in Fig. 18-1 is representative of the roller mills used by the coating and ink industries to grind (disperse) pigments into vehicles (1, 2). Four-roll and five-roll mills are also manufactured, but except in ink preparation this extension of the three-roll principle has not taken hold in the paint industry. Other variants of the three-roll mill are in commercial use. For example, in Europe a single-roll mill in which the feed nip is formed by forcing a vaned bar against a rotating roll has met with considerable favor (2). The mill base to be dispersed is fed into a hopper above the roll and is carried against the bar by the rotation of the roller. Multiple-bar assemblies have been designed where a recess between the two bearing surfaces provides paths to relieve pressure and return oversize particles to the hopper. In this chapter, however, only the conventional three-roll mill will be considered. The general principles developed for the three-roll mill apply equally well to alternative roller mill designs. All are smearer-type equipment.

DESCRIPTION OF THREE-ROLL MILL OPERATION (1-5)

As shown in the schematic design of Fig. 18-1, in a three-roll mill pigment grinding is initiated by charging the mill base to the space between the feed and center rolls, which are rotating inwardly at different speeds. End plates machined to fit the contours of the rolls prevent the mill base from spilling out the sides. As a result of the inward rotation of the rolls, the mill base is dragged into the region of the feed nip. Because of the ever-narrowing space, a major portion

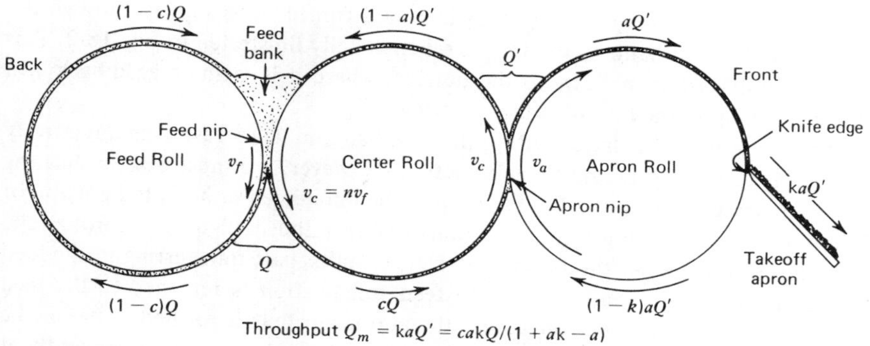

Fig. 18-1 Schematic diagram of a three-roll mill, showing flow path and transfer fractions for the mill base as it progresses from the feed bank to the takeoff apron.

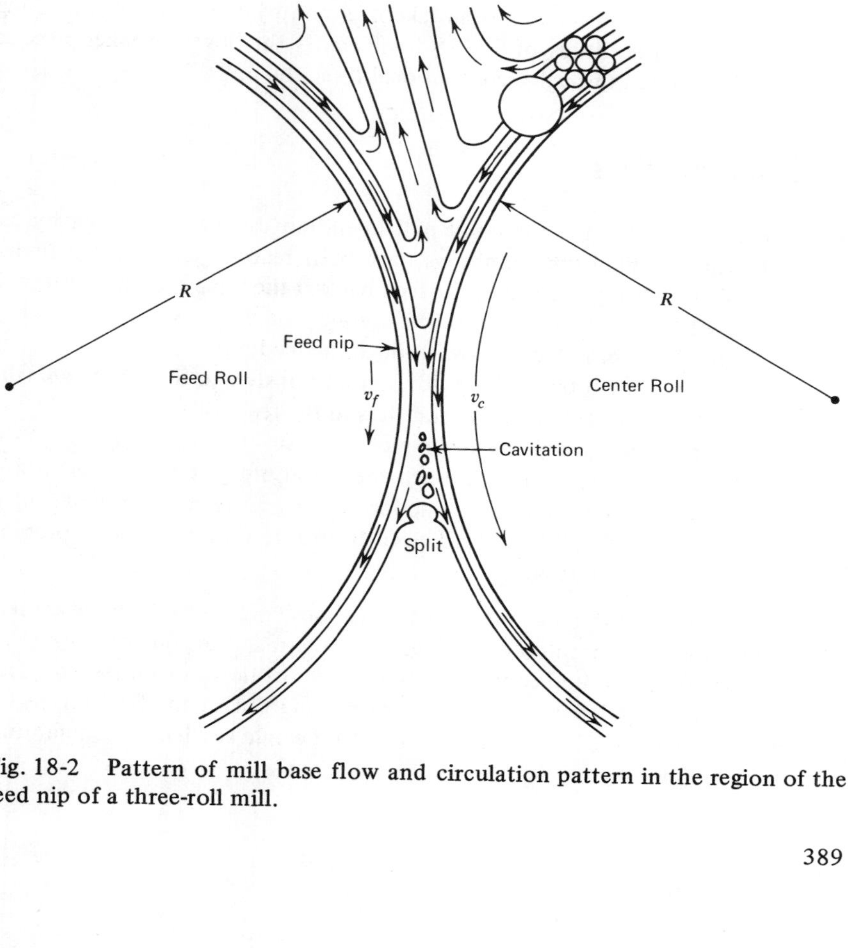

Fig. 18-2 Pattern of mill base flow and circulation pattern in the region of the feed nip of a three-roll mill.

of the feed material is necessarily rejected and forced back and up through the center of the feed bank to the top, as schematically illustrated in Fig. 18-2. This rejected mill base then flows to the outside, where it is again dragged back into the nip region by the inward rotation of the rolls.

This pattern of circulation within the feed bank, in itself, results in a relatively intense type of mixing and shearing action. However, the most drastic shearing action takes place through the feed nip as the unrejected or accepted portion of the mill base is continuously forced under tremendously high shear through the feed nip to emerge on the other side. Here it splits, part transferring to the feed roll and part to the center roll. The feed roll portion is returned to the feed bank; the center roll portion enters the apron nip that is formed between the center and apron rolls. At the apron nip (the rolls, as before, rotate inwardly at different speeds) the mill base undergoes an even more intense shearing action than that to which it was subjected at the feed nip. When it emerges at the top of the apron nip, a second split takes place. Part of the mill base transfers to the center roll (and returns to the feed bank), and part transfers to the apron roll, where it is in turn transferred to the takeoff apron by the knife edge pressing against the apron roll. Study of Fig. 18-1 will clarify the flow that takes place as the mill base is passed through a conventional three-roll mill.

MATERIAL BALANCE

A material balance for the three-roll mill operation can be established by assuming (*a*) that equilibrium conditions have been reached and (*b*) that undispersed mill base is being charged to the feed bank at the same rate that dispersed mill base is being discharged to the takeoff apron.

Let Q = total rate of mill base flow through the feed nip

c = fraction of total feed nip flow Q that transfers to the center roll (the remaining fraction, $1 - c$, transfers to the feed roll)

Q' = total flow through the apron nip

a = fraction of total flow through the apron nip Q' that transfers to the apron roll (the remaining fraction, $1 - a$, transfers to the center roll)

k = fraction of mill base transferred from apron roll to takeoff apron by the apron knife edge

Rates of flow at different points on the three-roll mill (which are self-evident from inspection) have been noted on Fig. 18-1 in terms of the quantities Q, Q', c, a, and k. Under equilibrium conditions, the streams of flow moving into the feed nip region must equal the streams of flow moving out of the feed nip region (otherwise an accumulation or deficit of mill base would result). The equilibrium relationship is given by Eq. 1.

$$(1 - c)\,Q + kaQ' + (1 - a)\,Q' = (1 - c)\,Q + cQ \tag{1}$$

Note that under equilibrium conditions the mill base lost from the three-roll system kaQ' is just made up by the addition to the system of new mill base equal in amount to kaQ'. Solving for Q' in terms of the other quantities yields Eq. 2.

$$Q' = \frac{cQ}{1 + ak - a} \tag{2}$$

Substituting this value for Q' in the expression for the apron discharge rate kaQ' yields Eq. 3, which gives the three-roll mill discharge rate Q_m in terms of Q and the three transfer fractions c, a, and k.

$$Q_m \text{ (discharge rate)} = kaQ' = \frac{cakQ}{1 + ak - a} \tag{3}$$

Under good operating conditions the takeoff knife efficiency closely approaches a value of 100% ($k = 1.00$). Hence for optimum three-roll mill operation Eq. 3 can be written as Eq. 4.

$$Q_m = caQ \tag{4}$$

From Eq. 4 it is evident that the rate of production for a three-roll mill depends primarily on the rate of mill base flow through the feed nip and on the fractions of mill base being transferred to the center and apron rolls, respectively.

Fractional Transfer c of Mill Base to Center Roll

For a feed and a center roll rotating at the same speed, it can logically be argued that a 50/50 split in mill base between the two rolls may be anticipated ($c = 0.50$). It might also be expected that for two rolls rotating at different speeds more mill base would transfer to the roll with the higher rim velocity.

Let v_f = rim (peripheral) velocity of feed roll
v_c = rim velocity of center roll
$n = v_c/v_f$ = ratio of center roll to feed roll rim velocity ($v_c = nv_f$)

As a starting assumption, for the case of differential speeds, assume that mill base transfer to the feed and center rolls is in proportion to their peripheral velocities. Then the transfer fraction c is given by Eq. 5.

$$c = \frac{v_c}{v_c + v_f} = \frac{nv_f}{nv_f + v_f} = \frac{n}{1 + n} \tag{5}$$

Careful experimentation with mill base transfer shows this to be a reasonable approximation to what actually occurs. However, the empirical expression based on actual experimental data given by Eq. 6 has been developed to express the relation between c and n to a higher degree of accuracy (5).

$$c = \frac{n^2 (n + 3)}{(n + 1)^3} \tag{6}$$

Equation 6 indicates that somewhat more mill base is transferred than is shown by Eq. 5. Thus, when $n = 3.0$, Eq. 6 gives a transfer fraction of 0.84 versus 0.75 for Eq. 5 (both give a transfer of 0.50 when $n = 1.0$).

Since higher production rates are obtained with higher roll speed ratios, the trend in manufacture has been to design three-roll mills with increased values for n (say from $n = 2$ to $n = 3$). However, this increase is subject to the limitation of the critical rim velocity of the apron roll, at which velocity the mill base starts to fly off because of centrifugal force. Approximate roll speeds of a typical high-speed three-roll mill of current manufacture (12-in. diameter by 30-in. long rolls) are 35 rpm (feed roll), 105 rpm (center roll), and 315 rpm (apron roll), corresponding to an n value of 3.0. The fraction a of mill base transferring to the apron roll is similar in its transfer behavior to the fraction c transferring to the center roll.

Rate of Volume Flow Q Through Feed Nip

The mill base flowing through the feed nip can be visualized as a ribbon of material of width L (the effective length of the rolls) and of thickness x (the feed nip clearance) which is being extruded from between the rolls at a velocity corresponding to the average of the rim velocities of the feed and center rolls.

$$Q = \frac{V}{t} = \frac{Lx \cdot (v_c + v_f)}{2} \tag{7}$$

Since $v_c = nv_f$, Eq. 7 can be expressed as Eq. 8.

$$Q = \frac{Lxv_f \cdot (1 + n)}{2} \tag{8}$$

For coating and ink calculations it is convenient to work with Q expressed in gal/hr, L in in., v_f in rpm, the roll diameter in in., and x in mils. It has also been shown both theoretically and experimentally that the average velocity through the feed nip is actually about 1.15 times higher than the value given by a simple averaging of the rim velocities of the feed and center rolls. Taking this correction into consideration and introducing appropriate conversion factors, we can rewrite Eq. 8 as Eq. 9.

$$Q = 4.7 \cdot 10^{-4}\ DLx \cdot \text{rpm}(1 + n) \tag{9}$$

where Q = volume of mill base flow through feed nip (gal/hr)
L = effective length of rolls (in.)
D = roll diameter (in.)
x = feed nip clearance (mils)
rpm = revolutions per minute for feed roll

Rate of Mill Base Flow Through Mill

Equation 9 gives the mill base flowing through the feed nip. The production rate for the mill as an entire unit is given by substituting this value for Q in Eq. 4.

$$Q_m\ (\text{gal/hr}) = 4.7 \cdot 10^{-4}\ DLx \cdot \text{rpm}(1 + n)\ ca \tag{10}$$

PROBLEM 18-1

Calculate the production rate Q_m for a 12 X 30 in. three-roll mill if the feed nip clearance is maintained at 6.0 mils. The mill rolls are geared for the following revolutions per minute: feed roll 35, rpm, center roll 105, rpm, apron roll 315 rpm.

Solution. Substitute the problem data in Eq. 10. The values for c and a are obtained by calculation from Eq. 6.

$$Q_m = 4.7 \cdot 10^{-4} \cdot 12 \cdot 30 \cdot 6 \cdot 35(1 + 3) \cdot 0.84^2 = 100 \text{ gal/hr}$$

Although Eq. 10 serves to calculate a production rate (a quantitative figure), it fails to give any clue as to the quality of the resulting dispersion. However, it can be presumed that the quality of the dispersion must be related to the work done on the mill base during its passage through the mill. This subject is now considered.

POWER INPUT TO THREE-ROLL MILL

A basic engineering study has shown that the power demand of a three-roll mill is related to a number of key variables, as shown by Eq. 11 in terms of dimensionless groups (6). Although this study was carried out on three-roll mills of the floating center roll type, the data are applicable also to the more conventional type of three-roll mill where the center roll is fixed, as schematically illustrated in Fig. 18-3.

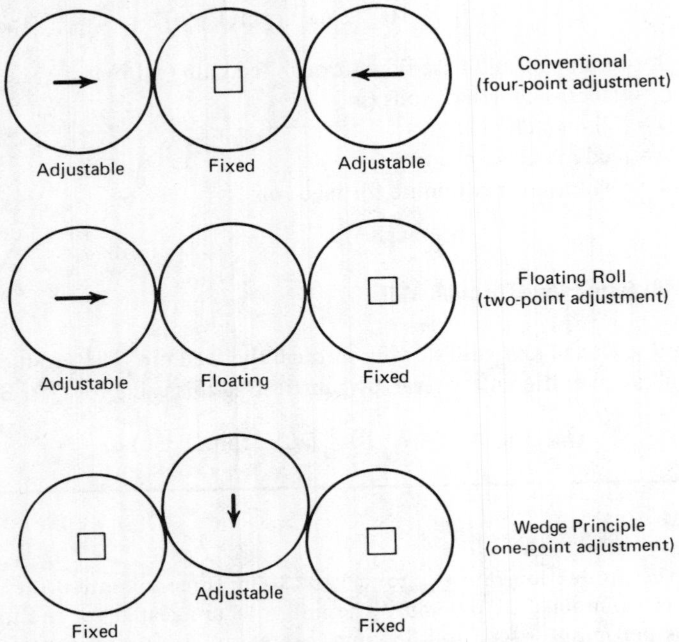

Fig. 18-3 Mechanical arrangements for forcing rolls together on a three-roll mill.

Power number	Froude number	Reynolds number

$$\frac{P}{Lv^3 D\rho} \cdot \left(\frac{v^2}{Xg}\right)^{1/3} \cdot \left(\frac{Xv\rho}{\eta}\right)^{2/3} = f(n, c) \quad (11)$$

where P = power (dyne-cm/sec)
v = peripheral roll velocity (cm/sec)
L = effective roll length (cm)
D = roll diameter (cm)
ρ = density (g/cm^3)
X = sum of feed and apron nip clearances (cm)
η = mill base viscosity (poises)
g = gravitational acceleration (cm/sec^2)

The function $f(n, c)$ in Eq. 11 is given by Eq. 12.

$$f(n, c) = n(n + 1)^{1/3} (n + c)^{1/3} \left(\frac{1.08n^2}{c^{1/3}} + 2.56\right) \quad (12)$$

The fraction c is taken as expressing the average fractional transfer of mill base to the faster roll in its passage through the mill. As given, Eqs. 11 and 12 are quite formidable. However, this author has found that the function $f(n, c)$ in Eq. 12 can be replaced by the simple approximate expression given by Eq. 13 with little loss of accuracy (a mill base transfer of 0.75 was assumed in deriving Eq. 13).

$$f(n, c) = 4n^3 \tag{13}$$

The sum of the two nip clearances is given by the expression $X = x(2c + n - nc)/c + n - cn)$.

By making these simplifying assumptions and converting to units with which the coating engineer is more familiar, there is derived Eq. 14, which expresses the horsepower demand of a three-roll mill in terms of L, D, x, n, rpm, ρ, and η.

$$P\text{ (horsepower)} = 7.0 \cdot 10^{-9} \cdot Ln^3 \left(\frac{D^8 \cdot \text{rpm}^5 \rho\eta^2}{X}\right)^{1/3} \tag{14}$$

where L = effective length of rolls (in.)
D = diameter of roll (in.)
n = ratio of center roll to feed roll speed; also ratio of apron roll to center roll speed
rpm = revolutions per minute of feed roll
ρ = density of mill base (lb/gal)
η = mill base viscosity (poises)
X = sum of gap clearances for feed nip and apron nip $[X = x(2c + n - nc)/(c + n - cn)]$ (mils); if the feed split is considered proportional to the rim velocities $[c = n/(1 + n)]$, then $X = 1.5x$

PROBLEM 18-2

Calculate the power demand of a 16 X 40 in. three-roll mill that is grinding a mill base having a viscosity of 50 poises and a density of 10.4 lb/gal. The mill rolls are geared to give the following revolutions per minute: feed roll 80 rpm, center roll 160 rpm, apron roll 320 rpm. Assume a 4-mill clearance for the feed nip gap and a transfer fraction c of 0.74.

Solution. Calculate a value for the sum of the feed and apron nip clearances X as given by the expression under Eq. 14.

$$X = \frac{4(2 \cdot 0.74 + 2 - 2 \cdot 0.74)}{0.74 + 2 - 2 \cdot 0.74} = 6.36 \text{ mils}$$

Substitute this value for X and the other given data in Eq. 14.

$$P = 7.0 \cdot 10^{-9} \cdot 40 \cdot 2^3 \left(\frac{16^8 80^5 \cdot 10.4 \cdot 50^2}{6.36}\right)^{1/3} = 87 \text{ horsepower}$$

Comment. This calculated power demand of 87 horsepower is probably somewhat higher than would be specified for a three-roll mill of this size in commercial manufacture.

Work Input per Unit Volume of Mill Base Dispersion

By dividing Eq. 14 by Eq. 10 and making appropriate unit conversions, there is obtained Eq. 15, which expresses the work input per gallon of processed mill base. The sum of the feed and apron gap clearances X is taken as equal to $1.5x$.

$$\frac{\text{Work (ft-lb)}}{V(\text{gal})} = \frac{26n^3(D^5 \cdot \text{rpm}^2 \rho\eta^2)^{1/3}}{x^{4/3}(1+n)\, c^2} \tag{15}$$

PROBLEM 18-3

Calculate the work done on each gallon of mill base that passes through the three-roll mill of Problem 18-2.

Solution. Substitute the given data in Eq. 15.

$$\frac{\text{Work}}{V} = \frac{26 \cdot 2^3(16^5 80^2 \cdot 10.4 \cdot 50^2)^{1/3}}{4^{4/3}\,(1+2) \cdot 0.74^2} = 1.1 \cdot 10^6 \text{ ft-lb/gal}$$

Both of the foregoing problems assumed an unchanging or effective average viscosity value for the mill base during its passage through the three-roll mill. That this viscosity may be difficult to determine is shown by the next problem.

PROBLEM 18-4

Assuming that all the work done on the mill base (specific heat $S = 0.35$) of Problem 18-3 goes to raising its temperature, estimate the temperature rise that can be expected for the mill base as it passes through the three-roll mill. Disregard heat loss to the outside environment.

Solution. The mechanical equivalent of heat is 778 ft-lb/Btu (British thermal unit). Using this conversion expression, calculate the temperature rise by Eq. 16, where H is the heat input in Btu, W is the weight of the material being heated, S is the specific heat of the material, and ΔT is the temperature rise.

$$H = W \cdot S \cdot \Delta T \tag{16}$$

$$\frac{1.1 \cdot 10^6}{778} = 10.4 \cdot 0.35 \cdot \Delta T; \qquad \Delta T = 388 \text{ F}^\circ$$

From the foregoing problem it is obvious that water cooling of the mill rolls is essential to reduce the extreme temperature rise that occurs because of the mill work input. The problem also points up the fact that the heat generated by a three-roll mill exerts a major influence on mill base viscosity. Hence good sense must be exercised in selecting a proper (equilibrium) viscosity to be entered into the derived equations.

Also to be noted is that the temperature rise due to work input tends to self-limiting, since the increase in temperature from this work source reduces the mill base viscosity. This in turn reduces the work input required, reducing still further the temperature increase. Eventually the rise in temperature will taper off to give an equilibrium temperature.

USEFUL EQUATIONS APPLYING TO THREE-ROLL MILLS

Except for the mill base viscosity η, the mill base density ρ, and the feed nip clearance x, all the quantities in Eqs. 10, 14, and 15 are uniquely determined by the three-roll mill design (roller dimensions, roller rpm's). Hence, for any given three-roll mill in the plant, these three equations can be simplified by inserting the appropriate mill constants and reducing them to expressions involving only viscosity, density, and feed nip clearance. Furthermore, since density is involved only to the one-third power and since a density value for a mill base intended for a three-roll mill grind does not vary significantly from batch to batch (say from an arbitrary value of 10.4 lb/gal), further simplification can be achieved by substituting 2.18 ($= 10.4^{1/3}$) for $\rho^{1/3}$ in the equations in question. This type of simplification has been carried out for a 16 × 40 in. three-roll mill with roll speeds as follows: feed roll 80 rpm, center roll 160 rpm, apron roll 320 rpm.

$$Q_m \text{ (gal/hr)} = 40 \cdot x \qquad \text{(from Eq. 10)} \qquad (17)$$

$$P \text{ (horsepower)} = 12 \left(\frac{\eta^2}{X}\right)^{1/3} \qquad \text{(from Eq. 14)} \qquad (18)$$

$$\frac{\text{Work}}{V} \text{ (ft-lb/gal)} = 560{,}000 \left(\frac{\eta}{x^2}\right)^{2/3} \qquad \text{(from Eq. 15)} \qquad (19)$$

From these reduced equations for a specific mill, the controlling influence of mill base viscosity and feed nip clearance becomes self-evident. Unfortunately, neither quantity is readily amenable to measurement. As already mentioned, the mill base viscosity decreases as it passes through the mill because of inevitable temperature increase. Moreover, non-Newtonian flow can be anticipated for most mill base dispersions (with possible dilatancy). Exact measurement of the feed nip clearance under actual operating conditions is also beset with difficulties. The tremendous forces acting across the mill roll tend to introduce strains that are likely to vitiate any attempt to arrange for a presetting of the mill roll gap that will hold for the mill in actual operation. However, by indirect methods it can be demonstrated that the nip clearance x between two rolls is approximately related to the average viscosity η, the roll length L, and the force F (in lb) pushing the rolls together by Eq. 20.

$$x = \frac{KL\eta}{F} \tag{20}$$

As a first approximation, the average mill base viscosity can be taken as the viscosity midway between the entering and exit viscosities. K is a constant that is specific for the roll mill in question. For a 16 × 40 in. three-roll mill a value of $K = 28$ is applicable, giving Eq. 21.

$$x = \frac{28L\eta}{F} = \frac{1120\eta}{F} \tag{21}$$

Substituting this value for x in Eqs. 17, 18, and 19 gives Eqs. 22, 23, and 24, applying to a 16 × 40 in. three-roll mill.

$$Q_m \text{ (gal/hr)} = \frac{45{,}000\eta}{F} \qquad \text{(from Eq. 17)} \tag{22}$$

$$P \text{ (horsepower)} = (\eta F)^{1/3} \qquad \text{(from Eq. 18)} \tag{23}$$

$$\frac{\text{Work}}{V} \text{ (ft-lb/gal)} = 48 \left(\frac{F^2}{\eta}\right)^{2/3} \qquad \text{(from Eq. 19)} \tag{24}$$

Practical Application of Equations

A number of equations have been derived that with proper interpretation can be instructive and very helpful to coating engineers in understanding the complex interactions taking place during a pigment grind in a three-roll mill.

For example, Eq. 22 indicates that the mill base production rate can be tripled by tripling the mill base viscosity (F remaining unchanged). Equation 23 in turn tells us that this can be accomplished by only a 44% (0.44) increase in power input ($3^{1/3} = 1.44$). However, Eq. 24 reveals that, in passing through the mill, 52% less work [$0.52 = 1.00 - 0.48$, where $0.48 = (1/3)^{2/3}$] would be done on the mill base per gallon, which means that the quality of the discharged mill base dispersion would suffer greatly. To maintain quality, the ratio F^2/η in Eq. 24 must remain unchanged. Hence, if the viscosity is to be tripled, F must be increased by 73% (0.73), since $3^{1/2} = 1.73$. Coming back to Eq. 22, we see that, to increase production while maintaining quality, a tripling of viscosity must be accompanied by a 73% increase in F (a factor of 1.73). Substituting these data in

Eq. 22 reveals that tripling the viscosity does increase the production rate, but to maintain quality (the same work input per gallon as for the original production) the production increase is only 73% (0.73 from 1.73 = 3/1.73), since it must be accompanied by a simultaneous 73% increase in F. The power demand increase (Eq. 23) under these conditions is also 73% [0.73 from 1.73 = $(3 \cdot 1.73)^{1/3}$].

Another problem involves estimating the change in mill base quality to be anticipated as a result of varying the feed nip clearance. Continuing with the premise that the grind quality is related to the work done on a unit volume of mill base dispersion, we can work out an answer along the following lines (viscosity assumed constant). Let the nip clearance be changed by a factor f (i.e., if the nip clearance is doubled, $f = 2.0$). As a result of this change, the mill throughput is also changed f times (Eq. 10). This change in nip clearance is brought about by changing the force F (which pushes the rolls together) by a factor of $1/f$ (Eq. 20). The work input per gallon, due to the fx nip clearance (from Eq. 15), is altered by a factor of $1/f^{4/3}$. Changes in throughput (quantity) and work input per gallon of mill base (quality) can then be calculated corresponding to the f change in nip clearance. Such changes have been calculated and plotted in Fig. 18-4 for a range of f values.

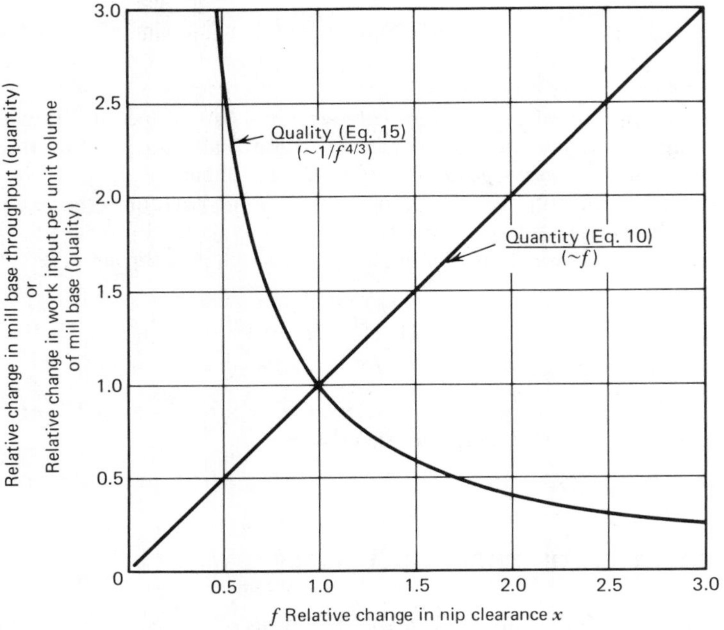

Fig. 18-4 Plot of relative change in mill base throughput (quantity) and in work input per unit volume of mill base (quality) as functions of the relative change in the feed nip gap clearance.

As expected, increasing the nip clearance by backing off on the force pushing the rolls together reduces the quality of the mill base dispersion. Thus an increase in nip clearance from 4.0 to 6.0 mils ($f = 1.5$) can be expected to reduce the mill base dispersion quality by a factor of 0.58 ($= 1/1.5^{4/3}$), equivalent to a 42% reduction in quality. At the same time there is an increase in production rate by 50% ($f = 1.5$). On the other hand, a 25% decrease in nip clearance (say from 4.0 to 3.0 mils, giving an f factor of 0.75) increases the mill base dispersion quality by a factor of 1.48 ($= 1/0.75^{4/3}$). However, this nearly 50% increase in quality is accompanied by a 25% reduction in output (f value of 0.75). From this it is seen that quality and quantity are inverse functions of the nip clearance. Hence, in commercial production, a nip gap must be established that permits a reasonable compromise between a satisfactory production rate and an acceptable mill base quality. This conclusion is borne out by experimental studies comparing fineness of grind with mill output, in which a more or less linear relationship was found between maximum particle size and production output.

Problem 18-5 considers a procedure for correlating results obtained on a laboratory mill with commercial production (7).

PROBLEM 18-5

From an equation proposed by Hummel, it has been calculated that six (5.9) passes of a given mill base composition through a 4 X 8 in. three-roll laboratory mill ($n = 3$, feed roll rpm = 33.3) are required to achieve the same grind as is given by a commercial 12 X 30 in. three-roll mill ($n = 3$, feed roll rpm = 30) for the same nip clearance. Show that the same result is also obtained on the basis of equivalence of work input per gallon of finished mill base.

Solution. From Eq. 15, which gives work input per unit volume (ft-lb/gal), it can be shown that one pass through the commercial mill provides 5.8 times as much work input per gallon of mill base as does one pass through the laboratory mill.

$$5.8 = \left(\frac{12}{4}\right)^{5/3} \cdot \left(\frac{30}{33.3}\right)^{2/3}$$

Comment. To compensate for the lesser work input into the laboratory mill, it is necessary to resort to multiple (six) passes.

CONSTRUCTION OF THREE-ROLL MILLS

The tremendous forces involved in the operation of three-roll mills demand unusually robust equipment. Furthermore, the rolls must be accurately ground so that the roll faces are parallel under running conditions, with an unchanging nip clearance over the entire length of the roll. This calls for water circulation through the rolls to minimize temperature variations over the roll surface that might otherwise cause grinding irregularities. The roll wall must not be so thin that distortion occurs under pressure, or so thick that the water cooling becomes ineffective. To obtain parallel rolls at running temperatures of 140 F (60 F) or over and also under very high pressure application, it is necessary to machine the rolls to have a slight degree of camber (convexity).

The rolls are normally manufactured from chilled cast iron having satisfactory hardness and abrasion resistance. The surface of the roll is preferably ground to a matt finish, since a smooth (shiny) finish promotes slippage rather than the desired mill base adhesion.

OPERATION OF CONVENTIONAL THREE-ROLL MILL (FIXED CENTER ROLL)

The procedure in starting a three-roll mill with a fixed center roll is as follows. First slightly slacken off the adjustable front and rear rolls from the center roll and release the takeoff apron (or knife). Add the mill base to the feed nip, start the mill, and then in turn bring up the feed (rear) roll, the apron (front) roll, and the takeoff apron. Then make adjustments until the mill base is flowing through the mill at a satisfactory rate.

Until all settings are satisfactory, dispersed mill base should be returned to the feed nip. When starting with a cold commercial mill, it may take some 30 min before equilibrium conditions are established. During this initial period, adjustments will be necessary, since the temperature increase that occurs affects both the mill base viscosity and the roll diameter. At equilibrium conditions the gap adjustment at the feed nip is usually such as to produce an apron nip clearance that falls at some fixed value between 0.4 and 2.0 mils (10 to 50 μ) (1).

The adjustment of the takeoff apron knife edge is also critical. When the three-roll mill is operating correctly, the mill base will flow onto the apron with the same film thickness from one side to the other. This can be quickly verified by removing the mill base from the region of the takeoff knife edge with a rapid sweep of a spatula and observing the restart of the flow for thickness variations. The exact setting of the knife requires considerable skill. Too much pressure causes excessive wear; too little pressure reduces output. The angle of contact of the knife edge against the roll should be 45° to the tangent at the point of contact (2).

Modern three-roll mills are equipped with many sophisticated improvements such as hydraulically adjusted rolls that maintain preset pressure ratios, an hydraulically operated apron takeoff knife, forced feed lubrication, self-aligning rollers, and constant monitoring of temperature. With care the rolls of these modern mills can have a long work life (several years before reconditioning becomes necessary).

MAJOR USES OF THREE-ROLL MILL

The three-roll mill provides a true smearer-type grind. This grind takes place at high shear rates, as indicated by Eq. 25.

$$\text{D (shear rate)} = \frac{105R \cdot \Delta\text{rpm}}{x} \tag{25}$$

where R = roll radius (in.)
Δrpm = difference in roll speeds (rpm)
x = nip clearance (mils)

PROBLEM 18-6

Calculate the shear rate in the feed nip of the mill described in Problem 18-2.

Solution. Substitute the problem data in Eq. 25.

$$\text{D} = \frac{105 \cdot 8 \cdot (160 - 80)}{4} = 16{,}800\ \text{sec}^{-1}$$

Comment. It is estimated that the shear rate at the apron nip of this mill is on the order of 80,000 sec^{-1}.

Because of the high shear rate the three-roll mill is especially adapted to the processing of hard-to-grind pigments and/or for attaining the ultimate in dispersion excellence. In the ink industry it is more or less standard for letterpress and lithographic ink production (2). In the coatings industry it is commonly reserved for dispersing hard-to-grind organic pigments.

The three-roll mill has the advantage of handling viscous systems. However, the throughput of this mill is only low to moderate, and overall coating manufacture calls for a three-stage process involving the preparation of a well mixed premix for charging to the three-roll mill and a carefully arranged letdown of the dispersed mill base to the finished coating. The high quality of the modern precision three-roll mill makes it a relatively high-priced piece of equipment. Hence it finds its place mainly as a specialty mill uniquely designed to handle refractory pigments or produce fine dispersions in viscous systems. As such it should be distinguished from the more common and less expensive high-output, general purposes mills designed to handle only easy-to-disperse pigments.

MILL BASE COMPOSITIONS FOR THREE-ROLL MILLS (8–10)

By substituting Eq. 20 in Eq. 10, there results Eq. 26, an expression that reveals the influence of nine key variables on the production rate of a three-roll mill.

$$Q_m \text{ (gal/hr)} = \frac{4.7 \cdot 10^{-4} KDL^2 \text{rpm}(1 + n)\, ca\eta}{F} \tag{26}$$

For any given mill, five of these quantities (K, D, L, rpm, and n) are set by the original mill design. The transfer fractions c and a are also more or less fixed, since, as shown previously, they are closely calculated from a knowledge of n. Collecting these constants into an overall constant K' we can rewrite Eq. 26 as Eq. 27, which gives the production rate for any particular three-roll mill in terms of K', F, and η.

$$Q_m \text{ (gal/hr)} = \frac{K'\eta}{F} \tag{27}$$

The coating and ink engineer has control of K' (the mill specifications) only at the time of the original mill purchase. Occasionally multispeeds are specified at this time to achieve a certain flexibility in the K' value. Other than this, subsequent production rates will depend almost exclusively on the average mill base viscosity and the force F pushing the rolls together. Inspection of Eq. 27 suggests that, by backing off on F, production can be increased, but as was demonstrated in the preceding section a rise in production achieved in this way is accompanied by a loss in quality. Hence the key to high production in a three-roll mill lies in formulating mill base compositions with the maximum viscosities commensurate with the ability of the mill to absorb the tremendous stresses and strains that derive from handling such high-viscosity mixtures. Whereas alternative mills (ball mills, sand mills) are limited in their ability to handle viscous mill base compositions, no such restriction applies to the three-roll mill, and for a highly viscous mill base (ink pastes, solventless coatings) these mills are ideally suited.

Since tremendous shear forces are generated within a three-roll mill, less dependence need be placed on using solvents to wet pigment agglomerates and on exposing their surfaces for subsequent wetting by vehicle binder. Hence mill base compositions for three-roll mills are characterized not only by high viscosity but also by vehicles of high solids content.

Mill Base Vehicle

Mill base vehicles intended for use on a three-roll mill usually contain from zero to a maximum of 40% volatile content. Bodied linseed oil (12-poise viscosity) is a typical grinding vehicle with zero volatile content. A 70% solids alkyd (20-poise viscosity) is representative of a grinding vehicle with volatile solvent present. The introduction of higher volatile contents into these mill base compositions generally leads only to poorer dispersions and/or reduced production rates.

Pigment Content of Mill Base

The pigment content of the mill base is made as high as is commensurate with convenience in handling on the mill (the mill base dispersion must exhibit a nominal degree of flow to the feed bank and from the takeoff apron). A comparison of the pigment/vehicle ratios recommended for 18 different pigments with bodied linseed oil (for three-roll milling) versus their corresponding rub-out oil absorption values reveals that from about 1.5 (TiO_2, chrome yellow) to 3.0 (carbon black) times as much bodied linseed oil is required for proper roll mill dispersion as is needed to just wet out the pigment on a spatula rub-out. Presumably, the alternative Gardner-Coleman method of determining oil absorption should provide a vehicle/pigment ratio only slightly lower than that which would actually be satisfactory for a three-mill grind.

In the case of a 70% solids alkyd vehicle, 65% by weight of a TiO_2 pigment (low-oil-absorption anatase) has been shown to provide an excellent mill base dispersion.

To recapitulate, in formulating mill base compositions for roll mill grinding, specify vehicles of high solids content (say from 70 to 100%) which have high viscosities (say in the range of 20 to 100 poises). To these vehicles charge as much pigment as possible just short of interfering with a satisfactory flow of the mill base from the premix to the mill and from the apron takeoff to the letdown tanks.

Mill Base Tack

Unless the mill base adheres to the surface of the rotating rolls, it will not be dragged into the roll nip. In turn, the particle agglomerates admixed with the mill base will fail to enter the nip region. (11, 12)

Equation Relating Tack to Crushing Force. Figure 18-5 illustrates in schematic form the forces acting on a spherical particle of radius r that is wedged between two inwardly rotating rolls which are smeared with mill base vehicle. Let C be the normal (crushing) force, which acts in a line running through the roll and particle centers. Let T be the tangential (frictional or drag) force, which acts tangentially at the points of contact of roll and particle. Since the spherical

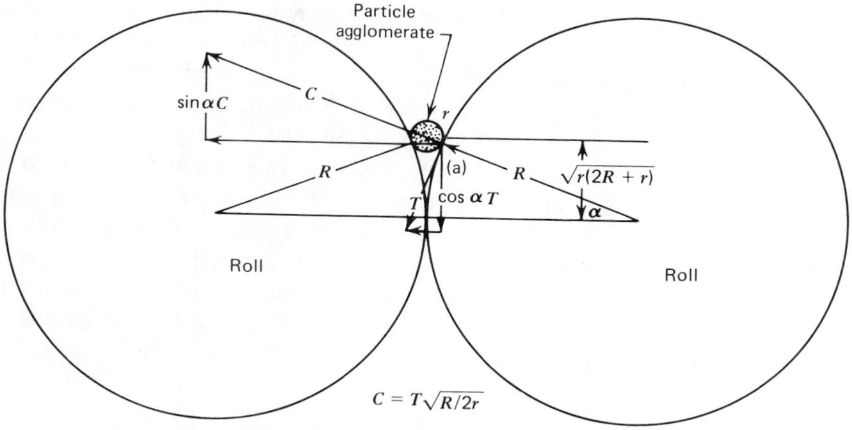

Fig. 18-5 Schematic diagram illustrating the forces acting on a pigment agglomerate lodged between two inwardly rotating rolls of a three-roll mill.

particle is symmetrically placed between the two rolls, the horizontal components of forces C and T acting on the particle on each side cancel out. On the other hand, under equilibrium conditions and for either point of contact, the upward vertical component of force C (= $\sin \alpha C$) is balanced by the downward vertical component of force T (= $\cos \alpha T$) as given by Eq. 28.

$$\sin \alpha C = \cos \alpha T \tag{28}$$

Equation 28 can be expressed alternatively by Eq. 29.

$$\frac{T}{C} = \frac{\sin \alpha}{\cos \alpha} = \tan \alpha \tag{29}$$

From inspection of Fig. 18-5 it can be seen that $\tan \alpha$ can be expressed in terms of R and r as Eq. 30.

$$\tan \alpha = \frac{\sqrt{(R + r)^2 - R^2}}{R} = \frac{\sqrt{r(2R + r)}}{R} \tag{30}$$

Since r for any actual situation is negligible in size compared with R, the expression $2R + r$ can be expressed simply as $2R$. Introducing this simplifying approximation in Eq. 30, substituting this value for $\tan \alpha$ in Eq. 29, and rearranging terms gives Eq. 31.

$$C = T\sqrt{R/2r} \tag{31}$$

Interpretation of Tack Equation. Although Eq. 31 strictly applies only to the case of a single pigment particle lodged between two inwardly rotating rolls, the form of the equation can probably be more generally interpreted. Thus it would appear that a force *C* acting to shear a particle agglomerate caught between two rolls is more or less proportional to the frictional force *T* dragging on the agglomerate. From this it can be argued that a higher degree of tack should act to accelerate the onset of the mill base shearing action (i.e., shearing can be expected sooner and at a point further out from the roll nip). In addition, for a given particle size *r* and tack condition *T* it is apparent that the force *C* increases in proportion to the square root of the roll radius. Hence larger rolls have a built-in advantage over smaller ones in developing shear for a given amount of tack. Finally, the fact that *R* and *r* appear as a ratio, means that a larger roll can handle a proportionately larger particle agglomerate for a given mill base tack factor. In turn, this indicates that larger roll mills have less tendency to reject (and hold back) oversize particles.

Effect of Tack on Rupture of Agglomerates. Vehicle tack also significantly influences the extent to which pigment agglomerates are broken up within the pigment vehicle mixture. For example, consider a pigment agglomerate caught between two mill base streamlines, as schematically shown in Fig. 18-6. Such a situation could conceivably arise in the nip region, where abrupt reversals in streamline direction are perfectly normal (see Fig. 18-2). If a pigment agglomerate is strongly adherent (i.e., if the individual particles are tightly bonded together) and if the mill base vehicle exhibits little tack, the pigment agglomerate will simply rotate like a ball bearing under the streamline forces. If the agglomerate particles are loosely held together and the vehicle is highly tacky (sticky),

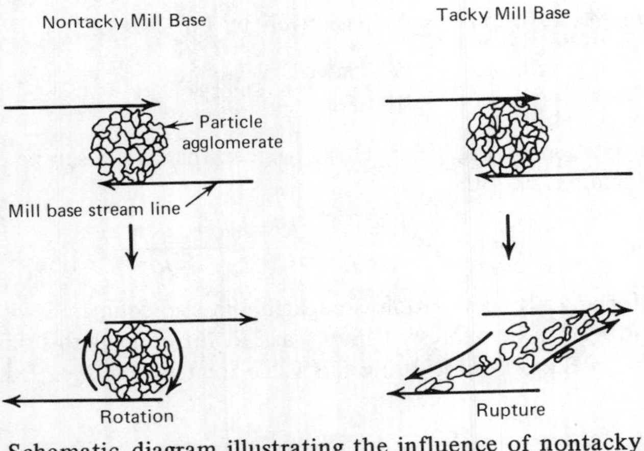

Fig. 18-6 Schematic diagram illustrating the influence of nontacky and tacky mill base compositions on the breakup of pigment agglomerates in a three-roll mill.

however, the agglomerate will be torn apart, a portion of the ruptured agglomerate moving off with one streamline and the other portion moving off with the oppositely directed streamline. Although this is a more or less idealized situation, it serves to dramatize the tug-of-war that is continually taking place between the vehicle tack forces and the agglomerate bonding forces during the grinding operation in a three-roll mill. To a marked extent, the overall effectiveness of the dispersion process is closely related to the tack exhibited by the mill base vehicle. As more data are obtained, it may even be possible to add a "rupture efficiency factor" to the equations previously developed in which vehicle tack is one of the controlling variables.

The foregoing analysis serves to show that tack plays an important role, and studies of mill base production rates reveal that a higher production throughput tends to be obtained with a tackier mill base. It is also a consensus that a significant degree of tack is necessary for a satisfactory transfer of mill base to the faster roll on emergence from the roll nip. In general the viscous mill base compositions that are recommended for use on the three-roll mill have a fully adequate tacky structure. Only with less viscous compositions are grinding aids (tackifiers) normally to be considered helpful in improving production.

Mill Base Premix

The importance of providing a thoroughly mixed premix for charging to a three-roll mill cannot be overemphasized. Since the premix is invariably a viscous composition, a heavy-duty (intensive) type of premixing equipment such as a dough mixer, a W & P (sigma blade) mixer, or a change can mixer is required. Sufficient vehicle to wet down the entire pigment charge (at about the end of 2 or 3 min of intensive mixing) is added first to the mixer. The amount of vehicle for this purpose should be determined by experiment ahead of time, or it may be estimated from previous experience. The entire pigment charge is then loaded on top of this portion of the vehicle, and the power turned on (the ratio of pigment to vehicle at this stage is commonly 2:1). As the rotation of the mixer blade commences, the initial wetting of the pigment mass (with accompanying air elimination) produces a ball-like condition of high structural strength, which in turn leads to the development of high shearing forces within the pigment/vehicle mixture. This intensive mixing is continued until the ball-like mass subsides to a smooth, flowing paste (a matter of a few minutes). The remaining vehicle is then slowly added to bring the premix into the viscosity range satisfactory for charging to the three-roll mill.

This manner of preparation provides even-textured, lump-free, well-wetted mill base premixes that are essentially free of entrapped air.

PARTICLE SIZE VERSUS NIP CLEARANCE

Pigment Lag or Holdback

The average feed nip clearance of a three-roll mill during a normal grind probably runs around 2 mils (51 μ). The average opening in the 325-mesh sieve which is commonly used in particle size analysis is about 44 μ. Allowable pigment retain on a 325-mesh screen is commonly 0.5%. Hence it can be anticipated that the larger pigment agglomerates in a mill base premix will have diameters on the order of magnitude of the gap clearance between the feed and center rolls.

Whereas the retention of oversize pigment agglomerates on a screen is easily visualized, the holdback or retention of oversize particles in the feed bank of a three-roll mill is not as readily apparent. Reference to Fig. 18-2 will clarify the pattern of flow that tends to classify particle agglomerates. First consider the large particle agglomerate located next to the center roll. Normally a particle travels along the streamline that passes through its center. Since the streamline through the large particle agglomerate reverses its flow in the nip region, rejection and return of the large agglomerate to the top of the feed bank takes place. Consider next the cluster of small particle agglomerates. Some of these will also travel along streamlines that lead to rejection. However, the small particle agglomerates directly adjacent to the center roll will almost certainly travel along the streamlines that descend directly into the feed nip. This preferential acceptance of a portion of the smaller particle agglomerates by the feed nip leads to a continuing accumulation of large particle agglomerates in the feed bank. If not recognized and allowed for, this accumulation can lead to a serious "pigment lag," in which oversize agglomerates continue to collect in the feed bank until possibly more than 99.5% of the mill base has passed through the mill. Up to this point an excellent pigment dispersion may have been obtained at the apron takeoff (most of the oversize and large refractory particle agglomerates have been preferentially rejected at the feed nip). However, as the final 0.5% of mill base with its heavy concentration of oversize agglomerates feeds down into the nip, there is passed through the mill the residual portion of the mill base (tail ends) that can ruin the otherwise excellent 99.5% of the dispersion which has gone before. If anticipated, this final portion can, of course, be diverted to some other, less demanding production run, introduced into a subsequent grind, or otherwise used. However, a technically sound method of overcoming pigment lag is to formulate a more viscous mill base that will permit opening up the nip clearance to a wider gap, thus reducing in turn the degree of classification within the feed bank.

Note that reducing the nip clearance during the run in an effort to reduce the oversize may actually aggravate the classification process.

REFERENCES

1. Schaffer, Martin H., "Dispersion and Grinding," Unit 16, Federation Series on Coating Technology, Federation Societies for Paint Technology, Philadelphia, Pa., 1970.
2. Askew, F. A., *Printing Ink Manual,* Society of British Printing Ink Manufacturers, W. Heffer & Sons, Ltd., Cambridge, 1969.
3. Leopold, C. H., "Roller Mills," *Off. Dig.,* **27,** No. 369, 682 (1955).
4. Taylor, J. H. and A. C. Zettlemoyer, "Production Rates in Three-Roll Mills," *Paint Manuf.,* **27,** 299, August (1957).
5. Zettlemoyer, A. C. and J. H. Taylor, "Effect of Flow Properties in Production on Roll Mills," *Off. Dig.,* **32,** No. 424, 648 (1960).
6. Maus, L., W. C. Walker, and A. C. Zettlemoyer, "Dispersion Studies–Correlation of Roll Mill Variables," *Ind. Eng. Chem.,* **47,** 696 (1955).
7. Baker, C. P. and J. F. Vozzella, "The Roll Mill, Pebble Mill, and Kneader as Dispersion Equipment," *Off. Dig.,* **23,** No. 319, 467 (1951).
8. Hoback, W. H., "Practical Aspects of Pigment Dispersion," *Off. Dig.,* **23,** No. 316, 255 (1951).
9. Shurts, R. B., "More Production from Your Present Equipment," *Natl. Paint Varn. Lacquer Assoc. Sci. Sect. Circ.* No. 753, 1951.
10. New York Production Club, "Roll Grinding Study," *Natl. Paint Varn. Lacquer Assoc. Sci. Sect. Circ.* No. 629, 1941.
11. Beakes, H. L., "Theory of Grinding," in J. J. Mattiello (Ed.), *Protective and Decorative Coatings,* Vol. 4, Wiley, New York, 1944, pp. 80–94.
12. New York Club, "A Study of Pigment Dispersion: Part II," *Am. Paint J. Conv. Daily,* p. 12, November 5 (1948).

BIBLIOGRAPHY

Barkman, A., "Equipment and Methods for Manufacturing Paint," *Off. Dig.*, **22,** No. 308, 630 (1950).

Brasington, E. T., "Mixing and Dispersion of Printing Inks," *Am. Ink Maker*, **39,** 32, December (1961).

Heiberger, P., "Chemical Engineering Aspects in Paint Manufacture," *Paint Varn. Prod.*, **42,** 21 (Nov. 1952).

Hurdelbrink, M. W., "Mixers and Mixing Operations," *Off. Dig.*, **27,** No. 369, 677 (1955).

Shurts, R. B., and P. Rosa, "A New Principle One-Point Adjustment Three Roll Mill," *Natl. Paint Varn. Lacquer Assoc. Sci. Sect. Circ.* No. 759, 1952.

Sonsthagen, L. A., "Milling as a Factor in the Cost of Making Paint," *Paint Technol.*, **21,** 347 (1957).

Weil, R. E., "The Selection, Installation and Operation of Three Roll Mills," *Am. Ink Maker*, **40,** 20, April (1962).

Weil, R. E., "Three Roll Mill: Installation and Operation," *Paint Ind. Mag.*, **76,** May (1961).

Zettlemoyer, A. C., and R. R. Myers, "The Rheology of Printing Inks," in F. R. Eirich (Ed.), *Rheology*, Vol, 3, Academic Press, New York, 1960, pp. 145–188.

19 *Ball and Pebble Mills*

A ball or pebble mill consists essentially of a cylindrical container, mounted horizontally and partially filled with either pebbles or ceramic or metallic balls (the grinding medium) (1-5). To this are charged the components of the pigment/vehicle mixture (the mill base). Grinding (dispersion) of the mill base is accomplished by rotating the ball mill and its contents about the horizontal axis of the mill at a rate sufficient to lift the pebbles or balls to one side and then cause them to roll, slide, and tumble (cascade) to the lower side, as shown in Fig. 19-1.

To prevent slippage or back-sliding of the ball charge along the lower mill wall as the mill rotates, resort is commonly made to baffling, either with narrow metal bars in ball mills or ridged ceramic blocks in pebble mills, to assist the lifting action.

Because of this cascading action, pigment particles are caught between the tumbling balls and subjected to both impact and intensive shear. This highly turbulent mixing action on the mill base trapped in the ball interstices imparts the required dispersion effect.

In the paint and ink industries, a pebble mill is conventionally defined as any mill in which a ceramic grinding medium is used (flint pebbles, porcelain balls) with the inside surface of the mill lined with a nonmetallic liner (burrstone, porcelain block, rubber). On the other hand, the designation ball mill is generally reserved for any mill in which the grinding medium is metallic (chilled iron balls, cast nickel alloy balls), the inside mill surface being an alloy steel or some other special metallic liner. Since the same principles of physical dispersion hold for both pebble and ball mills, they are treated as a unit with the understanding that the use of the term "ball mill" includes the term "pebble mill."

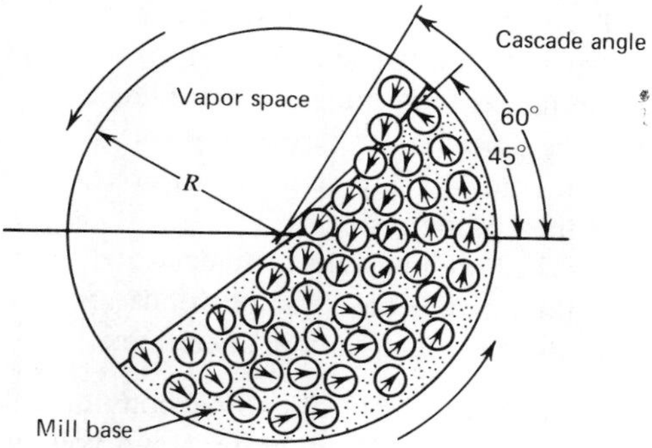

Fig. 19-1 Schematic diagram illustrating the cascading pattern of ball media in a conventional ball mill.

ADVANTAGES AND DISADVANTAGES OF BALL MILLS

Ball mills offer many advantages to the paint or ink manufacturer for dispersing pigments, including hard-to-disperse ones, into vehicles.

1. No premixing of pigment into vehicle is normally required. This is a distinct economic advantage, since many grinding processes require a premixing of the mill base before introduction into the grinding equipment. In the case of ball mills, this need is eliminated. The vehicle is normally charged directly to the mill, followed by the pigment charge.
2. There is essentially no volatile loss and contamination. Since ball milling is carried out in a closed system, volatile loss is limited strictly to the losses that occur during the charging and discharging operations. Contamination by moisture or other foreign material is also eliminated by the closed nature of the ball mill operation.
3. No special skilled supervision or attention to the process is required, since ball mill equipment is relatively safe and simple. Hence labor costs for ball mill operation are minimal.
4. Maintenance costs are low.
5. Ball mills are widely adaptable to the grinding of most paint dispersions and of all pigments, whether soft or hard, fine or coarse. Only highly viscous systems are not amenable to ball mill grinding.
6. Standardization among ball mills is readily accomplished. This is important from the standpoint of interplant processing. Plant production closely duplicates the results obtained with a laboratory mill.
7. Ball mills afford a high ratio of finished coating to dispersed mill base (on

the order of 5:1 up to 8:1) in contrast to lesser ratios for sand mills and high-speed disk dispersers (on the order of 2:1 up to 3:1) (6).

8. Ball mills have the inherent advantage of providing a significant physical size reduction of oversize particles. This generally acts favorably to upgrade pigment opacity and/or color development (it evens out particle size distribution in the direction of the optimum size).
9. The technology of ball milling is well understood, permitting the routine formulation of optimum mill base compositions and the scheduling of the most feasible conditions for the grinding operation.

Disadvantages include the occasional difficulty encountered in emptying the ball mill and a lack of flexibility in scheduling time and volume factors. Thus ball mills must usually run for a fairly long period (several hours) to achieve an acceptable degree of dispersion, and the charge volume is automatically determined by the mill size.

In view of the many advantages enumerated above, it is understandable that ball milling is a common dispersion technique in the paint industry. It has been reported that, of the paint sold in the United States in 1950, half was prepared in ball mills. Although this fraction has since been significantly reduced because of the keen competition offered by the newer types of dispersion equipment, it is almost a certainty that the ball mill will always be important in the manufacture of coatings and inks.

PHYSICAL FACTORS AFFECTING THE DISPERSION EFFECTIVENESS OF A BALL MILL (1-12)

The physical factors controlling the dispersion effectiveness of a ball mill are grouped under three headings in Table 19-1. For the most part, the influence

Table 19-1 Physical Factors Controlling the Dispersion Effectiveness of a Ball Mill

Ball mill
Size (diameter or radius)
Speed or rotation (rpm)
Ball charge
Load size (relative volume)
Ball composition (density)
Ball size (diameter or radius)
Ball shape (spherical, cylindrical, other)
Mill base charge
Load (relative volume)
Viscosity
Density
Composition (pigment/binder/solvent ratio)

of these controlling variables has been so well elucidated that conditions for optimum ball mill operation are quickly and routinely specified.

Size of Ball Mill and Optimum Speed of Rotation

Since the optimum number of revolutions per minute (rpm) for a ball mill is directly related to the ball mill radius, these two factors are grouped together for study.

Consider for a moment the path traveled by a typical ball in a half-full ball mill. At a low angular velocity (low rpm) this ball will be gradually lifted up on one side of the mill to a higher and higher level. Finally a point is reached where it no longer has any support from below. At this point it joins its neighbors in a cascade of balls falling and tumbling down over each other along the sloping surface of the ball charge to the low side of the mill. This process is continuously repeated, as shown in Figs. 19-1 and 19-2.

If the angular velocity of the mill is increased, a point will be reached where the ball no longer cascades but tends to be thrown into the vapor space as it reaches the outer surface of the ball charge. Such cataracting action is depicted graphically in the center diagram of Fig. 19-2. With still further increase in angular velocity, centrifuging will finally take over as the centrifugal force acting on the ball presses and holds it firmly against the inner surface of the mill.

Both cataracting and centrifuging conditions in a ball mill are to be avoided. Centrifuging obviously gives no dispersive action at all (the ball charge as a unit is completely immobilized). Cataracting induces excessive wear of both the lining of the mill and the balls and represents an inefficient dispersion condition.

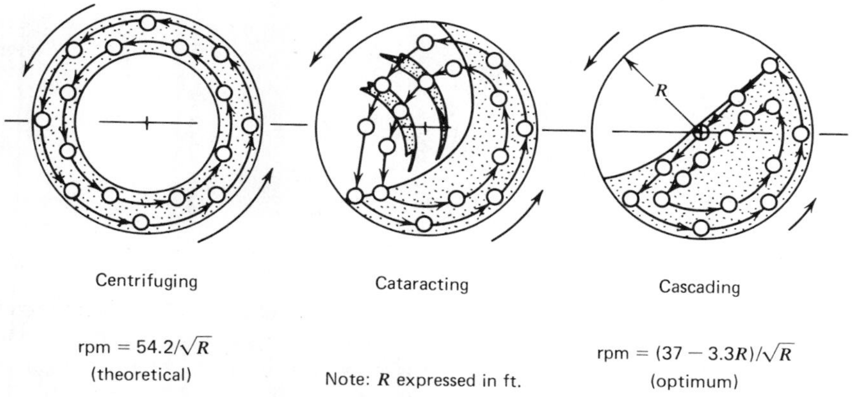

Fig. 19-2 Schematic illustration of centrifuging, cataracting, and cascading conditions for ball media in a conventional ball mill.

Cascading, however, is a highly efficient process for producing a fine pigment dispersion and is the goal in any ball mill paint grinding operation.

Experienced operators of ball mills can closely estimate the most favorable rpm for a ball mill by listening to the noise developed by the balls. A steady rumble is interpreted as an acceptable grinding condition. A high noise level is associated with unwanted cataracting as the balls hammer against and wear down the mill lining. Conversely, too low a noise level indicates an excessive mill base viscosity with an arrested ball motion (referred to as a choked mixture).

Conditions that lead to unwanted centrifuging can be computed from simple physical principles, as illustrated by Problem 19-1.

PROBLEM 9-1

A ball of mass m moves with constant velocity v along a circular path of radius R in a vertical plane. This corresponds to a ball lodged on the inside surface of a rapidly turning ball mill, as shown in Fig. 19-3. Calculate the angular velocity (rpm) just necessary to maintain this ball in orbit at the top of its circular trajectory.

Solution. To just maintain the ball in orbit at the top of its trajectory, it is necessary that the centrifugal force ($F = mv^2/R$) acting radially outward on the ball be exactly equal to the centripetal gravitational force ($F = mg$) acting radially inward on the ball. These conditions are stipulated in Eq. 1.

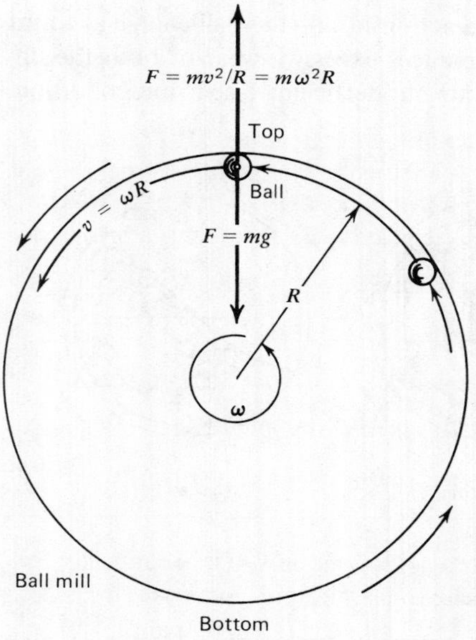

Fig. 19-3 Diagram illustrating the path and forces acting on a sphere that is traveling in a circular orbit in a vertical plane.

$$F\text{ (centrifugal)} = \frac{mv^2}{R} = F\text{ (gravitational)} = mg \tag{1}$$

Solving for v_c, the critical velocity just necessary for continued orbiting at the top, gives Eq. 2.

$$v_c = \sqrt{gR} \tag{2}$$

The tangential velocity in terms of rpm is given by Eq. 3.

$$v = \frac{2\pi R}{60} \cdot \text{rpm} \tag{3}$$

Substituting this value for v in Eq. 2 and solving for the critical rpm_c leads to Eq. 4.

$$\text{rpm}_c = \frac{60}{2\pi R}\sqrt{gR} \tag{4}$$

Using 32.2 ft/sec^2 for g (to be consistent in unit with R expressed in ft) and simplifying gives Eq. 5.

$$\text{rpm}_c = \frac{54.2}{\sqrt{R}} \tag{5}$$

This is the number of rpm just necessary to maintain the ball in circular motion at the top of its orbit. A lower angular velocity will result in either cataracting or cascading rather than centrifuging.

Optimum cascading conditions are based on practical production experience rather than on theoretical considerations (as were used for centrifuging). Yet there is a relationship between the two. An analysis of manufacturers' recommendations for optimum ball milling conditions shows that the rpm for the optimum grinding conditions (rpm_o) is related to the critical rpm just necessary to produce centrifuging (rpm_c) by Eq. 6 (13).

$$\text{rpm}_o = (0.68 - 0.06R)\,\text{rpm}_c \tag{6}$$

When the value for rpm_c given by Eq. 5 is substituted in Eq. 6, there results a second expression for rpm_o in terms of the ball mill radius (R must be expressed in ft).

$$\text{rpm}_o = \frac{37 - 3.3R}{\sqrt{R}} = \frac{37}{\sqrt{R}} - 3.3\sqrt{R} \tag{7}$$

Values for rpm_c and rpm_o are tabulated in Table 19-2 and graphed in Fig. 19-4 for ball mill radii ranging from 1 to 4 ft. Centrifuging and cascading conditions are schematically shown in Fig. 19-2.

Table 19-2 Centrifuging rpm_c and Cascading rpm_o Values for a Range of Ball Mill Radii

Ball Mill Dimensions		Critical Centrifuging, rpm, rpm_c (Eq. 5)	Optimum Cascading rpm, rpm_o (Eq. 7)
Radius (ft)	Diameter (ft)		
0.50	1.00	76.6	50.0
1.00	2.00	54.2	33.7
1.50	3.00	44.3	26.2
2.00	4.00	38.4	21.5
2.50	5.00	34.3	18.2
3.00	6.00	31.3	15.6
3.50	7.00	29.0	13.6
4.00	8.00	27.1	11.9

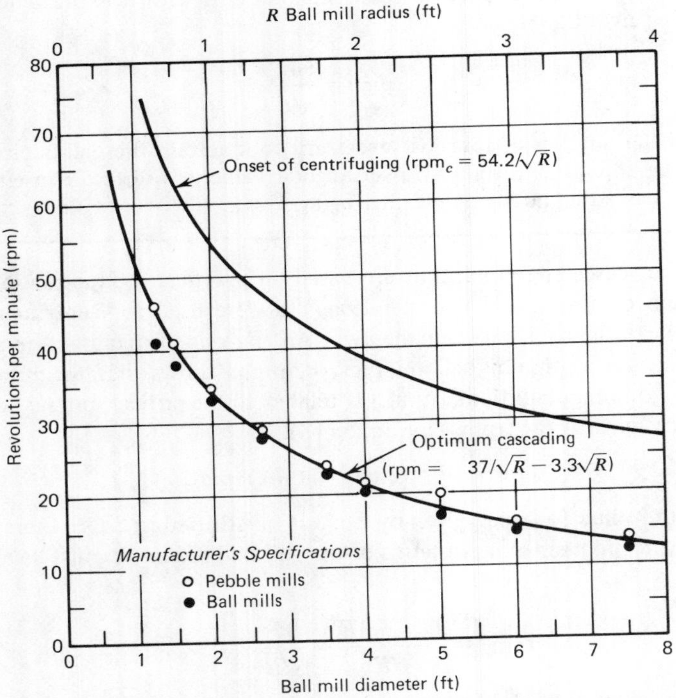

Fig. 19-4 Plot of the critical centrifuging rpm's and optimum cascading rpm's versus ball mill radii for conventional ball mills.

The values of rpm_o that provide the most favorable grinding conditions for a ball mill of a given radius R as calculated from Eq. 7, tabulated in Table 19-2, and graphed in Fig. 19-4 are representative of normal practice. This is indicated in Fig. 19-4, where optimum rpm's for cascading grinding, as listed in one manufacturer's literature, have been plotted for pebble mills (circles) and ball mills (dots) (13). Adjustments may be required for any actual production run to compensate for particular conditions. Thus for a hard-to-grind pigment a cascade angle of 60° may prove more efficient than the more common cascade angle of 45° (see Fig. 19-1). The higher angle is obtained by slightly increasing the ball mill angular velocity.

Optimum Ball Charge to Ball Mill

Charging the ball mill to the halfway mark with balls gives the most efficient loading arrangement. This optimum loading complies with theoretical considerations. At this level, cascading freely occurs across the full diameter of the mill. Furthermore, by loading to the halfway mark, the power input to the mill attains a maximum value. This is clearly revealed by reference to Fig. 19-5, which graphs the relative theoretical power input for mill loading ranging from an empty to a full mill. Note that the power input peaks at the 50% gross volume loading and tapers off to zero for both the 0 and 100% loadings. Since dispersion effectiveness is directly related to power input, it follows that any ball charge

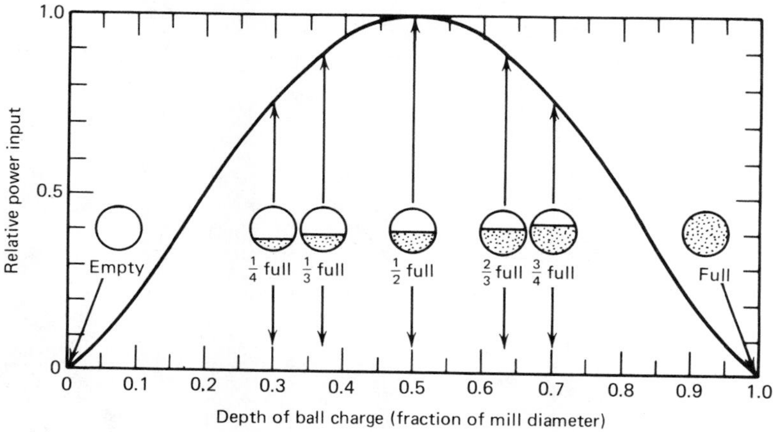

Fig. 19-5 Plot of theoretical power requirements for conventional ball mill versus ball mill loading.

which fills the mill to less or more than the halfway mark results in the corresponding decrease in grinding output. For various reasons (say to schedule a larger volume of mill base in the production run) lower ball loadings are occasionally specified. However, reduction of the ball charge, from, one half to one third, automatically introduces at least a 10% loss in potential grinding output. Loadings lower than 30% of the ball mill volume introduce a slippage of the ball charge (as a unit) along the inside ball mill surface with attendant excessive wear.

In industrial practice the ball charge to a ball mill commonly ranges from one third (metallic balls) to one half (ceramic balls). In the case of ceramic balls, some operators elect to slightly exceed the halfway point to compensate for wear, but this increase should probably not exceed 5%.

Power Required to Operate the Ball Mill with Optimum Half-Full Ball Charge

The following computation method for obtaining the power required to operate a commercial ball mill of radius R is predicted on the assumption that the ball mill is half filled with balls, there is a cascade angle of 45°, and the mill base just covers the top layer of balls. Let the average density of the ball/mill base mixture be ρ'. As the mill turns with uniform velocity at the optimum rpm value, cascading takes place as the balls are continuously rotated into the cascading zone. The energy input required to do this can be thought of as equivalent to that required to continuously move a wedge of balls (with mill base) from the lower half to the top half of the sloping cascade surface (see Fig. 19-6). The dif-

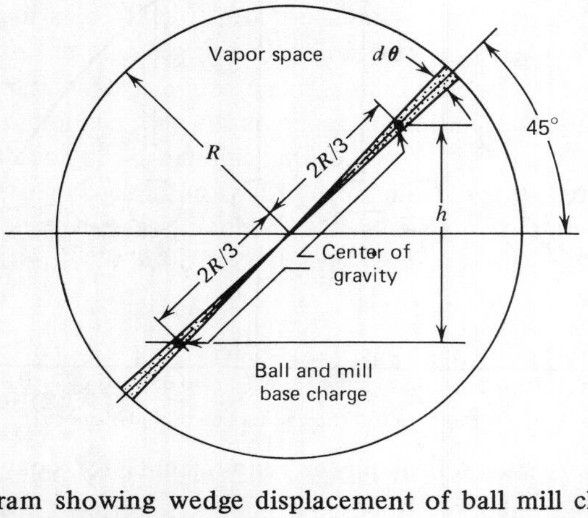

Fig. 19-6 Diagram showing wedge displacement of ball mill charge during rotation of conventional ball mill.

ferential volume of this narrow wedge is equal to $\pi R^2 L \, d\theta / 2\pi$ where L is the effective inside length of the ball mill. In turn, the differential weight dW of the load in the wedge section is given by Eq. 8.

$$dW = \frac{\rho' R^2 L \, d\theta}{2} \tag{8}$$

From physical principles it can be demonstrated that the center of gravity of the differential wedge lies at a distance $2R/3$ from the mill axis. In moving this wedge from the lower to the higher cascade level, this center of gravity is displaced a vertical distance h equal to $2(2R/3) \sin 45° = 0.943R$. The corresponding work input dE is equal to the product $dW \cdot 0.943R$. Substituting the value for dW given by Eq. 8 in this work input expression gives Eq. 9 for the differential energy required to produce this lifting effort.

$$dE = \frac{0.943 \rho' R^3 L \, d\theta}{2} \tag{9}$$

The number of differential wedges rotated into position for cascading each second is given by Eq. 10.

$$\text{Differential wedges per second} = \frac{2\pi \, \text{rpm}}{60 \, d\theta} \tag{10}$$

The power required to lift this number of wedges per second through a height of $0.943R$ is given by Eq. 11.

$$P = \frac{0.943 \rho' \pi L R^3 \, \text{rpm}_o}{60} \tag{11}$$

For optimum cascading the rate of mill rotation is related to the mill radius R by Eq. 7. Substituting this value for rpm_o in Eq. 11, dividing by 550 to convert ft-lb/sec to horsepower, and simplifying the numerical values yields Eq. 12 (density must be expressed in lb/ft^3, and R and L in ft).

$$P(\text{horsepower}) = \frac{\rho' L (3.3 - 0.3R) R^3}{1000\sqrt{R}} \tag{12}$$

This equation gives the theoretical horsepower requirements for a ball mill of radius R and length L when running at optimum rpm with the mill half-filled with mill base and balls (cascade angle of 45°).

Equation 12 can be rewritten as Eq. 13, where $f(R)$ is equal to $(3.3 - 0.3R)R^3/1000\sqrt{R}$.

$$P(\text{horsepower}) = \rho' L \cdot f(R) \tag{13}$$

Table 19-3 Values for the Function $f(R) = (3.3 - 0.3R)R^3/1000\sqrt{R}$ For a Selected Range of R Values

Ball Mill Dimensions		
Radius R (ft)	Diameter (ft)	$f(R)$
0.5	1.0	0.00056
1.0	2.0	0.0030
1.5	3.0	0.0079
2.0	4.0	0.0153
2.5	5.0	0.0252
3.0	6.0	0.0374
3.5	7.0	0.052
4.0	8.0	0.067

Values for the $f(R)$ corresponding to a range of R values are listed in Table 19-3.

PROBLEM 19-2

A 5 X 6 ft ball mill (diameter is always given first) is charged to the halfway mark with 1.0-in. porcelain balls (density 2.4 g/cm^3). To this is added mill base (1.8-g/cm^3 density) in sufficient amount to just cover the top layer of balls. Calculate the horsepower required to operate this mill at the optimum rpm.

Solution. Consider tne balls to occupy 60% and the mill base 40% of the volume of the total charge. The average density of the total charge is then 2.16 g/cm^3 (= 0.60 · 2.4 + 0.40 · 1.8). The density in American units is 135 lb/ft^3 (= 2.16 · 62.4). Substitute this and the given values for R (= 2.5) and L (= 6) in Eq. 12.

$$P = \frac{135 \cdot 6.0(3.3 - 0.3 \cdot 2.5) \cdot 2.5^3}{1000\sqrt{2.5}} = 20.4 \text{ horsepower}$$

Comment. This problem could also have been solved by using the function $f(R)$ given in Table 19-3.

$$P = 135 \cdot 6.0 \cdot 0.0252 = 20.4 \text{ horsepower}$$

Inspection of Eqs. 12 and 13 shows that the power demand of a ball mill is proportional to the average density of the ball media/mill base mixture. For example, if the ceramic balls in Problem 19-2 were replaced by steel balls, over 2.5 times as much power input would be required.

BALL MEDIA DIAMETERS AND BALL MILL FIXTURES

Ceramic ball media (flint pebbles, porcelain balls) are normally supplied in a range of sizes from $\frac{1}{4}$ to $3\frac{1}{4}$-in. in diameter (13). Steel balls are commonly supplied in ball diameters ranging from $\frac{1}{16}$ to $1\frac{1}{4}$ in. (14). The approximate number of balls n per pound can be calculated from the expression $n = 53/\rho d^3$, where ρ is the ball density (g/cm^3), and d is the ball diameter (in.). Thus the number of $\frac{1}{4}$-in. porcelain balls ($\rho = 3.5$ g/cm^3) in a pound is 970 ($= 53/3.5 \cdot 0.25^3$).

A ball mill is normally equipped with a manhole for loading the charge from the top (facilitated by a slope chute such as an 8-in. diameter stainless steel tube that reaches from the floor above to the surface of the ball charge within the mill), an air relief cock, an emptying port with a standby slotted cover for discharge, and a sampling vent opening with an easily removable vent plug. Venting should be carried out some 30 min after starting a new batch and periodically thereafter as necessary (venting contributes to grinding efficiency). Also, since flowing material or moving machinery generates static electricity, it is necessary to have grounding at all points, as well as grounding of the ball mill operators.

TYPES OF GRINDING MEDIA

There are three general classes of grinding media: flint pebbles, porcelain balls (regular and high density), and metallic balls (steel and other metals).

Ceramic Media

Naturally occurring flint pebbles ($\rho \sim 2.6$ g/cm^3), representing the oldest grinding media, are still quite popular. The best grade of these pebbles, such as those collected from Normandy Beach (France), are exceptionally tough and long wearing. However, because of their natural origin they are imperfectly shaped (making cleaning difficult), and their off-color may in turn impart a gray cast to light-colored dispersions.

Regular grade synthetic porcelain balls ($\rho \sim 2.3$ to 2.4 g/cm^3) are probably the type most commonly used today. They are pure white, highly vitrified, ceramic spheres that strongly resist chipping or cracking in service. More recently, higher density (and therefore more efficient) porcelain balls ($\rho \sim 3.3$ to 3.5 g/cm^3) with an approximate 85%w Al_2O_3 content (Borundum, Diamonite, Arlcite) have been introduced. These high-density balls, which are also more expensive, are harder and more abrasion resistant than regular porcelain, and hence ball wear is correspondingly less. However, the smaller amount of abraded material that mixes with the dispersion is also more abrasive.

Ceramic ball wear loss may range from 0.3 to 0.5%w/100 hr for flint pebbles to 2 to 3%w/100 hr for porcelain balls (15).

Metallic Media

Because of the far higher density of metallic media, metal balls provide faster grinding at smaller ball diameters than their ceramic counterparts. At the same time their use demands sturdier mill construction and a far higher power input. High-carbon/high-manganese steel balls ($\rho \sim 7.8$ to 7.9; Rockwell C hardness 60 to 70) are most commonly used. Stainless steel balls are occasionally used but only for special situations (acid resistance, nonmagnetism) where their higher cost can be justified. Relatively expensive cast nickel alloy balls (whitish metal) are sometimes used where metallic staining must be minimized. On the other hand, where metallic contamination is not considered objectionable, resort is often made to relatively cheap, low-carbon steel (forged) balls.

Metallic ball wear loss may range from 0.1 to 0.3%w/100 hr for chrome/maganese steel to 0.3 to 0.7 %w/100 hr for high-carbon steel (15).

BALL SIZE, DENSITY, AND SHAPE

Ball Size and Ball Density

The smallest ball size commensurate with proper cascading and proper drainage of the dispersed mill base should be used. Fortunately, when a ball size is once established as best for a given dispersion, it does not vary appreciably for a different mill size. For example, a ball size that proves successful for a laboratory ball mill dispersion also proves applicable to a scaled-up production run.

Balls of minimum diameter are specified for the simple reason that, per revolution of the mill, they provide the maximum number of impact or shearing contacts and, for a given volume of space, they offer the maximum area for dispersive action. The smaller void spaces between balls of lesser diameter also automatically limit the size of the mill base agglomerates that can exist in the interstitial space and resist dispersion. To specify a ball size and density which deliver an impact or shearing action far in excess of that needed for effective dispersion leads only to wasted energy and unnecessary ball wear.

Selection of the ball diameter and density that will provide the most favorable conditions for grinding in a ball mill is tied in with the density and viscosity of the mill base being processed. For example, the difference between the ball and mill base density $\Delta\rho$ determines the actual force acting on the balls that causes them to cascade. Thus the buoyant effect exerted by the mill base (acting upward) is proportional to the mill base density. The gravitational pull (weight) on the ball charge (acting downward) is proportional to the ball density. Hence the overall or net force acting downward is proportional to the difference in the two densities as given by Eq. 14.

$$\Delta\rho = \rho_{\text{ball}} - \rho_{\text{mill base}} \tag{14}$$

This relationship serves to explain why high-density alumina balls (3.5 g/cm^3), as compared with porcelain balls (2.3 g/cm^3), tend to impart a much more forceful dispersive action than would be predicted from the ratio of their densities (1.5X = 3.5/2.3). Actually a true comparison must take into account the flotation or buoyant effect exerted by the mill base on the immersed balls. Assume that the average mill base charged to a ball mill has an approximate density of 15 lb/gal (1.8 g/cm^3). Using this density value and correcting for the buoyant effect of the mill base, we find that a high-density alumina ball actually exerts about 3.4 times the impact and shearing action of a porcelain ball:

$$3.4\text{X} = \frac{3.5 - 1.8}{2.3 - 1.8}$$

Table 19-4 lists the four main factors controlling the forcefulness of the grinding action. For a given mill base, a set of values must be selected which will give a proper balance among these four factors and provide the most favorable set of processing conditions.

Optimum Mill Base Viscosity Related to Ball Size and Ball Density. Until recently the selection of variables from Table 19-4 was based largely on practical experience. Now results of experimental work are available that permit a selection on a more rational basis. The reported work is predicted on the argument that ball wear is the key to establishing optimum grinding conditions. It postu-

Table 19-4 Factors Controlling the Magnitude of the Dispersive Action in a Ball Mill

Factor	Low Impact, Mild Shear	High Impact, Strong Shear
Ball factors		
Ball diameter	Small	Large
Ball density	Low	High
Mill base factors		
Mill base density	High	Low
Mill base viscosity	High	Low

lates that excessive ball wear is indicative of wasted energy, whereas no ball wear is indicative of retarded and/or incomplete dispersive action. Between these two extremes there must exist a borderline region of nominal ball wear equivalent to optimum milling. From extensive testing this critical region of ball wear has been established as ranging between 5 and 10% per year.

On the basis of these data and also of information reported in a recent article (16) this author has developed a simple expression (Eq. 15) that approximately relates ball size and density to optimum mill base grinding viscosity.

$$\eta \text{ (optimum KU)} = 10\rho + 72d \tag{15}$$

where η = viscosity (Krebs units, KU)
ρ = ball density (g/cm^3)
d = ball diameter (in.)

The first edition of this book proposed a more complex expression for this situation, but Eq. 15 appears to be equally satisfactory. It provides a starting point for experimentation when no other background data are available.

PROBLEM 19-3

A mill base of 15-lb/gal density is to be ground in a ball mill, using $\frac{3}{4}$-in. diameter high-density alumina balls. Calculate the mill base viscosity that provides optimum grinding conditions for the problem specifications.

Solution. The density of the alumina balls is 3.5 g/cm^3. Substitute this value and the value of 0.75 in. for the ball diameter in Eq. 15.

$$\eta \text{ (KU)} = 10 \cdot 3.5 + 0.75 \cdot 72 = 89 \text{ KU}$$

Ball Media Shape

Ceramic grinding media are supplied in three major shapes: spherical, nodular or irregular, and cylindrical. There is considerable controversy concerning the effect of the ball (media) shape on the particle size distribution that is provided. Evidence has been presented that suggests that the more varied the type of media contact that is possible, the narrower is the grind size distribution (17). This is indicated schematically in Fig. 19-7, where the point, line, and face contacts afforded by cylindrical grinding media are shown to yield a narrower particle size distribution than do spherical grinding media (where essentially only point contacts are possible). Irregularly shaped media yield distributions intermediate to those obtained with spheres and cylinders. Incidentally, irregular media tend to suffer from considerable self-attrition, which introduces a sig-

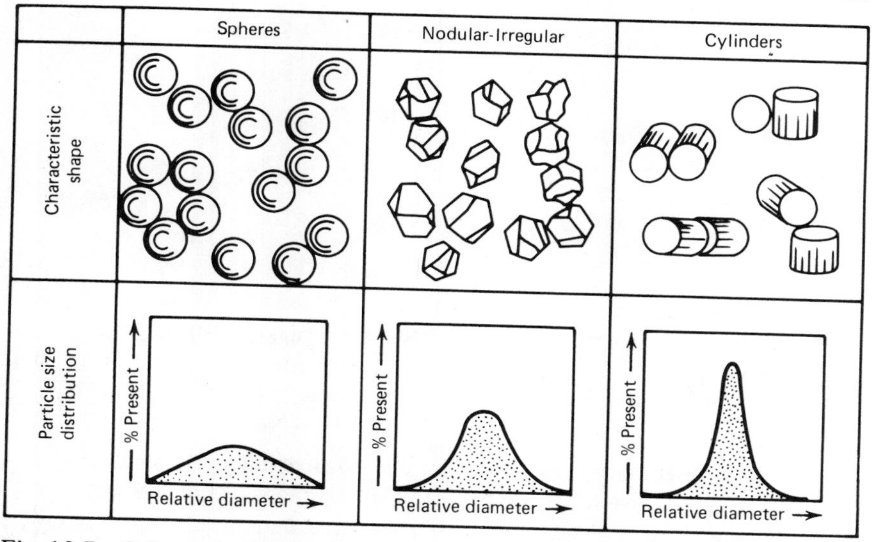

Fig. 19-7 Schematic illustration showing relationship between the grinding media shape and the particle size distribution provided.

nificant amount of contamination (at least until the irregularities have been smoothed off to give a more rounded shape).

Mixed Ball Sizes

It has occasionally been suggested that a mixture of small and large ball sizes might be advantageous. However, practical experience has shown that the larger balls accelerate the wear of the smaller ones. Hence, if the smaller balls are providing an adequate dispersion, the introduction of the larger balls simply expands needless energy in excessively abrading the smaller balls and the mill lining.

VOLUME OF MILL BASE CHARGE

Before considering the several factors that determine the proper volume of mill base to be charged to the ball mill, it is helpful to review first the geometry which applies to the situation. In the first place, the total ball charge, on being loaded into the mill, will assume the shape of an elongated segment volume of length L. Above this segment there will lie a complementary segment corresponding to the vapor space region.

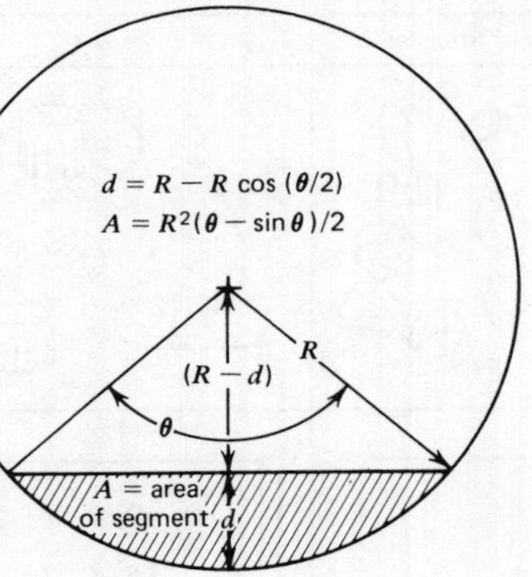

Fig. 19-8 Diagram illustrating geometry of segment area in relation to total circular area.

A typical segment area is illustrated by the crosshatched area of Fig. 19-8. Let the depth of this segment section be d, and let the chord of this segment be subtended by the angle θ. Values for the segment area A in terms of the mill radius R and the angle θ are given by Eqs. 16 and 17.

$$d = R - R\cos\frac{\theta}{2} \tag{16}$$

$$A = \frac{R^2(\theta - \sin\theta)}{2} \tag{17}$$

From these two equations there has been calculated and tabulated in Table 19-5 the percentage areas occupied by the segment section for various depth values (expressed as fractions of the circle diameter).

PROBLEM 19-4

The distance from the top of a 6 × 8-ft ball mill to the surface level of a ball charge is 3.3 ft. Calculate the volume of the vapor space above the ball charge.

Solution. The depth of the vapor space expressed as a fraction of the mill diameter is 0.55 (= 3.3/6.0). The mill cross-sectional area is 28.3 ft². Of this, 56.3% is occupied by the vapor space segment (see the value opposite 0.55 in Table 19-5). The volume of the vapor space above the ball charge is then 128 ft³ [= (56.3/100) · 8.0 · 28.3]. The gross volume occupied by the ball charge is the difference between this volume and the total mill volume.

In Problem 19-4 the term "gross" has been used to describe the overall ball volume, since a significant amount of space between the balls (interstices or

Table 19-5 Percentage of Total Circular Area Occupied by Segment Section for Various Values of Segment Depth Expressed as the Fraction $d/2R$ (see Fig. 19-8)

Depth ($d/2R$)	Area (%)	Depth ($d/2R$)	Area (%)
0.00	00.0	0.55	56.3
0.05	1.9	0.60	62.6
0.10	5.2	0.65	68.8
0.15	9.4	0.70	74.8
0.20	14.2	0.75	80.5
0.25	19.5	0.80	85.8
0.30	25.2	0.85	90.6
0.35	31.2	0.90	94.8
0.40	37.4	0.95	98.1
0.45	43.7	1.00	100.0
0.50	50.0		

interstitial space) is void volume. As defined, gross volume embraces both the actual and the void volume of the ball charge.

When mill base is added to the ball charge in the mill, the mill base proceeds to fill into the voids between the balls. It is important in ball mill calculations to estimate the relative percentages of ball and mill base volume when the two are mixed together in the ball mill.

Consider the case where all the balls are spherical and uniform in size. From geometry it can be shown that with tight tetrahedral packing the balls occupy 74.0% of the overall volume, with 26.0% of interstitial space left over for mill base occupancy (see Fig. 19-9). With loose cubical packing the balls occupy only 52.4% of the available space, with 47.7% left for mill base occupancy. On the basis of experimental evidence and with uniform, commercially manufactured balls, it is found that the balls actually occupy about 60% of available volume, leaving 40% of void space for mill base occupancy. The latter percentages, which apply to practical situations, are seen to fall between the values computed for the tightest and the loosest theoretical packing arrangements. This 60/40 apportionment of ball charge to mill base, where the two are present as a mixture, is used hereafter for any ball mill calculation.

It was established previously that, by charging the ball mill to the halfway mark with balls, maximum grinding efficiency is obtained. The question now arises as to the correct volume of mill base to be used with this optimum ball loading.

An amount less than that required to fill the ball voids obviously results in more or less direct ball-to-ball contact with no cushioning action of interspersed mill base. Not only does this produce unnecessary wear, but also a portion of the grinding action is rendered entirely unproductive.

Packing	Ball Volume	Void Volume
Cubical (loosest)	52.4% ($\pi/6$)	47.6%
Tetrahedral (tightest)	74.0% ($\sqrt{2}\ \pi/6$)	26.0%

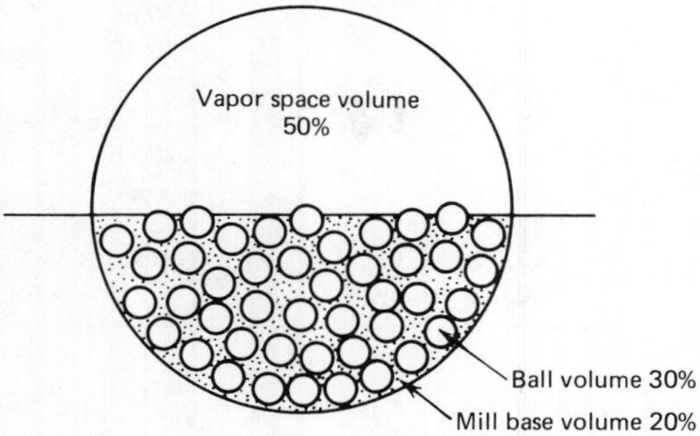

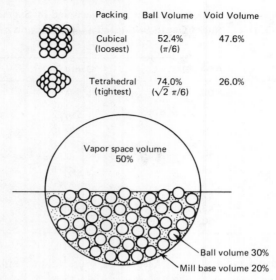

Fig. 19-9 Schematic diagram illustrating ideal charging conditions for a conventional ball mill. Volume relationships are shown for selected packing arrangements.

On the other hand, an amount of mill base greater than that necessary to cover the balls means that the excess remains undispersed until such time as it is afforded the opportunity to diffuse into and displace some of the mill base in the active ball region. In turn this displaced mill base must then await its chance to reenter the grinding zone to complete its dispersion. Eventually a uniform dispersion will result, but with too much excess mill base the dispersion process is markedly retarded.

Charging conditions for optimum dispersion effectiveness are shown in Fig. 19-9, where the ball mill is half filled with balls with sufficient mill base present just to cover the ball charge. For these ideal conditions the respective volumes are 50% vapor space, 30% ball volume, and 20% mill base volume. In practice, somewhat more mill base is added to form a cushioning puddle at the bottom of the cascade slope.

Before considering the effect of excess mill base, certain pertinent volume relationships should be clarified. Thus for a ball mill half-filled with balls with 25% excess mill base present, note that the total mill base volume (45%) includes both interstitial (20%) and excess (25%) mill base volume. The ball charge *gross* volume (50%) is still equal in amount to the actual ball volume (30%) plus the interstitial mill base volume (20%).

From an analysis of dispersion rates (for a ball mill half-filled with balls), there has been derived the empirical expression given by Eq. 18, relating dispersion time to mill base volume.

$$\log \frac{t_2}{t_1} = 1.8(M_2 - M_1) \tag{18}$$

where t = time required to attain a given degree of dispersion
M = volume of mill base expressed as fraction of total ball mill volume

The subscripts refer to two sets of conditions. The nomogram of Fig. 19-10 provides a graphical solution to Eq. 18.

PROBLEM 19-5

In a plant ball milling operation, 5% excess mill base is used, with the ball charge loaded to the halfway mark. If a satisfactory dispersion is obtained in 8.0 hr under these conditions, estimate the time required to achieve the same degree of dispersion when the total mill base volume is doubled.

Solution. For a half-full ball charge, the interstitial mill base volume is 20% of the ball mill volume. Together with 5% excess mill base, the plant operation is then normally carried out with 25% of total mill base volume present. Doubling this gives a 50% total mill base volume, equivalent to a 30% excess of mill base (= 50 − 20). Either a numerical or a graphical solution can be used to determine

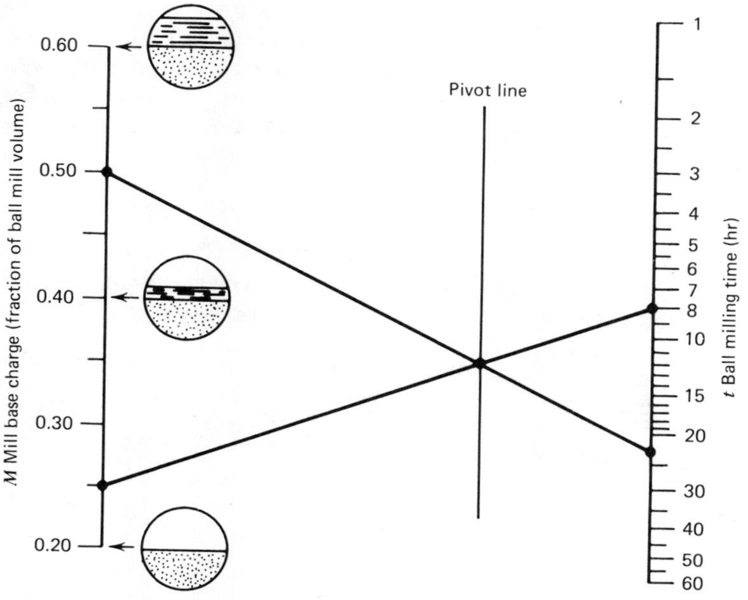

Fig. 19-10 Nomogram relating ball milling time to mill base volume when the ball mill is half-filled with the ball media.

the time necessary to attain the required degree of dispersion with the doubled mill base charge.

Numerical Solution: Substitute the two sets of data in Eq. 18.

$$\log \frac{t_2}{8.0} = 1.8(0.50 - 0.25); \qquad t_2 = 22.6 \text{ hr}$$

Graphical Solution: Connect 0.25 as located on the M scale of Fig. 19-10 with the value of 8.0 hr on the "ball milling time" scale by a straight line. Mark the intersection of this line with the pivot scale line. Next draw a straight line from 0.50 as located on the M scale through this pivot scale intersection point, and continue this line to intersect the "ball milling time" scale. Read off the required time of 22.5 hr at the intersection point.

In this problem it is important to observe that when the mill base volume is doubled it takes nearly three times as long to attain the same degree of dispersion.

Effect of Ball Mill Size on Rate of Dispersion

Other things being equal, faster dispersion rates are attained through the use of larger ball mills. The reason is explained by the following theoretical argument. When operating at maximum efficiency, the dispersing action within a ball mill is directly proportional to the power input to the mill. In turn, the power input is related to the mill radius by Eq. 12. Also for most effective dispersion, the mill must be filled halfway with the ball and mill base charge. Hence the volume of charge being acted on is related to the mill radius by Eq. 19.

$$V\,(\text{ft}^3) = \tfrac{1}{2}\pi R^2 L \tag{19}$$

Division of the power input by the charge volume (Eq. 12 by Eq. 19) yields Eq. 20 for the power input to the mill per unit volume of total charge (ball media and mill base). This ratio provides a relative measure of the dispersion rate that is possible for a ball mill of radius R (half filled with balls; cascade angle 45°).

$$\frac{P}{V}(\text{horsepower/ft}^3) = 6.36 \cdot 10^{-4}\ \rho'(3.3 - 0.3R)\sqrt{R} \tag{20}$$

Since the density ρ' is in units of lb/ft^3, Eq. 20 can also be expressed as Eq. 21.

$$\frac{P}{W}(\text{horsepower/lb}) = 6.36 \cdot 10^{-4}(3.3 - 0.3R)\sqrt{R} \tag{21}$$

Table 19-6 Theoretical Relative Dispersion Times for Ball Mills of Varying Diameter (Eq. 21)

Ball Mill Dimensions		Power Input per Unit Weight of Total Charge (horsepower/lb)	Relative Dispersion Time (ratio)
Radius (ft)	Diameter (ft)		
0.5	1.0	0.00141	1.00
1.0	2.0	0.00191	0.74
1.5	3.0	0.00221	0.64
2.0	4.0	0.00243	0.58
2.5	5.0	0.00256	0.55
3.0	6.0	0.00266	0.54
3.5	7.0	0.00268	0.53
4.0	8.0	0.00268	0.53

In Eq. 21 W refers to the weight of the ball media/mill base mixture in lb. From Eq. 21 there have been calculated and listed in Table 19-6 dispersion efficiencies, and from these the relative dispersion times for ball mills ranging in diameter from 1 to 8 ft. Inspection of Table 19-6 reveals that the reduction in dispersion time is most pronounced in going from small to intermediate-sized mills. Also it can be roughly estimated that, for a given required degree of dispersion, a production mill will grind a mill base in about half the time required by a laboratory ball mill (other things being equal).

These theoretical conclusions are in line with industrial findings, where it is reported that a 4-ft diameter production mill usually grinds in half the time required for a 1-ft diameter laboratory jar (18), or that a 7-ft diameter ball mill produces a given dispersion in less than one-third the time for a $\frac{2}{3}$-ft diameter laboratory mill or in about one-half the time for a 2.5-ft diameter production mill (19). A proposed rule-of-thumb is that the time required to achieve a given degree of dispersion on a production size unit (4 ft diameter and above) is approximately 50 to 60% of that required on a laboratory ball mill (1 ft in diameter or less) (20).

TEMPERATURE INCREASE DURING BALL MILL OPERATION

Dispersion (grinding) in a ball mill generates heat, which can be controlled as necessary by water jacketing around the mill barrel (water introduced through a water joint designed for continuous turning). However, some degree of temperature increase is not detrimental provided that the accompanying viscosity reduction is anticipated and allowed for in scheduling the milling operation.

In general a reduction in viscosity acts favorably to promote the dispersion process. However, the heat generation also creates a pressure buildup that should be released periodically during the grinding operation. Factors affecting temperature buildup are considered in Problem 19-6.

PROBLEM 19-6

Assuming that the heat generated in a ball mill is confined solely to heating the contents of the ball mill (grinding media and mill base), calculate the temperature increase that can be expected for each hour of operation of a 6-ft diameter ball mill half-filled with $\frac{3}{4}$-in. porcelain balls and mill base. Assume an overall specific heat of 0.27 for the ball mill contents.

Solution. Equation 21 gives the horsepower that is required per pound of ball mill content to operate a ball mill of radius R at an optimum rpm value when half-filled with ball media and mill base. By using appropriate conversion factors, this expression can be rewritten in terms of the rate of heat input H/t (Btu/hr) per pound as given by Eq. 22.

$$\frac{H/t}{W}\,(\text{Btu/hr-lb}) = (5.34 - 0.486R)\sqrt{R} \qquad (22)$$

For a 6-ft diameter ball mill ($R = 3.0$ ft), the hourly heat input per pound of content is 6.72 Btu [= (5.34 − 0.486 · 3) · 1.73]. The resulting hourly temperature increase for the ball mill contents is then 25 F° (= 6.72/0.27).

Comment. The hourly temperature rise for a 6-ft diameter mill in actual practice, although considerable, is far less than the calculated 25 F° because of heat loss to the surroundings.

Because of the small size of a laboratory ball mill, the temperature rise is slight. This fact must be taken into account when scaling up from laboratory to production units, since the temperature buildups in the latter can be significantly higher.

OPTIMUM SOLVENT/BINDER RATIO FOR BALL MILL GRINDING VEHICLE (21-29)

An optimum solvent/binder ratio for the ball mill grinding vehicle can be established from a Daniel flow point determination. This procedure is based on the premise that the binder portion of the vehicle is in reality a dispersing agent that acts to deflocculate wetted pigment suspended in solvent. Accordingly, only sufficient binder should be present in the solvent/binder mix to effect a complete particle separation and give a satisfactory flow. More than this amount of binder is excessive and leads to an undesirable increase in viscosity. At the minimum binder level, both the lowest viscosity and the highest pigment volume content consistent with adequate dispersion are simultaneously obtained.

The flow point procedure is carried out by preparing blends of solvent and binder over a range of solvent/binder ratios (say 10/90, 15/85, 20/80, 25/75, 30/70, 40/60, and 50/50). In turn each blend is mixed with a given weight of pigment (usually 20 g) until a characteristic flow and point is reached.

Details of Daniel Flow Point Determination

The solvent/binder mixture is added to the pigment in a 100-ml glass beaker with continuous kneading and rubbing with a sturdy fire-polished glass rod until a heavy, smooth paste is obtained. More of the solvent/binder mixture is then gradually added until the mass can be stirred without significant resistance. At this point the glass rod is quickly lifted out of the beaker, and the runoff material clinging to the rod (which is held vertically) is observed. A characteristic flow end point is reached when a thin, even film of material remains on the rod and when the last few drops, falling at 1- to 2-sec intervals, appear to break off with an elastic snap-back. This flow end point corresponds roughly to an 80-KU viscosity as measured on the Stormer viscometer.

A plot of the amount of each blend required to reach the characteristic flow end point versus the composition of the blend (see Fig. 19-11) gives a U-shaped curve, the minimum point of which corresponds to the required grinding mill base composition.

The mixture at this point is critical and represents a three-way balance in the selection of solvent, binder, and pigment. With this optimum mill base premix, the ball mill will be used most efficiently and will provide the maximum volume output of dispersed pigment. The solvent/binder ratio is also critically balanced in that more solvent will tend to induce flocculation, whereas more binder will increase the premix and mill base viscosity, both undesirable departures from the optimum blend at the U-curve minimum point.

Some difficulties occasionally arise with this simple hand-mixing procedure. For example, hard pigment agglomerates and ultrafine pigments may resist the hand shearing action, as indicated by a gritty consistency. Then a mechanical grind may become necessary. However, the method is generally applicable to most of the pigments in common use by the paint industry. The method is highly practical and represents a proven approach to establishing an optimum premix for charging to a ball mill.

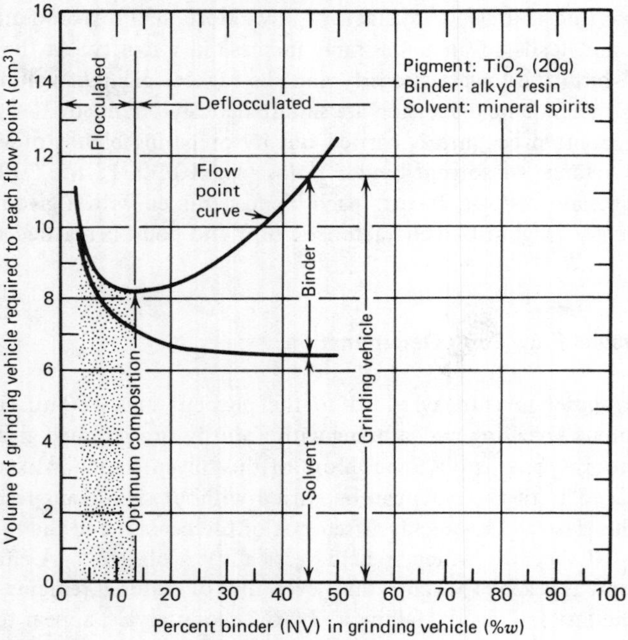

Fig. 19-11 Graph illustrating the results of a typical Daniel flow point determination. Optimum vehicle composition indicated by low point of the flow point *U* curve.

PROBLEM 19-7

Table 19-7 specifies the pigment and the vehicle composition for an olive drab enamel. The medium-oil-length (OL) alkyd contains 50% binder on a volume basis. Flow point determinations for the four pigments specified for the enamel, using mixtures of the medium-OL alkyd and mineral spirits (MS), give the results tabulated in Table 19-8. From these data formulate the premix composition providing optimum dispersing conditions for a conventional ball mill.

Solution. Inspection of the flow point data reveals that 20% binder content is optimum for calcium carbonate and red oxide pigment, whereas 15% binder content is best for chrome yellow and carbon black. The volume and composition of the grinding vehicle required for the quantities of pigment listed in Table 19-7 are computed from these flow point data.

A total of 32.1 gal of grinding vehicle is required, of which 5.69 gal must be binder nonvolatile (NV). The binder NV is furnished by 11.4 gal (= 5.69/0.50) of the medium-oil-length alkyd. At the same time this also provides 5.69 gal

Table 19-7 Composition of Specified Olive Drab Enamel

Pigment Composition (lb)	
Chrome yellow	110
Red oxide	35
Calcium carbonate	145
Lampblack	35
Vehicle Composition (gal)	
Alkyd, medium OL	100
Alkyd, long OL	30
Mineral spirits	60

Table 19-8 Flow Point Determination by Daniel Method (Problem 19-7)

Mixture of Medium-OL Alkyd and MS (cm^3)			Volume of Mixture Required to Reach Flow Point (20 g Pigment) (cm^3)			
Alkyd	MS	Binder in Mixture (%v)	Chrome Yellow	Red Oxide	Calcium Carbonate	Lampblack
60	40	30	10.3	10.8	20.8	49.2
50	50	25	10.1	10.2	18.8	46.8
40	60	20	9.4	<u>9.9</u>	<u>17.6</u>	44.4
30	70	15	<u>9.0</u>	9.9	17.8	<u>42.0</u>
20	80	10	9.9	10.5	18.4	45.4
			Volume of Vehicle per Unit Weight of Pigment at Minimum Flow Point (Underlined Value)			
		cm^3/g:	0.450	0.495	0.88	2.1
		gal/lb:	0.054	0.060	0.105	0.252

Pigment	Required Grinding Vehicle: Volume (gal)	Required Grinding Vehicle: Binder (%)	Required Binder Volume (gal)
Chrome yellow 110 lb · 0.054 gal/lb =	5.9	15	0.89
Red oxide 35 lb · 0.060 gal/lb =	2.1	20	0.42
Calcium carbonate 145 lb · 0.105 gal/lb =	15.3	20	3.06
Lampblack 35 lb · 0.252 gal/lb =	8.8	15	1.32
Total	32.1		5.69

of MS. The remainder of the 32.1 gal of grinding vehicle, equal to 20.7 gal (= 32.1 - 11.4), is provided by the addition of mineral spirits.

An analysis of many flow point determinations reveals that most pigments exhibit optimum grinding properties with vehicles of 20% binder content. Hence, as a shortcut, this one vehicle composition (80% solvent/20% binder) can be scheduled for evaluation with good prospects that a fully satisfactory grind will be obtained. Only if a shortage of solvent prevents such a formulation or seeding-out occurs on reduction with letdown solvent will it be necessary to reformulate with higher binder content. When the solvent content of a paint composition is so low that a grinding vehicle with at least 65% solvent cannot be formulated, it is probable that the use of alternative grinding equipment will be more economical.

Grinds made with mill base vehicles containing a high percentage of solvent (say 85% or more) are commonly called slush grinds.

In general the Daniel flow point method for formulating mill base compositions appears to give reliable results. However, though affording the highest grinding efficiency, the binder concentration may be too low to act as an efficient lubricant for the ball media, and the ball wear may prove intolerable from an economic standpoint. Furthermore, the use of low binder content in the grind may lead to "pigment shock" on subsequent letdown of the mill base dispersion. In practice these undesirable effects may be avoided by using a somewhat higher binder fraction than is called for by the minimum flow point composition and scheduling a longer grinding time (15).

PRACTICAL CONSIDERATIONS

Up to this point the chapter has focused on establishing conditions for maximum dispersion rates. In actual plant operation, however, other factors may also be taken into consideration.

Thus it may prove more economical to use low-density pebbles rather than high-density alumina balls, which, although more efficient, are at the same time more expensive. Or when time is not a critical factor (a ball mill can be run overnight with minimum attention), it may be more economical to schedule a single ball mill run with 30% excess mill base than two concurrent runs, each with 5% excess mill base. It is true that to reach a given degree of dispersion the 30% excess run will take more time than the overall time required for the two more efficient 5% excess runs. However, if the extra time is scheduled for the night period, the operational cost involved may be relatively minor.

The ball mill operator must also keep in mind and possibly anticipate the change in mill base viscosity that will occur because of the nominal rise in temperature experienced by the ball mill during operation (even with water cooling). Also it is possible that a mill base composition, while optimum for ball mill grinding, may be unsatisfactory for subsequent reduction with letdown vehicle. Here a compromise may be necessary to avoid an uncontrollable flocculation due to a mixing of vehicles which are excessively disparate either in their composition or in their temperature, or in both.

One feasible method for avoiding loss of grind on letdown is to add some of the letdown vehicle to the mill after a required dispersion has been obtained and then continue the grind (actually a semimixing operation) for an hour or so. This procedure acts to stabilize the extended mill base to the point where pigment shock-out is no longer a problem. The additional vehicle introduced also helps to drain the mill (22).

Although the ball mill has the advantage of accepting the mill charge ingredients without premixing, this does not rule out the possibility of resorting to premixing to reduce milling time or to secure some other advantage. The use of a relatively homogeneous premix charge has been claimed to save as much as 75 to 80% in milling time (22).

The usual method of discharging mill base is to allow it to drain through a slotted discharge gate with or without the use of suction or pressure. Ball mills are normally designed to withstand maximum pressures of 15 psi (13). Hence pressures used to facilitate mill base discharge should be well within this limit (say ~5 psi). Also the ball diameter should be considerably larger than the slot width to avoid plugging the flow path to the outside.

A good method of cleaning is to charge a small quantity of solvent into the mill and run it for not more than 30 revolutions before dumping. This short cycle can be repeated as necessary. Cleaning for longer time periods results in excessive wear of the grinding media (13).

The objective of this chapter has been to discuss the fundamental principles of ball milling. It is the job of the paint or ink engineer to bring good common sense to bear on any actual ball mill operation and to use the theoretical equations and concepts derived in the chapter as guides for practical plant production runs.

REFERENCES

1. Brown, H. M. "Ball Mills," *Off. Dig.*, **20**, No. 284, 668 (1948).
2. Appell, F. "Optimum Conditions of Operating Ball Mills," *Off. Dig.*, **22**, No. 303, 315 (1950).
3. Bulletin P-290, "Jar, Ball, and Pebble Milling–Theory and Practice," U. S. Stoneware Company, Akron, Ohio, 1962.
4. Paul O. Abbe, Inc., *Optimum Conditions of Operating Ball Mills,* Little Falls, N.J.
5. Monograph Series No. 101, "A General Consideration of Paint Paste Formulation," and Monograph Series No. 102, "A General Review of Grinding Media and Mill Considerations," Patterson Foundry and Machine Company, East Liverpool, Ohio, 1962.

6. Gaynes, N. I., "Ball Mills Are Not Dead," *Am. Paint J.*, Reprint, November 4 (1968).
7. Garlick, O. H., "Ball and Pebbie Mills," *Paint Varn. Prod.*, **49,** 53, March (1959).
8. Baker, C. P. and P. Virtue, "Factors Affecting Economic Optimum Paint Milling Time," *Off. Dig.*, **29,** No. 385, 178. (1957).
9. Redd, O. F., "A Physical Basis for Mill Selection," *Off. Dig.*, **24,** No. 324, 29 (1952).
10. Mountsier, S. R., "Review of Some Mixing and Grinding Equipment," *Off. Dig.*, **23,** No. 315, 233 (1951).
11. Baker, C. P. and J. F. Vozzella, "Effect of Temperature on the Dispersion of Titanium Dioxide in Alkyd Resin/Mineral Spirits by Means of a Pebble Mill," *Off. Dig.*, **21,** No. 294, 433 (1949).
12. Jebens, R. H., "Ball Mill Operation," *Paint Varn. Prod.*, **50,** 53, March (1960).
13. Paul O. Abbe, Inc., *Handbook of Ball Mill and Pebble Mill Operation,* rev. ed., Little Falls, N.J.
14. Epworth Manufacturing Company, *Paint Machinery Handbook*, South Haven, Mich.
15. Nylén, P., and E. Sunderland, *Modern Surface Coatings*, Wiley-Interscience, New York, 1965.
16. Rahter, J. M., "Selecting Grinding Media for Pebble Mills," *Paint Varn. Prod.*, May (1973).
17. Conley, R. F., "Effects of Crystal Structure and Grinder Energy on Fine Grinding of Pigments," *JPT*, **44,** No. 567, 67–84, April (1972).
18. Paul O. Abbe, Inc., *Grindings and Mixings*, **5,** No. 1.
19. Parfitt, G. D., *Dispersion of Powders in Liquids*, 2nd ed., Wiley, New York, 1973.
20. Schaffer, Martin H., "Dispersion and Grinding," Unit 16, Federation of Societies for Paint Technology, Philadelphia, Pa., 1970.
21. Daniel, F. K. and P. Goldman, "Evaluation of Dispersion by a Novel Rheological Method," *Ind. Eng. Chem.*, **18,** 26–31, January 15 (1946).
22. Cooper, E. K., "Formulating for Production," *JPT*, **39,** No. 515, 752–762, December (1967).
23. Ensminger, R. I., "Techniques for Efficient Pigment Dispersion Operations," *Mod. Paint Coatings*, pp. 29–35, May (1975).
24. Daniel, F. K., "A System for Determining the Optimum Grinding Composition of Paints in Ball and Pebble Mills," *Natl. Paint Varn. Lacquer Assoc. Sci. Sect. Circ.* No. 744, 1950.
25. Shurts, R. B., "The Determination of Proper Paint Grinding Formulations and Ball and Pebble Mills," *Natl. Paint Varn. Lacquer Assoc. Sci. Sect. Circ.* No. 745, 1950.
26. Shurts, R. B., "More Production from Your Present Pigment Dispersion Equipment," *Natl. Paint Varn. Lacquer Assoc. Sci. Sect. Circ.* No. 753, 1951.
27. Baker, C. P. and J. F. Vozzella, "The Roll Mill, Pebble Mill, and Kneader as Pigment Dispersion Equipment," *Off. Dig.*, **23,** No. 319, 467 (1951).
28. Doorgeest, T., "Selection and Optimum Use of Dispersion Equipment," *Paint Manuf.*, pp. 31–38, May (1968).
29. Askew, F. A., *Printing Ink Manual*, Society of British Printing Ink Manufacturers, Heffer & Sons, Ltd., Cambridge, England, 1969.

20 *Modified Ball Mills (Attritors and Vibration Mills)*

Conventional ball milling based on the cascading of the ball media in a horizontally mounted rotating drum is necessarily restricted to batch processing and a processing rate dependent on the ball mill radius (see Chapter 19). In recent years modified mill designs that rely on alternative methods for tumbling the balls about (agitation) and that free the milling process from one or both of these restrictions have been developed. Some of these new types of mill equipment are briefly reviewed here. The mill base formulation principles outlined in Chapter 19 for conventional ball mills still apply. However, judicious adaptation will be required to meet the altered processing conditions.

ATTRITOR MILLS (SZEGVARI)

The attritor consists essentially of an upright, stationary, cylindrical grinding tank fitted with a centralized, vertical, rotating shaft to which is attached (at right angles) a distribution of metallic fingers. These fingers protrude into the ball media/mill base mixture which fills the attritor tank during the milling operation. The rotation of these fingers, which sweep through the tank contents, vigorously agitates the ball charge, which in turn provides the necessary shear and impact to effect the dispersion of the pigment into the mill base vehicle. Starting with this basic concept, the process has been designed for either batch or continuous operation (1).

Continuous attritors, which are relatively narrow in diameter, are supplied in two basic types, type C to handle medium to high viscosities (agitator rotation between 100 and 500 rpm) and type H to handle low to medium viscosities (agitator rotation between 500 and 1400 rpm). Each type is made in a selection of sizes ranging from laboratory models (1.5 horsepower) to large production units nearly 12 ft high (75 to 100 horsepower).

The grinding media are selected to suit the mill base to be processed and are generally ceramic or steel balls with a $\frac{1}{4}$-in. diameter or less (C type) or a $\frac{3}{16}$-in. diameter or less (H type).

From data abstracted from one study on the dispersion of a red organic pigment in an alkyd vehicle using a small production attritor (2) this author developed Eq. 1, which approximately relates ball diameter d (in.) and attritor agitator rotational speed n (rpm) to the time t required to reach a given degree of dispersion.

$$t = \frac{k \cdot d}{\sqrt{n}} \tag{1}$$

where t = time to reach a given degree of dispersion
k = constant for the particular attritor being used and the mill base being processed
d = ball media diameter (range only from $\frac{1}{8}$ to $\frac{1}{2}$ in.)
n = agitator speed (rpm)

From this expression it is seen that, by decreasing the ball media diameter from $\frac{1}{2}$ to $\frac{1}{8}$-in. (but not beyond) and/or by increasing the impeller rotational speed, faster dispersion times are attained. However, below the $\frac{1}{8}$-in. ball size the efficiency trend reverses, and dispersion time became intolerably long.

The processing rate is controlled by a pumping system which meters with accuracy the mill base slurry that is continuously introduced into the bottom of the attritor. The fineness of grind obtained on exiting at the top is also a function of the dwell time within the attritor (the time between entering and leaving the grinding tank).

The attritor type of agitation is claimed to accelerate the dispersion process manyfold (10X to 100X) over a comparable conventional ball mill operation.

Batch-type attritors are available in sizes ranging from a research model (750 cm^3; $\frac{1}{2}$ horsepower) to a large production unit (300 gal; 20 to 40 horsepower). These batch units of wider tank diameter are generally equipped with a pumping system that permits continuous circulation of the mill base through the attritor during the dispersion operation. The rotation of the batch agitator unit is slower than that of the continuous unit, since the metallic fingers extend further into the ball media/mill base mixture in the wider batch tank.

The high dispersion efficiency of the attritor is due to the fact that the rotation of the agitator fingers maintains practically all of the ball media in constant motion, in contrast to the conventional ball mill, where most of the balls are largely static during the rotation of the mill. Since no limitation is placed on the angular velocity of the revolving shaft (with its protruding fingers), the attritor is capable of handling much higher viscosities than can a conventional ball mill. Water jacketing is provided with production attritor units.

When suitable premixing tanks are used, a 5-gal attritor can readily match the output of a conventional ball mill of 50-gal working capacity. Both are capable of processing difficult-to-disperse pigments.

VIBRATION MILLS

High-Speed (Quickee) Laboratory Ball Mill

The high-speed (Quickee) ball mill is an adaptation of a commercial paint shaker to a laboratory grinding operation. The ball mill consists of a stainless steel cylinder (capacity from 200 to 800 cm^3) that is clamped to the shaking arm of a commercial (Red Devil) paint shaker (3). Normally this paint shaker is used to redisperse the pigment settlement that occurs in cans of over-the-counter paints during storage in stores. In adapting the paint shaker to a grinding mill, the container has been made more rugged and use is made of grinding media ($\frac{1}{8}$-in. steel balls) (4).

The mill base to be dispersed is charged to the container (vehicle first and pigment second) and thoroughly stirred with a spatula to break up any lumps larger than $\frac{1}{4}$ in. The $\frac{1}{8}$-in. steel balls are then added, and the entire mass is stirred to uniformity. The closed system is clamped to the arm of the paint shaker, and the shaker started. Processing is effected by the vibration of the shaker, which shakes the grinding media about to provide the grinding action arising from the innumerable ball collisions. Under correct conditions an easy-to-disperse pigment should be reduced to less than a 5-μ grind in 3 to 4 min; a hard-to-grind pigment, to this same fineness in 30 to 60 min (4).

It has been found by experience that excellent results are produced when the grinding media and mill base occupy 70% of the container volume (30% solid media; 40% mill base). Steel balls $\frac{1}{8}$ in. in diameter are the preferred grinding media, since $\frac{1}{4}$-in. balls process too slowly (the time is 3X as long) and $\frac{1}{16}$-in. balls, although affording faster results, are difficult to clean and handle.

Good correlation between the high-speed laboratory mill and factory ball mills is obtained. Thus a grinding time of 1.0 min in the high-speed laboratory unit provides substantially the same grinding effect as 1.0 hr in a production ball mill (20% solid media consisting of $\frac{5}{8}$-in. steel balls; 30% mill base). Minor adjustments of the mill base viscosity may be required because of the smaller mass of the $\frac{1}{8}$-in. steel balls (slightly less viscosity than the plant mill bases).

The consistency of the mill base should be such that the operator hears the sound made by the balls bouncing briskly off the sides of the container.

Commercial or Production Vibration Mills

Although production vibration mills using ball media are widely employed for the comminution of dry or wetted materials by impact crushing in, for example, the refractory and ceramic industries, there apparently has been almost no application of the vibratory principle for the commercial manufacture of paints and inks.

Commercial vibratory mills are capable of exerting tremendous impact forces. The collisions provided by the vibrating ball media can reduce such materials as ores (bauxite, alumina, columbite), clays, talc, graphite, frits, glazes, and even abrasives from gravel-sized feed to a fineness under 10 μ. To date the major utilization, by far, of these vibratory mill has been for the pulverizing and reduction of coarse particulate matter to ultra fine sizes (as required). However, certain models have recently been designed to provide submicron wet grinding of pigment agglomerates. These vibration mills are stated to be ideally suited for the manufacture of coatings and inks. However, little work has been carried out to establish any background for their use. Presumably, with time, more attention will be directed to the investigation of vibration milling using ball media for paint and ink preparation. The vibratory equipment is available and awaits further study by the engineer.

Palla Vibration Mill (Continuous). The Palla vibration ball mill is a scaled-up adaptation of the Quickee ball mill principle. It consists mainly of two horizontal, equidirectional grinding tubes braced rigidly, one above the other, in two webplate frames. These grinding tubes are thrown into circular vibrations by an unbalanced weight drive mounted centrally between them. Flexible hoses connect the vibrating tubes with stationary containers for introducing the feed material and discharging the finished product. At the discharge end of each tube is a slotted plate that retains the grinding media but allows the dispersed material to pass through (5).

The largest Palla vibratory mill (125 horsepower demand) can operate with a vibratory acceleration seven times that of gravity if necessary. The entire vibrating system (loaded mill tubes, crossplates, and vibrating drives) is supported on rubber buffer pads.

The grinding cylinders range in diameter from about 8 in. (~4 ft long) for the smallest unit to about 25 in. (~14 ft long) for the largest unit.

Although this author knows of no coatings manufacturer presently using the Palla mill for the dispersion of pigments in vehicles, it is widely employed for the mechanical grinding of minerals. Presumably the potential use of this vibratory mill in the coatings and ink industries is yet to be exploited.

SWECO Vibro-Energy Mill (Batch Type). The SWECO mill is characterized by a three-dimensional vibratory action that is communicated to a ball charge in a chamber mounted on high-tensile steel springs. The claim is made that a SWECO mill can grind (comminute) a greater variety of materials to a finer degree (0.5 μ or less) than can any other mill and to an exact size distribution as required. These mills are furnished for wet grinding in working capacities ranging from 1.0 pt to 182 gal. Since the major portion of the energy input into the mill is directed to grinding (there is little waste motion), they are far more efficient than conventional ball mills. The grinding chamber is in the form of a vertical cylinder or annulus, and usually the hard grinding media are $\frac{1}{2}$-in. sintered alumina cylinders. One 3-gal model (D3 mill) is specifically manufactured for the deagglomeration and dispersion of pigments in coatings and inks (6).

Laboratory runs made on a number of coating systems ranging from enamels to primers demonstrate that the time to reach a Hegman 7 grind is from 4 to 12 hr in this vibratory mill (Model M-18 SWECO Vibro-Energy Mill) versus 48 to 96 hr in a conventional ball or pebble mill (7). Although there appears to be no significant use of the SWECO mill in the coatings and ink industries to date, it is technically sound and certainly merits further investigation for commercial use.

REFERENCES

1. Company brochures, "The Attritor" and "Continuous Attritors," Union Process, Inc., 1925 Akron Peninsula Road, Akron, Ohio 44313.
2. Doorgeest, T., "Selection and Optimum Use of Dispersion Equipment," *Paint Manuf.*, pp. 31–38, May (1968).
3. Brochure B/4: Dispersion, "MG-8600 Quickee Ball Mill" and "MG-8600-B Red Devil Paint Conditioner," Gardner Laboratory, Inc., Bethesda, Md. 20014.
4. Orwig, B. R., "The Quickie: A High-Speed Ball Mill for the Paint Laboratory," *Off. Dig.*, pp. 830–836, September (1954).
5. Company brochure, "PALLA Vibration Mills," Humboldt Wedag, 1 Huntington Quadrangle, Huntington Station, N.Y. 11746.
6. Company brochures, "SWECO Vibro-Energy Mills" and "SWECO Laboratory Report on Fine Grinding," SWECO, Inc., 6033 East Bandini Boulevard, Los Angeles, Calif. 90051.
7. Sjogren, J. K. and J. J. Kasevniak, "Paint Formulation Conducted in the Vibro-Energy Mill," Laboratory Report, Southwestern Engineering Company, Vibro-Equipment Division, 4800 Sante Fe Avenue, Los Angeles, Calif. 1961.

21 *Sand, Bead, and Shot Mills*

Dispersion of pigments in vehicles using rounded sand particles as the grinding media has been practiced since 1952. Although the original sand mill called for a 24- to 40-mesh Ottawa sand, it was not long before a variety of competing synthetic media became available (glass beads, ceramic beads, metallic shot) which, although more expensive, offered improved processing characteristics.

Sand grinding can be thought of as an extension of the ball mill principle wherein use is made of tiny balls, beads, or shot. Since the largest beads that are used in sand or bead mills closely approach the dimensions of the smallest balls used in ball mills, there is really no sharp differentiation between bead and ball mills in the region where the two tend to overlap. However, the Ottawa sand that is commonly specified for sand mills is a 20- to 30-mesh grade corresponding to a particle diameter of about 0.7 mm or 0.028 in ($\sim\frac{1}{32}$ in). Synthetic bead media for bead mills are normally supplied in a range from 0.7 to 3.0 mm ($\sim\frac{1}{32}$ to $\sim\frac{1}{8}$ in). Hence sand or bead mills will be defined as dispersion units using spherical grinding media in this general range. Ball mills, on the other hand, will be considered as dispersion units that normally use grinding media above this range (above $\frac{1}{8}$-in diameter). Actually this distinction is somewhat artificial in that some mills are designed to operate with media diameters over a wide range and may be considered as either bead or ball mills, depending on the size of the media used in the grinding operation.

DESCRIPTION OF THE SAND GRINDING PROCESS

Basically the sand grinding process (a pigment dispersion process) consists in pumping a homogeneous mixture of a pigment/vehicle slurry (the mill base) through a cylindrical bank of sand which is being subjected to intense agitation (see Fig. 21-1). During passage upward through the agitated sand zone, the mill base is caught and ground between the sand particles—a strong shearing action which effects the dispersion of the pigment into the vehicle. On emerging from the active sand zone, the dispersed mill base overflows through an exit screen sized to permit free flowthrough of the pigment dispersion while holding back the sand particles.

The agitation of the sand particles is produced by flat disk impellers which revolve at high rates of speed (peripheral velocities on the order of 2,000 ft/min) within the sand grinder. Sand particles and mill base adjacent to the impeller surfaces pick up the impeller motion through viscous resistance and as a result are slung outward against the confining walls of the sand grinder. An approximate flow pattern for the overall turbulent flow that ensues is depicted in Fig. 21-1 and may be grossly described as a rolling double-doughnut motion which provides an excellent dispersing effect, especially in the regions adjacent to the impeller surfaces and between the outside edges of the impeller and the container walls.

At this point it is instructive to calculate the magnitude of the force acting on a particle of sand at the impeller periphery. Assume that a sand particle located at the outside edge of the impeller disk moves at the same velocity as the impeller velocity v. If the impeller radius is r, the centrifugal force F, is equal to mv^2/r. Compared with the gravitational force acting on the particle, the centrifugal force is quite large, as can be calculated from Eq. 1 (g is the acceleration of gravity).

$$\frac{\text{Centrifugal force}}{\text{Gravitational force}} = \frac{mv^2/r}{mg} = \frac{v^2}{rg} \qquad (1)$$

For example, let the impeller peripheral velocity be 2000 ft/min and the impeller radius be 4.0 in. Then the centrifugal force acting on the sand particle is equal to 104 times its own weight [$= (2000/60)^2/(4/12) \cdot 32.2$]. It is this forceful action on the sand particle which compensates for the latter's small size and leads to the generation of strong shearing forces within the sand mass.

Although the sand grinder imparts a strong and highly efficient dispersing action, it confers only a mild disintegrating action.

Type of Sand

In the original patent describing the sand grinder, issued in 1952 (U.S. 2,581,414), a 20- to 40-mesh Ottawa-type sand was specified as the grinding medium. This corresponds to a particle diameter range of about 0.024 ± 0.008 in. It is notable that subsequent evaluation of alternative grinding media has failed to uncover

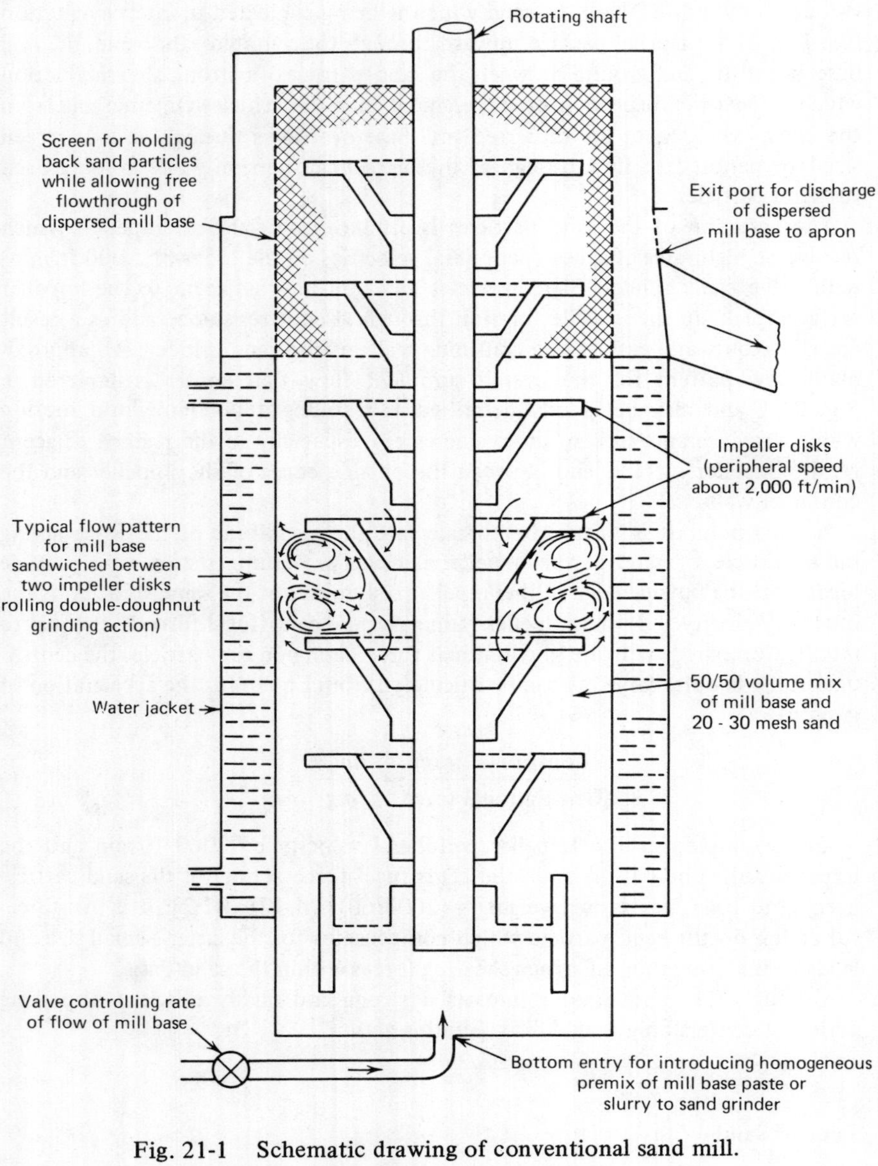

Fig. 21-1 Schematic drawing of conventional sand mill.

any more suitable *natural* material. Hence from the standpoint of cost, availability, uniformity, size, density, abrasion resistance, grinding efficiency, and acceptable suspension this grinding medium as originally selected is still commonly specified for use.

However, synthetic spherical beads (glass, ceramic, steel) that greatly widen the choice of potential grinding media are also available on the market.

SELECTION OF BEAD MEDIA

Aside from cost, three major factors determine the selection of bead media–size, density, and chemical composition. In Fig. 21-2 are plotted values for representative commercial bead media in terms of bead diameter, weight, and density. Inspection shows that a considerable range is offered for all three variables. The bead weight can be estimated by interpolating between the weight lines slanting slightly downward from left to right.

Bead Size

Every sand and ball mill has a screening mechanism for separating the grinding media from the mill base at the exit from the grinding chamber. This screening may take the form of slots, holes, or other orificial configurations. The general practice is to select a bead size such that its diameter exceeds the maximum dimension of the opening by at least 50%, since otherwise plugging of the screening device soon becomes troublesome. For example, a commonly used nickel screen, with a width of 0.014 in, normally calls for a media with a diameter of 0.021 in (1).

Two other factors also should be considered in bead selection. In the first place, a bead size of specified diameter actually represents a distribution of diameters clustering around the specified diameter. Hence the smallest diameter in this distribution may also be a secondary controlling dimension. In the second place, beads wear, and since their diameters become reduced with time, the beads must eventually be replaced or a screen with smaller openings must be introduced. Fortunately, it takes considerable wear to reduce the volume of a bead to the extent where its smaller diameter becomes of critical concern. For example, if an initial bead diameter were 3 times the maximum screen opening dimension, nearly 90% (87.5%) of the bead volume would have to be worn away before reaching the point where the diameter was 1.5 times the maximum screen opening dimension. An intolerably reduced bead size is normally manifested by the onset of excessive screen plugging.

Smaller beads have the advantage of providing more points of grinding contact. If a random packing factor of $\phi = 0.639$ is assumed for beads in a given volume

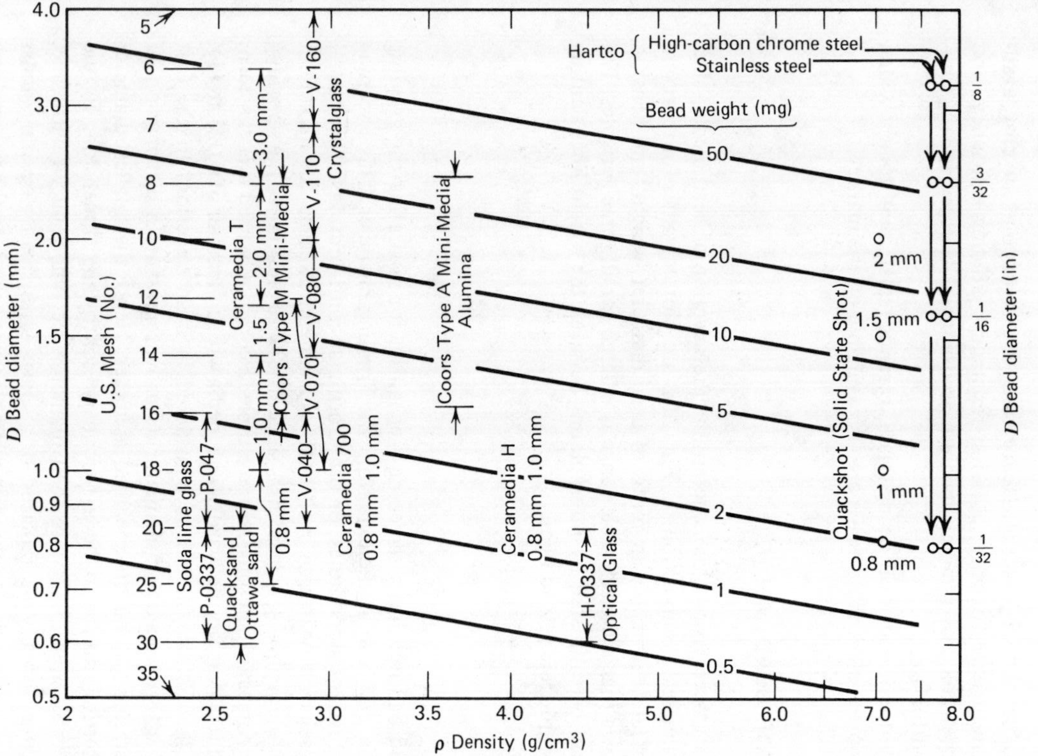

Fig. 21-2 Plot of data for some commercial sand, bead, and shot grinding media in terms of the diameter, density, and weight of their respective spherical particles. Commercial suppliers: Ceramedia, Quacksand, Quackshot (Quackenbush Company, Box 607, Palatine, IL 60067); Mini-Media (Coors Porcelain Company, Golden, CO 80401); H-, P-, and V-series of Ballotini glass spheres (Potters Industries, Inc., 377 Route 17, Hasbrouck Heights, NJ 07604); Hartco steel balls (Morehouse Industries, Inc., 1600 Commonwealth Avenue, P.O. Box 3620, Fullerton, CA 92633).

(with approximately 4.6 contacts per bead), the number of bead contacts that are provided when the beads are packed in 1.0 in.^{3} of space is given by the expression $5.6/D^3$, where D is the bead diameter expressed in inches. Thus the number of contacts for $\frac{1}{8}$-in. diameter beads in 1.0 in.^{3} is 2900, for $\frac{1}{16}$-in. diameter beads the number of contacts is 23,000, and for $\frac{1}{32}$-in. diameter beads the number of contacts is 180,000. Obviously, it is best to use the smallest possible bead size provided that (*a*) it supplies the dispersive energy required for the grinding action and yet (*b*) is not so small as to plug the screen or unduly retard the screening of the bead media from the mill base at the screen surface.

Present evidence indicates that mixing bead sizes offers no special advantage; rather, beads should be graded as closely as possible to the optimum size for the particular screen opening (1).

Bead Density

A low bead density, such as is provided by sand particles, has the advantage of minimizing the weight of the bead media/mill base mixture in the grinding chamber. This is beneficial, since dense mixtures induce bead abrasion and mill wear. A low bead density also reduces the tendency for the bead media to settle and pack at the bottom of the mill. It should be noted that these considerations apply primarily to conventional sand or bead mills (vertical design), since the newer mills with a horizontal grinding chamber are quite free from these restrictions (2, 3).

Although a low-density bead has the advantage of less weight to handle and process, it fails to provide the more energetic dispersing action provided by higher density beads. Hence high-density bead media are generally employed to process difficult-to-disperse pigments and/or large agglomerates. With dense media it is almost always necessary to match the high density with a higher mill base viscosity; otherwise the dense beads are subject to packing, which in turn leads to excessive wear and impaired dispersion. Since the synthetic denser beads are more costly than the less dense natural sand media, it follows that they are generally reserved for particularly difficult grinding operations (refractory pigments) and/or for viscous compositions. For example, steel shot is commonly employed for dispersing carbon black in very viscous ink pastes. Here any discoloration or iron contamination is not objectionable (4).

Chemical Composition

All bead media, of whatever composition, are extremely resistant to crushing (1). Therefore the stresses developed during a normal grinding operation are incapable of breaking them apart. The presence of any broken beads in a mill base is usually indicative of some malfunctioning or break down of the mill itself. Wearing of the impeller disk to the point where deep grooves are formed can also sometimes fix beads in a position where they can be fractured (impeller disk replacement is indicated).

Sand and the synthetic grinding media (mostly glass beads) that fall in the lowest density range are relatively clear products. Since they are also not very abrasive, they do not cause excessive mill wear and they normally yield invisible and nonabrasive wear products. Beads in a somewhat higher density class, ranging from high-strength glass beads (~ 2.75 g/cm^3) to ceramic mullite pellets (~ 2.85 g/cm^3), exhibit only moderate wear. The wear products of the mullite are opaque, whereas those of the glass beads are invisible.

Still higher density bead media include alumina oxide (~ 3.6 to 3.8 g/cm^3) and zirconium oxide (~ 5 g/cm^3) pellets (both of which are very abrasive) and high-density glasses (~ 3.8 to 4.5 g/cm^3). Of the three, only the glass beads provide invisible wear products.

The highest density media are metallic spheres (steel shot, chrome-iron spheres). Although only moderately abrasive, they tend to introduce coloration (gray-off) of the mill base.

Sand is generally more resistant to wear than glass grinding media. Most glass beads have about the same hardness and wear rate.

From the foregoing discussion it is concluded that the grinding medium with the lowest particle diameter, the lowest density, and the lowest cost that does an adequate milling job should be given first consideration. Secondary factors affecting the choice are the closeness of the bead grading to the specified media size, the conformity of the bead to true sphericity, the abrasiveness of the media and its wear products, the expected wear life, and the visual appearance of the wear products.

IMPELLER UNIT

Flat annular disks (steel of 40 Rockwell C hardness or nylon) are the basic impellers for the sand grinder. Bars, blades, and protruding rods are not suitable, since they tend to wear excessively when the sand grinder is operating at optimum efficiency from a dispersion standpoint.

VOLUME RATIO OF SOLID SAND PARTICLES TO MILL BASE

A 1 : 1 volume ratio of solid sand particles to mill base, corresponding to a solid sand volume fraction of 0.50, is considered optimum for sand grinding. If the average sand particle has a diameter of 0.028 in. and the sand volume fraction v is 0.50, the average standoff distance s between two sand particles is 60 μ. This is calculated from Eq. 5 based on the following derivation.

Consider a packing of spheres having an average radius r in a cubical space L^3, as shown in Fig. 21-3. In the schematic illustration the front spheres are shown as uniformly spaced, but they could just as well be randomly spaced. Let the average standoff distance between two spheres be s (the distance from the outside face of one sphere to the outside face of an adjacent sphere). The average distance between the centers of two spheres is then $s + 2r$, as shown in Fig. 21-3. Let V be the volume of all the spheres packed in the cubical space L^3. Then the ball fractional volume is given by Eq. 2.

$$v = \frac{V}{L^3} \tag{2}$$

Consider next two cubical volumes L_1^3 and L_2^3 that contain the same number of spheres (the same volume V). In the larger cubical space the spheres are of necessity more widely separated. In fact, in enlarging the cubical space, the center-to-center separation distance between two spheres automatically keeps pace with the increase in any other linear dimension of the cube (say a side L). Since $L = \sqrt[3]{V/v}$ from Eq. 2, there results Eq. 3 showing the proportionality between the two linear dimensions for two cubical volumes of ball volume fractions v_1 and v_2.

$$\frac{s_2 + 2r}{s_1 + 2r} = \frac{\sqrt[3]{V/v_2}}{\sqrt[3]{V/v_1}} \tag{3}$$

Since the volume of spheres contained in either cubical volume is V, Eq. 3 reduces to Eq. 4.

$$\frac{s_2 + 2r}{s_1 + 2r} = \sqrt[3]{v_1/v_2} \tag{4}$$

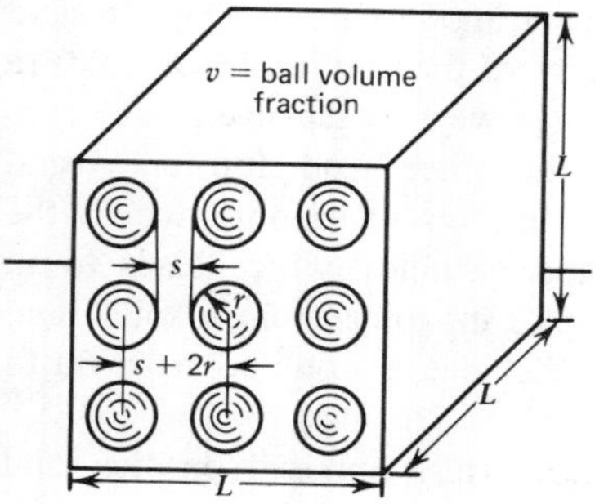

Fig. 21-3 Schematic drawing illustrating packing of spheres in cubical volume and applicable spacing dimensions.

The packing factor for most sand and bead media can be taken as $\phi = 0.639$, corresponding to a condition where the balls are in contact ($s = 0$). Substituting this value for s_1 in Eq. 4 and solving for s_2 (or simply s) yields Eq. 5.

$$\frac{s}{r} = 2(\sqrt[3]{0.639/v} - 1) \tag{5}$$

To help in visualizing the significance of this equation the standoff distance versus the bead volume fraction for beads for varying radii is plotted in Fig. 21-4.

PROBLEM 21-1

Calculate the standoff distance for Ottawa sand when mixed with mill base at an optimum solid sand volume fraction of 0.50. Assume the average diameter of the rounded sand particles to be 0.028 in.

Solution *Numerical:* Substitute the given data in Eq. 5, and solve for s.

$$s = \frac{0.028}{2} \cdot 2(\sqrt[3]{0.639/0.5} - 1) = 0.00238 \text{ in.} \ (60\ \mu)$$

Graphical: Read off the value 60 μ on the plot of Fig. 21-4 for Ottawa sand at a v value of 0.50.

From practical experience an average standoff distance of 60 μ as calculated in Problem 21-1 has been found to provide effective dispersion conditions. At a higher sand fraction the sand/mill base mixture starts to show signs of dilatancy, since the more tightly packed sand particles must deviate more sharply from straight-line motion in order to pass each other as they are whirled about by the impeller rotation. For example, at a solid sand fractional volume of 0.60 (see Fig. 21-4), the standoff distance is 15 μ, representing a tremendous reduction in sand particle maneuverability which in turn result in a jamming rather than a strong shearing action. Hence, at practical operating speeds, too high a sand fraction tends to result in a "freezing" or rigidity of the sand/mill base mixture with attendant slippage at the disk surfaces (high wear), excessive power consumption, and poor dispersion efficiency. On the other hand, too low a sand fraction also results in a marked reduction in milling efficiency, due to the increased gap separation between adjacent sand particles, which leads to an intolerable loss in shearing action. For example, in going from a solid sand volume fraction of 0.50 to 0.40 the standoff distance is doubled from 60 to 120 μ (see Fig. 21-4).

It has been estimated that the average shear stress imposed on the sand

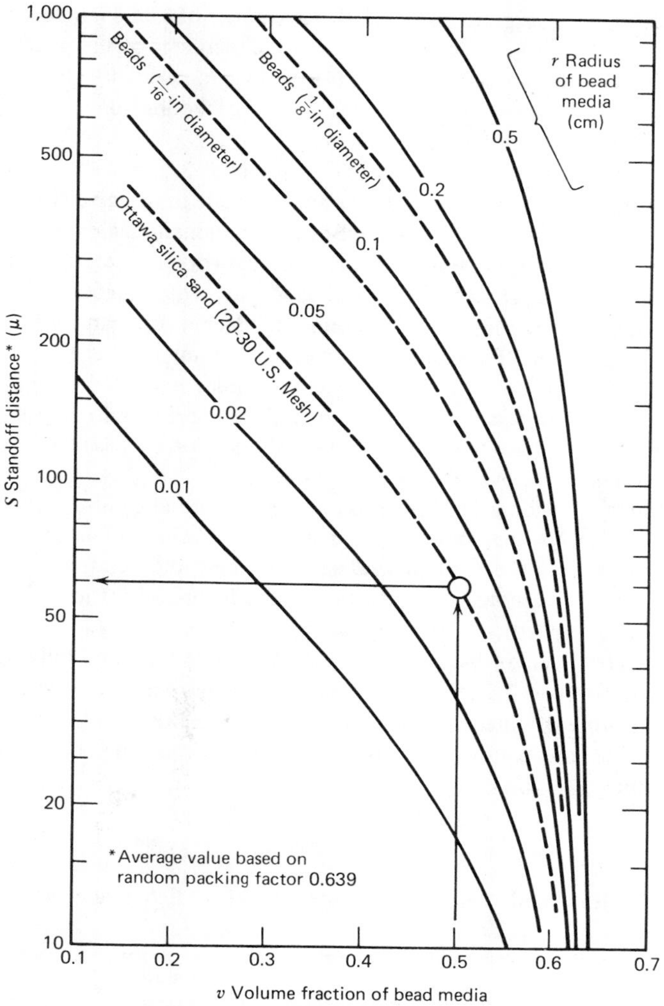

Fig. 21-4 Graph giving standoff distance s as a function of bead radius r and volume fraction v of the bead media, based on an assumed packing factor of 0.639.

particles at the rotating disk surface during normal sand grinding conditions is on the order of 30,000 dynes/cm^2. In turn, the stress exerted by a sand particle on a pigment agglomerate is roughly proportional to the ratio of the sand particle cross-sectional area to the agglomerate cross-sectional area. Hence smaller agglomerates of pigment that are caught between sand particles are more strongly stressed than larger agglomerates. If the diameter of the average sand particle (in the 20- to 30-mesh range) is 700 μ and the diameter of the pigment agglomerate is 7 μ, the stress exerted on this pigment agglomerate by the sand particles is roughly 10,000 times ($10{,}000 = 700^2/7^2$) that acting on the sand particles themselves because of the impeller rotation.

This stress multiplication is sufficient actually to fracture friable pigment particles in the 5- to 10-μ range. It is also sufficient to disperse strongly bonded agglomerates in the 20- to 80-μ range. For larger pigment agglomerates, however, where the stress multiplication becomes minimal (or even less than 1 when the pigment agglomerate is larger than the sand particle), acceptable dispersion is markedly delayed or may never be achieved. Hence the sand grinder is most effective for providing dispersions of small to medium-sized agglomerates. From this it follows that the efficiency of a sand grinder is related to the initial size of the particle agglomerates present in the mill base premix. For optimum operation, then, the mill base fed to the sand grinder should be reasonably well mixed and free from large or oversized agglomerates.

A secondary consideration concerns the possible presence of ultrafine pigments (say carbon black) which absorb large quantities of vehicle. The adsorbed vehicle cushions the stressing action of the sand particles and retards the overall dispersing action. In compensating for such a condition, adjustment of the vehicle viscosity and composition may be quite critical.

The variables thus far discussed, that are applicable to a sand mill, as tabulated in the top section of Table 21-1, are seen to fall within a very narrow range of values for optimum sand grinder operation. On the other hand, the variables of viscosity, mill base composition, processing time, and temperature, considered next, are not so readily bracketed.

Table 21-1 Range of Major Variables Controlling Efficiency of Sand Grinder Operation

Variable	Value
Grinding medium	20–30-mesh Ottawa sand
Type of impeller	Annular disk
Impeller velocity	Peripheral velocity of 2000 ± 100 ft/min
Mill base/sand ratio	Equal volumes of mill base and sand particles
Mill base premix	Homogeneous, finely mixed, free from large or oversized agglomerates
Mill base viscosity	Normally 3–15 poises (60–95KU) at room temperature
Temperature	110–150 F
Mill base vehicle NV	15–35%
Mill base pigment percentage (by weight)	From 8–20% for carbon black and certain color pigments up to 60–70% for TiO_2 and most inert, coarse extender pigments

MILL BASE FORMULATION

The Daniel flow point procedure for establishing an optimum mill base composition for ball mills, discussed in Chapter 19, applies equally well to sand mill grinding. Hence a mill base suitable for ball milling is usually suitable also for sand mills. In general the maximum amount of solvent that is commensurate with maintaining the stability of the dispersion on letdown to the final coating should be used in the mill base. Not only does the solvent rapidly wet out the pigment but in addition the lower solvent viscosity permits a desirable high loading of pigment in the mill base composition.

The critical effect of the mill base composition on production rate is illustrated in Fig. 21-5 for the dispersion of a titanium dioxide pigment in an alkyd vehicle.

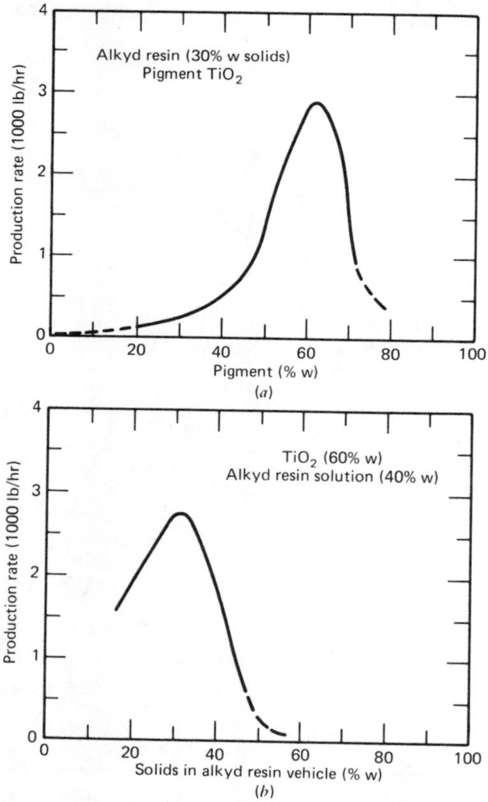

Fig. 21-5 (*a*) Plot of sand mill production rate versus percent weight of titanium dioxide pigment in a mill base formulated with a 30%w solids alkyd resin vehicle. (*b*) Plot of sand mill production rate versus percent solids (nonvolatile) in an alkyd resin vehicle based on a fixed 60%w of titanium dioxide in the mill base composition.

In Fig. 21-5*a* is plotted the production rate for mill bases formulated with varying contents of titanium dioxide pigment in an alkyd resin vehicle of 30%w nonvolatile content. Note that the production rate peaks at about 60%w pigment content. In a similar manner, when the titanium dioxide pigment content is fixed at 60%w and the percent solids (nonvolatile) of the alkyd resin solution is varied, as shown in Fig. 21-5*b*, a similar peak in production rate is exhibited at 31%w alkyd solids content (5). It is obvious that small percentage deviations from optimum mill base composition result in drastic reductions in acceptable sandmill throughput rates.

Figure 21-6 gives recommended formulation regions for mill bases intended for sand mill processing (6). Inspection shows that the nonvolatile content (NV) of the mill base vehicle ranges from 15 to 35%w, with the lower limit applying to easily dispersed pigments and the upper limit to difficult-to-disperse pigments. From this graph tentative mill base compositions for experimental

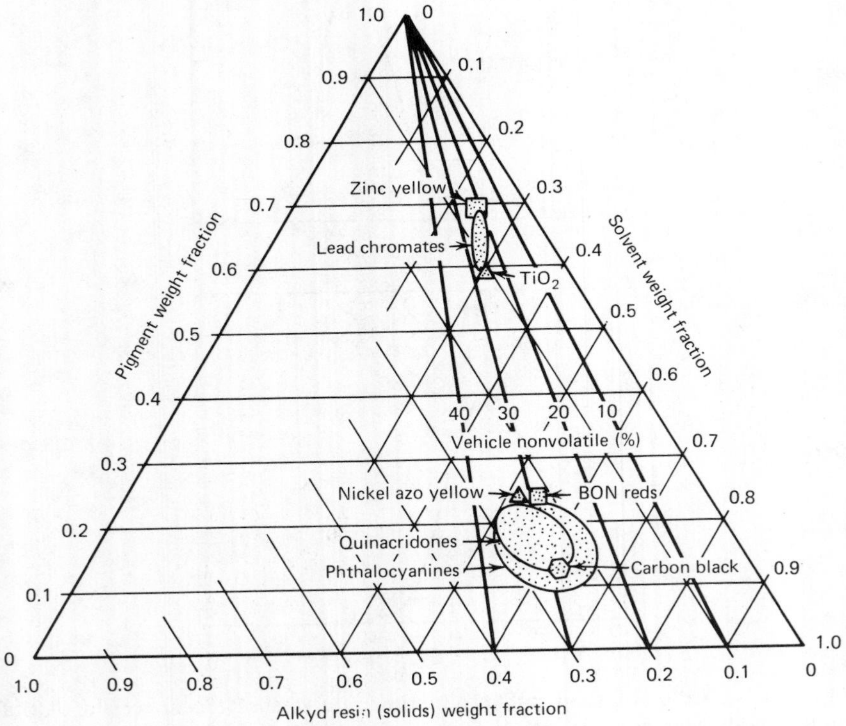

Fig. 21-6 Triangular graph giving practical formulating regions for mill base compositions to be charged to a sand or bead mill. The weight compositions are formulated with an alkyd vehicle.

evaluation can be formulated with reasonable assurance of approximating acceptable compositions for sand mill use. Once formulated, a mill base should be checked to determine whether its viscosity falls in an acceptable range (say from 3 to 12 poises or 60 to 90 KU), and minor adjustment made as necessary. Note that the organic color pigments and carbon black (high-oil-absorption pigments) tend to cluster in a pigment weight fraction region ranging from 0.08 to 0.22, as opposed to the inorganic pigments (low- to moderate-oil-absorption pigments), which tend to cluster in a pigment weight fraction range from 0.58 to 0.70.

When density is removed as a contributing formulating factor and the mill base compositions are plotted on a basis of fractional volume, the differentiation between high-oil-absorption pigments (organic colors and carbon black) and low-oil-absorption pigments (inorganic pigments) is still exhibited, although to a lesser extent, as shown in Fig. 21-7*a*. Changing to another vehicle (from alkyd

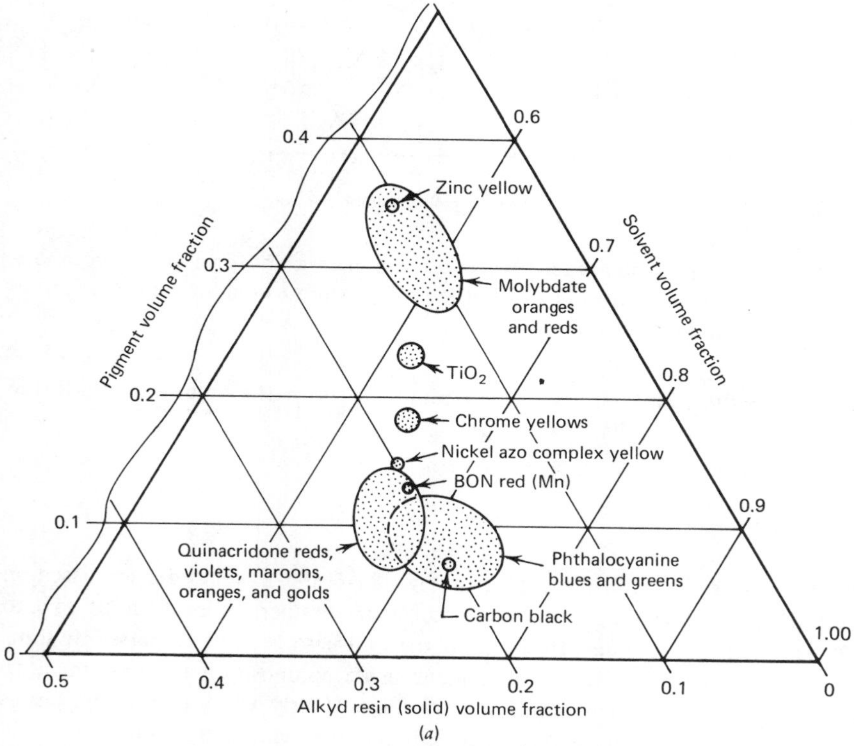

Fig. 21-7 (*a*) Triangular graph giving practical formulating regions for mill base compositions to be charged to a sand or bead mill. The volume compositions are formulated with an alkyd vehicle.

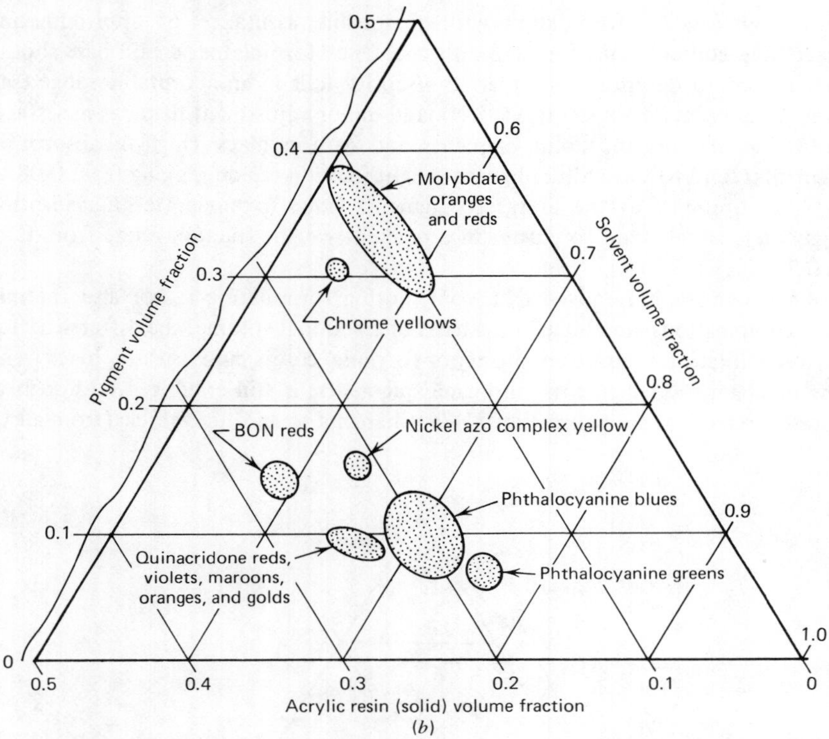

Fig. 21-7 (*b*) Triangular graph giving formulating regions for mill base compositions to be charged to a sand or bead mill. The volume compositions are formulated with an acrylic vehicle.

resin to acrylic resin) results in only minor shifts in the volume composition variables, as shown in Fig. 21-7*b* (6).

Effect of Temperature

The mill bases shown in the graphs of Figs. 21-6 and 21-7*a* and *b* are based on conventional vehicles and mill base processing temperatures of about 110 to 120 F in the sand grinder. However, if the mill base is to be dispersed at some lower or higher temperature, adjustment in its composition must be made to compensate for the altered mill base viscosity. Fairly viscous mill base pastes can be handled by a bead mill provided that the temperature within the mill is raised sufficiently to reduce the mill base viscosity to a suitable working level at the given processing temperature. In one laboratory study on carbon black, the small bead (shot) mill was run in the range of 222 to 240 F, using $\frac{1}{32}$-in. steel shot grinding media (4).

In many plants the sand mill temperature operating range is from 120 to 150 F (a moderate flow of cooling water is used). Raising the temperature to this level permits higher pigment loadings and higher pigment throughputs because of reduced viscosity (6). High temperatures are usually advantageous in that they assist pigment wetting and/or permit a higher binder fraction to be used in the mill base composition.

For a given pigment/binder/solvent system, there must be one mill base composition that provides optimum dispersion conditions yet permits satisfactory letdown. A logical approach to such an optimum can usually be accomplished by using the general concepts that have been presented. At no time should the importance of temperature be disregarded or overlooked. Moreover, it must be remembered that for any operating temperature too low a viscosity results in poor mill base circulation, sand slippage, and excessive wear, and excessively high viscosity restricts the sand movement, leading to a "choked" grinder. In either case markedly reduced dispersion results.

Although sand mills are very versatile, pigments that do not wet well and/or contain large, hard aggregates should not be routinely scheduled for sand milling. In such cases use should be made of alternative equipment such as roller or conventional ball mills.

DESIGN OF SAND AND BEAD MILLS

Significant advances have been made in the design of sand and bead mills during recent years, ranging from the setting up of specifications for miniature laboratory sand mills to the design and manufacture of highly sophisticated production mills that can handle either sand or bead media. These are briefly reviewed.

In general, the peripheral (outside edge) speed of the rotating disk that is working within the mill, whether a laboratory or commercial unit, is adjusted to about 2000 ft/min (1000 cm/sec). This is roughly half the peripheral speed that is considered optimum for the high-speed disk disperser considered in Chapter 22.

Also, a throughput rate that has been found optimum for one unit (say a laboratory sand mill) can be used to calculate a reasonably acceptable throughput rate for another unit (say a production sand mill) in terms of equivalent dwell times ($t = S/Q$, where t is the dwell time, S is the milling chamber size, and Q is the throughput rate).

Miniature Sand Mills

A simple batch-type miniature sand mill using 20- to 30-mesh Ottawa sand was reported in 1967. It consists of a 200-cm^3 tall form beaker (polyethylene or stainless steel), within which a 1.625-in. phenolic fiber disk mounted on a precision shaft is rotated at either 5000 rpm (rim velocity 2130 ft/min) or 8000 rpm (rim velocity 3390 ft/min) by a $\frac{1}{2}$-horsepower motor (7, 8). The amount of sample required ranges from 3 g for high-oil-absorption pigments ($\overline{OA} > 50$) to 50 g for low-oil-absorption pigments ($\overline{OA} \approx 10$). The vehicle requirement per batch is normally about 60 cm^3, and the sand requirement about 100 g. A Hegman grind of 7 is achieved with this equipment on most pigments in 5 to 10 min. The procedure is fast, economical, and reproducible and closely correlates with factory production. A modification of this procedure is described in detail in ASTM D3022-72, "Color and Strength of Color Pigments by Use of a Miniature Sand Mill."

A somewhat similarly designed laboratory sand mill was used in a 1967 cooperative testing program aimed at studying dispersibility differences in a controlled energy sand mill (3-in diameter nylon impeller disk; 4-in diameter polyethylene container; impeller rotational speed variable from 425 to 3200 rpm) (9).

In a later investigation (1968) a commercial, water jacketed, stainless steel sand mill (quart size) was employed (10). Milling was carried out based on a rotor speed of 2380 rpm and a sand charge of 540 g (Ottawa 30- to 40-mesh). This study showed that a viscosity of about 90 KU is suitable for providing a good throughput for starting evaluations, and also that the grind reading for a completed dispersion based on inorganic pigments corresponds roughly to the largest particle size in the original pigment mixture (a 15-μ pigment gives a grind of 6H; a 10-μ pigment a grind of 7H; and a 0.03-μ pigment a grind of >8H).

Production Sand, Bead, and Shot Mills

Significant advances have been made in the design of production mills (8- to 60-gal grinding chamber capacity) using small grinding media.

Vertical Mills. In the original sand mill the screen holding back the sand media at the top of the mill was an open type that often became encrusted with dry mill base, especially with high-volatility solvent. However, most of the newer sand and bead mills have a submerged type of screen that avoids this type of plugging. Many current production units are also tightly sealed to permit operation under pressure (in excess of 100 psi in one commercial unit) and also to accept high-viscosity mill bases (several hundred poise) (11). Presumably, corresponding changes in formulation will be required to take advantage of these extreme conditions.

High-Speed Shot Mill (Schold). The continuous high-speed shot mill is somewhat similar to the sand mill (although more rugged) in that the milling unit is a relatively narrow, upright, cylindrical tank equipped with a centralized, rotating, vertical shaft that supports a series of uniformly spaced, stainless alloy, circular platforms (10- to 100-gal models) (12). As with the bead mill, these platforms communicate the rotary motion of the agitator shaft to the steel ball media/mill base mixture.

The Schold shot mill operates under internal pressure and hence can process mill bases with relatively high viscosities and/or bases that exhibit thixotropic behavior. The mill also features a variable speed pump and a unique filter that rotates with the shaft. Since this filter is designed for complete submergence in the liquid, it prevents both drying of the product and entrainment of air.

This Schold high-speed shot mill works best with a charge of small grinding media, which may be steel (or ceramic) balls. The recommended ball media diameters range from 0.08 down to 0.02 in. The mill base viscosity and the required throughput rate dictate the size and density of the grinding media.

SWMill. The SWMill sand grinder (U.S. patent 3,055,600) is a batch-type unit consisting essentially of an outwardly flared, water-cooled, open, cylindrical tank (with removable cover) equipped with a centralized, vertical shaft (entry through bottom) that rotates a single circular rotor platform located at the bottom of the tank (13). The standard grinding media supplied with the mill are tiny ceramic balls (density 2.8 g/cm^3) of $\frac{1}{16}$-in. diameter, which are placed above the rotor. SWMills are supplied in sizes ranging from a 62-gal unit (10 horsepower) to a 300-gal unit (60 horsepower), which process, respectively, 25- and 200-gal batches of mill base.

No premixing of the mill charge is required, but the loading sequence is important. The liquid portion must be added before starting the mill (about 5% of solvent is withheld for mill washing). The mill is then started, and the pigments are chaiged, the more difficult-to-disperse pigments being added first. The viscosity of the resulting mill base composition (slurry) during processing must be such as to allow good media flow. This is indicated by a vortex that exposes a top portion of the impeller assembly as the ball media/mill base mixture is whirled about the tank by the rotary motion imposed by the rotating platform underneath. Overall processing time is usually less then 1 hr.

Discharge of the finished product (which is commonly stabilized by a last minute addition of vehicle) is accomplished while the mill is still running. This is done by raising a screen guard inside the tank and opening an exit valve to allow the dispersed mill base to discharge through a screened opening at the bottom of the tank.

The claim is made that the SWMill disperses about 30 times faster than a conventional ball mill of comparable capacity (batch processing time about as fast as it is physically possible to charge the raw materials to the mill and move the finished product away for final disposition).

As with conventional ball mills, the optimum fraction for the binder portion of the mill base vehicle (nonvolatile fraction) usually lies between 0.20 and 0.30. A rough starting weight fraction w for the pigment portion of the mill base can be obtained from the expression $w = 1.00 - 1.45 \cdot \overline{OA}$ (for $\overline{OA}$ values between 0.05 and 0.55), where $\overline{OA}$ is the mean or equivalent oil absorption for the pigment mixture (expressed as a fraction). Beyond an $\overline{OA}$ value of 0.55 the pigment weight fraction should be held to about 0.20 (20%). As an example, the mean oil absorption value for the pigmentation of a green enamel is 0.26 (26 lb oil/100 lb pigment mixture). The calculated weight fraction for the pigment in the mill base mixture is then 0.62 ($= 1.00 - 1.45 \cdot 0.26$) or 62%w. This should provide a viscosity between 85 and 115 KU. However, if insufficient motion is noted on observing the batch flow pattern, additional solvent and/or vehicle should be added incrementally until good flow is established.

To avoid overgrinding, a check should be made every 10 min in attaining a required coarse grind (0 to 3H reading), every 20 min for a medium grind (3 to 5H reading), and every 30 min for a fine grind (6H and higher). On reaching the specified grind, any scheduled stabilizing vehicle is added and the unit is run for a few more minutes, after which the dispersed mill base is discharged.

Horizontal Mills. Probably the most revolutionary change in the design of sand and bead mills has been simply to rotate the grinding unit assembly through an angle of 90° (from an upright or vertical alignment to a horizontal one). At least two grinding systems based on horizontal positioning of the sand/bead grinder are being manufactured (2, 3). Many advantages accrue from this changed position. For example, in the event of the unit being stopped, the horizontal position enables it to be restarted without difficulty, regardless of the product being processed, its viscosity, or the type of grinding media being used. The density of the grinding media is also not a limiting factor. Thus the use of steel balls, which may be contraindicated in an upright sand mill because of ball settlement to the bottom, is no problem with the horizontal design. The ball media remain uniformly distributed along the length of the horizontal unit at all times. In this new design, maximum utilization is made of the grinding chamber (as much as 80 to 85% of the grinding container is filled with grinding media). These horizontal units are very compact and are easily assembled and disassembled from the side.

As with the newer type of vertical sand mills, horizontal bead mills are completely sealed (no introduction of air or loss of volatiles), and they feature effective enclosed dynamic screening arrangements.

Horizontal mills are manufactured in a number of sizes based on grinder container volumes of less than 1 qt (laboratory model) to 50 gal (production unit). The degree of dispersion attained depends on the rate at which the mill base is pumped through the unit, the selection of grinding media (density, diameter), and the impeller rotational speed.

The excellent construction of these grinding units and the abrasion resistance of the working faces are substantiated by the fact that a 4-gal capacity mill processed nearly 900 products (totaling over 500,000 lb in weight) during a 22-month period before replacement of agitator disks was required (3).

The claim is made that the horizontal sand or bead mill outproduces the vertical mill by a factor of 2 or 3, based on the same volume capacity.

PRODUCTION RATES AND ECONOMIC CONSIDERATIONS

A plot of production rates for dispersed mill base versus sand mill size, as shown in Fig. 21-8*a*, reveals a possible fourfold variation in throughput. Thus the lowest feasible production rate Q (gal/hr) for a conventional vertical sand mill is numerically equal to about 3 times the chamber size S of the sand mill ($= 3S$), whereas the highest production rate is about 12 times the chamber size ($= 12S$). An intermediate or typical production rate can be taken as equal to $6S$, as given by Eq. 6, where S is the chamber size of the sand mill in gal.

$$Q\ (\text{gal/hr}) = 6S\ (\text{gal}) \tag{6}$$

Production rates double and half this mean rate represent approximate limiting extremes.

A plot of the power P (horsepower) required to operate sand mills, as shown in Fig. 21-8*b*, reveals a significantly smaller power demand per unit volume of processing space as sand mill size is increased. Equation 7 closely expresses the relationship between the power demand (horsepower) and the sand mill size.

$$P\ (\text{horsepower}) = 5.5\sqrt{S\ (\text{gal})} \tag{7}$$

This expression states that the power required to operate a sand grinder varies as the square root of the sand mill size. Thus a 16-gal unit demands only twice as much power input as a 4-gal unit ($2 = \sqrt{16/4}$). From this it follows that large sand mills are far more efficient than smaller units from a power demand standpoint. This fact can be expressed in the form of an equation by dividing Eq. 6 by Eq. 7, yielding Eq. 8, which gives the volume V (gal) of dispersed mill base produced by 1.0 horsepower-hr of energy for an average sand mill operation.

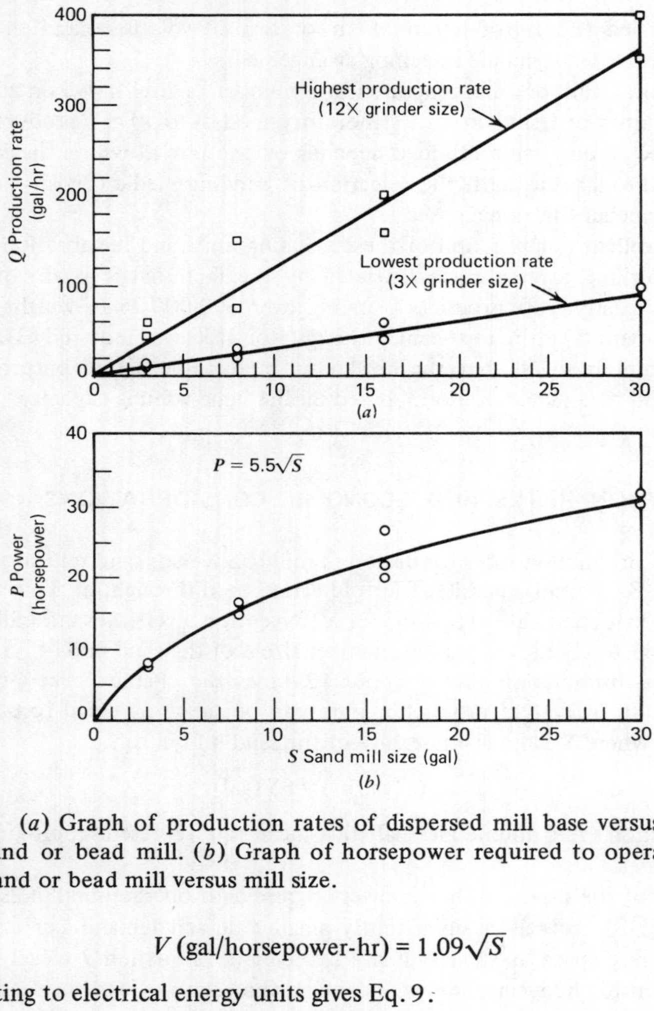

Fig. 21-8 (*a*) Graph of production rates of dispersed mill base versus chamber size of sand or bead mill. (*b*) Graph of horsepower required to operate a continuous sand or bead mill versus mill size.

$$V\,(\text{gal/horsepower-hr}) = 1.09\sqrt{S} \tag{8}$$

Converting to electrical energy units gives Eq. 9.

$$V\,(\text{gal/kW-hr}) = 1.45\sqrt{S} \tag{9}$$

PROBLEM 21-2

Assuming an electrical rate charge of 2¢/kW-hr, calculate the cost of processing a 500-gal batch of mill base through a 16-gal sand grinder under average processing conditions.

Solution. Calculate the volume of dispersed mill base that can be processed by 1.0 kW-hr of energy from Eq. 9.

$$V \text{ (gal/kW-hr)} = 1.45\sqrt{16} = 5.8$$

From this information compute the energy cost for processing a 500-gal batch of mill base.

$$\frac{500}{5.8} \cdot 0.02 = \$1.72$$

ADVANTAGES AND DISADVANTAGES OF CONTINUOUS SAND AND BEAD MILLS

Advantages

Sand and bead mills offer the coatings manufacturer many advantages for dispersing pigments into vehicles.

1. Relatively low initial investment, low power consumption, small space requirements, light factory construction, and minimal maintenance costs are attractive economic features.
2. No special skilled supervision or attention is required because of the safety and simplicity of the sand mill equipment. The setting of the valve that determines the mill base flow into the sand grinder chamber is the central control for the sand mill operation. Once the mill is running at a prescribed rate, minimal attention is necessary (several units can be supervised by a single operator).
3. Flexibility in volume output is afforded by the continuous nature of the process. High production rates can be attained by use of a proper mill base premix.
4. The sand mill is especially effective for providing ultrafine dispersion of pigments having clean bright colors and for yielding coatings with improved gloss, maximum tinting strength, and, frequently, improved suspension properties.
5. Reproducibility is good once the throughput rate has been established at the optimum level.
6. With most mill bases, rapid and thorough cleanup can be accomplished by flushing the sand unit and pipelines with a suitable solvent and/or solvent/vehicle mixture. Cleaning cycles should be kept as short as possible, especially when running with solvent alone, since otherwise excessive mill and media wear occurs.

Disadvantages

The sand mill also has several disadvantages or limitations.

1. The charge to the sand mill must be a homogeneous premix, since sand grinding is essentially a refining operation.
2. Screen blinding can occasionally become troublesome.
3. The sand mill is not equipped to process large or oversize agglomerates.
4. Cleanup between drastic color changes (say from dark red to bright yellow) is difficult, calling in some cases for a change in the media loading.
5. Dilatant (shear thickening) mill bases process poorly in a sand or bead mill and normally should be avoided.

Sand and bead mills have gained widespread acceptance by the coatings industry, especially for the manufacture of architectural paints.

REFERENCES

1. Company brochure, "Selecting Sandmill Medias," Quackenbush Company, Palatine, Ill., 1977.
2. Company brochure, "Supermill by Premier," Premier Mill Corporation, 1140 Broadway, New York 10001.
3. Maag, Theodor, "Dyno-Mill: An Advance in Dispersing Technology," *Chem. Rundsch.*, **28**, No. 10, (1975) (U.S. distributor: Impandex Inc., 260 West Broadway, New York 10013).
4. Garret, M. D. et al., "Dispersion of Carbon Black by Modified Sand Mill Technique," *Am. Ink Maker*, November (1969).
5. Company brochure, "Bugs? In Your Bead Mill," Quackenbush Company, Palatine, Ill. 60067, 1970.
6. Progress Report 7C (revised), "Sand Grinding of Pigment Colors," E. I. du Pont Company, Pigments Division, Wilmington, Del.
7. Orwig, B. R., "Color, Strength, and Dispersibility of Pigments by the Sherwin-Williams Miniature Sand Mill," *JPT*, **39**, No. 504, 14–18, January (1967).
8. Publication B/4, "Sherwin-Williams Miniature Sand Mill," Gardner Laboratory, Bethesda, Md., 1978.
9. Boyer, Wilbur, (Chairman), "Studies on a Dispersibility Test Using a Controlled Energy Sandmill," *JPT*, **39**, No. 514, 677–685, November (1967).
10. Goodman, Barnard, (Chairman), "Evaluation of Pigment-Vehicle Relationships Using a Laboratory Sand Mill," *JPT*, **40**, No. 527, 591–594, December (1968).
11. Company brochure, "Morehouse Pressure Mill System," Morehouse Industries, Inc., Fullerton, Calif. 92633.
12. Company brochure, "High-Speed Shot Mills," Schold Machine Corporation, Chicago, Ill. 60638.
13. Company brochure, "SWMill," Epworth Manufacturing Company, Inc., 1400 Kalamazoo Street, South Haven, Mich. 49090.

BIBLIOGRAPHY

Boose, D. G. "Development and Use of the Sand Grinder," *Offi. Dig.*, **30**, No. 398, 250 (1958).

Brownlie, G. "Performance of Titanium Dioxide Pigments in a Sand Grinder," *JOCCA*, **43**, 737 (1960).

Chicago Boiler Company, *A Primer of Sand Grinding*, Chicago, Ill., 1961.

E. I. du Pont Comapny, *Sand Grinding Process for Dispersing Pigments*, Wilmington, Del., 1961.

Lore, M. B., "Sand Grinding," *Am. Ink Maker*, **39**, 33, February (1961).

Quackenbush, I. C. "Dispersion by Fine Media through Hydraulic Action–the Sand Grinder," *Offi. Dig.*, **31**, No. 408, 77 (1959).

Quackenbush, I. C., "The Sand Grinder," *Paint Varni. Prod.*, **49**, 39, May (1959).

Robe, Karl, "Forte of Sealed Sand Mill," *Chem. Process*, August (1966).

Schlapfer, L. A., Jr., "Sand Grinding Flexographic and Rotogravure Inks," *Am. Ink Maker*, July (1965).

Wahl, E. F., "Dispersion with High Velocity Media," *JPT*, **41**, No. 532, 345–352, May (1969).

22 *High-Speed Disk Disperser*

The present abundant supply of both readily dispersible pigments (jet-milled, surface-treated) and ready-made colorant dispersions, together with the availability of compact and efficient heavy-duty power sources, has led to the development of high-speed disk-type impeller dispersers that can premix, grind (disperse), and let down in a single piece of equipment.

DESCRIPTION OF HIGH-SPEED DISK DISPERSER

The high-speed disk disperer consists essentially of a circular saw-blade-type impeller mounted on a shaft rotating at high speed that is vertically centered in an upright cylindrical tank. In its simplest form the impeller blade is a flat, disk-shaped plate. Starting from this basic design, the disk has been subject to numerous modifications, with the notching along the impeller rim varying among the manufacturers of impeller equipment. Although the serrations along the edge of the impeller may be cut according to different designs–turned upward, downward, inward, and outward at varying angles; punched with holes or otherwise gouged to induce special attrition or cavitation effects (even lined with links of chain)–the main impeller disk itself is generally left flat and continuous. Attempts to interrupt the continuity of the impeller from the shaft to near the impeller rim have led only to a decrease in dispersion efficiency. Figure 22-1 illustrates a typical impeller disk design.

Manufacturers of dispersion equipment feature many special edge designs, such as a pick design that is claimed to provide cutting in addition to dispersing

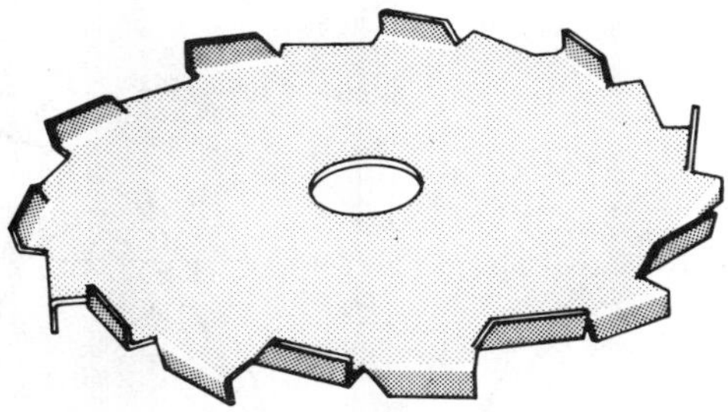

Fig. 22-1 Schematic drawing of a representative high-speed disk impeller blade.

(1), a high-vaned circular blade that is said to promote turbulence at the expense of laminar flow (2), and a multi-flat-ring assembly at the edge of the flat disk blade (fortified with slanted adjacent flat teeth) that allegedly provides more attrition than the conventional sawtooth blade (2). Further modification of the multi ring design has even been introduced that involves an arrangement for a Venturi configuration between adjacent rings (3, 4). Despite this proliferation of ideas, the majority of the disk impellers used in the coatings industry closely follow the more or less standard design shown in Fig. 22-1.

SIZE, POSITIONING, AND SPEED OF DISPERSER BLADE

Figure 22-2 illustrates the normal positioning of the disperser blade in the mixing/dispersing tank in terms of the impeller diameter *D*. The given geometry has been found quite satisfactory in practice. For any particular run, however, it is usually necessary to make adjustments within the specified *D* values until optimum dimensions have been established. When the high-speed impeller disperser is operating properly, a rolling-doughnut type of circulation is induced by the rotation of the impeller blade. No splashing or surging should be evident, and within the hole of the doughnut a portion of the impeller blade should be visible, although some of these dispersers operate with the lower portion of the shaft partially covered. The swirling pattern should be such that a particle located at the outside edge of the mill base (top surface) will spiral down to the bottom of the vortex before completing one revolution around the mixing tank. The cylindrical tank itself should be free of baffles and blind corners (preferably with a dished bottom and beveled edges).

Experience has shown that the peripheral velocity of the impeller blade must reach about 4,000 ft/min or higher for satisfactory pigment dispersion. The impeller shaft is generally centrally located, although some off-centering has been found helpful when loading the pigment charge. On the other hand, Daniel

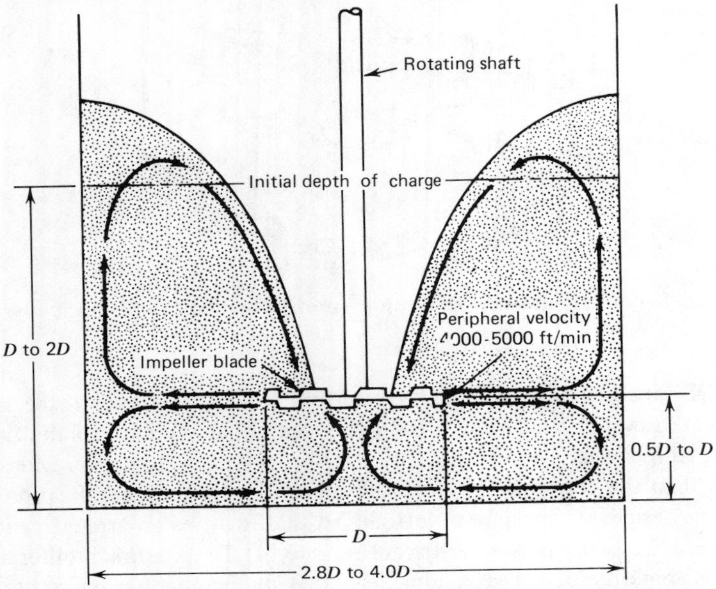

Fig. 22-2 Correct positioning of disk impeller blade and optimum dimensions for impeller container in terms of impeller disk diameter D.

recommends pigment loading with the shaft at the center. A shift is then made to a position midway between the center and the side of the tank for most efficient dispersing (3).

MILL BASE RHEOLOGY IN A HIGH-SPEED DISK DISPERSER

The high-speed disk disperser is in the unhappy position of having to straddle both smashing and smearing types of dispersion. This means that it cannot truly function effectively as either a smasher or a smearer but must settle for half measures. However, from practical experience it is concluded that the high-speed disperser has the best chance of operating with relatively good efficiency if it is considered primarily as a smearer (5). As a smasher it leaves much to be desired. However, if a low-viscosity system must be accepted (say a water system with low pigment solids content and low viscosity), then smasher principles should be applied although not too much effective dispersing action should be expected.

Dispersion of pigment in a mill base by smearing is accomplished by laminar flow. Any introduction of turbulence simply upsets the smearing operation, and the pigment particles tend to swirl around each other instead of being dragged across each other.

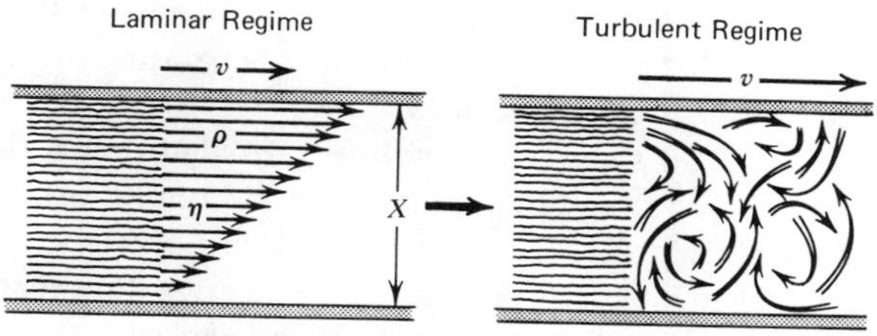

$$\mathrm{Re} = \frac{\text{kinetic (toppling) component}}{\text{viscous (antitoppling) component}} = \frac{v \cdot \rho \cdot X}{\eta}$$

Fig. 22-3 Schematic diagrams illustrating conditions of laminar and turbulent flow. A Reynolds number of 2000 is commonly taken as the critical value that separates the laminar from the turbulent regime in pipelines.

Fluids in flow can move in two distinct patterns. If the shear stress producing the flow provides a sufficiently low shear rate, layers of liquid slide over each other in an orderly manner to give laminar flow (also referred to as viscous, Newtonian, or streamline flow). As the shear rate is increased, however, a critical point is reached where the motion suddenly becomes chaotic. The former orderly laminar glow is disrupted, and in its place there is generated a swirling chaos of eddies and vortices. This condition of turbulence is referred to as turbulent or hydraulic flow. The two regimes of tubulent and laminar flow are depicted in Fig. 22-3.

Critical Reynolds Number

From the equations of fluid motion, Reynolds reasoned that the onset of turbulence should depend on a critical value for the dimensionless quantity given by Eq. 1.

$$\text{Re (Reynolds number)} = \frac{\rho v X}{\eta} \tag{1}$$

where ρ = density (g/cm^3)
v = velocity (cm/sec)
X = characteristic linear dimension (cm)
η = viscosity (poises)

For flow conditions in pipes and similar circular conduits the change from laminar to turbulent flow occurs at Re $\approx$ 2000, and this value will also be assumed as applying to the mill base flow in a high-speed disperser. However, a Reynolds number as high as 10,000 has been mentioned in the literature as applying to rotors (6). In piping systems the characteristic linear dimension is taken as the pipe diameter; for the high-speed disk disperser the characteristic linear dimension X will be taken as the distance separating the disperser blade from the bottom of the dispersing tank (5).

Inspection of Fig. 22-3 shows that swirling, turbulent flow is fine for mixing, with pockets of mill base from different locations becoming nicely intermingled. But (and this is most significant) these intermingled pockets still represent sections of undispersed mill base. Conversely, the laminar flow pattern may fail to provide efficient mixing, but the drag of one layer over the other tears the clumps of pigment apart and effective dispersion is attained. Obviously, the more viscous the mill base vehicle, the more effective is the tearing apart (dispersion) of the pigment aggregates. As shown in Fig. 22-3, the coating or ink engineer should mentally bracket mixing with turbulence and dispersion with laminar flow (with a Reynolds number of about 2000 separating the two regimes). Inspection of the Reynolds number (Eq. 1) shows that it is composed of a kinetic component (ρvX) that tends to induce turbulence and a viscous component (η) that tends to promote laminar flow. The key to obtaining laminar flow in a high-speed disk disperser is to raise the viscosity component and minimize the kinetic component.

Conditions for Producing Laminar Flow in High-Speed Disperser Related to Viscosity

To estimate suitable conditions for laminar flow in a high-speed disk disperser, assume a mill base density of 1.0 g/cm^3 and a rim velocity of the circular blade of 4000 ft/min (2000 cm/sec), as schematically illustrated in Fig. 22-4. For laminar flow (dispersion conditions), the Reynolds number must be less than 2000. Therefore the condition for laminar flow is given by Eq. 2.

$$\text{(Condition for laminar flow)}\ \frac{\rho vX}{\eta} < 2000 \tag{2}$$

Substituting the assumed values for v and ρ in Eq. 2 and simplifying gives Eq. 3.

$$\text{(Condition for laminar flow)}\ X < \eta \quad \text{or} \quad \eta > X \tag{3}$$

Despite the several approximations and assumptions that have been introduced into the derivation of Eq. 3 there emerges a simple, informative expression for the conditions that will assure laminar flow in a high-speed disk desperser, namely, that the distance X (in cm) between the disk blade and the container bottom be less than the viscosity η (in poises) of the mill base (viscosity mea-

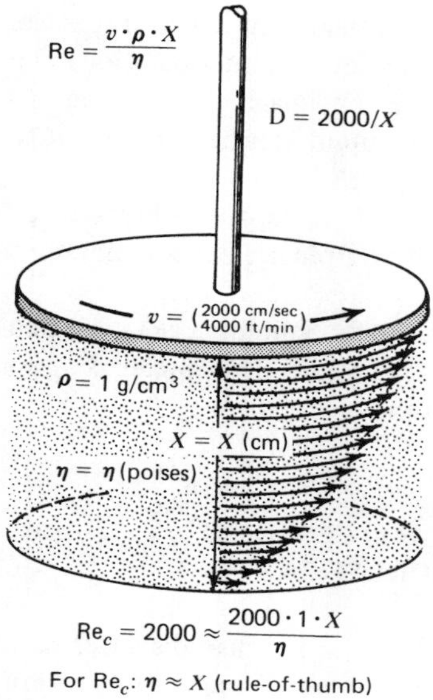

Fig. 22-4 Schematic drawing illustrating the laminar flow that takes place between the high-speed impeller disk and the bottom of the impeller container. For the given conditions, the viscosity η (poises) should be greater than the separation distance X (cm) to assure laminar flow under the impeller disk.

sured at applicable shear rate). For example, if the disk blade is placed 1.0 ft (30.5 cm) above the container bottom, the mill base should be formulated to have a viscosity of 31 poises or greater to provide laminar flow. This is more or less in line with industrial experience.

The flow pattern, of course, is not as simple as that depicted in Fig. 22-4. Mill base is obviously being centrifuged outward, and material is flying off in a tangential direction at the edges of the disk blade at a velocity of 4000 ft/min. Hence, imposed on the series of flat disks of mill base, which are grinding over each other in a laminar manner (running downward under the disk to the container bottom), there is also taking place an outward movement of mill base to the side of the container. At the container bottom, mill base moves in to replenish this overhead depletion.

Observations on High-Speed Disperser Design

There seems to be a trend by some disperser manufacturers to strive for higher peripheral velocities. However, examination of Eq. 2 shows this to be a questionable objective. This is also borne out by a report in the literature stating that above a critical speed grinding efficiency drastically decreases (7). If a mill base operation is close to the critical Reynolds number, boosting the rim speed from 4000 to 6000 ft/min may well topple the laminar flow into turbulence. As a result effective dispersion is replaced by a useless mixing operation. The thought of higher peripheral velocities may excite the imagination of the disperser manufacturer, but a better goal would be to build sturdier equipment to handle higher viscosities.

Note that Eq. 2 tells us that lowering the disk blade closer to the bottom can compensate (up to a point) for a reduced viscosity (laminar flow is maintained by this blade adjustment).

The region above the dispersion blade is essentially a holding area, so that loading it with mill base is merely a storage operation. This material must patiently await its turn for dispersion underneath the blade.

POWER REQUIREMENTS FOR MODEL HIGH-SPEED DISK DISPERSER UNDER IDEALIZED SHEAR CONDITIONS

The theoretical power requirements for a model high-speed disk disperser have been analyzed by this author (5), yielding Eq. 4 for the power expended per unit volume of mill base.

$$\frac{P}{V} = \frac{\eta\omega^2 R^2}{2X^2} \tag{4}$$

where P = power (dyne-cm/sec)
V = volume (cm^3)
ω = angular velocity (radians/sec)
η = viscosity (poises)
R = radius of impeller disk (cm)
X = distance between impeller blade and bottom of container (cm)

Equation 4 states that the rate of energy input to a high-speed disk disperser is proportional to the viscosity of the mill base and to the square of the impeller disk radius, and inversely proportional to the square of the separation distance between the disk blade and the bottom of the container. Again, high viscosity and a low setting of the disperser disk are shown as conducive to effective processing.

The interrelationships of the quantities given in Eq. 4 are graphically illustrated in Fig. 22-5. The boundary line *BB* separates the region of laminar from turbulent flow. Equation 4 applies only to the laminar flow region (shaded). Therefore viscosity lines below the boundary line *BB* actually refer to imaginary conditions. In this lower turbulent region a different set of equations would be required to express the several relationships. However, since the present interest is in achieving dispersion, it is necessary to consider only the upper laminar flow region.

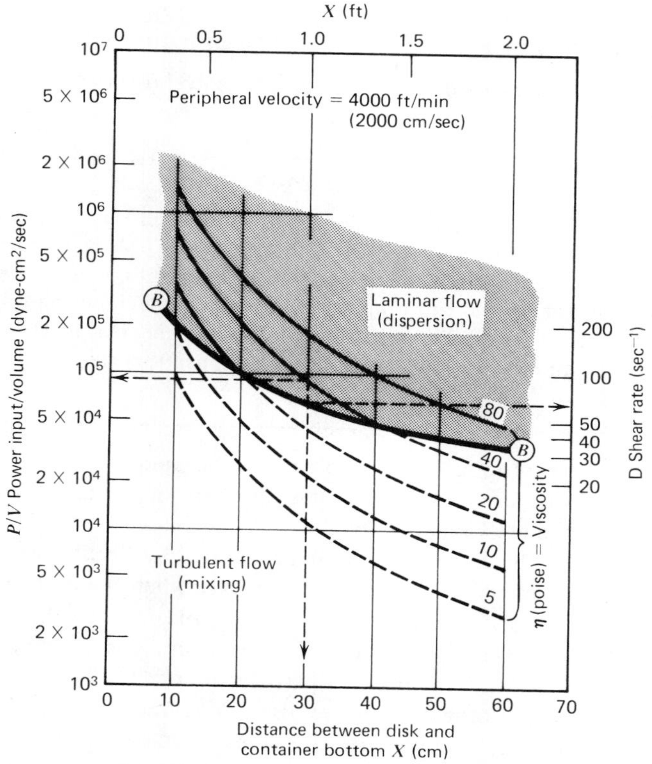

Fig. 22-5 Relationship of power input per unit volume to (*a*) mill base viscosity and (*b*) distance between impeller disk and bottom of container. Shear rate is also shown as a function of the distance X for the given conditions (curve *BB*).

PROBLEM 22-1

Calculate the theoretical power requirements for a high-speed disk disperser using a 2.0-ft diameter blade (rim velocity 4000 ft/min) positioned 1.0 ft above the dispersion tank. Assume a 40-poise viscosity for the mill base under operating conditions.

Solution. Equation 4 can be rewritten as Eq. 5 after conversion for the changed set of units.

$$P\text{ (horsepower)} = \frac{17 \cdot 10^{-10} \eta R^2 v^2}{X} \tag{5}$$

where η = viscosity (poises)
R = impeller radius (ft)

v = impeller peripheral velocity (ft/min)
X = distance from impeller blade to container bottom (ft)

Substitute the problem data in Eq. 5, and solve for the required power demand.

$$P\text{ (horsepower)} = \frac{17 \cdot 10^{-10} \cdot 40 \cdot 1 \cdot 4000^2}{1} = 1.1 \text{ horsepower}$$

Comment. This calculated theoretical power demand is unduly small, partly because the computation is based on 100% efficiency. More importantly, the calculation does not take into account the laminar flow occurring outside the boundardies set by the rim dimension, nor does it include the large amount of power expended in producing turbulent flow at the periphery of the disk. Hence in actual practice a power source many times this size (on the order of 10X to 20X) would be required.

A quantitative description of the flow conditions and a theory for the deaggolomeration of pigment aggregates at the periphery of the disk disperser blade have been published by Wahl et al. (6).

Based on a study by Daniel (3) and data reported by Christensen and Arnold (8), it has been estimated by this author that only when the mill base viscosity exceeds about 10 poises does it contribute significantly to the total power demand (5). Below 10-poise viscosity, the power input is largely expended in creating turbulence. Unless the pigment particles are practically "self-dispersing," this turbulence fails to provide effective dispersion. Above about 10 poises, the viscosity factor begins to contribute to the power demand, and above about 20 to 30 poises, viscosity becomes a controlling parameter as laminar flow takes over and dispersion rather than mixing becomes predominant (5).

Note in Fig. 22-5 that a shift from turbulent to laminar flow (from mixing to dispersion) can be accomplished by either increasing η or decreasing X. If η is fixed, the only recourse is to reduce X. Conversely, if X is fixed, the only recourse is to increase η.

SHEAR RATE IN A HIGH-SPEED DISK DISPERSER

Despite the appearance of tremendous activity in the high-speed disk disperser, the shear rates that are involved under the disk blade are quite moderate. The shear rate at the center (axis) is zero and becomes maximum at the disk edge. In the model disperser of Fig. 22-4, the maximum shear rate D is given by Eq. 6.

$$\text{D} = \frac{v}{X} \tag{6}$$

For a rim velocity of 4000 ft/min (2000 cm/sec) and a separation distance X of 1.0 ft (30.5 cm), the shear rate is only 66 sec^{-1} (=2000/30.5). This hardly compares with the very high shear rate of ~10,000 sec^{-1} that is set up when paint is brushed over a substrate. Plotted points for shear rate versus separation distance result in a curve that coincides with the line *BB* on Fig. 22-5 (rim velocity assumed equal to 4000 ft/min). Hence this line serves a dual function.

INFLUENCE OF MILL BASE VISCOSITY PROFILE (RHEOLOGY) ON DISPERSION EFFICIENCY

The question of the adjustment of mill base viscosity can be resolved by thinking in terms of what the high-speed disperser is trying to accomplish, namely, a relaying of the stressing action of the disperser blade (overhead) down through the body of the mill base to the bottom of the container. If the application of the stress thins out the mill base, communication between successive layers is weakened or lost. Hence, if the mill base viscosity becomes drastically reduced as the stress forces attempt to build up, there are no opposing forces against which the stress forces can work. In this case energy is spent mainly in whirling a disk about its axis, with undispersed mill base remaining unaffected as it observes this display of high-speed rotation by the overhead disk. Even the edge geometrics (vanes, teeth, serrations, notchings, lips, chains, gougings, and bendings) at the impeller blade periphery (which are put there mainly to aid in getting the top layers of mill base into circular motion) are helpless to assist if the mill base thins down to a watery consistency under the impelling action. From this discussion it becomes obvious that viscous resistance is the key to broad stress distribution. It is evident that marked pseudoplasticity defeats the purpose of high-speed disperser action. This tends to suggest that dilatancy is the answer, since a dilatant mill base offers increased opposition to flow as the stressing action of the disperser blade is increased. This is definitely valid. However, caution must again be exercised at this point, since a mill base can be made so highly dilatant that under strong shearing action it approaches infinite viscosity (approaches a rigid state). In this case the mill base is ripped apart into chunks of material (takes on the aspects of a soft solid), and shattering replaces laminar flow.

Somewhere between these undesirable extremes there is a compromise flow behavior that provides optimum laminar flow conditions. This condition is mild dilatancy—a conclusion that was also reached by Daniel (3), Weisberg (9), and Ensminger (10, 11) in their papers on mill base compositions for high-speed dispersers.

The Daniel wet point/flow point procedure is most helpful for evaluating dilatancy, as discussed in Chapter 17 on the grinding of pigment vehicles (12).

If a mill base is somewhat thixotropic, it may start to hang up on the sides of the dispersion tank, the circulation pattern in turn becomes unsatisfactory, and in extreme cases the impeller may even carve out a cavern in the mill base within which the impeller blade freely rotates with little or no contact with the immobilized mill base outside. This suggests that, when thixotropic agents are part of the mill base composition, it may be expedient to postpone their introduction until the main part of the pigment dispersing action has been completed. On the other hand, if the thixotropic additive requires time for its effect to become manifest, it is best to add the thixotrope at the beginning.

Dilatancy can be brought about either through the use of a dilatant vehicle (an approach that may not be very realistic) or by pigment loading (the practical route). As the pigment loading in a mill base formulation is increased, a state is eventually reached that is characterized by mild dilatancy. This is the target rheological condition for a mill base intended for high-speed disk dispersing.

EFFECT OF TEMPERATURE BUILDUP DURING PROCESSING

The temperature increase that arises during high-speed disk disperser processing is an adverse influence, and any step taken to hold down a rise in temperature (say by jacketing with cold water) is highly beneficial. A high-speed disk disperser is highly inefficient with possibly 5 to 10% of the power input expended in useful dispersive action. Most of the energy is diverted into creating turbulence. In turn this turbulent kinetic energy is soon dissipated as heat, resulting in an undesirable temperature increase. Since only viscosity can counter the onset of turbulence and since even a small rise in temperature markedly reduces viscosity, it follows that even after 10 to 15 min of processing the mill base viscosity may be so reduced in value that any further operation of the disperser unit becomes a waste of time and energy.

The dependence of temperature and viscosity on time of processing is illustrated in Fig. 22-6 (a plot of viscosity vs. temperature). The processing time trend is seen to move down the curve. When a point is reached on the curve corresponding to the critical Reynolds number, the laminar flow is toppled into turbulent flow and dispersion is replaced by mixing. Useful mixing is accomplished mainly during the early stages of the processing cycle. A good operator usually recognizes the critical stage in the operation where the temperature rise has brought about an intolerable reduction in viscosity. At this point the run should be terminated. As necessary, the run is continued after a cool-down period to complete the dispersion operation.

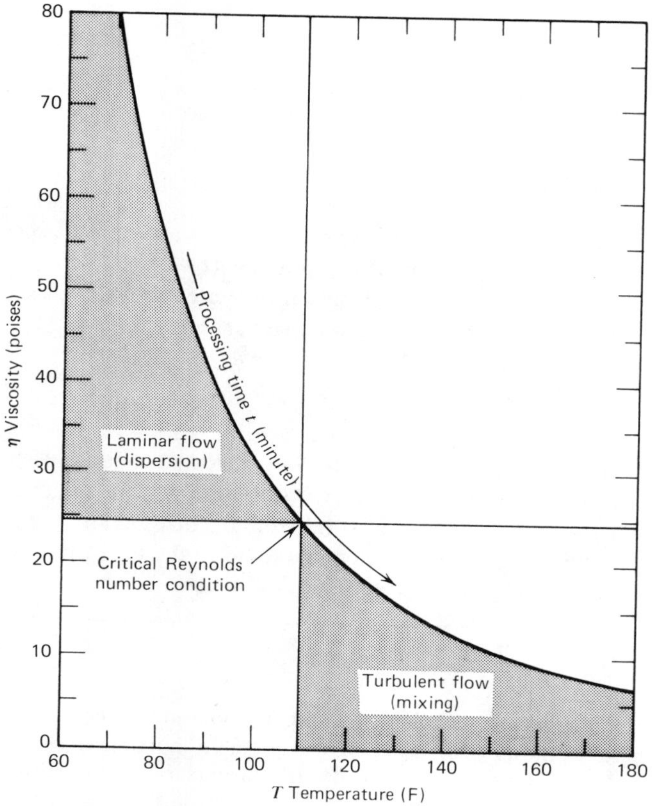

Fig. 22-6 Graph illustrating the rapid rise in temperature and the accompanying drop in viscosity that occurs with processing time. At some critical temperature the viscosity becomes reduced to a point where the laminar flow erupts into turbulent flow and dispersion by laminar flow (smearing) under the impeller disk is ended. Beyond this stage there is only excessive mixing.

DISPERSION RATE IN A HIGH-SPEED DISK DISPERSER

A mathematical analysis of the dispersion rate of pigments in a high-speed disk disperser has been developed, and data have been submitted to support the theoretical derivation (6). The rate of dispersion at any given time t is considered proportional to the fractional weight of pigment agglomerates above a specified size that are still present in the pigment portion of the mill base. The rate of dispersion in terms of the disappearance of aggolmerates above a fixed size ($>35\ \mu$ in the case of the reported experimental work) is then given by Eq. 7.

$$\frac{dw}{dt} = -k \cdot w \tag{7}$$

Equation 7 integrates to Eq. 8.

$$2.3 \log \frac{w_t}{w_0} = k \cdot t \tag{8}$$

where w_0 = initial weight fraction of pigment agglomerates above a specified size in the pigment portion of the mill base ($t = 0$)
w_t = fractional weight of pigment aggolmerates above a specified size in the pigment portion of the mill base at time t
t = dispersion time period
k = dispersion rate constant

Equation 8 is of such a form that, if the logarithm of the pigment agglomerate weight fraction above a given size (say >35 μ) is plotted versus time (on semilog paper). a straight line results. However, with time an end point is generally reached where the sloped line becomes horizontal, corresponding to a condition where the few remaining oversize agglomerates are of the type that is completely resistant to breakdown by the high-speed disk disperser.

PROBLEM 22-2

The amount of titantium dioxide pigment agglomerates over 35 μ in diameter in a mill base is initially 0.19%w of the pigment portion. After 8.0 min of dispersion in a production high-speed disk disperser, this weight fraction is reduced to 0.03%w (after which no further reduction is obtained). Assuming a logarithmic decrement decrease in pigment agglomerate weight fraction >35 μ with time, calculate the dispersion rate constant k.

Solution. Substitute the problem data in Eq. 8, and solve for the dispersion rate constant.

$$2.3 \log \frac{0.0019}{0.0003} = k \cdot 8.0; \qquad k = 0.23/\text{min}$$

In a series of pigmented alkyds the dispersion rate constant was found to vary from 0.06 to 0.70 for a laboratory dispersion unit and from 0.050 to 0.24 for a production unit, thus indicating the excellent response of this type of measurement to altered dispersion conditions (6). The same report also revealed that the dispersion rate in a high-speed disk disperser is roughly proportional to the surface of the impeller disk ($\sim R^2$), rather than its rim circumference ($\sim R$). This tends to verify the thesis that excellence of dispersion is a function of laminar flow under the disk blade, rather than of turbulent flow at the disk periphery.

In this regard it was shown that a reduction of laminar flow, produced by adding less than 1% of solvent to reduce viscosity, lowered the dispersion rate in one case by a factor of 5. This again confirms the key role of high viscosity in securing adequate dispersion.

FORMULATION OF NONAQUEOUS MILL BASE FOR A HIGH-SPEED DISK DISPERSER

As discussed in preceding sections, any mill base to be charged to a high-speed disk disperser should be quite viscous (yet flowable), with Newtonian or slightly dilatant flow behavior. A high-viscosity mill base can be formulated by using either a high-viscosity vehicle, a high pigment loading, or a combination of the two. From the standpoint of economy and efficiency, the use of high pigment loading is preferable, since it provides a maximum pigment charge per unit volume of dispersing equipment. Also, the somewhat tightly packed pigment particles in a highly loaded mill base tend to be more dilatant than loosely packed particles. Other things being equal, resorting to a highly viscous vehicle is undesirable, since such a vehicle is generally slow both in penetrating the interstices of the pigment aggolmerates and in wetting out the pigment particle surfaces to effect a pigment dispersion.

In view of these several facts, it is not unexpected that the trend in formulating mill bases for high-speed disk dispersers is toward highly loaded compositions that are based on medium-viscosity vehicles with low binder content. Experience has shown that 15% is about the minimum binder content that should be considered for the vehicle, since at lower values the mill base dispersion is unstable. Moreover, during subsequent letdown there is the definite possibility that difficulty will be experienced in maintaining the original degree of dispersion and even more difficulty in arriving at a compatible letdown composition. This suggests a range of about 20 to 35% w for the binder (nonvolatile) portion of the mill base vehicle. To this is added sufficient pigment to give a highly viscous mill base (above 30 poises at a shear rate of approximately 400 sec^{-1}).

Numerical Procedure for Estimating a Suitable Mill Base Composition

From an analysis of the results provided by a series of runs with a high-speed production disk disperser, Guggenheim developed the empirical expression relating vehicle solids (NV), vehicle viscosity (η), and Gardner-Coleman oil absorption ($\overline{OA}_c$) to an optimum vehicle to pigment weight ratio given by Eq. 9 (13).

$$\frac{W_v}{W_p} = (0.9 + 0.69\ \mathrm{NV} + 0.025\eta)\left(\frac{\overline{\mathrm{OA}}_c}{100}\right) \tag{9}$$

where W_v = weight of vehicle
W_p = weight of pigment
NV = nonvolatile binder (expressed as fraction)
η = vehicle viscosity (poises)
$\overline{\mathrm{OA}}_c$ = Gardner-Coleman oil absorption (lb/100 lb)

An inspection of Eq. 9 shows that this expression is premised primarily on a base value of $0.9\ \overline{\mathrm{OA}}_c/100$, to which are added two modifying terms, one that adjusts upward for the nonvolatile binder content, $0.69\ \mathrm{NV} \cdot \overline{\mathrm{OA}}_c/100$, and one that adjusts upward for the vehicle viscosity, $0.025\,\eta\,\mathrm{OA}_c/100$. Their relative effects are illustrated by the following problem.

PROBLEM 22-3

Calculate a mill base composition intended for high-speed disk dispersion based on a 35%w nonvolatile alkyd vehicle of 4.0-poise viscosity and a pigmentation with a Gardner-Coleman oil absorption value of $\overline{\mathrm{OA}}_c = 28$.

Solution. Calculate the W_v/W_p ratio using Eq. 9.

$$\frac{W_v}{W_p} = (0.9 + 0.69 \cdot 0.35 + 0.025 \cdot 4.0)(28/100)$$

$$= (0.9 + 0.24 + 0.10)(0.28) = 1.24 \cdot 0.28 = 0.347$$

The mill base composition is then calculated.

Binder $(0.35 \cdot 0.347) = 0.121$ (9.0%w)

Solvent $(0.65 \cdot 0.347) = 0.226$ (16.8%w)

Pigment = 1.000 (74.2%w)

1.347

From both laboratory experience and a study of the relevant literature Ensminger proposed that the mill base vehicle for high-speed disk dispersion should be formulated to a viscosity of at least A to F bubble viscosity (~1.0 poise) before admixture with the pigment, although in no case should the binder content of the vehicle be below 15%w (10). A suitable vehicle viscosity range usually falls between 1.0 and 4.0 poises (11). Sufficient pigment is then added to the vehicle to give a mill base viscosity of about 30 to 40 poises (11). This corresponds roughly to a volume pigment loading of 0.75 (=V/U).

PREPARATION OF LATEX COATINGS WITH A HIGH-SPEED DISK DISPERSER

The sequence for adding the components in the preparation of a latex coating is now more or less standardized (14). Normally the pigment dispersants, the coalescing agents, the antifoamers, and the thickener solution are charged first to the water in the tank and intimately mixed. Occasionally a minor portion of the latex vehicle may be introduced at this stage to assist in the pigment dispersing action, but the amount should be minimal.

The pigmentation is then added by centering the impeller disk close to the bottom of the tank and, with the disk rotating at relatively low speed, gradually sifting the dry pigment into the deep vortex that is formed. As the mill base becomes more viscous because of the initial incremental additions of pigment, the height of the impeller blade should be adjusted upward to provide a shallower vortex. After all pigment has been added, the rotation of the disk is increased to provide a rim velocity of about 5400 ft/min, which is considered optimum for dispersing the pigment in the mill base slurry. The grinding time should be kept to a minimum, with a temporary stopping of the dispersion operation if the temperature rises above about 110 F for flat latex grinds (because of low water content) or about 135 F for semigloss latex grinds. Completion of the dispersion stage is normally monitored by using a grind gage. In the case of flat latex mill bases, a reading of from 3 to 5H is considered acceptable (usually obtained in about 6 min). For semigloss latex mill bases a grind of 7H or higher must be reached.

On attaining an acceptable grind, the mill base slurry is let down with the bulk of the latex vehicle, together with any scheduled defoamer and/or additional water or thickener solution (to adjust the viscosity), using low-speed mixing. After thorough blending is accomplished, final adjustment of the alkalinity (pH) is made with a suitable amine or, commonly, ammonium hydroxide.

PRACTICAL CONSIDERATIONS IN USING A HIGH-SPEED DISK DISPERSER

Even with highly viscous mill base compositions, the relatively mild shearing action developed by a high-speed impeller disperser is incapable of grinding (dispersing) difficult-to-disperse pigments. However, most pigments can be obtained either in easily dispersed forms or as finely ground predispersions. Hence this restriction is not a serious limitation on the use of high-speed impeller equipment. It also is agreed that some heat buildup is a necessary adjunct of successful impeller operation. In fact, if too little heat is generated, it can be concluded that the impeller is being inefficiently operated. Hence, if highly volatile solvents or heat-sensitive vehicles (or pigments) are contemplated for inclusion in a mill base, some compromise may be necessary to resolve this conflict of volatility and/or heat sensitivity with a controlled heat development during the run.

Whereas the dispersion efficiency may be high at the beginning of high-speed impeller operation (low temperature, high viscosity, strong shear), the efficiency at the end becomes markedly reduced (high temperature, low viscosity, weak shear). This important observation should be kept in mind in working out techniques designed to provide high-efficiency processing (water cooling, multistage processing with intermediate cool-down periods).

Premixing of the mill base in the high-speed impeller disperser is preferably done at reduced speed, a step that should be carried out to yield a well-mixed slurry free from lumps. Operating the disperser at top speed may actually run counter to this objective, a slower speed being found more conducive to an initial smooth acceptance of the dry pigment powder. One premixing procedure consists in charging the mill base vehicle to the impeller tank, starting the impeller, and adding the pigment portion to the vortex as fast as is consistent with mill base mobility. Too fast a loading may produce lumping and sticking to the sides of the tank. During the loading operation (say after the addition of each bag of pigment), pigment adhering to the tank walls and the shaft of the disperser should be scraped down into the mill base below. After the loading is complete, the mill should be stopped momentarily and any residual pigment again scraped down and incorporated with the mill base mixture. Failure to observe this precaution frequently leads to a seedy or sandy dispersion. A final wash-down with thinner may be desirable at the end of the loading cycle. During the initial loading stage, a lower rpm also significantly reduces dusting. The dispersing stage, using the highest permissible rpm, follows the loading stage.

A second method of loading consists in alternate additions of vehicle (minor portions at first) and pigment (major portions at first) with the same precaution observed that all added pigment be well admixed with mill base vehicle by the end of the loading stage and before the high-speed dispersion step is started.

When pigment mixtures are involved, the general rule is to add first the pigments that are most difficult to disperse (say color pigments).

Highly viscous compositions (at room temperature) can be handled by a high-speed impeller disperser provided that the mill base or the composition in question is preheated to yield a flowable composition and the processing is carried out at this elevated temperature.

Excellent correlation is reported between laboratory and plant high-speed impeller dispersers (possibly better than for any other type of paint grinding equipment).

With time, the serrated disk impeller becomes eroded and the sharp edges and projections that provided points of attrition, shear, or cavitation become rounded off (streamlined) with some ensuing loss of efficiency. Hence, when a high degree of smoothness has been attained, replacement or resharpening of the disk becomes necessary.

Experience has shown that the low-solids-content vehicle so favorable for dispersing is also best for minimizing excessive heat development and for permitting faster pigment addition during the loading stage of the overall dispersion operation.

The ultimate dispersion possible with a high-speed impeller disperser is usually attained within a relatively short period after the loading stage (a time period possibly equal to or less than the loading period itself). Continuing the run beyond this minimum time period contributes little or nothing to the degree of dispersion.

From a critical analysis of numerous TiO_2 grinds made on a laboratory high-speed impeller disperser, Dowling has submitted a number of recommended conditions for obtaining an optimum mill base dispersion. These controlling variables are listed in Table 22-1, together with a scale-up of the conditions that should apply to a production unit for dispersing titanium dioxide pigment. Dowling's recommendations are in line with what has already been stated and can be considered reasonable guides to good industrial practice (15).

In operating a high-speed disk impeller, it is useful to have an ammeter or wattmeter connected into the power line feeding to the disperser motor so that an immediate check can be made on the motor power demand at any step of

Table 22-1 Suggested Operating Conditions for Obtaining an Optimum Mill Base Dispersion, Using a High-Speed Disk Disperser, Based on Titanium Dioxide Pigmentation

Variable	Laboratory Unit	Production Mill
Disk impeller diameter (in.)	3.0	D^a
Tank diameter (in.)	8.5	2.8*D*
Depth of mill base charge (in.)	6.0	2.0*D*
Impeller height above bottom (in.)	1.8	0.6*D*
Impeller speed (rpm)	4900	14,700/*D*
Impeller rim velocity (ft/min)	3860	3860
Vehicle binder content (NV %w)	25–35	25–35
Mill base pigment volume content (%v)	42–46	42–46
Milling time (min)	12	10–15

[a] *D* expressed in in.

the dispersion process. For efficient processing, the power consumption of the disperser unit should normally run close to (or at least not very far under) the rated capacity of the disperser motor. This observation applies to both the loading and the dispersing step. By experimenting with loading and dispersing schedules that call at all times for reasonably high power consumption, optimum processing conditions are generally achieve automatically.

HIGH-SPEED DISK DISPERSER EQUIPMENT

High-speed disk dispersers are commercially available in a wide range of sizes from compact laboratory units (<5-gal capacity; 1 to 2 horsepower; 1000 to 6000 rpm) to large production units that weight over 4 tons (~1000-gal capacity; 75 to 100 horsepower; 400 to 1100 rpm). The disk impeller blades generally range in diameter from 4 in. (laboratory mill) to 36 in. (production unit). The horsepower demand through this range is roughly proportional to the square of the disk impeller radius, a further confirmation that dispersion in a high-speed disk disperser is essentially a bulk effect (laminar flow under surface of disk) rather than an edge effect (turbulent flow along edge of disk).

Manufacturers of high-speed disk dispersers have extended the simple, single-impeller disk design to such mechanical elaborations as two impeller disks mounted on the same shaft, two impeller disks mounted on separate shafts with an overlapping of the two blades (16), and a slower moving, open-type impeller (not a disk) that is positioned coaxially just above the high-speed disk impeller (17) or, in another design, is mounted off to one side of the high-speed disk on a separate shaft (16). Despite the mechanical ingenuity exercised in developing these more complex constructions, the single-impeller disk, with variable speed capabilities, usually provides all the dispersing action that can be reasonably expected from this general type of equipment.

ADVANTAGES AND LIMITATIONS OF THE HIGH-SPEED DISK DISPERSER

Advantages

A high-speed disk disperser probably provides the simplest, quickest, and least expensive means of dispersing easy-to-disperse pigments in conventional vehicles. It also has the advantage of handling all phases of a coating preparation (loading, premixing, dispersion, letdown, and shading) in a single piece of equipment.

The high-speed disk disperser is also a good mixer and can be used to prepare premixes for charging to sand mills, high-speed stone mills, colloid mills, and the like and/or to carry out postmixing operations. It calls for a relatively low initial capital investment, and the only working part requiring replacement is the inexpensive impeller disk (11).

Although the disperser appears simple in operation, a trained operator can, by intelligent observation, select, arrange, and adjust the disperser disk and manipulate the loading sequence to remarkably improve the disperser efficiency and output within it physical limitations.

Limitations

The main drawback of the high-speed disk disperser is its inability to disperse hard or tough agglomerates. This restricts its use to soft-textured, easily-dispersed pigments such as the better grades of titanium dioxide. Even when only a small fraction of the pigmentation consists of oversize particles, hard agglomerates, or foreign contaminants, it is not possible to obtain a clean grind. A second and less obvious limitation (since it is not reflected on a grind gage reading) is the inability of the high-speed disk disperser to fully disperse the overwhleming number of refractory white hiding and color pigment agglomerates in the lower size range (less than 3 μ). For example, as much as one fourth of the pigment value (opacity, tint strength) may be lost through incompletion of the dispersion process (3, 12). In the case of inexpensive pigments this loss may be defended from an overall economic standpoint, but with expensive pigments such waste is difficult to justify on any grounds. In summary, the high-speed disk disperser is an excellent mixer but is equipped only to disperse easy-to-disperse pigments.

REFERENCES

1. Company brochure, "Impellers by Morehouse," Morehouse Industries, Inc., 1600 West Commonwealth Avenue, Fullerton, Calif. 92633.
2. Company literature, "Dispersers," Big "H" Equipment Corporation, 610 Worthington Avenue, Harrison, N.J. 07029.
3. Daniel, F. K., "High Speed Dispersers: Operating and Design Principles," *Paint Varn. Prod.,* May (1970).
4. U.S. Patent 3,486,741.
5. Patton, Temple C., "Theory of High-Speed Disk Impeller Dispersion," *JPT*, **42,** No. 550, 626–635, November (1970).
6. Wahl, E. F. et al., "Dynamic Analysis and Mathematical Modeling of the High-Speed Dispersion System," *JPT*, **44,** No. 564, 98–107, January (1972).
7. Schaffer, Martin H., "Dispersion and Grinding," Unit 16, Federation Series on Coatings Technology, Federation of Societies for Paint Technology, Philadelphia, Pa., 1970.
8. Christensen, V. J. and B. Arnold, "Power Requirements for Disc-Type Impellers," *Off. Dig.*, **37,** No. 491, 1640 (1965).
9. Weisberg, H. E., "A Rheological Study of High-Speed Pigment Dispersion," *Off. Dig.*, **36,** No. 478, 1261 (1964).

10. Ensminger, R. I., "Pigment Dispersions in Synthetic Vehicles," *Off. Dig.*, **35,** No. 456, 71 (1963).
11. Ensminger, R. I., "Techniques for Efficient Pigment Dispersion Operations," *Mod. Paint Coatings*, pp. 29–35, May (1975).
12. Daniel, F. K., "Determination of Mill Base Compositions for High-Speed Dispersers," *JPT*, **38,** No. 500, 534–542, September (1966).
13. Guggenheim, S., "Data on Pigment Dispersion with High Speed Impeller Equipment," *Off. Dig.*, **30,** No. 402, 729 (1958).
14. Staff Article, "The Use of High-Speed Impellers in Emulsion Paint Manufacture," *Paint Technol.*, **32,** No. 10, 34, 36, October (1968).
15. Dowling, D. G., "The Behavior of Titanium Dioxide in High Speed Impeller Dispersion Mills," *JOCCA*, **44,** 188 (1961).
16. Company literature, "Fundamentals of High-Speed Dispersion" and Technical Bulletin No. 1, "Advantages of Dual Shaft Dispersers," C. K. "Bud" Myers Engineering, Inc., 8376 Salt Lake Avenue, Bell, Calif. 90201.
17. Company brochure, "Co-Axial Variable Disperser," Schold Machine Corporation, 7201 West 64th Place, Chicago, Ill. 60638.

BIBLIOGRAPHY

Moore, C. W., "Jet Milled Pigments," *Off. Dig.*, **22,** No. 304, 373 (1950).

Schiesser, R. H. et al., "Pigment Dispersion in High-Speed Impeller Mixers," *Off. Dig.*, **34,** No. 446, 265 (1962).

Wade, W. G. and B. A. Taylor, "Dispersion of Micronized Pigments," *Paint Manuf.*, **30,** 355 (1960).

23 *High-Speed Stone and Colloid Mills*

The high-speed stone mill and the colloid mill represent true smearer-type dispersing equipment. The first commercially important colloid mill was developed in 1920 in England. The high-speed stone mill was designed at a much later date. They differ essentialy only in the contour of their rotor/stator gap configurations.

HIGH-SPEED STONE MILL

The modern high-speed stone (Carborundum) mill, which has evolved from the older and outmoded slow stone mill, consists essentially of two accurately shaped Carborundum stones working against each other, as schematically illustrated in Fig. 23-1a. One stone (the stator) is held stationary, while the other stone (the rotor) is rotated at high speed (3600 to 5400 rpm), with only a small gap separating the two stone surfaces. Dispersing action by viscous laminar flow takes place between the parallel faces of the stones as mill base is fed by gravity or under pressure (15 to 25 psi) to the flat restricted space between the stones. The spacing separating the rotor and stator faces is adjustable from positive contact to a multimil distance.

Description of Operation

The mill is adjusted for a run by first bringing the stones together to establish minimal contact and then immediately separating them to provide a clearance

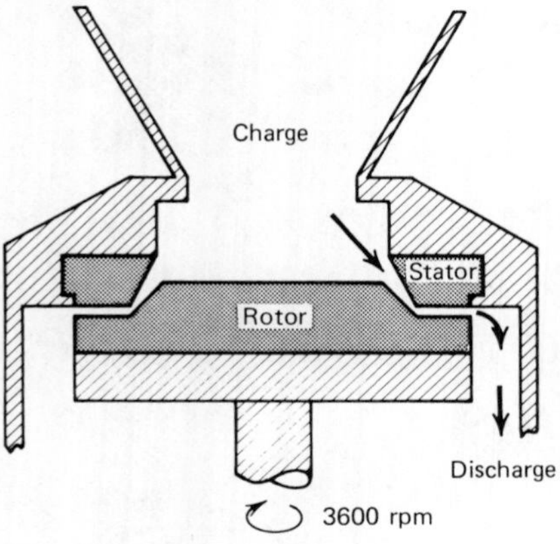

(*a*)

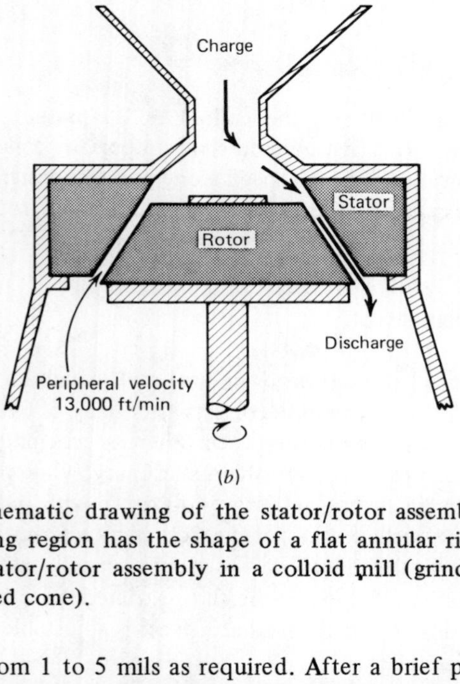

(*b*)

Fig. 23-1 (*a*) Schematic drawing of the stator/rotor assembly in a high-speed stone mill (grinding region has the shape of a flat annular ring). (*b*) Schematic drawing of the stator/rotor assembly in a colloid mill (grinding region has the shape of a truncated cone).

(gap setting) of from 1 to 5 mils as required. After a brief period of operation, further adjustment becomes necessary to compensate for the stone expansion due to the heat generated by the intensive shearing action. Since this heat expansion can close the gap between the stones by as much as 3 to 4 mils in a 15- to 20-min period, an initial gap setting (on a dial) of, say, 2 to 3 mils may have to be increased to an eventual gap setting of 5 to 7 mils to maintain the actual spacing between the stones at the desired 2- to 3-mil level.

It is imperative that any mill base fed to the high-speed stone mill be initially well mixed to a uniform composition. It takes only a simple calculation to show that the average dwell time of mill base between the stones is on the order of a fraction of a second. Hence the dispersing action provided by the mill is substantially a refining rather than a gross mixing operation. Although a certain amount of mixing does take place in the region above the gap entrance, dependence should not be placed on this minimal mixing action for achieving a premix dispersion.

PROBLEMS 23-1

Calculate the average dwell time of an increment of mill base in the rotor/stator milling region of a high-speed stone mill and the distance the increment travels, based on a 4-mil stone clearance and a rotor speed of 3600 rpm. Assume the milling area to be that of an annular ring (1.5-in. inside and 2.0-in. outside radius) and the throughput under these conditions to be 10 gal/hr.

Solution. Consider the mill base as flowing through a slot 4 mils thick (stone gap clearance) and 11 in. long (circumference of a circle having a radius of 1.75 in., an average of the inside and outside radii for the milling area). For 10 gal (2310 in.^{3}) of mill base to flow through the slot area of 0.044 in.^{2} (= 0.004 · 11) in 1.0 hr (3600 sec), an average velocity of 14.5 in./sec must be attained (= 2310/3600 · 0.044). Since the width of the annular ring (mill area) is 0.5 in., the average dwell time is then approximately 0.035 sec (= 0.5/14.5).

At 3600 rpm (60 rev/sec) the mill base will be rotated on the average a distance of 23 in. (= 60 · 11 · 0.035). This extended distance is due to the spiral path that the increment of mill base takes as it passes through the milling region.

The shear rate in the milling zone is very high. For example, the average shear rate in Problem 23-1 is 165,000 sec^{-1} (= 60 · 11/0.004).

Strong Influence of Stone Grit Size on Mill Base Fineness of Grind

The size of the Carborundum grit used in fabricating the stones (ranging from 36 to 100 grit size) provides a major control of the fineness of grind that can be achieved by high-speed stone milling. This is shown in Fig. 23-2, where fineness-of-grind values have been plotted versus grit sizes for the stator and rotor stones, respectively. The mill base feed stock was a 50% solids vehicle in all cases with the pigmentation composed of a mixture of white hiding and fine-particle-size extender pigment. The mill base viscosity was maintained at 103 KU, and the clearance of the stones was adjusted to a 2-mil separation. The data were based on sealed stones (used stones permeated with paint, as opposed to new or unsealed stones which exhibit some degree of porosity). That this sealing factor can

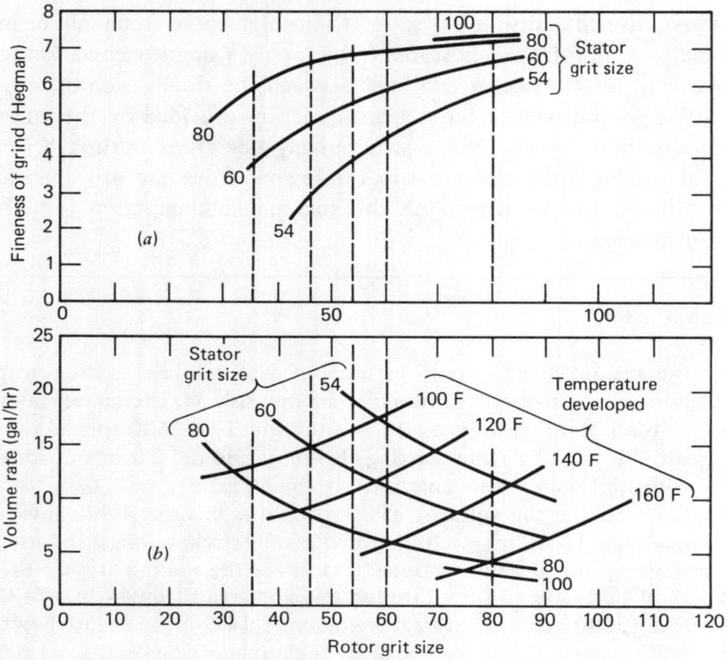

Fig. 23-2 (*a*) Graph showing the effect of the stator and rotor grit size on fineness of grind. (*b*) Graph showing the effect of the stator and rotor grit size on volume throughput and temperature development in the mill.

be significant is shown by the fact that under identical feed conditions a 36 rotor/80 stator grit combination when new (unsealed) gave a 1 to 2 fineness of grind, whereas after several runs (sealed stone) a 6 to $6\frac{1}{2}$ grind was obtained.

Throughput Versus Quality

Reference to Fig. 23-2 shows that an improved fineness of grind is invariably accompanied by a reduction in volume throughput. Also, as might be expected, this slower production rate gives rise to an increase in the temperature developed in the mill base during its passage between the stones. A careful study of Fig. 23-2 provides a better understanding of the importance of grit size selection for the stator and rotor stones and the part this selection plays in establishing optimum conditions for practical production runs.

With daily usage the life of a stone is 4 to 12 months. Changing or replacing stones takes only a few minutes.

Mill Base Compositions for High-Speed Stone Mills

Since the high-speed stone mill is a smearer-type grinding mill, it calls for a fairly high-viscosity mill base, which is formulated using a vehicle with a relatively high binder (nonvolatile) content, ranging, say, from 45 to 53%w. Suggested amounts of vehicle in terms of pigment oil absorption values, together with suggested vehicle NV contents, are listed in Table 23-1.

The mill base charged to the high-speed stone mill must be well mixed. In attaining this goal it is helpful if the supplied vehicle has a higher NV content than is called for by the mill base premix. For example, assume that the supplied vehicle has a 70%w NV content and that this must be let down to 45%w for the mill base composition. Then, in preparing the premix (say on a high-speed disk disperser), it is most efficient to start the premixing using all of the letdown solvent but only part of the vehicle (as supplied). The higher solvent content of the starting vehicle promotes rapid wetting of the pigment, after which the remaining vehicle can be added and the total premix processed to a uniform mixture for charging to the high-speed stone mill.

A graph showing the effects of percent vehicle solids (nonvolatile) and mill base viscosity on fineness of grind and throughput rate for titanium dioxide pigment is given in Fig. 23-3. Also shown are some graphed data for Prussian blue and carbon black. It is seen that a 40%w solids content for the mill base vehicle is indicated for attaining a fine grind with a TiO_2 pigment. However, for the other two pigments, higher vehicle solid contents are necessary.

Mill Base Viscosity

Reference to Fig. 23-3 shows that higher mill base viscosities result in higher fineness-of-grind values, and this trend generally holds. Provided that sufficient

Table 23-1 Suggested Amounts of Vehicle and Vehicle Nonvolatile Contents for Premixes to be Charged to High-Speed Stone Mill

Pigment	Vehicle Demand ($f \cdot \overline{OA}$) f Values	Vehicle Nonvolatile (%w) Content
Carbon black	2.2	45
Chrome yellow	3.3	45
Chrome green	4.0	51
Toluidine red	4.0	53
Para red	4.0	47
Watchung red	4.2	48
Milori blue	4.4	50
Phthalocyanine blue	5.5	48

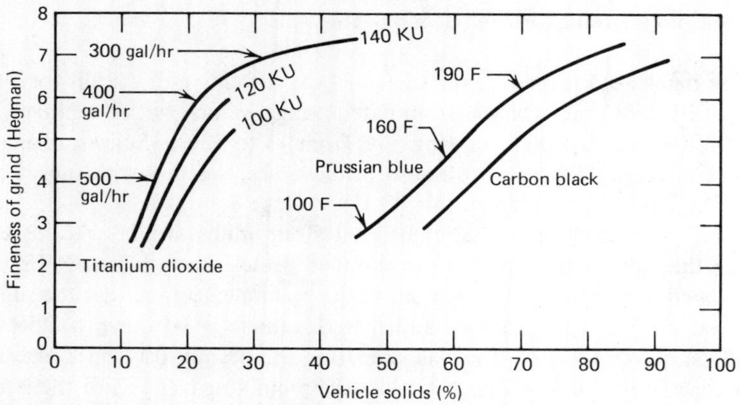

Fig. 23-3 Graph showing the effect of vehicle solids (NV) and mill base viscosity on fineness of grind, volume throughput, and temperature development in a high-speed stone mill, based on a selection of pigments.

vehicle solids are present in the mill base composition to disperse and stabilize the pigment, it is found that mill base viscosity, stone grit selection, and gap clearance are the major factors controlling the fineness of grind of high-speed stone mill dispersion. The mill base viscosity must not be made too high; otherwise, inadequate flow to and unsatisfactory flow from the grinding region result. Also the temperature rise becomes excessive. As in the case of the three-roll mill, a certain degree of mill base tack is desirable to pull the agglomerates apart rather than having them roll around as coherent clusters within the stator/rotor gap.

For low-oil-absorption pigments, formulation of the mill base to viscosities on the order of 100 to 140 KU is normally satisfactory; for high-oil-absorption pigments, lower viscosities in the range of 80 to 100 KU are more common.

Three passes through a laboratory high-speed stone mill provide a dispersion about equivalent to that given by one pass through a production factory unit.

Summary

Table 23-2 summarizes most of the main factors controlling the quality and quantity of mill base dispersion obtained by high-speed stone milling. Because of the many variables involved it is obvious that optimum conditions for any specific production run must be established on a more or less empirical basis. However, from the foregoing discussion it is possible to derive reasonable guides for good practice.

Table 23-2 Summary of Factors Controlling the Quality and Quantity of Mill Base Dispersion Obtained by High-Speed Stone Milling

Factor	Effect on Fineness of Grind	Effect on Production Rate
Stone grit size		
Fine (80–100)	Improves	Reduces
Coarse (36–46)	Reduces	Increases
Stone gap clearance		
Narrow (1–2 mils)	Improves	Decreases
Wide (4–6 mils)	Reduces	Increases
Mill base viscosity		
Low (70–90 KU)	Reduces	Increases
High (100–140 KU)	Improves	Decreases
Rotor speed		
Increase	Improves	Increases
Mill base vehicle solids		
Low (20–30%)	Reduces	—
High (40–85%)	Increases	—
Type of feed		
Gravitational	—	Normal
Pressure	—	Increases

1. Select the proper combination of stone grit size to fit the type of paint being produced. For example, select fine grit stones for the production of gloss paints.
2. Maintain vehicle solids at a high level consistent with the achievement of the final paint properties desired. For white enamels this is about 45% NV.
3. Use the highest mill base viscosity commensurate with adequate flow and reasonable throughput rate. A viscosity value of 100 KU can be taken as a reasonable starting point for initiating experimental work where no other data are available.
4. Provide a well-mixed and uniform mill base premix for charging to the high-speed stone mill.
5. Maintain a proper clearance of the stones at all times. This is usually the tightest setting that will hold the mill base temperature at about 130 to 140 F under equilibrium processing conditions (usually given by a 2- to 3-mil gap clearance).

COLLOID MILL

Although the colloid mill is intended primarily for the preparation of emulsion-type colloidal dispersions, it is also suitable for providing fine-particle-size pigment dispersions. The colloid mill differs from the high-speed stone mill in that the gap configuration for the former is that of a truncated cone, as schematically illustrated in Fig. 23-1*b*. Two types of colloid mills are available, based on the nature of the materials used to fabricate the stator and rotor.

The first type consists of an all-metal stator/rotor assembly that is commonly constructed from Invar, a high-nickel alloy that has the lowest temperature coefficient of expansion of all industrial metals. Hence heating of the product and unit during the grinding operation does not affect the gap clearance. The working faces of the stator and rotor are lapped to provide complete smoothness. If the faces become gouged or nicked, the desirable laminar flow between the stator and the rotor is disrupted by eddy currents, and impaired dispersion results. It is usual to protect the working surfaces by lining them with a thick layer of an extremely abrasion-resistant material (e.g., Stellite).

The second stator/rotor combination consists of Carborundum stones. In one stone design three different layers of grit size are featured in progressing from the top of the bottom of the stone. Thus a mill base entering the grinding region is continually being subjected to a finer grind in moving down through the milling region. With Carborundum stones it is necessary to make gap adjustments for heat expansion during the processing operation.

Colloid mills are normally charged by feeding a well-mixed premix into an open funnel at the top. However, the colloid mill can be adapted to operate as a closed system by using accessory pumping equipment. Jacketing the main housings allows the use of cooling water as necessary.

The clearance between the stator and the rotor is commonly 2.0 mils (51 μ) but is adjustable from as high as 0.125 in. (3.18 mm) down to 0.0010 in. (1.0 mil or 25 μ). The gap clearance should not be used as an index of the grind to be expected. For example, a separation gap of 3.0 mils (75 μ) can yield a particle dispersion having an average particle size as small as 2 to 3 μ. As with the high-speed stone mill, widening the gap increases the throughput but also lowers the quality of the grind.

The colloid mill will handle any well-mixed premix that is flowable. Even flowable powders produced by premixing fine particulate matter (such as finely divided calcium oxide, magnesium oxide, or antimony oxide) with suitable vehicles (e.g., mineral or natural oils) can be charged to a colloid mill to yield soft pastes, as outlined in U.S. Patent 3,951,849 (1976).

Commercial colloid mills range in size from small or laboratory units (3-in. rotor; 30-gal/hr capacity; 5-horsepower motor) to large production mills (15-in. rotor; 1500-gal/hr capacity; 50- to 100-horsepower motor).

Mill Base Composition and Viscosity for Colloid Mills

The composition and viscosity of a mill base intended for dispersion in a colloid mill should be formulated along the same general principles as those outlined for the high-speed stone mill. For both types of mills it is essential that efficient premixing be attained before charging the mill base to the dispersion unit. This places a limitation on the indiscriminate use of these mills but is also the key to their successful operation.

High-speed stone and colloid mills are refining dispersion units that can serve to produce good to excellent quality pigment dispersions based on all but the most difficult-to-disperse pigments.

BIBLIOGRAPHY

Company literature, "Colloid Mills by Premier" and "What Is a Colloid Mill?" Premier Mill Corporation, 224 Fifth Avenue, New York 10001.

Foy, W. L., "A Preliminary Study of Fine Pigment Dispersion on the Morehouse Mill," *Off. Dig.*, **23,** No. 322, 751 (1951).

Morehouse, G. H., "The Principle of Grinding Paint on a Carborundum Stone Mill," *Off. Dig.*, **21,** No. 291, 199 (1949).

Pirrone, F. B., "The Morehouse Mill," *Off. Dig.*, **27,** No. 369, (1955).

Schaffer, Martin H., "Dispersion and Grinding," Unit 16, Federation Series on Coatings Technology, Federation of Societies for Paint Technology, Philadelphia, Pa., (1970).

Taylor, J. J., "Fine Pigment Dispersion," *Off. Dig.*, **26,** No. 353, 445 (1954).

24 *High-Speed Impingement Mills*

The high-speed impingement mill, sometimes referred to as a kinetic dispersion mill, is a true smasher, since it disperses pigment agglomerates primarily by impact. As schematically depicted in Fig. 24-1, a pigment/vehicle mixture that is sucked in at both top and bottom of the mill head is slung outward by rotating slots (the rotor) against a relatively close-fitting surrounding slotted collar (the stator). As a result of this forceful impact, breakup and attrition develop. The suspended pigment agglomerates, emerging at high velocity from the rotor slots, impinge with a shattering effect against the stator wall and against each other. This leads to a smasher-type agglomerate disintegration that provides the required pigment dispersion.

The rotor peripheral velocity is maintained at about 8,700 ft/min for most mills (corresponding to a rotational speed of 5,200 rpm for a large production unit and 16,000 rpm for a small laboratory mill). Since all the mechanical dimensions of the mill are preset by the original mill design, only the factors of vehicle solids, fractional pigment volume, mill base viscosity, order of addition, and running time remain to control optimum dispersion conditions.

VISCOSITY AND VEHICLE COMPOSITION

Since the dispersing action of the high-speed impingement mill depends primarily on smashing impact, it follows that the higest efficiency is obtained by using low-viscosity pigment/vehicle mixtures. A relatively viscous mill base premix introduces so much viscous resistance in passing through the rotor slots that the

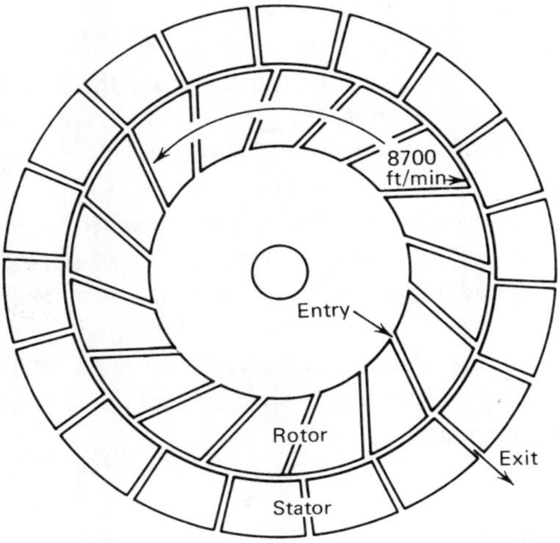

Fig. 24-1 Schematic drawing of the milling head of a high-speed impingement (kinetic dispersion) mill.

resulting low emergent velocity of the premix would fail to develop any useful impact or attrition action. This immediately suggests the use of solvents alone for the vehicle. However, since post flocculation occurs with an all-slovent vehicle, a proper vehicle composition, although mainly solvent, must include sufficient binder (say 15 to 25%w) to ensure that flocculation does not occur on letdown after pigment dispersion has been effected.

The limit of pigment loading is determined by the viscosity that develops in the mill base composition. The loading of pigment must not be so high as to give an excessively viscous mixture during the milling cycle, since this would act to nullify the impact and attrition action.

ORDER OF ADDITION

Pigment dispersion in a high-speed impingement mill is a batch operation. No premixing is required. The vehicle (low solids content) is placed in the mill tank before starting the milling action. After the rotor is started, pigment is fed to the mixture in the tank as rapidly as the pigment can be entrained into the main stream in a smooth manner. At all times it is imperative that the mill head be completely covered by the mill base both to ensure efficient processing and to avoid the introduction of air.

Successful operation of the high-speed impingement mill is based to a large extent on acquired experience. Presumably a technique starting with a vehicle that is rich in solvent (say only 10 %w binder solids), followed by additions of binder-rich vehicle at later stages in the grinding operation, is the type of manipulation that enables one company to routinely use the high-speed impingement mill for the successful preparation of a given product, whereas another company fails to obtain an acceptable product with this type of mill.

The batch grinding time is generally less than 25 min, and since process efficiency declines significantly with time, there is little to be gained by prolonging the operation beyond this time limit. The high-speed impingement mill has the advantages of high-volume production and the use of a single unit for the total coating preparation (premixing, dispersing, letdown, and shading). However, its utilization is more or less limited to dispersing the more easily ground pigments, since the smashing forces generated are inadequate to break up and disperse difficult-to-disperse pigments.

COMMERCIAL UNITS

High-speed impingement (kinetic energy) mills are commercially available in sizes ranging from a laboratory mill designed for in-plant testing and small-volume operations (2- to 6-qt processing capacity; 3- to 5-horsepower motor) to large factory units (250- to 1000-gal processing capacity; 250-horsepower motor).

The laboratory high-speed impingement mill should not be confused with a dispersion unit tradenamed the Dispersator, which also sucks in pigment/liquid mixtures through openings at the bottom and/or top of the milling unit and then slings the mixture outward through slots. Up to this point the two have some common mechanical features, but whereas in the impingement mill the out-moving mill base is flung through deeply grooved slots to smash against a solid stator, in the Dispersator the out-moving mill base is flung only through shallow slits into the slower moving mill base surrounding the rotating Dispersator head.

BIBLIOGRAPHY

Company brochure, "High-Speed Impingement Dispersion Mill," Dispersion Equipment Corporation, 9 Hackett Drive, Tonawanda, N.Y. 14150.

Company literature, "Premier Dispersator," Premier Mill Corporation, 224 Fifth Avenue, New York 10001.

Kew, C. E., "Dispersion by High Velocity Impingement (The Kady Mill," *Off. Dig.*, **31**, No. 408, 70 (1959).

Schaffer, Martin, H., "Dispersion and Grinding," Unit 16, Federation Series on Coatings Technology, Federation of Societies for Paint Technology, Philadelphia, Pa., 1970.

25 *Assessment of Pigment Dispersion*

The degree of pigment dispersion attained in a coating composition can be assessed either directly by some visual observation involving the particle size distribution (as by a grind gage) or indirectly by evaluating some performance aspect of the coating (as by measuring color development or hiding power). In the first case the paint engineer working with practical systems generally considers the absence of oversize particles a reasonable index to a satisfactory dispersion. A research worker would be more interested in observing and measuring the actual particle size distribution within the paint film. In the second case, where the degree of pigment dispersion is assessed in terms of the relative excellence of peformance of some aspect of the coating system, a laboratory worker commonly uses a simple finger rub-up test and bases his or her judgment of the dispersion on a visual color comparison. A research investigator would be more disposed to use sophisticated instrumentation to measure such dispersion-related properties as color development, gloss, opacity, and viscosity. All these testing schemes have their particular fields of usefulness.

MAXIMUM SIZE OF PIGMENT PARTICLES

Sieve Analysis (1)

Pigments intended for paint and ink production are normally supplied in a particle size range such that all or substantially all of the particle pass through a 325-mesh sieve (this sieve size screens out particles 45 μ or greater in diameter).

The 325-mesh sieve represents about the finest screen opening that can be fabricated commercially on a routine basis. ASTM test method D185-72, based on this sieve size, specifies a standard procedure for determining the percentage of coarse particles in a pigment sample for either dry pigments or for mixtures of pigment in vehicles.

In the case of dry pigments the pigment sample and the dry sieve are weighed to the nearest milligram. Water is then run through the pigment (wetted with alcohol) which rests on the sieve screen. By slightly shaking the sieve and by gently stroking with a camel's-hair brush, the bulk of the pigment is assisted through the meshwork of the screen. An end point is reached when a sample of the wash water fails to reveal the presence of any pigment particles. Particles of pigment adhering to the brush are then washed into the sieve, the sieve with its retained oversize particles is dried, and from the several weighings the percentage of particle retain is calculated.

For metallic pigments, denatured alcohol is used in place of water. For water-soluble pigments, oil pastes, or paints, kerosene is used. For pigmented lacquers, a mixture of equal parts of ethyl acetate, benzene, and denatured alcohol is substituted for the water. For carbon black in pellet form, a water solution containing a dispersant specific for carbon black is used. In general, the sieve test is employed primarily to determine the coarse or oversize retain for dry pigment samples.

An analysis of 24 specification coatings reveals that for gloss and enamel paints a retain of less than 0.5 to 1.0% pigment on the 325-mesh sieve is specified for the pigment ingredients, and that for flat and lusterless paints from 1.0 to 2.0% retain on the 325-mesh sieve is permitted.

ASSESSMENT OF PIGMENT DISPERSION BY A PHYSICAL OBSERVATION OR BY MEASUREMENT OF THE PARTICLE SIZE DISTRIBUTION

Fineness of Grind

The degree to which a pigment is dispersed in a vehicle is routinely assessed by the use of a "fineness-of-grind" gage or, more simply, a grind gage (2, 3). The fineness of grind provides a direct measure of the coarsest pigment particles present in a pigment dispersion. In practice, mill bases are commonly ground to comply with some specified fineness-of-grind requirement.

ASTM test method D1210-64 specifies a conventional grind gage procedure to be used by the coatings industry. A single-channel type of gage is schematically illustrated in Fig. 25-1. The gage consists of a hardened steel block which is first ground flat and smooth on one face (approximate dimensions 6.4 × 17.1 × 1.3

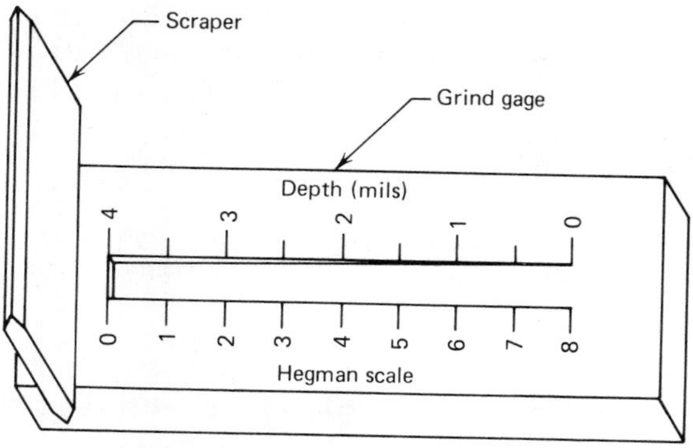

Fig. 25-1 Sketch of a conventional coating grind gage and scraper for measuring fineness of grind of pigment dispersions.

cm). One or two shallow and uniformly tapered channels (each about 0.5 × 5.25 in.) are then machined into the block surface, with the tapering channels running lengthwise from a 4.0 mil depth at one end to zero depth at the other end. The scraper is a wedge-shaped blade with the scraping straight edge rounded off to a radius of 0.01 in., as schematically shown in Fig. 25-1. Grind gages are commercially available in several designs featuring either one or two tapered channels, channel widths from 0.5 to 2.0 in., and depths ranging from 0 to 1.0 mil (grindometer for the ink industry), through 0 to 4.0 mils (standard for the paint industry), to 20.0 mils (for the peanut butter industry) (4).

The test procedure consists in placing a puddle of the test coating in the deep end of the channel (with the gage resting on a flat horizontal surface). With the rounded straight edge of the scraper, the coating is drawn down the length of the channel with uniform motion and with sufficient pressure to wipe clean the flat top face of the gage. The thin, elongated wedge of paint remaining in the tapered channel is then immediately inspected for projecting particles (viewed from the side at a grazing angle). The point on the gage where the coating first shows a substantially speckled appearance is noted and is either recorded as an actual depth (mil or micron) or expressed in terms of an arbitrary scale (such as the Hegman scale shown on the gage diagram of Fig. 25-1) (5). Incidentally, one investigator has pointed out that the actual depth of the coating film is only about 0.65 that of the channel depth dimension (6).

The Hegman (H) scale, sometimes referred to as the National or North Standard (NS) scale, is in common use by the coatings industry today. Although the New York Production Club submitted a paper urging the adoption of a Production Club (P) scale ranging from 0 to 10 in place of the Hegman scale (7), this change has not been widely accepted. More recently it has been proposed that the grind gage readings be recorded directly in mil or micron depth. This is certainly logical and obviates any argument as to the choice of scales which lack fundamental significance. The algebraic expressions relating the three scales are given by Eqs. 1, 2, and 3, where H = Hegman scale reading, P = Production Paint Club reading, and x = channel depth (mils).

$$\mathrm{H} = 0.8\mathrm{P} = 8 - 2x \tag{1}$$

$$\mathrm{P} = 1.25\mathrm{H} = 10 - 2.5x \tag{2}$$

$$x = \frac{8 - \mathrm{H}}{2} = \frac{10 - \mathrm{P}}{2} \tag{3}$$

Figure 13-1 relates the Hegman scale to pigment particle dimensions in graphical form.

The ink industry is less tolerant of particle coarseness than the paint industry, mainly because ink films are generally thinner than paint films. For this reason the standard ink fineness-of-grind gage (grindometer) is designed to run linearly from a maximum depth of 1.0 mil to zero depth. This corresponds to a NPIRI (National Printing Ink Research Institute) scale running from 10 to 0. In commenting on the ink fineness-of-grind (ink grittiness) reading using the grindometer, the ASTM specification covering this test notes the the preferred method of reporting the result is in terms of the film depth in microns rather than the NPIRI scale (3). The relation between the NPIRI scale reading and the film depth x in mils and μ, respectively, is given by Eq. 4.

$$\text{NPIRI scale reading} = 10x\ (\text{mils}) = 0.394x\ (\mu) \tag{4}$$

Grindometer and Hegman gages are commonly used by both the paint and ink industries.

Some years ago a survey showed that values reported for similar particle patterns on the Hegman gage varied widely among operators. To overcome this serious disagreement, standard particle distribution patterns were established for grind readings ranging from 0 to 4 mils. Typical particle distribution patterns for these standardized grind gage readings are depicted in Fig. 25-2. With such pictorial guides, reproducibility among operators is now within ±1.0 Hegman unit.

In evaluating the efficiency of a dispersion operation using a grind gage, it has been found that a straight-line relationship normally results when log fineness-of-grind (μ) is plotted versus log dispersion (milling) time. Occasionally two intersecting straight lines are required to fit the overall data (6).

An analysis of 24 specification coatings reveals that specified minimum Hegman readings range from 3H to 4H for flat paints, concrete paints, primers, undercoats, and the like to 7H and above for gloss coatings and enamels.

Two early papers (1946) reported cooperative studies that served to establish correct procedures for using the Hegman grind gage (8, 9). Many subsequent articles have contributed useful refinements. One article (1970) reported a

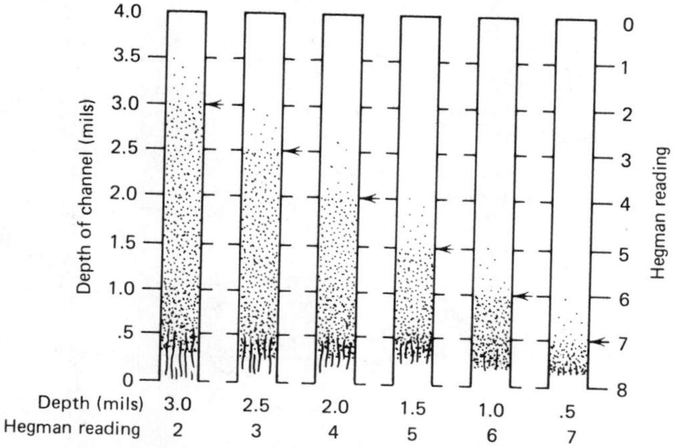

Fig. 25-2 Schematic diagrams illustrating grind gage readings corresponding to standardized particle distribution patterns.

photographic procedure (illumination type and intensity, viewing angle, operator technique) that reproducibly operates a grind gage and immediately takes an instant (Polaroid) picture, providing a precise and permanent record (10). Another article (1969) describes an elaboration of the grind procedure (special illumination and magnification) in which an actual count is made of the protruding particles within the limits of selected graduations along the grind gage (2-in wide path). With this count technique it is possible to readily differentiate among the relative dispersibilities of pigment (titanium dioxide samples in the article) and/or evaluate the efficiency of different dispersion processing procedures (11). A third important contribution outlines a method for examining with greater precision the agglomerates present on a grind gage drawdown that are smaller than 25 μ (greater than 6H) (12). By using a narrow incident beam and a correct geometry, and by proper focusing with a microscope, images of protruding agglomerates can be observed that are not detectable with the naked eye. With this technique it is possible to closely estimate the agglomerate population through the 2- to 25-μ range, which can be critical in affecting such properties as gloss or flatting. Thus a Hegman reading was $7\frac{1}{2}$H for three samples of TiO_2 pigment, yet the sizes of the largest agglomerates viewed through the microscope were 2, 6, and 12 μ, giving 20° gloss values of 75, 65, and 55, respectively.

Despite the refining of the grind gage procedures, two inherent physical shortcomings remain. In the first place, many of the present day fine-particle-size pigments require dispersion to a size that is below the resolution limits of the fineness gage. In the second place, the grind gage gives a picture of only the coarser fraction of the pigment dispersion. Presumably a general reduction in the size of all agglomerates is taking place as the largest ones are being broken up and dispersed.

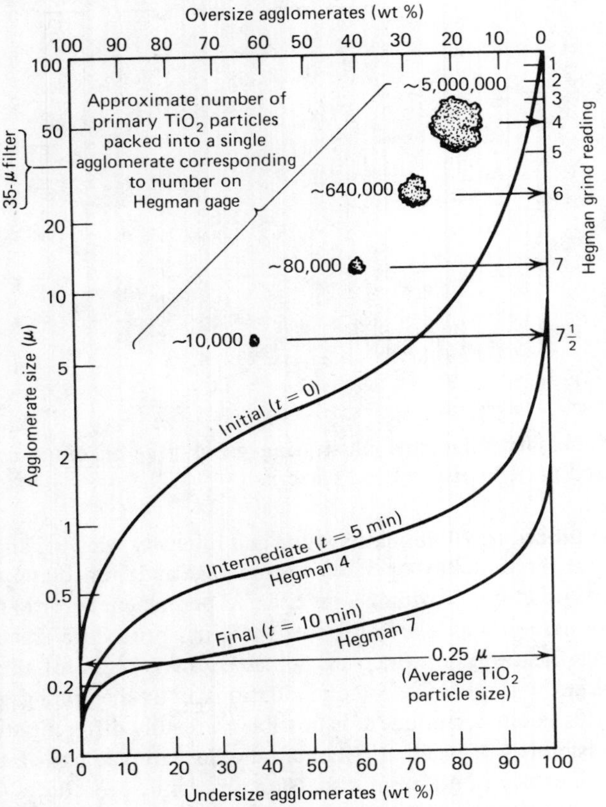

Fig. 25-3 Distribution of titanium dioxide agglomerates corresponding to the initial, intermediate, and final stages in the dispersion of TiO_2 pigment particles in an alkyd vehicle by a high-speed disk disperser.

However, the fineness-of-grind test is incapable of providing data on the overall reduction in the particle size distribution that is occurring. Figure 25-3 serves to provide a visual picture of the relative particle dimensions and the particle size distribution patterns that apply to a titanium dioxide pigment during its dispersion (three stages) in a high-speed disk impeller. It is obvious that a Hegman reading, even with the refinements discussed in the preceding paragraphs, provides only an index of the very coarsest fraction, and even this capability disappears below about 2 to 5 μ. However, the fineness-of-grind test is a quick, inexpensive, and reproducible procedure and as such is an indispensable tool of the coatings industry.

An alternative method for rating degree of dispersion is to measure the fraction of coarse pigment above a fixed size (say above 35 μ) that is present in the dispersion. This technique was used by Wahl et al. in their study of the dispersion capabilities of a high-speed disk disperser (13). This report notes that the deagglomeration process has the twofold purpose of (*a*) deagglomerating all materials larger than 10 to 100 μ (which cause surface defects) and (*b*) deagglomerating all optically active pigment particles to their optimum size of about 0.25 μ. In this experimental work the deagglomeration process was monitored in terms of the pigment fraction greater than 35 μ that was left after filtration of the mill base (at successive stages) through a nylon screen with a 35-μ opening (Millipore filter apparatus). Differences in pigment dispersibility were found to be readily apparent when the log of the retained oversize fraction (>35 μ) was plotted versus dispersion time.

Visual Observation of Dry Paint Film

Simple visual observation of a dry paint film is widely used to assess degree of pigment dispersion. A thin coating of specified thickness is applied to a transparent substrate (say a glass panel). Uniform thickness is commonly achieved by spinning the panel with its applied film in a horizontal position for a fixed time period. Other methods call for spraying, dipping, and the use of an applicator blade as specified in ASTM D823-53 (1970). After drying, the paint film is compared with a set of standard panels (14). The degree of dispersion is judged by observing the relative number of detectable imperfections that are present, such as minute protrusions (nibs) that break the film surface and/or specks that are discernible with transmitted light.

Other Direct Techniques

Direct procedures for assessing degree of dispersion, other than the fineness-of-grind test and visual observation of a dry film, are generally concerned with establishing the total particle size distribution. This much more involved task calls for elaborate and time-consuming techniques suitable only for research investigations. The degree of difficulty becomes understandable when it is considered (*a*) that optically active pigments become most efficient only when dispersed to their optimum particle sizes and (*b*) that this size range is at or below the wavelength of light.

For wet coatings, sedimentation techniques (both gravitational and centrifugal) and electronic particle counting (Coulter counter) can be used to establish the particle size distribution (15, 16). However, these methods usually require a pretest reduction in either particle concentration or viscosity, or both. In diluting the pigment dispersion to comply with such requirements the initial particle size distribution tends to be upset.

For dry paints, resort is normally made to microscopic techniques for directly measuring degree of pigment dispersion. Contact microradiography can be used to study the particle size distribution in the body of the paint film, but the poor resolution afforded by this method (which is limited by the grain size of the photographic film) restricts it to the measurement of particles greater than about 2 μ (14, 15). Other constraints on contact microradiography are the necessarily selective X-ray absorption between the vehicle and pigment components and the application of a very thin film and/or dilution of the coating to a low pigment volume concentration.

The second and most important method for directly assessing the degree of pigment dispersion is by direct microscopic observation. For pigments other than coarse extenders this necessitates the use of an electron microscope to secure the resolving power required to distinguish the primary pigment particles. Even with access to such instrumentation, many practical problems arise regarding the preparation of the sample for viewing and the interpretation of the results that are obtained. This subject is well covered in survey papers by Hornby and Murley (14) and Murray (17). Since this research procedure is of only peripheral interest to the coatings engineer, it will not be discussed further in this book.

ASSESSMENT OF PIGMENT DISPERSION BY OBSERVING OR MEASURING SOME PROPERTY RELATED TO DISPERSION

Excellence of pigment dispersion and optimum paint performance properties are essentially synonymous. In fact, the two are so intimately related that paint properties such as color strength, paint opacity, paint hiding power, surface gloss, and rheological behavior are routinely used to measure degree of pigment dispersion.

Color Development

The degree of dispersion in a tinted or colored paint can be readily assessed in a qualitative manner by resorting to subjective testing techniques. Thus, if a coating containing colored pigment is inadequately dispersed (deficient processing) or if the paint contains pigment that became severely flocculated because of a questionable postgrinding or letdown operation, this unsatisfactory condition can be demonstrated by manually working a portion of the applied film and comparing this portion with an unworked portion. Ideas differ on how the manipulation should be carried out to produce worked and unworked areas for comparsion, but the following procedures appear to be commonly employed.

Finger Rub-up. A drawdown or sprayout is made on a nonporous substrate. After a portion of the solvent has been released from the coating film, the fore-finger is applied to the slightly wet paint and a circular rub-up is made, using a rotary finger motion over a small circular area. A color comparison is then made between the circular rub-up region and the surrounding unworked area. If the color pigment has been dispersed to its ultimate particle size, no significant color difference will be observed. However, if the colored pigment has not been properly dispersed or if flocculation has occurred, the rub-up area will exhibit a more intense color because of the strong dispersing action of the finger rub-up.

Brushing versus Pouring. A vigorous brushout is made over a nonporous substrate, and after a portion of the solvent has been released, two or three blobs of coating are plopped on the brushed-out area. Again a comparison is made between the unworked paint blobs and the worked (brushed) surrounding area. A significantly lighter coating blob is indicative of faulty color pigment dispersion.

Flocculation Number. The foregoing type of testing can be made quantitative by resorting to photometric reflectance measurements. In one such procedure a wide drawdown of the test coating is made and allowed to dry. A small additional amount of the coating is then applied to a section of the dry drawdown and brushed until almost dry. Reflectance measurements are then made over the brushed and unbrushed areas, and from these data a calculation is made to establish the extent of the pigment agglomeration or flocculation. A flocculated or incompletely dispersed color pigment will brush darker than the unbrushed portion; a flocculated or incompletely dispersed white hiding pigment will brush lighter than the unbrushed portion. Equation 5, where the subscripts refer to the unbrushed and brushed conditions, can be used to derive a flocculation number.

$$\text{Flocculation number} = \frac{100(K/S)_U}{(K/S)_B} \tag{5}$$

The K/S ratio of a paint is calculated from the paint reflectance R_∞ by Eq. 6.

$$\frac{K}{S} = \frac{(1 - R_\infty)^2}{2R_\infty} \tag{6}$$

A flocculation number greater than 100 is indicative of white hiding pigment flocculation or incomplete dispersion; a number less than 100 is indicative of color pigment flocculation, incomplete color pigment dispersion, or impaired color acceptance.

Thus, starting with the reflectance measurements, the two K/S ratios for the unbrushed and brushed paint areas can be calculated and used to compute the flocculation number.

The scattering coefficient S of a white hiding pigment is high, and its absorption coefficient K is low. Conversely, the S value for a colored pigment is low and its K value is high. From this it can be shown that in a mixture of white and colored pigments a well-dispersed white pigment imparts high scattering (high tinting strength and high reflectance), corresponding to a low K/S value. On the other hand, a well-dispersed colored pigment provides a high degree of absorption (low reflectance), corresponding to a high K/S value. This is the physical basis for the flocculation number.

Other Dispersion-Related Properties

Practically any paint performance property can be related to the degree of pigment dispersion in the original coating. Even electrical properties are occasionally used to measure the state of dispersion (conductivity is greater for the flocculated state) (18), although in general optical and rheological properties are most commonly employed for this purpose (19).

In a paper reporting on the dispersion of organic pigments with modern dispersing equipment, Herbst used four performance properties (color shade, color hue, viscosity, and gloss) to assess the degree of dispersion (20). Gloss is a surface phenomenon that relates closely to dispersion, since undispersed particle agglomerates and flocculates that protrude through a smooth film interfere with the specular reflection characteristic of high gloss. The close relationship between pigment dispersion and viscosity has been made the basis for many testing programs aimed at rating dispersibility. For example, Seivard and Downey measured the relative abilities of dispersants to provide optimum dispersion of pigments in latex coatings by plotting log mill base viscosity versus level of dispersant (gloss, contrast, and photomicrographs were also used) (21). In a study by Richards and Bovenizer, weathering properties were found to be closely related to the state of dispersion (22). Interestingly, they found that overgrinding (more intensive grinding than was necessary for complete dispersion) led to loss of weather resistance. Presumably this was due to the contamination introduced by the excessive milling, as indicated by a loss in paint reflectance. Overgrinding of flatting pigments (such as diatomaceous silica) can also occur because of unnecessary fracturing of the flatting particles, as reported by Brody (23). Here the dispersed state was measured in terms of loss of gloss and sheen (as well as by fineness-of-grind).

The number of papers relating degree of dispersion to performance properties is practically endless. The papers cited in the foregoing paragraph are simply representative of current technology.

Among the suitable optical performance tests that are adaptable to assessing the degree of pigment dispersion are the following:

ASTM D332-64 (1975) "Tinting Strength of White Pigments"
ASTM D387-60 (1972) "Mass Color and Tinting Strength of Color Pigments"
ASTM D2805-70 (1975) "Hiding Power of Paints"

REFERENCES

1. Patton, Temple C., "Sieving," *Pigment Handbook*, Vol. III (T. C. Patton, Ed.), Wiley-Interscience, New York, 1973.
2. ASTM Standard Test Method D1210-64, "Test For Fineness of Dispersion of Pigment-Vehicle Systems," 1970.
3. ASTM Standard Test Method D1316-68, "Fineness of Grind of Printing Inks by the Production Grindometer."
4. Instrument Brochure B/4, "Dispersion: Fineness-of-Grind Gages," Gardner Laboratory, Inc., Bethesda, Md. 20014.
5. Doubleday, D. and A. Barkman, "Reading the Hegman Grind Gage," *Off. Dig.*, **22,** No. 300, 598 (1950).
6. Doorgeest, T., "Selection and Optimum Use of Dispersion Equipment," *Paint Manuf.*, pp. 31–38, May (1968).
7. Baker, C. P. and J. F. Vozzella, "Evaluation of Production Club Fineness Gage and Paint Film Characteristics Based on Particle Size Distribution of Pigment Dispersions," *Off. Dig.*, **22,** No. 311, 1076 (1950).
8. Baltimore Paint and Varnish Production Club, "Measurement of the Fineness of Grind," *Off. Dig.*, No. 163, 631–633, December (1946).
9. New York Paint and Varnish Production Club, "A Study of Pigment Dispersion: Part I. The Measurement of Fineness of Grind," *Off. Dig.*, No. 163, 633–643, December (1946).
10. Houston Society for Paint Technology, "Development of A Photographic Fineness of Grind Instrument," *JPT*, **42,** No. 551, 736–739, December (1970).
11. Armstrong, W. G., "Some Ideas on Dispersing Titanium Dioxide in High Speed Dispersers," *JPT*, **41,** No. 530, 179–184, March (1969).
12. Golden, Hubert J., "Microscopic Method for Observing Pigment Dispersions," *JPT*, **45,** No. 576, 54–57, January (1973).
13. Wahl, E. F. et al., "Dynamic Analysis and Mathematical Modeling of the High-Speed Dispersion System," *JPT*, **44,** No. 564, 98–107, January (1972).
14. Hornby, M. R. and R. D. Murley, "The Measurement of Pigment Dispersion in Paints and Paint Films," *Prog. Org. Coatings*, **3,** 261–279 (1975).
15. ASTM Special Technical Publication No. 500, Paint Testing Manual, American Society for Testing and Materials, Philadelphia, Pa.
16. Kinsman, Shepard, "Coulter Counter (Measurement of Particle Size)," *Pigment Handbook*, Vol. III (T. C. Patton, Ed.), Wiley-Interscience, New York, 1973.
17. Murray, Jay Whitney, "The Transmission Electron Microscope (Its Role in Pigment Technology)," in *Pigment Handbook*, Vol. III (T. C. Patton, Ed.), Wiley-Interscience, New York, 1973.
18. Parfitt, G. D., *Dispersion of Powers in Liquids*, 2nd ed., Wiley, New York, 1973.
19. Sheppard, I. R. and G. Cope, "Methods Available for Dispersion Testing," *JOCCA*, **46,** 220 (1963).
20. Herbst, W., "Dispersion of Organic Pigments with Modern Dispersion Equipment," *JPT*, **45,** No. 579, 39–50, April (1973).

21. Seivard, L. L., and W. W. Downey, "Pigmentation and Formulation Variables Affecting Gloss in Latex Paints," *JPT*, **40,** No. 522, 293–300, July (1968).
22. Richards, D. P. and G. W. Bovenizer, "Titanium Dioxide: Relation of Dispersion Process to Weathering Properties," *JPT*, **44,** No. 572, 90–96, September (1972).
23. Brody, Donald E., "Performance of Flatting Pigments in High-Speed Dispersion," *JPT*, **40,** No. 525, 439–445, October (1968).

BIBLIOGRAPHY

Jefferies, H. D., "The Dispersion of Titanium Dioxide Pigment and the Influence of the Degree of Dispersion on Some Film Properties," *JOCCA*, **45,** No. 10, 681 (1962).

Patton, T. C., "Paint Rheology and Pigment Dispersion," in *Pigment Handbook*, Vol. III, (T. C. Patton, Ed.), Wiley-Interscience, New York, 1973.

Stieg, Fred B., "High-Speed Dispersion Problems and Titanium Pigment," *JPT*, **47,** No. 603, 43–49, April (1975).

26 *Mill Base Letdown*

During the development stage of any pigmented coating, the formulator is normally required to formulate three distinct but related compositions. The first composition involved is the mill base (designed to yield an optimum pigment dispersion); the second is the coating proper (designed to give satisfactory application); and the third is the converted coating (designed to provide a durable and serviceable film over the substrate to which it is applied). Although this is the sequential history of a paint or ink, the engineer normally thinks in the reverse direction, being concerned initially in deciding on the type of coating best suited to fulfill the demands of a given specification. He or she then works backwards by formulating this coating so that it can be properly applied, and still further backwards by formulating a mill base that will yield a uniform pigment dispersion for the paint or ink involved.

The present discussion is concerned with the letdown step in this cycle of events, where the completed mill base dispersion is let down or reduced with further solvent and/or binder to give a coating that can be satisfactorily applied to a substrate.

CONVENTIONAL (NONAQUEOUS) LETDOWN (1-3)

Graphical Presentation

For better visualization of the place of this letdown step in the preparation of a conventional paint, a flow pattern has been graphed in Fig. 26-1. It gives in turn the volume relationships for the three compositions involved in the develop-

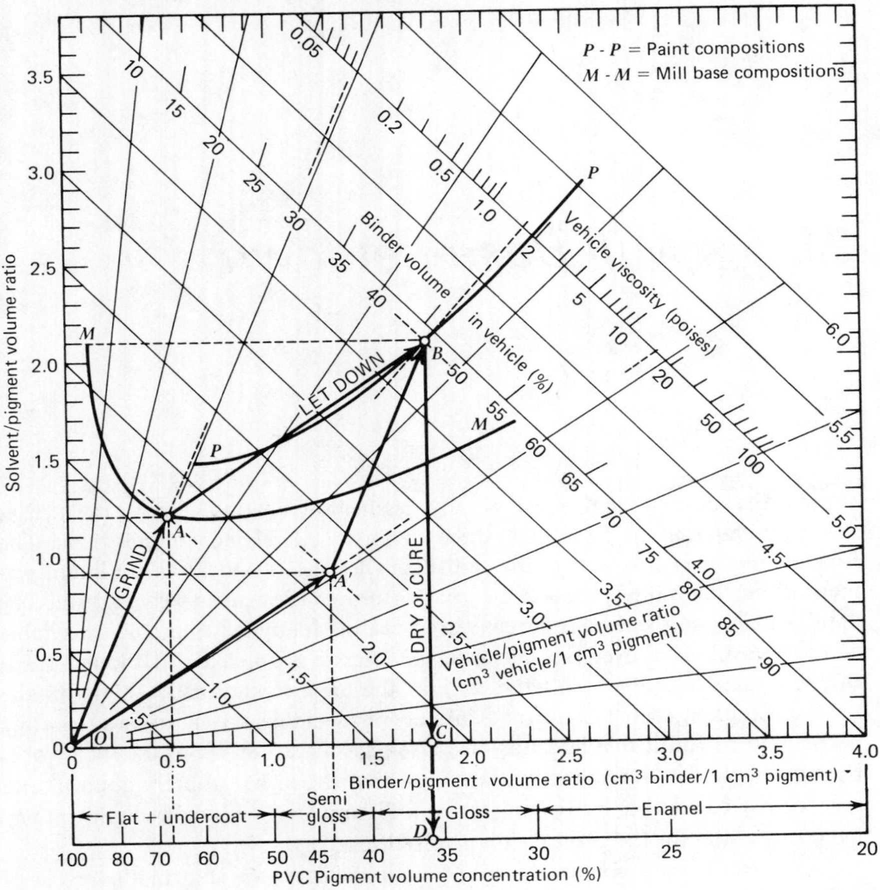

Fig. 26-1 Plot of flow path for the preparation of a typical gloss paint, giving volume interrelationships and vehicle viscosity.

ment of a typical gloss paint prepared from white hiding pigment, mineral spirits, and a long-oil alkyd vehicle (sand mill grind).

Although the manner of plotting is more or less self-evident, a few comments may help in clarifying the interrelationships presented in this graph. All compositional data are presented on a basis of volume. Solvent/pigment volume ratios (cm^3 solvent/1 cm^3 pigment) are plotted on the left vertical scale as ordinate values. Binder/pigment volume ratios are plotted on the bottom horizontal scale of the graph as abscissa values. Also given at the bottom is a supplementary horizontal scale on which corresponding pigment volume concentration (PVC) values are plotted. Running downward (to the right) at an angle of 45° are a series of equally spaced parallel lines giving vehicle/pigment volume ratios (vehicle volume = binder volume + solvent volume). Radiating outward from the origin of the plot are lines corresponding to the varying percentages of binder by volume in the vehicle. The intersections of these radiating lines with the viscosity scale give the corresponding viscosities for these vehicle compositions. Hence, by reference to these several scales, any point on the graph can be immediately related to (*a*) solvent/pigment volume ratio, (*b*) binder/solvent volume ratio, (*c*) vehicle/pigment volume ratio, (*d*) PVC content, (*e*) percentage volume of binder in vehicle, and (*f*) vehicle viscosity. For example, Table 26-1 lists values for these six quantities for the coating compositions *A*, *A'*, *B*, and *C* as read from the graph (see dotted lines).

The five volumetric quantities are invariant (remain fixed on the chart and are independent of the nature of the vehicle). The viscosity, however, depends on the nature of the binder/solvent combination. Hence the viscosity scale will be graduated differently for each new binder/solvent selection. The viscosity scale given on the graph of Fig. 26-1 is one that holds for a typical long-oil alkyd reduced with mineral spirits.

The heavy curves labeled *MM* and *PP* delineate compositions that are suitable for formulating mill base compositions (sand grinding) and paint compositions (brushing application), respectively, for coatings prepared from a long-oil alkyd, mineral spirits, and hiding pigment.

The *MM* line is the locus of a series of Daniel flow points, determined according to the procedure outlines in Chapter 19. From the standpoint of economy, the minimum of mill base vehicle that can be used to secure an effective pigment dispersion is indicated, for a minimum quantity permits a maximum volume of dispersed mill base to be processed through a given volume capacity of milling equipment. In line with this argument, reference to Fig. 26-1 reveals that for a given gloss paint a mill base vehicle having about a 20% binder content (by volume) provides this minimum vehicle demand. However, a mill base vehicle having 30% binder content is only very slightly less efficient and for reasons to be explained later is preferred.

Table 26-1 Volumetric Relationships and Vehicle Viscosity for Coating Compositions *A*, *A'*, *B*, and *C* as Read from Fig. 26-1

Quantity	*A*	*A'*	*B*	*C*
Solvent/pigment volume ratio	1.2	0.9	2.1	0
Binder/pigment volume ratio	0.5	1.3	1.8	1.8
Vehicle/pigment volume ratio	1.7	2.2	3.9	1.8
Pigment volume concentration (PVC)	67%	44%	36%	36%
Percentage binder volume in vehicle	29%	59%	46%	100%
Viscosity of vehicle (poises)	0.09	16	1.6	∞

The *PP* line connects a series of points corresponding to coating compositions that can be satisfactorily applied to a substrate by brushing. Note that this compositional curve tends to approach the 2-poise viscosity line (radiating outward from the origin of the graph) as the PVC content of the paint decreases (enamel range), whereas it departs sharply from the 2-poise line as the PVC content increases much above 45% PVC (flat paint range).

The flow pattern for the preparation of a gloss paint starts at the graph origin *O*, a point corresponding to an all-pigment mass with no solvent or binder present. To this pigment there is then added mill base vehicle having the composition designated by point *A*. The flow path for this addition process runs from point *O* to *A*, at which point the mill base grind is carried out. After the grind step at *A*, the mill base dispersion is reduced with letdown vehicle to give the paint composition at *B*. The flow path for this letdown step is from point *A* to *B*. From a parallelogram relationship (*OABA′O*) it can be shown that the letdown composition involved in this mill base reduction is given by point *A′* (a corner opposite to that of the mill base composition). The spatial geometry afforded by this parallelogram figure is quite helpful in visualizing and estimating the compositional relationships that are involved in proceeding from *O* to *B*.

The flow path for the paint, on application to a substrate, takes a direction vertically downward from *B* to *C*(*D*) as solvent is lost by surface evaporation and substrate absorption. Whereas the initial rate of solvent loss is generally very rapid, the final rate of loss may be quite slow, resulting in a significant amount of residual solvent being retained in the film even after a month or more of exposure. Hence it should be recognized that the initial rapid movement away from *B* on path *BC*(*D*) is considerably retarded by the time a final approach is made to *C*(*D*).

It is suggested that Fig. 26-1 be carefully studied so that the spatial interrelationships of the several graphed quantities are well understood. This understanding is a prerequisite to the following discussion, which makes use of abbreviated and abridged forms of this basic graph.

Of the several stages in the processing and application of a coating, the letdown step is probably least understood. Certainly the technical aspects of this reduction step have been neglected in the technical literature. A primary reason, doubtless, is that the makeup of a letdown composition is normally determined by what must be added to the mill base (designed for dispersion) to give the final paint (designed for satisfactory application). However, resorting to such a routine calculation of a letdown composition by difference, with no thought as to the technical implications involved, can frequently lead to a serious downgrading of the paint or even its outright ruination. Actually, equal technical importance should be assigned to the letdown step, and this consideration may well call for some degree of compromise in the mill base and paint compositions to achieve a more favorable letdown composition.

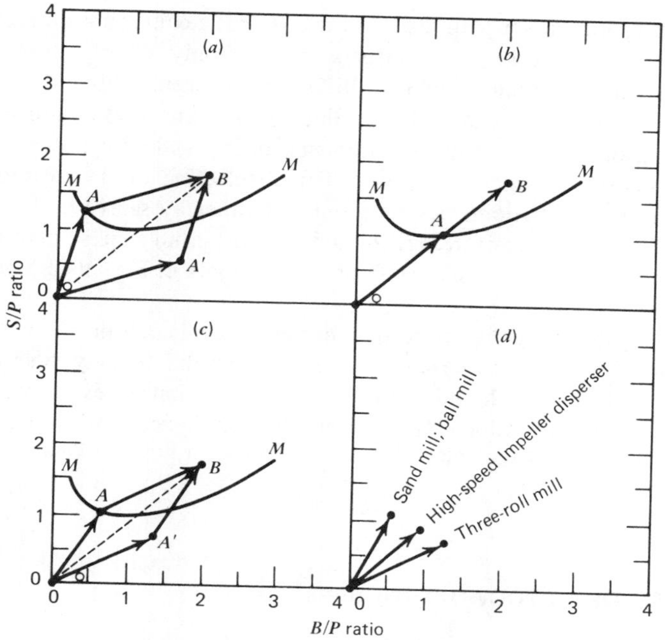

Fig. 26-2 (*a*, *b*, *c*) Alternative preparation routes for gloss paint made on sand mill. (*d*) Location of mill base compositions for four types of grinding equipment.

Reference to the parallelograms of Figs. 26-2*a*, *b*, and *c* shows that several paths are available in proceeding from *O* to *B*. In interpreting these graphed figures, bear in mind that mixing difficulties normally stem from vehicle dissimilarity. Hence, with few exceptions, the more dissimilar the two vehicles (mill base and letdown compositions), the greater is the chance that trouble will develop. Preparation of a paint via path *OAB* of Fig. 26-2*a* is highly economical in that the composition at point *A* corresponds to the minimum amount of mill base composition that can be used to disperse the paint pigment satisfactorily. However, since point *A′* is widely separated from point *A*, it follows that the letdown vehicle composition at *A′* is also widely different from the mill base composition at *A*, a disparity that could well introduce a reduction problem in the letdown step.

Paint preparation via path *OAB* of Fig. 26-2*b* is completely free of any letdown difficulty, since the letdown and mill base vehicles are identical in composition. However, an uneconomical mill base dispersion process for hybrid dispersers (see Table 17-1) is involved here because an excessive vehicle demand is necessary to carry out the dispersion process.

Paint preparation via path *OAB* of Fig. 26-2*c* is a compromise which provides both a reasonably economical dispersion step (only slightly more vehicle is demanded than for point *A* of Fig. 26-2*a*) and a considerably reduced disparity between the mill base and letdown vehicle compositions at *A* and *A′* (thereby minimizing any possible reduction problems during the letdown step).

Other considerations entering into a final selection of the related mill base and letdown compositions may be prompted either by safety factors (a temporary shift of *A* to the left in Fig. 26-2*a* during the grind step is potentially much more dangerous than a similar shift of *A* to the left in Fig. 26-2*c*) or for other reasons.

Although the foregoing discussion has been predicated on the use of a sand mill grind, the principles developed are equally valid for any type of coating preparation. Figure 26-2*d* indicates the initial dispersion stages for paint preparation using four conventional types of grinding equipment. Note that in the case of the three-roll mill the roles of points *A* and *A′* in Figs. 26-2*a* and *c* are essentially reversed.

SOURCES OF LETDOWN TROUBLES

Basically, letdown difficulties generally derive from a gross dissimilarity between the mill base and letdown vehicles. The term "colloidal shock" has been loosely used to describe the chaotic condition that arises when two widely different vehicle phases are abruptly brought into intimate contact. The following discussion considers the mechanisms whereby this dissimilarity causes reduction difficulty during the letdown step. Occasionally, as will be shown, dissimilarity can prove helpful.

Vehicle dissimilarity can be present not only as a disparity in composition but also as a difference in vehicle viscosity, surface tension, temperature, vehicle micelle development, and other vehicle aspects. Furthermore, the mill base is significantly different in that it is carrying a pigment dispersion.

The objective of the paint preparation is primarily the achievement of a uniformly mixed paint. This implies the development of not only a well-dispersed pigment system but also a well-dispersed binder system. It is recognized that some degree of nonuniformity can be tolerated and that frequently slight nonuniformity is purposely developed in a coating to attain certain desirable end characteristics. Thus some degree of binder agglomeration (micelle development) can serve to give a certain degree of nonpenetration, and some degree of pigment flocculation is helpful for minimizing hard settling. However, when this tendency to nonuniformity gets out of hand (leading, say, to seeding), coatings become seriously if not irrevocably downgraded.

Binder (Resin) Precipitation

A common source of unwanted binder (resin) precipitation is a faulty letdown technique wherein the dilution tolerance of the binder resin is exceeded at some stage in the letdown process. To visualize this situation, refer to Fig. 26-3*a*, which schematically graphs the preparation route for a gloss paint (three-roll mill grind). The mill base composition for this gloss paint is an alkyd vehicle (67% solids in xylene) that can tolerate dilution with mineral spirits (MS) down to a 27% solids content. Since the gloss paint specifies an alkyd vehicle of 45% solids content, this paint composition, of itself, is well on the safe side of the MS tolerance limit.

If the letdown procedure is not carried out properly, however, localized regions of precipitation (resin "kick-out") can develop where binder vehicle and letdown vehicle (in this case MS solvent) are inadvertently brought together to give a mixed vehicle in the resin precipitation range (0 to 27% solids content).

To exaggerate an *improper* letdown technique, let the pigmented vehicle be slowly added to all of the solvent in the letdown tank. At the start of this letdown procedure the resin content of the solvent in the tank is zero. As pigmented mill base vehicle is added, the binder content of the letdown solvent rises, and from 0 to 27% resin precipitation takes place. Any strong agitation here only acts to accelerate the precipitation process. Above 27% the process is reversed, and solution rather than precipitation occurs. Eventually, all the resin is brought back into solution, giving the 45% vehicle solids solution called for by the paint composition.

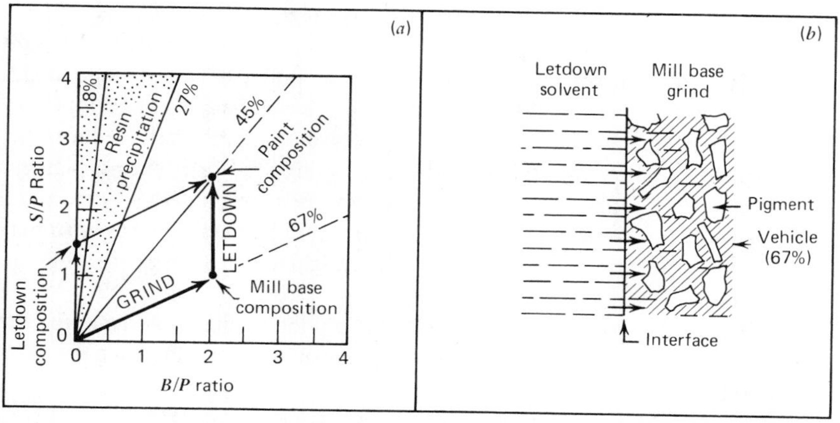

Fig. 26-3 (*a*) Plot of preparation route for a gloss paint, using a three-roll mill. (*b*) Penetration of interface by letdown solvent to extract binder from pigment particles.

Now, in the foregoing sequence of events, it was tacitly assumed that no irreversible side effect takes place. But this is exactly what cannot be assumed, for during the temporary binder precipitation period havoc can be visited on the pigment system, leading to irrevocable agglomeration. To avoid this potentially disastrous situation, resort must be made to any one of several alternatives. In the first place, the order of addition can be reversed; thus solvent can be slowly added to the pigmented binder vehicle under strong agitation. With this technique the approach to the 45% solids content of the paint vehicle is made from the 67% side (mill base composition), thus avoiding the dangerous 0 to 27% precipitation region. Strong mixing agitation can and should also be used to ensure that the small incremental additions of solvent are immediately smeared over the bulk of the pigmented binder vehicle. In this way the letdown is prevented from ever building up to a dangerous percentage at any localized point.

Finally, by simply formulating the mill base and letdown vehicles so that both are on the safe side of the dilution tolerance limit, resin precipitation is again avoided.

In summary, to avoid or minimize letdown problems where resin precipitation is a factor, add the letdown solvent or solvent-rich vehicle to the mill base dispersion slowly, with vigorous mixing agitation, and preferably formulate both mill base and letdown vehicle to lie on the safe side of the dilution tolerance limit of the resin in question.

Binder Extraction

Extraction or stripping of binder from a pigment arises in situations where a major amount of solvent or solvent-rich letdown vehicle is inadvertently brought into contact with a minor amount of binder-rich mill base dispersion. In this situation the bulk solvent penetrates the binder-rich mill base that envelopes the pigment particle and by diluting the mill base vehicle drastically reduces the binder dispersion efficiency. Pigment stability is impaired or lost, and pigment flocculation ensues. Since the net effect is a movement of binder away from the pigment particle into the bulk solvent, the action can be described as a stripping effect. To see how this undesirable condition might occur in practice, refer to Fig. 26-3*a* and *b*.

In this particular case assume an unlimited mineral spirits tolerance for the mill base alkyd resin (this assumption is equivalent to assuming that the resin precipitation region does not exist).

As before, to exaggerate an *incorrect* letdown technique, let the pigmented vehicle be slowly added to all of the solvent in the letdown tank. No resin precipitation occurs in this case, but an equally dangerous situation arises; each small incremental addition of mill base dispersion is immediately drowned in solvent, a condition that permits the solvent to effectively strip binder from around the dispersed particles. As a result of this binder loss, pigment flocculation ensues, with the possibility of serious, irreversible pigment agglomeration. Note that a more vigorous mixing action under these addition conditions only accentuates the stripping effect.

The solution to this problem is to reverse the addition procedure, namely, add the letdown solvent slowly with vigorous mixing action to the mill base dispersion. In this way the solvent is never afforded the chance to build up a concentration sufficient to effectively strip or extract mill base from the pigment particle.

An additional helpful technique is to add binder to the letdown solvent and, by thus initially satisfying the solvent demand for binder, lessen its tendency to strip binder from the dispersed particles. This idea is actually equivalent to narrowing the paint preparation parallelogram (see Fig. 26-3*a*) by bringing the points corresponding to the mill base and letdown vehicle compositions closer together.

Solvent Extraction

When a solvent-rich vehicle is brought into contact with a solvent-impoverished vehicle, cross migration and interdiffusion immediately ensue and soon blend these two disparate vehicle phases into a single uniform vehicle composition of intermediate solvent content. Although both the solvent and binder components of a vehicle enter into this cross-migration and diffusion process, it is almost entirely the *more mobile solvent* component that accounts for the bulk of the transfer across the interface and the subsequent blending of the two phases. Hence, in interpreting any letdown behavior, the process should be visualized as taking place in terms of a rapid movement of solvent from the solvent-rich vehicle to the solvent-starved vehicle. To a major extent, the success of the letdown process depends on control of this solvent transfer, which must be carefully directed to avoid potentially dangerous conditions during the letdown operation.

An interesting application of the above mobility principle has been reported by Garrison (4). He shows that of two solvents in a homologous series the slower diffusing one (the solvent with the higher molecular weight) invariably produces significantly less flocculating action (say when tinting paste is incorporated with a white base enamel containing the solvent). Thus the higher molecularweight solvent diffuses at a slower rate into the protective resin layer surrounding the color pigment, and hence there is less chance that a given particle will be free to coalesce with another particle. In any homologous series the rate of diffusion varies inversely with the square (approximately) of the solvent molecular weight

M. Consider two isoparaffinic solvents E ($M = 120$; $\eta = 0.0085$ poise; $\gamma = 20$ dynes/cm; b.p. = 240 to 286 F) and L ($M = 165$; $\eta = 0.0154$ poise; $\gamma = 23$ dynes/cm; b.p. = 372 to 406 F). The overriding influence on mobility here is the diffusion rate, and from the inverse square relationship solvent E should diffuse nearly twice as fast as solvent L ($1.9 = 165^2/120^2$). Such a diffusion ratio provides a fairly accurate picture of the respective rates of incursion of the two solvents into a protective binder surrounding pigment particles.

From this discussion the important conclusion can be drawn that the tendency toward flocculation during a coating letdown can be minimized by the use of high-molecular-weight solvents (high-boiling solvents). This suggests in turn that, when commercial hydrocarbon solvents are involved, those with low initial boiling points should be avoided (other things being equal). As a general rule, when flocculation presents a possible problem, relatively high-boiling solvents that have a narrow boiling-point range (say 370 to 400 F) should be specified.

Note that such danger spots are related, not necessarily to a vehicle condition in itself, but rather to the ill effects that such a vehicle condition can visit upon the pigment that it suspends. Thus, in the previous case of binder extraction, the flooding of binder-rich mill base with solvent is not necessarily bad per se. Rather, the harmful effect derives from the deleterious influence of this solvent-flooded vehicle on the pigment dispersion. Temporarily stripped of binder-rich vehicle, the pigment is induced to flocculate, completely negating the work of the prior dispersion step.

Pigment agglomeration within a mill base dispersion by solvent extraction is, in a sense, the reverse of pigment flocculation due to binder extraction. The difference is clarified by Figs. 26-3*b* and 26-4*b*. In the case of binder extraction (Fig. 26-3*b*) the pigment is initially dispersed in the binder-rich vehicle (say by a three-roll mill); in the case of solvent extraction (Fig. 26-4*b*) the pigment is initially dispersed in the solvent-rich vehicle (say by a sand grinder). In both cases, however, the same letdown trend is observed, namely, a rapid solvent migration from the solvent-rich to the solvent-starved vehicle that largely accounts for the subsequent mixing and letdown operation. From Fig. 26-3*b* it can be seen that the solvent flooding taking place with letdown solvent during this operation can readily strip binder from pigment dispersion in the binder-rich phase (at least locally). It will now be shown from Fig. 26-4*b* that pigment agglomeration (seeding) can be induced when pigment dispersed in a solvent-rich phase is let down with binder-rich vehicle.

Figure 26-4*a* schematically graphs the preparation route for a gloss paint (sand mill grind). The mill base for this gloss paint is an alkyd that has been reduced with solvent to a 25% solids content. The letdown vehicle is the unreduced alkyd (60% solids). Figure 26-4*b* shows a section of the interface between

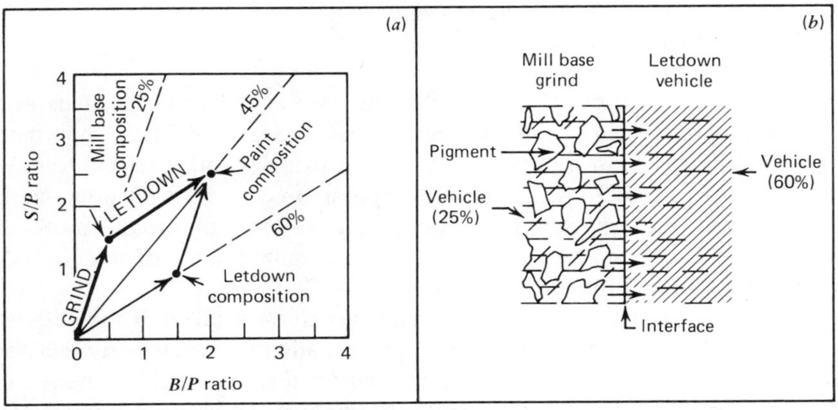

Fig. 26-4 (*a*) Plot of preparation route for a gloss paint, using a sand grinder. (*b*) Diffusion of mill base solvent into binder-rich letdown vehicle, leaving pigment particles behind.

these two vehicle phases at the beginning of the letdown operations. The following sequence of events now take place. In accordance with the general principle that the low-viscosity phase transfers most readily across an interface, the solvent of the solvent-rich mill base vehicle immediately commences to diffuse into the binder-rich letdown vehicle. But (and this is the key consideration) this migration of solvent does not necessarily include the transfer of pigment. In fact, the pigment is normally left stranded behind in the solvent-rich phase, a phase that continually shrinks in volume as solvent is extracted by the solvent-starved letdown vehicle. As a result the stranded pigment particles are squeezed and compressed ever closer together in the shrinking mill base until they finally come into contact and agglomerate, a compacted condition that is recognized as pigment seeding in the letdown paint.

The occurrence of such seeding can almost always be avoided by simply resorting to a proper reduction sequence, namely, an order of addition in which the letdown vehicle is slowly added to the mill base under conditions of vigorous agitation. When this is done, the mill base becomes capable of continuously accepting small increments of binder-rich letdown vehicle, and because of the slow addition no buildup of letdown vehicle is threatened at any point in the letdown operation. As a result solvent extraction is avoided, and the danger of pigment seeding is automatically eliminated.

An additional safeguard can also be arranged by formulating the mill base and letdown vehicles to more nearly match each other in composition. This is again equivalent to narrowing the paint preparation parallelogram of Fig. 26-4*a*.

Importance of Properly Allocating Mixed Solvents in the Preparation of a Paint

The proper allocation of solvents to the mill base and letdown portions of a paint composition (when a mixed solvent system is specified) is vitally important. A guiding principle is to formulate in a manner that will retard solvent transfer from one vehicle phase to the other. This appears to contradict the intermixing objective of the letdown step, but the idea here is simply to slow down solvent transfer to render it manageable and to avoid localized spots of unduly high transfer that lead to difficulty.

In carrying out this principle in practice, advantage is taken of the observation that the so-called stronger solvents are generally more tenaciously retained and more resistant to displacement from a binder than are weaker solvents. In the case of hydrocarbon solvents, relative strength can be assessed in terms of a kauri-butanol (KB) value. Thus the relative strengths (in increasing order) of three common hydrocarbon solvents are odorless mineral spirits or OMS (23 to 28), mineral spirits or MS (33 to 48), and xylene (98). On this basis OMS should diffuse at a much slower rate into a xylene/binder vehicle than xylene into a OMS/binder vehicle (other things being equal). The foregoing is highly empirical, but it serves as a rough rule-of-thumb approach for allocating solvents and for avoiding mixing conditions that have been shown to give pigment agglomeration and seeding in practice.

Figure 26-5 charts safe and (potentially) unsafe solvent combinations for the mill base and letdown vehicles, respectively. Also indicated are areas of possible pigment agglomeration when one solvent is used throughout (the same solvent for both vehicle phases, as previously discussed). For a mixed solvent system the general rule is to formulate the weaker of two solvents into the solvent-rich phase. To reverse this order is to invite the probability of serious pigment trouble. The one exception arises when the binder of a low-solvent mill base has a limited dilution tolerance. In such a case it becomes necessary to resort to alternative stratagems to achieve an acceptable letdown, such as adjusting the letdown and mill base vehicles to more nearly match each other in binder/solvent composition.

PROBLEM 26-1

A 16% mill base dispersion of flatting silica pigment in a high-solvent-content alkyd vehicle (minaral spirits) exhibits seeding when let down with the same type of vehicle (but with low solvent content). Suggest a method for overcoming this seeding deficiency.

Solution. Replace a portion of the mill base MS with OMS (say 50% substitution). In this way the solvent-rich vehicle is formulated with a weaker solvent, and seeding is avoided.

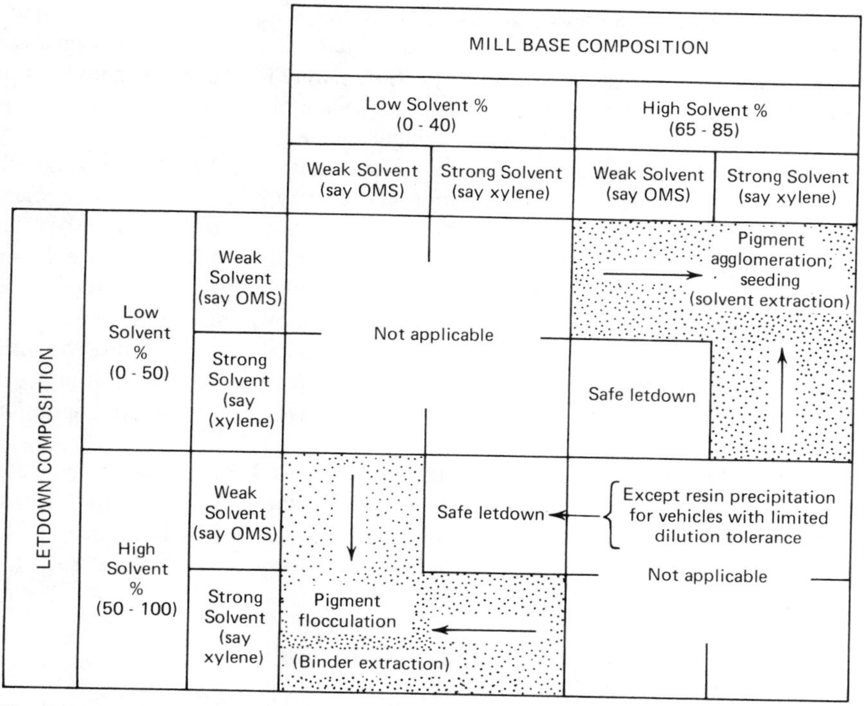

Fig. 26-5 Chart giving safe and unsafe combinations of two solvents for the mill base and letdown vehicles, respectively.

Binder Aggregation

Resin precipitation due to exceeding the dilution tolerance of a resin during the letdown stage has already been pointed out as a source of letdown trouble. The present section considers a less obvious latent danger that derives from blending vehicles of widely different binder content.

Binder molecules are invariably large in size and in solution tend to agglomerate in groupings (presumably symmetrical) termed micelles. In extreme instances these agglomerations or micelles can grow to yield specks or gels that are visible to the eye. Although these binder agglomerates seldom develop to the visible state in coating vehicles, they still exist as smaller entities. Hence in paint preparation two types of dispersions are actually being dealt with, namely, visible pigment dispersions and the less apparent micelle or colloidal dispersions of binder.

Under strong conditions of agitation, these binder agglomerates are normally broken down to a dispersed form. On termination of the mixing action, they proceed to reform gradually. However, cases frequently arise where the agitation of the agglomerated binder may be insufficient to completely disperse the binder components. Instead, the mixing action merely serves to cut them up into tiny clumps. Such insufficient dispersing action may lead to problems. Thus, if the agitation process is one in which a highly viscous vehicle (characterized by high binder content and strongly agglomerated binder molecules) is mixed with a low-viscosity solvent, a casual mixing action may well result in a problem product in which tiny islands of agglomerated binder float about in a bath of solvent. In this case the original state of the binder has not been broken down, but rather has been sliced or shredded into metastable sections that can often be recognized as gel-like seeds on a drawdown (fineness-of-grind gage).

It is also important to recognize that, if the agitation for the process is insufficient, no amount of time expended on the mixing operation will improve the paint dispersion. Most letdown operations have such a minimal agitation requirement.

On the high side, it has occasionally been reported that excessive agitation can also introduce an undesirable effect. Thus undue breakdown of micellular structure in the presence of pigments can conceivably result in irreversible changes that in turn can be related to a downgraded paint.

PRACTICAL RECOMMENDATIONS FOR ESTABLISHING OPTIMUM LETDOWN CONDITIONS

Mechanical

1. Slowly add the letdown vehicle by small increments to the mill base under vigorous mixing conditions.
2. Avoid temperature and/or viscosity extremes between the letdown and mill base vehicles.

Compositional

3. Design the composition of the letdown vehicle so that it is not too different from that of the mill base vehicle.
4. Allocate the weaker solvent of any solvent mixture to the vehicle phase that is solvent-rich (except for a mill base having a binder with limited dilution tolerance).

The first recommendation specifies the correct order of addition when reducing mill base with letdown vehicle. The reverse order of addition (mill base to letdown vehicle) is at all times highly questionable. Even with the correct order of addition, the rate of addition of the letdown vehicle should be gradual (especially at the beginning of the reduction operation). Dumping bulk letdown vehicle into mill base is to be avoided, since such a procedure is likely to set up localized danger spots, especially if the dumping is followed by weak mixing action.

The second recommendation warns against bringing together letdown and mill base vehicles that are radically different in temperature and/or viscosity properties. Thus adding a viscous, cold letdown vehicle to a warm mill base of low viscosity is an invitation to trouble. As a general rule, it can be stated that the addition of any strongly thixotropic or near-gel composition to a thin base should always be avoided. The proper technique is to schedule a strong agitation of the viscous phase, followed by a gradual dilution of the viscous with the less viscous phase.

The third recommendation practically rules out the use of pure solvent or 100% binder for the letdown vehicle, suggesting instead a closer equalization of the letdown and mill base binder contents, at least for early additions of the letdown vehicle.

The fourth recommendation gives the proper allocation of the weaker and stronger solvents in a mixed solvent system and by implication suggests that the deliberate introduction of two solvents may be feasible in certain instances to achieve a safer letdown condition, with the weaker solvent being assigned to the solvent-rich phase.

LATEX PAINT LETDOWN (4-8)

In the preparation of a latex paint, the letdown of the aqueous mill base dispersion with polymer emulsion and thickener is a critical step that invariably involves some degree of pigment and polymer agglomeration. A minimal amount of such agglomeration is acceptable, as may be necessary to achieve a desirable thickening of the latex paint, but excessive agglomeration leads to impaired hiding and color development, can instability, and reduced durability (weathering).

Unlike a solvent paint, the latex mill base seldom contains binder (polymer emulsion) for the reason that the dispersion agitation necessary to incorporate the pigment into the mill base would act to break the polymer emulsion. Latex paint letdown also differs from solvent paint letdown in that only moderate agitation (over a limited time period) is called for, whereas vigorous agitation is specified for a solvent paint. The letdown procedure for a latex paint then consists essentially in intermixing two types of particles by mild agitation. The polymer particles in the letdown emulsion are tiny, discrete spheres with average diameters (depending on the emulsion type) ranging from about 0.1 to 10 μ.

On the other hand, the pigment particles in the mill base dispersion may be of varied shape and exhibit a more extreme range in dimension. The success of the letdown step depends on achieving an intimate intermingling of these two disparate particle systems to yield a stable and uniformly distributed overall particle suspension (the latex paint). As a firm rule, the polymer emulsion should always be added to the mill base dispersion.

In considering any aqueous suspension, careful thought must be given to such controlling factors as the pH of the system; the presence of electrolytes; the nature and amount of surfactants and stabilizers present; the particle pigment volume concentration; the size, shape, and distribution of the suspended particles; the influence of thickeners, plasticizers, and other additives present; the viscosity; and the suspension temperature. If two suspensions are to be mixed, it stands to reason that the closer they match each other in all of the foregoing respects, the less chance there is that antagonistic situations will arise to jeopardize the securing of a stable and satisfactory mixed particle suspension. Obvious incompatibilities involve the admixture of an anionic with a cationic emulsifier; the inadvertent introduction of polyvalent ions (say Ca^{2+}, Mg^{2+}, and Al^{3+} cations from hard water) into an unprotected anionic polymer emulsion; the admixture of pigments capable of contributing calcium ions with phosphate-type dispersants (calcium phosphate precipitation); the addition of a polymer emulsion to a hot mill base; and the admixture of suspensions with widely different pH values.

Despite the care given to the letdown stage some degree of both polymer and pigment flocculation appears to develop. Of the two it is probably more important to maintain the polymer particles in as high a state of dispersion as possible, because the ability of a polymer particle to deform into a continuous and useful film during the last stages of water removal is favored by small particle size and high deformability. Hence any flocculation of polymer particles into agglomerates (equivalent to a larger particle size) reduces the effect of the surface tension forces favoring deformation and renders the particles less deformable. Particle flocculation also severely limits the ability of the polymer particle to penetrate into chalked paint and hence aggravates this already unsatisfactory adhesion problem.

Since the letdown mixture of a mill base dispersion with a conventional latex invariably results in a latex paint that is deficient in viscosity, it is usually necessary to resort to a thickening agent (generally a water-soluble polymer or its alkaline salt) that is capable of boosting the latex consistency. Unfortunately, the thickening agents that have been developed to date for this function induce a more or less severe polymer particle flocculation as part of the thickening process. If this agglomeration becomes excessive, it leads to latex paint instability, such as creaming and polymer "kick-out." Also, as has already been mentioned, latex paints thickened by this flocculation mechanism exhibit an impaired film integrity (grain cracking, deficient bonding). However, as long as résort is made to thickening agents to achieve a proper application consistency, the best answer to the flocculation problem is to select the thickening system that induces a minimum amount of polymer agglomeration (visible as film hazing and reduced gloss of the dry latex film).

It is difficult to prescribe specific rules to follow in scheduling a latex paint letdown. Undoubtedly the main guiding principle is to match the letdown composition to the mill base dispersion composition in all possible respects and to avoid any severe mechanical agitation that would disrupt and coagulate the polymer emulsion.

The letdown procedure for a latex coating is discussed in Chapter 22 for a high-speed disk disperser.

REFERENCES

1. Daniel, F. K., "Some Observations on the Theory and Practice of Post-mixing," *Off. Dig.*, **33**, No. 438, 830 (1961).
2. Daniel, F. K., "The Influence of Solvents on Pigment Dispersion and Seeding," *Off. Dig.*, **28**, No. 381, 837 (1956).
3. Brook, C. H., "Phenomena in Paints and Varnishes," *Natl. Paint Varn. Lacquer Assoc. Sci. Sect. Circ.* No. 546, 247 (1937).
4. Garrison, R. A., "Color Development in Alkyd Paints," *Off. Dig.*, **36**, No. 469, 167 (1964).

BIBLIOGRAPHY

Carpenter, M. C., "Latex Paints for Industrial Finishes," *Off. Dig.*, **30**, No. 403, 845 (1958).

Hahn, F. J. and J. F. Heaps, "Heterogeneity in Latex Paints," *Off. Dig.*, **35**, No. 459, 366 (1963).

Heiberger, P., "Aqueous Coatings," *Off Dig.*, **29**, No. 385, 100 (1957).

Liberti, F., "Latest Vinyl Copolymer Formulating Techniques," *Off. Dig.*, **29**, No. 389, 560 (1957).

Scholl, E. C., "Recent Developments in Finishes Based on Water Dispersed Paint Systems," *Off. Dig.*, **32**, No. 423, 519 (1960).

Schwahn, C. O. and W. M. Sullivan, "Effective Filming of Polyvinyl Latex Paints," *Off. Dig.*, **30**, No. 405, 1122 (1958).

Tess, R. W. and R. D. Schmitz, "Use of Hexylene Glycol and Other Solvents in Styrene-Butadiene Latex Paints," *Off. Dig.*, **29**, No. 395, 1346 (1957).

27 Pigment Settling

Any discussion of pigment settling must start with a consideration of simple systems, since the slow downward migration of pigments through a paint vehicle can and does become enormously complex. Hence it is expedient to develop some concepts of what happens in simpler systems and then extend these concepts to generalized settling behavior in practical paint compositions.

SETTLING OF SINGLE SPHERICAL PARTICLE IN NEWTONIAN LIQUID

Figure 27-1 illustrates in schematic form the settling of a single spherical particle in a Newtonian liquid of infinite extent under the force of gravity. The viscous resistance encountered by a sphere moving through such a bath can be calculated by Stokes's law (Eq. 1).

$$F = 6\pi r \eta v \tag{1}$$

where F = viscous resistance (dynes)
r = radius of sphere (cm)
η = viscosity (poises)
v = velocity (cm/sec)

By using a correction factor, Eq. 1 can be modified to apply to a container of finite dimensions. However, the radius of the pigment particle is normally so small compared with the container radius that this correction can be safely ignored.

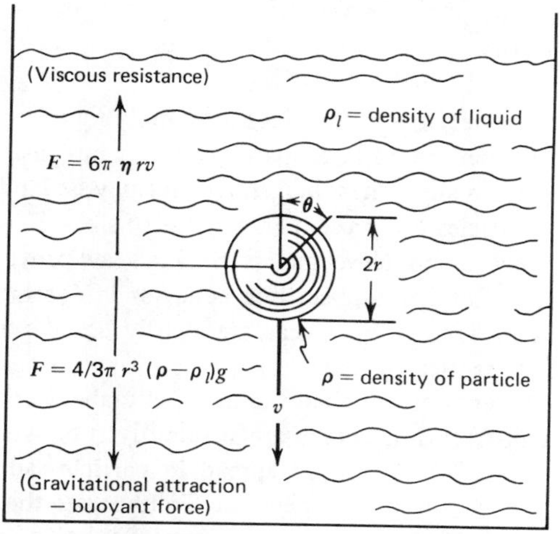

Fig. 27-1 Schematic diagram illustrating the forces acting on a single spherical particle of radius r settling with velocity v in a vehicle of viscosity η and density ρ_l.

Assume that the spherical particle has density ρ (the volume of the spherical particle is $\frac{4}{3}\pi r^3$). When placed in a liquid of density ρ_l, the weight of the particle (gravitational force) acts to pull the sphere downward through the liquid, whereas the buoyant force of the liquid acts to push the sphere upward. The net effect is normally a downward force ($\rho > \rho_l$) equal in value to F, as given in Eq. 2 ($g = 980$ cm/sec^2).

$$F = \tfrac{4}{3}\pi r^3(\rho - \rho_l)g \tag{2}$$

As the spherical particle falls through the liquid under the influence of gravity (Eq. 2), it is opposed by the viscous resistance of the liquid (Eq. 1). When these two forces are exactly balanced, the spherical particle falls at a constant equilibrium velocity v. Setting the two opposing forces equal to each other, corresponding to the constant velocity condition, yields Eq. 3.

$$6\pi r\eta v = \tfrac{4}{3}\pi r^3(\rho - \rho_l)g \tag{3}$$

Solving for v gives Eq. 4 expressing the settling rate for a single particle.

$$v = \frac{218r^2(\rho - \rho_l)}{\eta} \tag{4}$$

Inspection of Eq. 4 shows that the settling rate of a single particle is greatly accelerated by an increase in the particle size (proportional to r^2), and retarded by an increase in the liquid viscosity.

SETTLING OF SPHERICAL PIGMENT MIXTURES IN NEWTONIAN LIQUID

Actual paints depart widely from the conditions applying to Eq. 4. Thus millions of particles must be considered in a coating system, not just one; the pigment shapes involved are normally other than spherical (may be laminar, acicular, and so forth); and the particles are present in a range of sizes. Furthermore, coating vehicles are rarely Newtonian. Nevertheless, Stokes's law provides a basic expression that is helpful in interpreting settling behavior in coating systems, and by introducing modifying factors Eq. 4 can frequently be adapted to yield useful quantitative data for practical systems.

For example, pigments as supplied commercially always embrace a range of particle sizes, and distribution curves are invariably submitted by the pigment manufacturer to show the percentage spread in particle radii that can be expected for each type of commercial pigment. To illustrate the effect of multiple particle size on settling, consider the academic mixture of spherical particles given in Table 27-1 and graphed in Fig. 27-2. Any actual mixture, of course, consists of a continuous distribution, rather than the neatly segregated fractions given in the theoretical mixture. However, the academic mixture serves as an aid in understanding the nature of multiple particle settling.

Let a dilute dispersion of the spherical particle mixture of Table 27-1 (assume a particle density of $\rho = 3$ g/cm^3) be prepared in a Newtonian vehicle having a density ρ_l of 1 g/cm^3 and a viscosity of 1.09 poises. Assume that the particles are separated from each other by a sufficient distance so that no particle interaction takes place and that each particle settles out as an independent entity in the Newtonian vehicle. Under these conditions it is possible to calculate the time for each particle size to completely settle out to the bottom of a quart paint container (say with a maximum height of 10 cm) when the container is initially filled with a uniform mixture of the dilute dispersion of pigment in the vehicle.

From Eq. 4 the velocities of descent for each particle class has been calculated, and from this the time for each particle class to completely collect at the bottom of the quart container (see values given in Table 27-1 and Fig. 27-2).

Table 27-1 Settling Data for an Academic Pigment Mixture Composed of Five Fractions of Spherical Particles of Specified Radius

	Particle Radius of Fraction (μ)				
Property	1	2	3	4	5
Weight fraction (%)	10	20	40	20	10
Settling velocity (10^{-6} cm/sec)	4	16	36	64	100
Time to settle 10 cm (days)	29	7.3	3.2	1.8	1.2

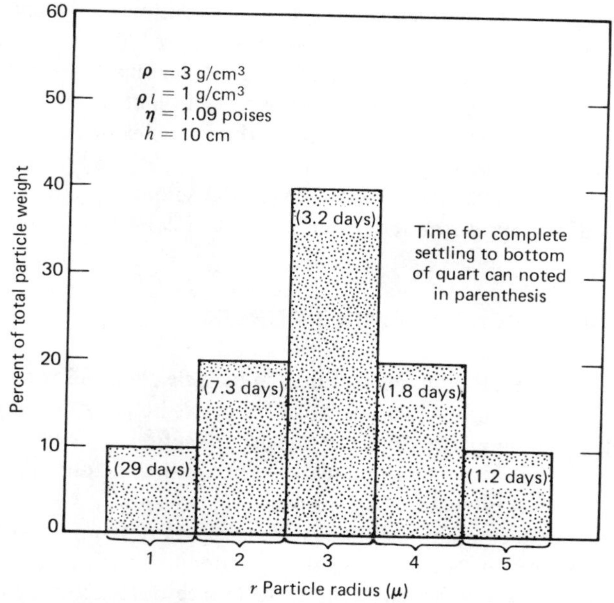

Fig. 27-2 Graph of an academic mixture of five fractions of spherical particles having radii of 1, 2, 3, 4, and 5 μ, respectively.

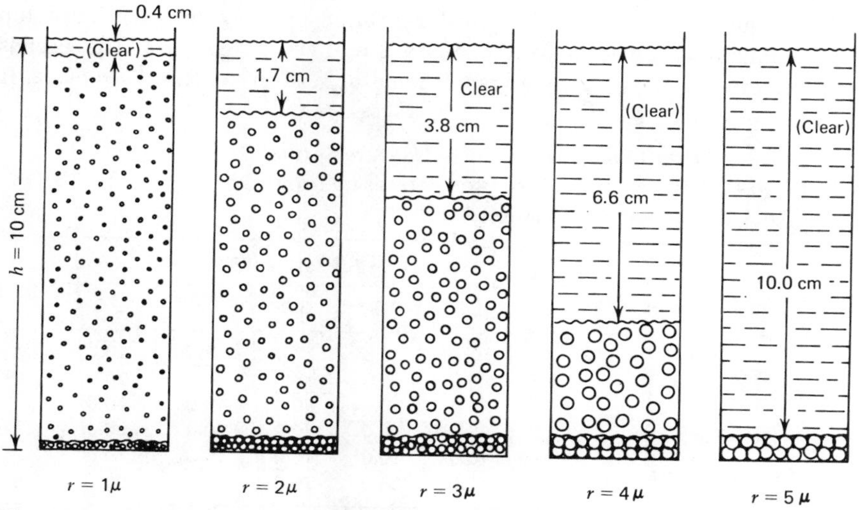

Fig. 27-3 Theoretical settling conditions at the end of 1.2 days for particles 1, 2, 3, 4, and 5 μ in radius, respectively. Vehicle viscosity 1.09 poises, and difference between pigment and vehicle density 2.0 g/cm^3.

Figure 27-3 schematically shows the settling situation at the end of 1.2 days. For clarity, each particle class has been drawn as a separate diagram, but to visualize the composite settling pattern imagine the five diagrams as superimposed on each other. It can be seen that the collection at the end of about 1 day is composed of a sediment consisting of all the particles in the 5-μ class, about two-thirds of the particles in the 4-μ class, and proportionally lesser percentages of particles in the smaller size classes. Hence the sediment collecting at the bottom is decidedly nonuniform under the given set of conditions.

CUMULATIVE WEIGHT SETTLING CURVES

Table 27-2 gives the percent weight for each particle class and for the mixture as a whole that has settled out in a quart can (effective height = 10 cm) for selected time periods. The cumulative percent weight collection of sediment at the bottom of the can has also been graphed in Fig. 27-4 as a function of time. The form of this graph is typical for pigment settling behavior.

A handy graphical relationship becomes apparent from inspection of Fig. 27-4. Note that, on the graph, tangent projections of the main curve have been made back to cut the "particle weight" scale (see dotted-line extensions). The segments that these projected dotted lines bracket on the weight scale correspond exactly to the weight proportions of the particles in the original particle mixture. Offhand, this might not have been expected, but it represents a helpful graphical device for estimating particle size distributions.

The argument applying to a formal derivation of this graphical relationship for a continuous distribution of particles is outlined in Fig. 27-5. The legends given in the four boxes on the graph should be read in numerical sequence. It

Table 27.2 Percent Weight of Pigment Mixture Settled Out in Quart Can (h = 10 cm) for Given Time Periods in Terms Of Both Total Settlement and Pigment Size[a]

Pigment Radius (μ)	Time Period (days) 1.2	1.8	3.2	7.3	29.0
1	0.4	0.6	1.1	2.5	10.0
2	3.3	4.9	8.8	20.0	20.0
3	15.0	22.5	40.0	40.0	40.0
4	13.3	20.0	20.0	20.0	20.0
5	10.0	10.0	10.0	10.0	10.0
Total	42.0	58.0	79.9	92.5	100.0

[a]Below the staggered line settlement has been completed.

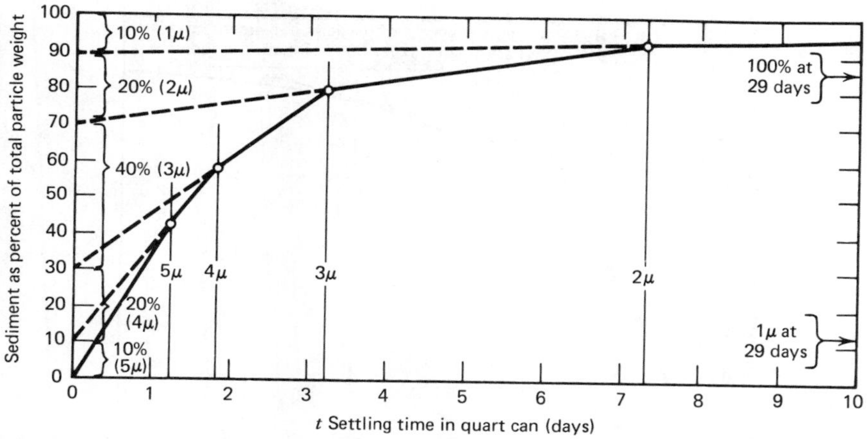

Fig. 27-4 Graph of percent sediment deposited (based on total particle weight) versus settling time. Noted particle radii correspond to times for their complete settlement. Settling curve (solid line) and tangent projections from this curve (dotted lines) give percentage composition of the particle mixture as read off the percentage sediment scale.

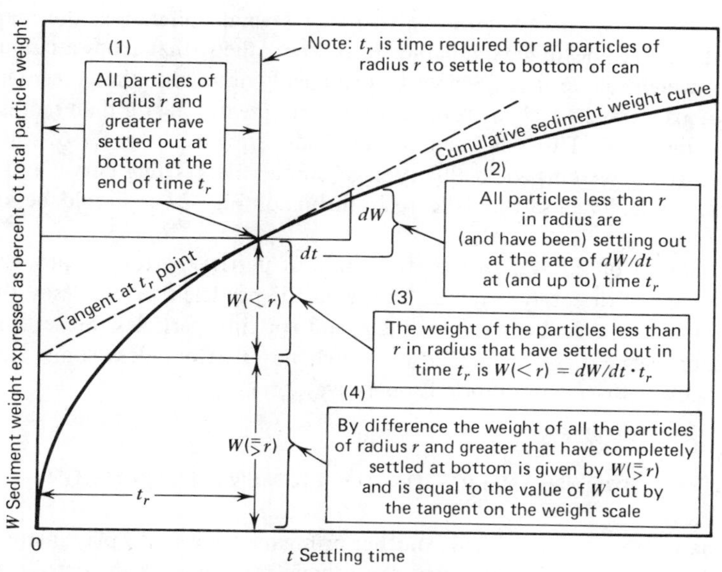

Fig. 27-5 Graphical derivation of equation relating weight fractions to cumulative weight settling curve.

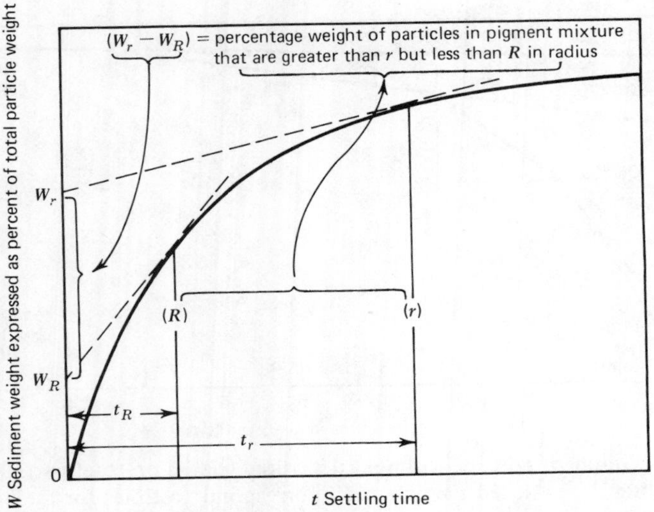

Fig. 27-6 Schematic diagram showing relationship of percent weight of particles in settled pigment mixture (that are more than r but less than R in radius) to the total settling time for particles of these radii.

is believed that, by following along with the several comments, the graphical relationships become understandable. It follows, then, that from a cumulative sediment weight curve it is possible to graphically determine the percentages by weight of particles in a pigment mixture that are greater in size than radius r but less than radius R. This has been schematically shown in Fig. 27-6. All freely settling particle systems yield this type of cumulative settling curve, and in interpreting unhindered settling the relationship of Fig. 27-6 should be kept in mind.

Examination of Eq. 4 shows that the rate of particle settling is proportional to the square of the particle radius. Hence, as the particle size decreases, the rate of particle descent declines very rapidly, and for tiny particles the settling rate may become essentially negligible (although never zero unless some other influence such as Brownian motion is operative).

EFFECT OF BROWNIAN MOTION ON PIGMENT SUSPENSION

As the particle size is decreased, another influence comes into play, namely, the effect of the molecular impacts of the vehicle molecules as they strike against the sides of the particle. The molecules of any vehicle are continually in a state of thermal agitation, and as they bombard a particle they move it about. Refer-

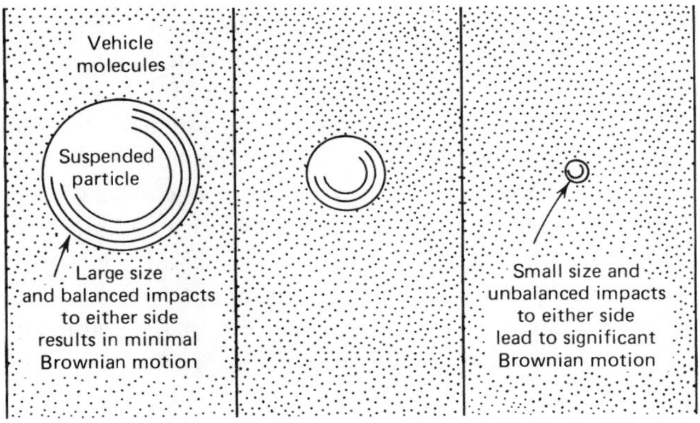

Fig. 27-7 Schematic diagram illustrating forces involved in Brownian motion.

ence to Fig. 27-7 shows that, as long as the particle is large compared with the vehicle molecules, the impacts received from any one direction will be essentially balanced by those received from the opposite direction, so that the total effect of these impacts on the particle is canceled out and the particle motion is left substantially unaltered by this molecular bombardment. Furthermore, the inertia of the larger particle is relatively great, so that it undergoes essentially no displacement from its normal descent in the vehicle. Conversely, as a particle becomes smaller in radius, the impact rate on opposite sides will tend to become unbalanced. The particle will receive statistically more impacts from one side than the other, causing it to move about. Furthermore, because of smaller particle size (mass), the effect of this unequal bombardment will become quite pronounced, and the particle will exhibit (under microscopic observation) a zigzag motion that is imposed on its otherwise vertically downward settling path.

This oscillatory or jiggling motion (first observed by Brown in 1826 and therefore referred to as Brownian motion) is exhibited by all particles that fall in a size range of about 1 μ and under.

The question now arises as to whether such Brownian motion is of sufficient magnitude to disrupt the normal descent of a single small particle under the influence of gravity in a vehicle and in effect alter the settling pattern so that the particle is jiggled into a state of suspension. For a single particle the answer is no. However, for a collection of particles the answer depends on a more complex situation.

At the beginning of the sedimentation cycle, the concentration of the particles in a dispersion is uniform throughout the vehicle. Provided that disturbing influences such as thermal convection currents and pigment flocculation are eliminated, the particles will start to descend and in the process of collecting at the bottom will tend to act on each other in much the same way as the vehicle molecules act on the individual particles. Thus there will be more particle impacts from the bottom than from the top for any given horizontal layer of particles. This unequal impact condition will tend to counteract the gravitational forces causing settling and in effect establish a vertical concentration gradient of particles. Since this upward force could possibly be a consideration in pigment settling in paint cans, it is necessary to inquire into whether the particles that are used in coating work are in a suitable size range to be suspended by this manifestation of Brownian motion.

Perrin and others derived an equation expressing the sedimentation equilibrium for a dispersion of small particles in suspension in a bath of liquid (Eq. 5).

$$2.3 \log \frac{c_0}{c_h} = \frac{mgh}{RT/N} \tag{5}$$

In Eq. 5 the pigment particle concentrations c_0 and c_h in the suspension apply to conditions at a base level 0 and a level h cm above this base level. The effective mass m of each pigment particle equals $\frac{4}{3}\pi r^3 \Delta\rho g$, where g is the acceleration of gravity (980 cm/sec^2), r the particle radius (cm), and $\Delta\rho$ the difference in density between the pigment and the vehicle ($\Delta\rho = \rho - \rho_l$). The temperature T is in absolute units ($C + 273$), R is the gas constant in energy units ($8.3 \cdot 10^7$ ergs/deg-mole), and N is Avogadro's number ($6.0 \cdot 10^{23}$ molecules/mole).

If the difference in the height of the two levels is taken as 1.0 cm and the temperature as 25 C, and r is expressed in microns (1.0 cm = 10,000 μ), Eq. 5 reduces to Eq. 6.

$$\log \frac{c_0}{c_{1.0}} = 4.3 \cdot 10^4 r^3 \Delta\rho \tag{6}$$

Experience has shown that for the conditions applying to Eq. 6 settling is just detectable but not measurable when $c_0/c_{1.0} \approx 2.0$. Taking this as a reasonable value that separates suspension from settling and inserting 2.0 in Eq. 6 yields Eq. 7, which gives the approximate particle radius (in terms of $\Delta\rho$) for this borderline condition.

$$r \text{ (in } \mu) = 0.02 \left(\frac{1}{\Delta\rho}\right)^{1/3} \tag{7}$$

PROBLEM 27-1

When $\Delta\rho$ between the spherical particles and the vehicle in a suspension is 2.0 g/cm^3, the particle radius at which settling just becomes detectable (from Eq. 7) is 0.016 $\mu = [0.02(\frac{1}{2})^{1/3}]$. Neglecting the effect of Brownian motion, calculate the time for a particle with radius 0.02 μ to settle a distance of 1.0 cm in a vehicle having a viscosity of 1.0 poise.

Solution. Substitute the given data in Eq. 4 to determine the velocity of descent of the particle in the vehicle. Note that, in substituting for the radius, r must be expressed in cm ($0.020\ \mu = 2.0 \cdot 10^{-6}$ cm).

$$v = \frac{218(2.0 \cdot 10^{-6})^2 \cdot 2}{1.0} = 1.74 \cdot 10^{-9} \text{ cm/sec}$$

The time to settle a distance $h = 1.0$ cm is calculated from the expression t (time) $= h/v$.

$$t = \frac{1.0}{1.74 \cdot 10^{-9}} = 5.7 \cdot 10^{8} \text{ sec} = 18 \text{ years}$$

Two conclusions can be reached from the foregoing discussion and the problem result. In the first place, it appears that Brownian motion cannot be expected to exert any significant effect on the suspension of pigment particles in most commercial coating systems, since Brownian motion starts to exhibit a suspension effect only on particles with radii less than about 0.016 μ. In the second place, at this particle radius and smaller ones settling is not a problem anyway, since, as indicated by the problem computation, it would take years for particles in this extremely minute size range to settle to the bottom of a quart can in a vehicle having a viscosity suitable for paint preparation. Hence Brownian motion cannot normally be expected to contribute any useful suspension properties to commercial coating compositions.

RATE OF SETTLING FROM DIFFERENT HEIGHTS

Another question that occasionally arises concerns the relative settling patterns of the same paint composition in containers of different heights. For example, does a paint settle out faster in a quart or a gallon container? Or, in other words, does the completion of a sedimentation cycle take the same time regardless of the container dimension?

Again a graphical device can be used to estimate relative settling starting from two different initial heights (horizontal dimensions have no influence on settling, since the settling movement is entirely in a vertical direction). This graphical method is illustrated in Fig. 27-8, where the settling of the same paint in a quart and a gallon can is considered. Assume that the paint settles out in such a manner that a well-defined demarcation boundary is maintained between a clear overhead phase and a sediment-filled lower phase. At the beginning of the settling cycle the boundary in both cases is identical with the top surface of the paint. As settling commences, however, a clear overhead phase starts to form and with time deepens, whereas the sediment-filled phase becomes compacted. The height of the boundary between the clear and sediment-filled phases for the gallon can has been plotted as a function of time in Fig. 27-8 and is represented by the curve that starts at a value of 16 cm ($t = 0$) and then descends to height values A, B, and C at the times noted thereafter. In the same way the height values for the boundary layer of the quart can have also been plotted as a function of time, starting with a value of 10 cm ($t = 0$) and proceeding to values a, b, and c at other noted time periods. Now it can be shown both theoretically and practically that there exists a simple correspondence among these points. Thus, if two or more straight lines are drawn from the origin of the graph through the two height/time curves, these curves will cut off segments on the straight lines that are proportional to each other. Thus on the graph of Fig. 27-8 the several straight lines radiating outward from the graph origin are cut into segments by the two settling curves, which have the following relationship:

$$\frac{16}{10} = \frac{OA}{Oa} = \frac{OB}{Ob} = \frac{OC}{Oc}$$

This graphical device can be put to many uses. For example, if the settling pattern for a quart can of paint is known, it is possible to routinely estimate the settling pattern for the same paint in a container of some other size, say a gallon or a pint can. Also, since the settling cycle is completed in a shorter time from a lesser height, it suggests that in designing any future procedure for assessing settling the test should preferably be carried out in a small container, say a half-pint can. It is even conceivable that a microtechnique using a microscope

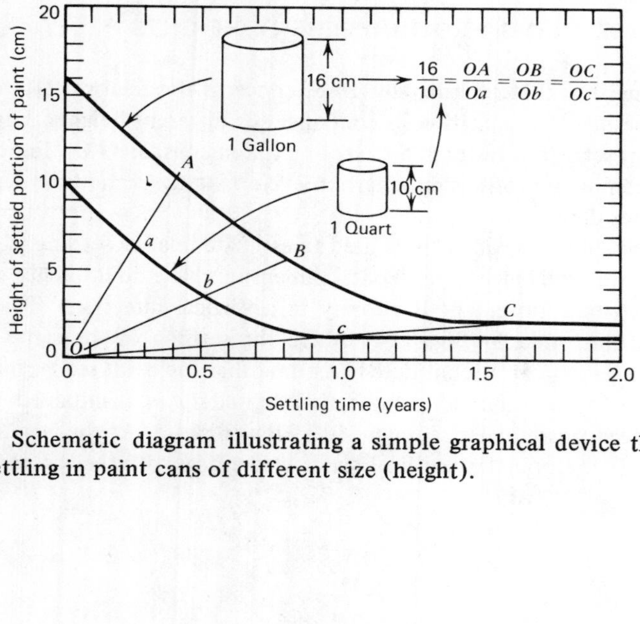

Fig. 27-8 Schematic diagram illustrating a simple graphical device that relates pigment settling in paint cans of different size (height).

might be worked out whereby the settling obtained in a week in a very shallow container could be extrapolated to give the settling that would occur in a quart can over a period of several months.

In the example used above for illustrating the simple ratio relationships holding for the quart and gallon containers, the point taken for observation was the boundary layer between the clear and sediment-filled phases. It was selected primarily because it was well defined and its height was readily measurable. However, any other two corresponding points that lend themselves to observation and/or measurement could just as well have been used, and the same type of proportionality would have been obtained.

CONSIDERATION OF COMPLEX SETTLING SYSTEMS

When a coating composition is allowed to stand undisturbed in a paint can (shelf storage), there ensues a gradual downward drift of pigment, and unless some factor opposes this downward trend a hard and substantially dry sediment that strongly resists redispersion will eventually form at the bottom of the can. As

Table 27-3 Considerations Related to Pigment Settling

Major Considerations

1. Pigment particles
   - Size distribution
   - Shape
   - Density
2. Vehicle
   - Viscosity profile
   - Density
3. Particle/vehicle interaction
   - Degree of dispersion (flocculation versus deflocculation)
4. Additives
   - Effect on interface activity and viscosity
5. Settling environment
   - Undisturbed versus disturbed (say during shipment)

Important Minor Considerations

1. The density and viscosity of a vehicle can become modified in the presence of suspended particles.
2. An upward vehicle velocity usually develops because of displaced vehicle (hence apparent settling is less than actual settling).
3. Since smaller particles are entrained and dragged downward by large particles, they tend to exhibit an accelerated settling.
4. Flocculation is encouraged by the closely spaced particles in a concentrated suspension.

has been pointed out and as is apparent from Eq. 4, both fineness of particle size and a more viscous vehicle reduce the rate of settling. However, resorting to such measures serves only to delay (not prevent) settling. Hence alternative mechanisms must be used to reduce or prevent pigment settling, namely, pigment flocculation and colloidal structure.

In the case of both pigment flocculation and colloidal structure, a loose network is introduced into the paint composition that imparts a low degree of rigidity to the pigmented system. These structures are weak and are readily broken down by mild agitation. However, when left undisturbed, these structures are capable of suspending pigment particles.

In actual coating systems a host of influences act to complicate the settling patterns. Major and minor controlling factors are listed in Table 27-3.

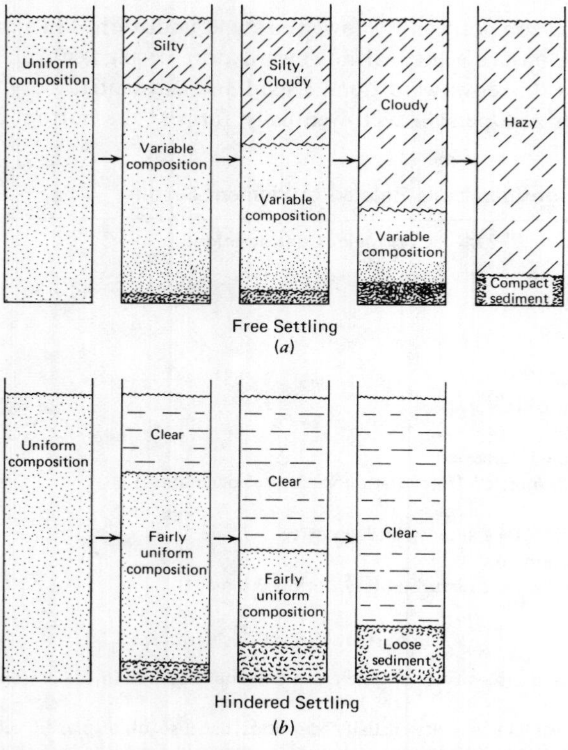

Fig. 27-9 (*a*) Schematic diagram showing the settling pattern exhibited by freely falling pigment particles. (*b*) Schematic diagram showing the settling pattern exhibited by pigments subject to hindered settling due to pigment flocculation.

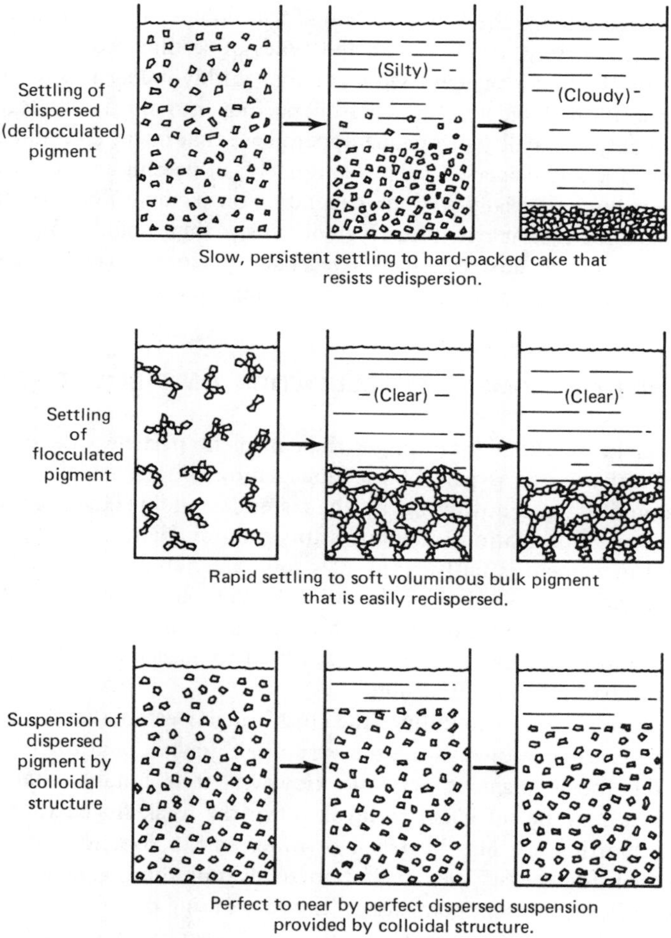

Fig. 27-10 Schematic illustration of three types of idealized settling patterns.

For purposes of discussion it is helpful to classify the settling habits of pigments into three fairly distinctive types, as schematically illustrated in Figs. 27-9 and 27-10. In Fig. 27-9 the difference between free and hindered settling is shown when colloidal structure is absent. The progressive nature of the settling patterns (proceed from left to right) merits careful study. In Fig. 27-10 the three types of settling are roughly compared.

It is believed that all three settling patterns are recognizable from practical experience. Of these the free settling type is obviously undesirable, since the terminal condition consists of a compacted pigment at the bottom of the can and reconstruction of the paint to a uniform composition is difficult if not impossible. Conversely, the two hindered types of settling provide a final settlement of loose or bulky pigment sediment that readily redisperses to yield a uniform composition. Incidentally, a person peering into the freely settled paint would probably see a silty or opaque upper phase that might give him or her the impression of a well-dispersed paint. Conversely, in peering into the paints with hindered settling, especially the one depending on pigment flocculation, the lay person might see a clear overhead layer of vehicle that would tempt him or her to reject the paint as unsatisfactory. Even paint personnel have been known to fall into these two errors in interpreting visual appearance in the can.

INFLUENCE OF PIGMENT FLOCCULATION ON PIGMENT SETTLING

Since pigment flocculates are larger than pigment particles, they settle more rapidly. Furthermore, as part of the flocculation process smaller particles are entrained and dragged downward by the faster descending larger particles. However, this same flocculation contributes to structure. Hence, although the initial downward movement is fairly rapid, it is soon arrested at a soft, bulky stage of settlement with a more or less clear overhead liquid phase normally present.

Flocculation and compact settling thus oppose each other. In fact, degree of pigment compactness has frequently been used to measure the degree of deflocculation resident in a test dispersion.

Under normal storage conditions the cross-bridging established by the pigment flocculates is capable of resisting the very weak gravitational force that acts to disrupt the fragile pigment network. However, it is notable that even mild disturbances may be sufficient to break down the weak flocculate bridgework. Thus transporting coatings by truck or other carrier over long distances may sometimes introduce sufficient agitation to disrupt the pigment structure and permit significant pigment compaction to occur during transit.

INFLUENCE OF COLLOIDAL STRUCTURE ON PIGMENT SETTLING

Although pigment flocculation contributes a good measure of pigment suspension, this departure from the dispersed state is at the same time self-defeating, since in proportion to the flocculation that takes place there result a concomitant loss in pigment hiding and an impairment in other properties such as durability. After all, the function of the grinding (dispersion) step is to effect a complete dispersion, and to undo this by deliberately reflocculating the pigment is a backward step. Some flocculation can occasionally be tolerated, but by far the best approach to achieving pigment suspension is to take advantage of the colloidal structure provided by so-called thixotropes in organic solvent systems and by thickeners in water systems. By incorporating these flow control agents into a coating composition, good pigment dispersion is retained and pigment suspension is assured. Figure 27-10 schematically illustrates the idealized patterns for the three representative types of pigment settling. The colloidal structure route for attaining pigment suspension is obviously preferable, and this technique is generally used today in the paint industry.

It never ceases to be amazing that the addition of a fractional percent of a thixotrope or thickener is capable of exerting an effect on viscosity that is all out of proportion to its very minor percentage level in the coating composition. However, it should be carefully noted that this increase in viscosity is manifested only in the regions of very low shear rate.

During paint application, where high shear rates are involved (brushing, roller coating, spraying), the colloidal additive might as well be absent as far as its effect on viscosity is concerned. However, during storage, where minute gravitational influences are acting to induce settling, a colloidal agent is capable of increasing the low-shear-rate viscosity many times. In fact, under sufficiently small stress the viscosity becomes infinite for all practical purposes (the coating is said to exhibit a yield value).

PIGMENT SETTLING SHEAR RATE

The shear rates developed by a falling spherical particle, in a liquid are highest next to the sphere surface (see Fig. 27-1). A surface shear rate can be calculated approximately by Eq. 8 where the angle θ is subtended by the topmost (or bottommost) point on the sphere and the point on the sphere where the shear rate is to be determined.

$$\mathrm{D} \approx \frac{3/2 \cdot \sin\theta \cdot v}{r} \tag{8}$$

A maximum shear rate is experienced when the angle $\theta = 90°$ ($\sin\theta = \sin 90° = 1.00$), as expressed by Eq. 9.

$$\mathrm{D} \approx \frac{1.5v}{r} \tag{9}$$

PROBLEM 27-2

Calculate the maximum shear rate developed by a spherical particle ($r = 1.0\ \mu$; $\rho = 3.0\ \mathrm{g/cm^3}$) in falling through a liquid of 1.0-poise viscosity ($\rho_l = 1.00\ \mathrm{g/cm^3}$).

Solution. Calculate the velocity of descent of the particle from Eq. 4.

$$v = \frac{218(0.0001)^2(3.0 - 1.0)}{1.0} = 4.36 \cdot 10^{-6} \text{ cm/sec}$$

Substitute this calculated value for v in Eq. 9.

$$\text{D (maximum)} = \frac{1.5 \cdot 4.36 \cdot 10^{-6}}{1.0 \cdot 10^{-4}} = 0.065 \text{ sec}^{-1}$$

From Problem 27-2 it is seen that pigment settling takes place at very low shear rates. It has been observed empirically that the amount of colloidal agent required to control sagging is normally more than adequate to provide satisfactory pigment suspension. This is fortunate from a testing standpoint, since a sag test can be carried out fairly quickly whereas a conventional pigment settling evaluation usually requires a significant time period (say 3 months of shelf storage) for the development of a suitably thick sediment for testing.

COLLOIDAL ADDITIVES

For effective performance a colloidal additive must be correctly incorporated into the coating composition. Since it is present in such minute amount, it must be well distributed, that is, it must be well dispersed. Even though present in small amount, it must be adequate to provide the required suspension effect. Furthermore the colloidal agent must be afforded ample opportunity to interact and swell in the vehicle. Manufacturers of colloidal additives usually recommend that their products reach at least a minimum temperature during their introduction, especially into organic solvent systems. Otherwise these agents act substantially as inerts until activated by a rise in temperature. These and other important considerations are carefully discussed in the literature of additive suppliers, and this technical information should be studied and observed in the use of these products.

RATING OF PIGMENT SETTLEMENT

Both qualitative and quantitative ratings have been used to assess pigment settling.

Qualitative Measurement

ASTM test method D869-48 (1974) specifies a procedure for evaluating the degree of settling of (traffic) paint in terms of a number scale running from 0

Table 27-4 Relation of an Arbitrary Number Scale to Qualitative Terms Used to Describe Pigment Settling

Number Rating	Penetration of Spatula through Sediment Under Own Weight (45 g)	Resistance Offered by Sediment to Movement of Flat-Tipped Spatula: Sidewise	Resistance Offered by Sediment to Movement of Flat-Tipped Spatula: Edgewise	Quality of Deposit Brought up by Spatula	Difficulty Encountered in Reincorporating Sediment
10		Perfect suspension			
8		Not significant[a]		Slight	
6	Complete	Definite		Coherent cake	
4	Incomplete	Difficult to move	Slight		Remixes easily
2		Very difficult	Definite		Can be remixed
0			Very firm cake		Cannot be reincorporated

[a]Some feeling of settlement.

(hard sediment that cannot be reincorporated into the paint system) to 10 (perfect suspension). The test is conducted manually by probing the sediment that has collected at the bottom of a pint can of the test paint over a 6-month undisturbed storage period. The probing is carried out using a square-edged, flexible steel spatula (13/16 × $4\frac{3}{4}$ in.). Terms descriptive of the nature of the settlement, as tubulated in Table 27-4, are then related to the 0 to 10 rating scale.

Quantitative Measurement

Several test procedures have been proposed for quantitatively evaluating settling. Most of them are based on mechanical devices that are capable of thrusting a probe down through the test paint under precisely controlled conditions. In the past this type of testing has tended to be somewhat cumbersome. Recently, however, a simplified gage has been made available that, when dismantled, can be slipped into the pocket of a brief case (1). A diagram of the gage is given in Fig. 27-11*a*, and a schematic illustration of the gage in place over a quart can of test paint at the end of a test run (resting on packed sediment) is shown in Fig. 27-11*b*. The test method consists in pushing a perforated disk (sketched in Fig. 27-11*b*) downward through the coating composition, using a stepwise procedure wherein weights are periodically added to the probe load at half-minute intervals. Typical curves developed from the loading schedule for diverse types of paints are shown in Fig. 27-12. The advantage of this empirical type of test is that it provides a permanent and reproducible record of paint settlement which is free from operator bias.

More recently an X-radiation procedure has been employed to study paint settlement. However, the technical competence and elaborate equipment required will undoubtedly restrict this method to the realm of research.

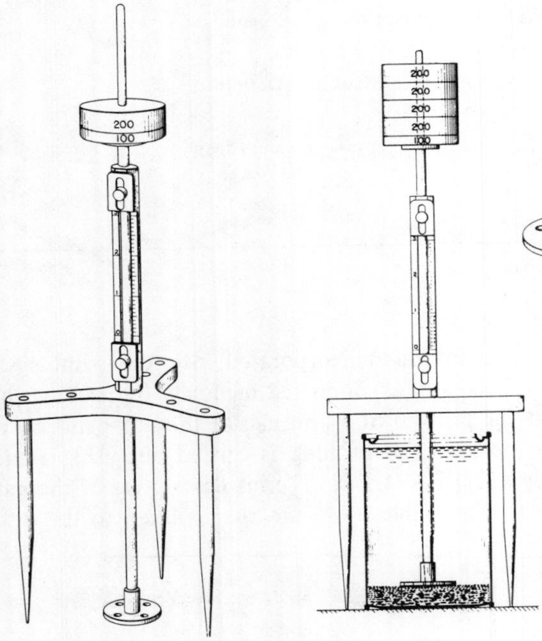

(*a*) (*b*)

Fig. 27-11 (*a*) Diagram of a simple pigment settling gage. (*b*) Diagram illustrating use of pigment settling gage. Full load on probe has failed to penetrate the hard, compacted pigment sediment at bottom of quart can, as indicated by benchmark on probe rod opposite gage scale.

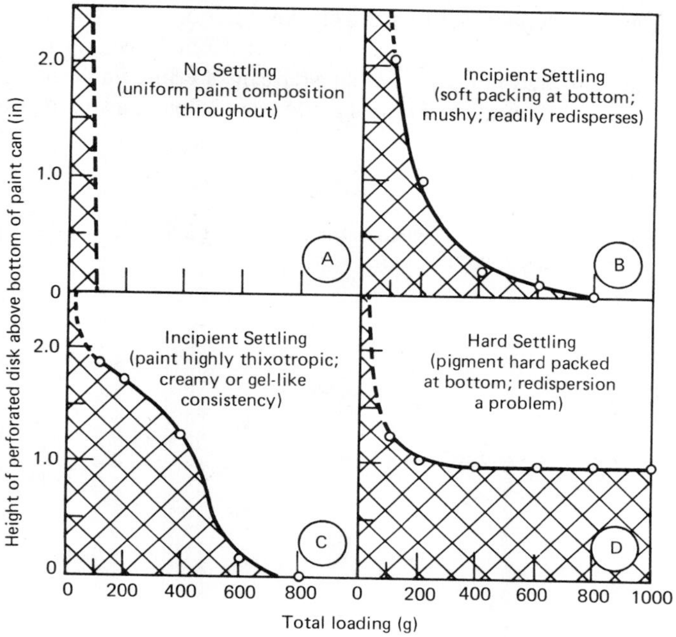

Fig. 27-12 Characteristic pigment settling curves obtained through use of pigment settling gage shown in Fig. 27-11.

ACCELERATION OF SETTLING

Pigment settling is normally a slow, gradual process. Hence, to expedite testing, attempts have been made to establish accelerated aging conditions that simulate long-term natural aging. Centrifuging techniques are virtually excluded for the reason that, by artificially increasing the downward force on the particles, they almost certainly disrupt the weak structural network suspending the pigment particles, and pigments that might be suspended indefinitely under gravitational influence alone would rapidly pack at the bottom under centrifuging conditions.

ASTM test method D1309-56 (1975) outlines an alternative approach to achieving accelerated settling conditions, namely, a temperature-cycling schedule wherein the test paint in a pint can is subjected to some two dozen alternate periods of cold and heat over a span of 2 weeks (−5 to −10 F cold period to 158 to 160 F heat period). This treatment is said to yield a final settling condition equivalent to that attained at the end of about 6 months of undisturbed shelf aging at room temperature.

In setting up accelerated settling conditions, it should be remembered that final settlement conditions are achieved sooner from lesser starting heights, suggesting that pint, half-pint, or even smaller cans are preferable to quart cans for accelerating the attainment of a final settled condition in a stored paint can.

REFERENCE

1. Patton, Temple C., "A Simple Pigment-Settling Gage and a Simple Anti-Sag Test," *Off. Dig.*, **29**, 10–25, January (1957).

28 *Leveling*

The application of a paint to a substrate by brushing introduces two rheological extremes. During the brushing application a shear rate on the order of 5000 to 20,000 sec^{-1} is encountered. Immediately after application, a shear rate of from about 200 to less than 0.01 sec^{-1} is involved (leveling). If the paint is applied to a vertical surface, another low-shear-rate effect due to gravitation is imposed on the system (sagging). Since leveling and sagging tend to be interrelated, they are treated in consecutive chapters.

COATING APPLICATION METHODS

Brushing

The brushing of a coating over a more or less smooth surface tends to leave brushmarks behind in the wake of the brush. These parallel furrows and ridges (striations) generally disappear in a short time because of the leveling of the wet coating. Since brushmarks are not esthetic in appearance and, more important, since they constitute sites of weakness (starting points for corrosion, cracking, and blistering), the attainment of acceptable leveling is an objective of prime importance.

The exact contribution of brush bristles to the brushing application is still imperfectly understood. A straightforward explanation that the furrows are merely indentations produced by the individual bristles as they pass over the wet coating is inadequate for the reason that the furrows formed are considerably wider than the bristles themselves (1). It has been argued that local clumping of the bristles takes place during the brushing operation and that these clumped bristles produce the wider furrows. This theory is partially substantiated by the observation that brushmarks do become wider as greater pressure is exerted on the brush during application. However, it fails to explain the fact that, when no change is made in the pressure, the furrows automatically become wider as the thickness of the coating is increased (1). Any brushmark theory must also explain the experimental observation that the number of brushmarks per unit width is substantially independent of the coating viscosity or the rate of application.

Spreading by Applicator Blade or Rod

It has been proposed that the drawdown of a coating by a coating applicator (straight-edge spreader or wire-wound rod) is quite similar in nature to brushing. Thus it is common experience that striations invariably result when an applicator with an overhanging, receding edge is used to draw down a coating over a smooth surface (a trailing edge gives little or no striation pattern). As with brushing, the striation spacing given by the spreader depends primarily on the coating thickness (controlled in turn by the applicator gap clearance) (1). Figure 28-1 reveals

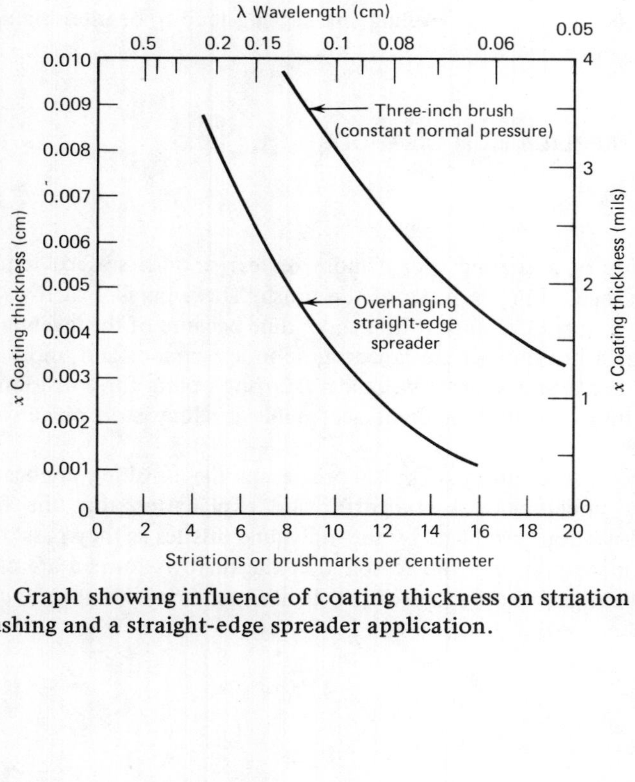

Fig. 28-1 Graph showing influence of coating thickness on striation spacing for both a brushing and a straight-edge spreader application.

the definite trend toward a more widely spaced striation pattern with increased coating thickness.

Roller Coating

A roller coater can be thought of as presenting an overhanging edge as it is rolled over a surface during the roller coater operation. Again tracks are formed in the wake of the receding face of the roller. However, they then tend to take the form of randomized pits or minute blotches to give a mottled appearance rather than the parallel striations typical of brushmarks or the tracks of spreading straight-edge applicators.

STRIATION THEORY

The obvious similarities among the tracks left behind in the wake of a coating application, be it from a brush, spreader, or roller coater, suggest that some common physical effect must be operating. Studies indicate that in all three cases the striation pattern is due to the unstable flow that is set up when the effluent coating is forced to split between the substrate and the receding applicator surface.

In the case of the overhanging applicator blade, a mathematical explanation using this theoretical approach has been developed to account for the striation effect (2). Presumably this same theory with modification should be equally successful in explaining the formation of the tracks left behind by brushing and roller coater operations.

In summary, it can be stated (qualitatively) that wider spaced striations are produced by thicker coatings and that these striations become more pronounced with increased overhang of the receding applicator edge. On the other hand, the striation pattern is apparently little or not at all affected by coating viscosity or by rate of application.

LEVELING

Tracks such as brushmarks are an inevitable consequence of a coating application by an applicator. Usually they are quite transient, and leveling may be so rapid that the striation effect is never observed. On the other hand, the striations may tend to persist, and when they actually fail to level out to a smooth surface before the conversion of the wet coating to a dry paint film they become objectionable.

Surface tension is the driving force that causes a liquid coating to level. In its effort to constrict the coating to a shape with a minimum of surface, surface tension levels the paint from a rippled, furrowed, striated, or ridged surface to a smooth, even surface. Any leveling contributed by gravitational force is usually of minor consequence and can be disregarded.

Rheology of Leveling

Leveling phenomena can be considered in terms of an idealized striation pattern such as is given by the cross section schematically shown in Fig. 28-2. The controlling linear dimensions are the wavelength λ, the average coating thickness x, and the striation amplitude a. In practice the curved profile of the surface presumably conforms to that of a circular circumference, and an elementary body of theory has been worked out based on this concept (3). However, a more rigorous theoretical derivation (1) based on a sine wave profile is now accepted as more accurately reflecting the experimental findings, despite the slight inaccuracy introduced by the sine wave profile as opposed to the circular profile.

The equations given by the two different approaches differ only in the value of the equation constants; the variables are still grouped to give the same overall unit dimensions. In this revised edition only constants based on the sine wave profile will be used, since it is felt they represent best the quantitative aspects of leveling.

Leveling Equation

The basic leveling equation is given by Eq. 1 (1).

$$\ln \frac{a_0}{a_t} = \frac{16\pi^4}{3} \cdot \frac{\gamma x^3}{\lambda^4} \cdot \int \frac{dt}{\eta} \tag{1}$$

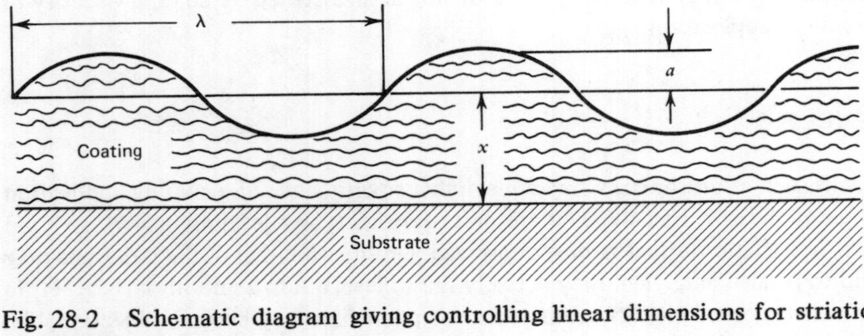

Fig. 28-2 Schematic diagram giving controlling linear dimensions for striation cross section.

where a_0 = initial amplitude (cm)
a_t = amplitude at time t (cm)
x = average coating thickness (cm)
λ = wavelength (cm)
γ = surface tension (dynes/cm)
η = viscosity (poises)
t = time (sec)

Changing to common logarithms and reducing the numerical values to a single constant gives Eq. 2.

$$\log \frac{a_0}{a_t} = \frac{226\gamma x^3}{\lambda^4} \cdot \int \frac{dt}{\eta} \tag{2}$$

The leveling equation is most revealing and points up the controlling influence of such factors as surface tension, viscosity, and striation geometry on the time required for a given degree of leveling to take place.

Surface tension usually falls within a narrow range of values from about 25 to 55 dynes/cm. For most practical purposes and in lieu of better data it can be assigned a routine value of 35 dynes/cm. Although surface tension is the compelling force behind leveling, it is an essentially constant quantity in the leveling equation.

On the other hand, the linear dimensions of thickness x and striation wavelength λ vary more widely. Moreover, since they enter into the leveling equation to the third and fourth power, respectively, their influence is enormously magnified. For example, other things being equal, doubling the coating thickness cuts the leveling time to one eighth; halving the coating thickness increases the leveling time eightfold. The wavelength is even more critical. Doubling the wavelength increases the leveling time sixteenfold; halving the wavelength decreases the leveling time sixteenfold.

Leveling Viscosity

Leveling viscosity refers to the viscosity conditions encountered during the leveling process. In most practical systems it is a viscosity that increases in value as leveling proceeds, and as the leveling end point is approached this viscosity increase generally becomes dramatic (becomes infinite).

Leveling Considered as a Stepwise Process

It is instructive to consider the leveling process in terms of successive and relatively small fractional reductions in the striation amplitude. If, for each fractional reduction, average values are assumed for the variables involved, Eq. 2 reduces to Eq. 3, where Δt is the time for the fractional amplitude reduction.

$$\Delta t = \log \frac{(a_0/a_t)\lambda^4 \eta}{226\gamma x^3} \tag{3}$$

For a half-reduction in amplitude for each successive fractional decrease, Eq. 3 simplifies to Eq. 4.

$$\Delta t = \frac{0.0133\lambda^4}{\gamma x^3} \cdot \eta \tag{4}$$

Consider now a typical leveling situation such as might be encountered in practice. For this purpose, the following common leveling conditions are assumed:

$a_0 = 0.0080$ cm
$a_t = 0.000050$ cm
$x = 0.010$ cm (4 mils)
$\lambda = 0.10$ cm (10 striations/cm)
$\gamma = 35$ dynes/cm

Reference to Fig. 28-1 shows that the selected value of 0.10 cm for the striation wavelength is representative. The selected average coating thickness of 0.010 cm (4 mils) is commonly applied to a substrate in commercial paint work. The surface tension of 35 dynes/cm is typical. The initial striation depth of 0.0080 cm (0.80 of the assumed coating thickness) may be somewhat large but is chosen deliberately to provide a marked reduction in the striation amplitude (a 160-fold reduction from 0.0080 to 0.000050 cm for complete leveling). The choice of 0.000050 cm (0.020 mil) for the leveling end point is based on the experimental observation that brushmarks and similar tracks formed in the wake of an applicator are no longer discernible when the amplitude reaches this value, or falls below it.

In this connection nine visual drawdown levelness standards that illustrate extremely poor to good leveling are commercially available (4). A wide range of striation amplitudes are covered, and both numerical and subjective ratings have been assigned to these levelness standards, as shown in Fig. 28-3. Note that a rating of 10, assigned to the region of excellence beyond the last visual standard (rating 9) corresponds to amplitudes that are at or below the selected practical leveling end point of 0.000050 cm.

Continuing now with the typical leveling situation and substituting the listed values for x, λ, and γ in Eq. 4, we obtain Eq. 5.

$$\Delta t \text{ (half-reduction in amplitude)} = 0.0038\eta \tag{5}$$

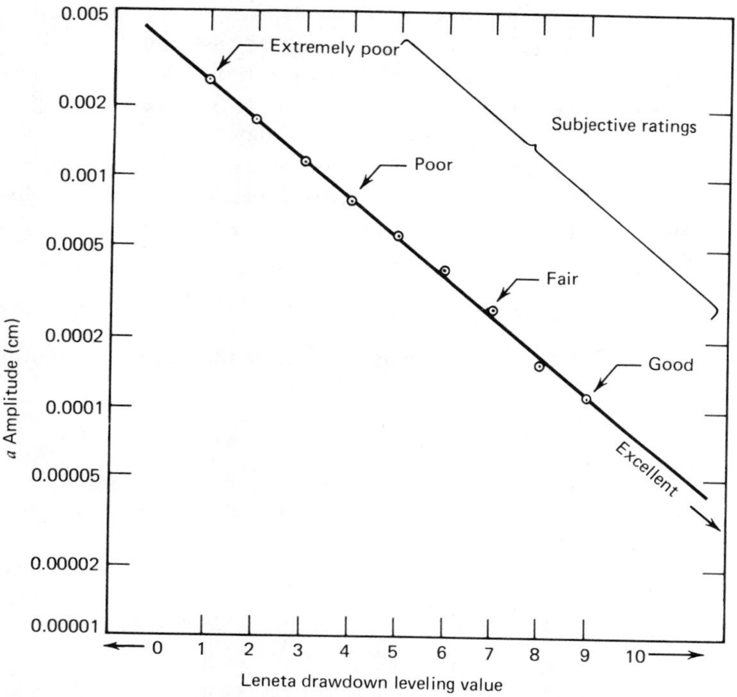

Fig. 28-3 Graph showing striation amplitudes corresponding to subjective ratings and Leneta drawdown leveling values.

To progress from an amplitude of $a_0 = 0.0080$ cm to $a_t = 0.000050$ cm (ratio of 160) requires 7.32 halvings (7.32 = log 160/log 2).

For conditions of constant viscosity, the total time to attain complete leveling is then given by Eq. 6.

$$t \text{ (complete leveling)} = 7.32 \cdot 0.0038\eta = 0.028\eta \tag{6}$$

For example, if the viscosity were a constant 10 poises, the time for complete leveling would be 0.28 sec; if 100 poises, 2.8 sec; if 1000 poises, 28 sec. These results could have been obtained, of course, by direct substitution in the basic leveling equation (Eq. 2). Thus for constant viscosity of 100 poises, the time is again calculated as 2.8 sec.

$$\log \frac{0.0080}{0.000050} = \frac{226 \cdot 35 \cdot 0.010^3}{0.10^4} \cdot \frac{t}{100}$$

$$t = 2.8 \text{ sec}$$

However, the purpose of introducting the stepwise procedure was to set the background for the computation of leveling times based on practical viscosity situations. The assumption of a constant viscosity throughout the leveling operation in unrealistic, since in practice the leveling viscosity is ever changing.

At any given moment the leveling viscosity is determined by (*a*) the viscosity profile of the coating at that time and (*b*) the shear stress at that given moment.

The viscosity profile of a coating is represented best by an expression discussed earlier in Chapter 16 and repeated here as Eq. 7 for ready reference.

$$\eta = \frac{\tau}{\mathrm{D}} = \left(\frac{\tau_0^n}{\mathrm{D}^n} + \eta_\infty^n\right)^{1/n} \tag{7}$$

For purposes of the proposed calculation scheme this equation is transformed algebraically to Eq. 8.

$$\mathrm{D} = \frac{(\tau^n - \tau_0^n)^{1/n}}{\eta_\infty} \tag{8}$$

The maximum shear stress exerted by the surface tension at any given moment is given by Eq. 9 (5, 6).

$$\tau = \left[\frac{(2\pi)^3 \gamma x a}{\lambda^3}\right] \cdot \left[\frac{\cosh(2\pi x/\lambda)}{(2\pi x/\lambda)^2 + \cosh^2(2\pi x/\lambda)}\right] \tag{9}$$

This equation can be rewritten as Eq. 10, where the function $f(x/\lambda)$ corresponding to the term in the second bracket of Eq. 9 can be read off the graph of Fig. 28-4.

$$\tau = \frac{248\gamma x a}{\lambda^3} \cdot f\left(\frac{x}{\lambda}\right) \tag{10}$$

The quantities γ, x, and λ remain substantially unchanged during the leveling operation. Hence the shear stress varies directly with the changing striation amplitude. Shear stress values developed at the leveling end point ($a = 0.000050$ cm) under varying conditions of coating thickness x and wavelength λ have been graphed in Fig. 28-5, based on a surface tension of 35 dynes/cm. If the yield value of the coating is greater than this shear stress, perfect leveling becomes unattainable. The dashed lines in this graph correspond to values for the shear stress before being modified by the function $f(x/\lambda)$; the solid lines correspond to values of τ for the full Eq. 10.

The yield value that a coating should have for proper leveling has been variously reported in the literature. Smith et al. indicate that a yield value of 0.5 dyne/cm^2 will result in barely discernible brushmarks, whereas a value of 20 dynes/cm^2 results in pronounced brushmarks (1). Quach and Hansen report that best leveling is obtained with yield values <10 dynes/cm^2, and poorest leveling with yield values >20 dynes/cm^2 (6). Probably the best correlation between

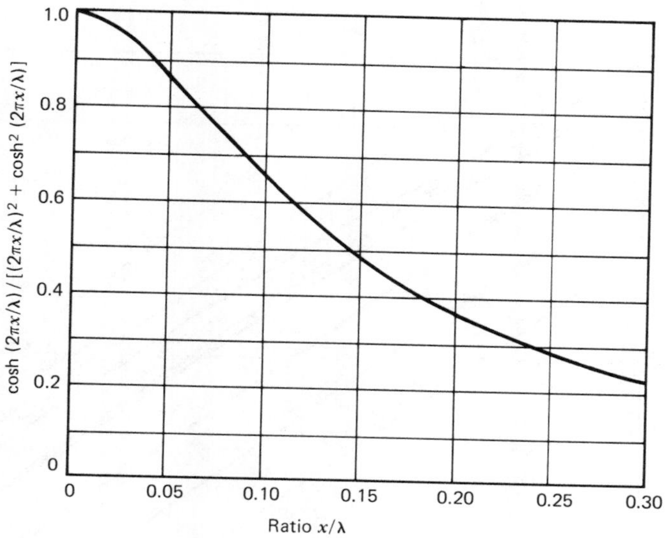

Fig. 28-4 Graph relating $f(x/\lambda)$ to x/λ, where $f(x/\lambda)$ is the function cosh $(2\pi x/\lambda)/[(2\pi x/\lambda)^2 + \cosh^2 (2\pi x/\lambda)]$.

yield value and satisfactory leveling is given by Sarkar and Lalk, who propose that the ideal paint coating should have a yield value >10 dynes/cm^2 before brushing and a value of <2.5 dynes/cm^2 immediately after brushing (7).

Conceivably it should be possible to set up a single expression relating Eqs. 2, 8, and 9 that would permit a direct calculation of the time needed for complete leveling. However, the necessary mathematical treatment is obviously formidable.

However, by resorting to the proposed stepwise procudure, Eqs. 3, 8, and 9 can be combined to provide an expression for the incremental leveling time for a single step, using average values. Then the several time increments can be added to give the total leveling time. The time increment obtained by combining Eqs. 3, 8, and 9 for a single step is given by Eq. 11. In this equation a step corresponds to a definite specified fractional change in amplitude (say a reduction by one half), and $\bar{a}$ is the average of the initial and final amplitudes.

$$\Delta t = \left[\frac{1.1 \log (a_0/a_t)\lambda \cdot f(x/\lambda)\eta_\infty}{x^2}\right] \cdot \bar{a} \cdot \left\{\frac{1}{[248\gamma x\bar{a} \cdot f(x/\lambda)/\lambda^3]^n - \tau_0^n}\right\}^{1/n} \tag{11}$$

This equation approaches exactness as the initial and final amplitudes come closer together (the fractional change becomes smaller). However, in the interests

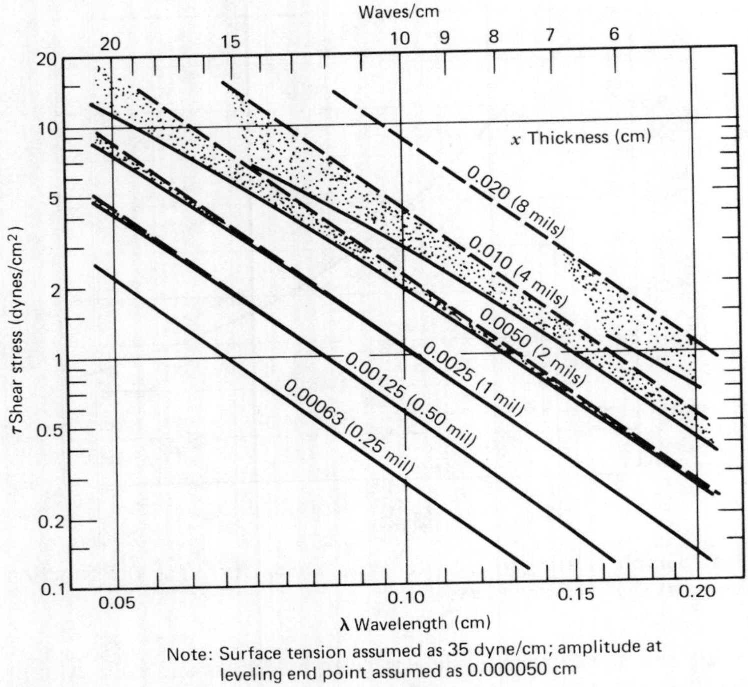

Fig. 28-5 Graph relating the maximum shear stress that can be exerted by a coating at the leveling end point to the coating thickness and striation wavelength. Leveling end point assumed to be an amplitude of 0.000050 cm, and surface tension taken as 35 dynes/cm. The dashed lines correspond to shear stress values unmodified by the function $f(x/\lambda)$; the solid lines correspond to the full Eq. 10.

of reducing the calculation work, it is convenient to work with a half-reduction in amplitude for each step.

Inspection of this equation shows that the leveling time is directly proportional to the viscosity at infinite shear rate η_∞. Also to be noted is the fact that, when the denominator of the term in braces equals zero, the time for leveling becomes infinite. From the latter observation there is derived Eq. 12, which gives the smallest amplitude that can be reached for a given set of leveling conditions.

$$a \text{ (minimum attainable)} = \frac{\tau_0 \lambda^3}{248\gamma x \cdot f(x/\lambda)} \qquad (12)$$

Equation 11 for an incremental time Δt appears somewhat involved, but inspection shows that by selecting a half-reduction in amplitude for each step in the leveling process ($a_0/a_t = 2.0$; $\log a_0/a_t = 0.301$) the time increment becomes a function of the average amplitude only. An example serves to illustrate the overall procedure and will show the unusual trend that a leveling viscosity takes during the leveling operation. For this purpose a coating has been selected with a viscosity profile that is characterized by a fairly high yield value of 10.0 dynes/cm^2, a representative viscosity at infinite shear rate of 2.0 poises, and a conventional value of 0.5 for the exponent n. This viscosity profile is shown as curve A in Fig. 28-6. The several conditions applying to the given leveling situation are as follows:

Viscosity Profile	Striation Geometry	Surface Tension
$\tau_0 = 10$ dynes/cm^2	$x = 0.010$ cm	$\gamma = 35$ dynes/cm
$\eta_\infty = 2.0$ poises	$\lambda = 0.10$ cm	
$n = 0.50$	$a_0/a_t = 2.0$	

Substituting these data in Eq. 11 yields the time increment required to halve the striation amplitude in terms of the average amplitude during this period (Eq. 13).

$$\Delta t = 431\,\bar{a}\left\{\frac{1}{[(56{,}700\bar{a})^{0.5} - 3.16]^2}\right\} \tag{13}$$

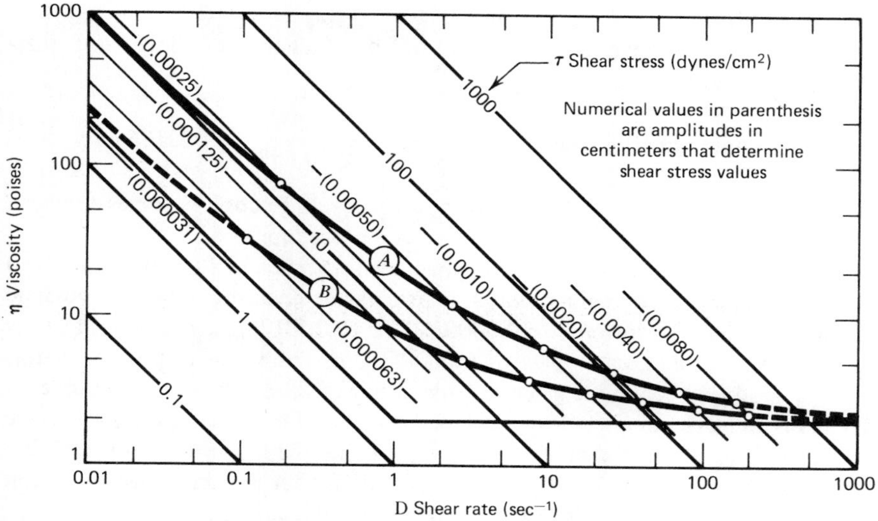

Fig. 28-6 Viscosity profiles for coatings A and B, which differ only in yield value (10.0 vs. 2.0 dynes/cm). Half-step reductions in striation amplitude indicated at applicable shear stress values.

Table 28-1 Time Increments for Successive Half-Reductions in Striation Amplitude As Leveling Proceeds and Total Time for Complete Leveling

Amplitude Half-Steps during Leveling, a (cm)	Average Amplitude per step, $\bar{a}$ (cm)	Time Increment for Half-Reduction in Amplitude (sec)	
		τ_0 = 10.0 (dynes/cm^2)	τ_0 = 2.0 (dynes/cm^2)
0.0080			
	0.0060	0.011	0.009
0.0040			
	0.0030	0.013	0.010
0.0020			
	0.0015	0.018	0.011
0.0010			
	0.00075	0.029	0.012
0.00050			
	0.000375	0.077	0.016
0.00025			
	0.000188	8.0	0.024
0.000125			
	0.000094	∞	0.051
0.000063			
	0.000047		0.39
0.000031			
Total		∞	0.523

In Table 28-1 are listed the time increments for successive half-reductions in amplitude from an initial value of 0.0080 cm to the lowest amplitude attainable (Δt calculated by Eq. 13). Inspection of these data reveals the fast reduction in the amplitude during the first stages of leveling, followed by the sudden slowing and then cessation of leveling during the last two stages.

The fact that complete leveling could not be attained could have been ascertained from Eq. 12.

$$a \text{ (minimum attainable)} = \frac{10.0 \cdot 0.10^3}{248 \cdot 35 \cdot 0.010 \cdot 0.653} = 0.000176 \text{ cm}$$

However, the nature of the cessation of the leveling action becomes apparent only from the results obtained by the stepwise procedure.

If the yield value τ_0 for the coating had been specified as 2.0 rather than 10.0 dynes/cm^2, complete leveling would be attainable, since the minimum possible amplitude would be lowered to 0.0000352 cm (0.000176/5), a value well below the practical leveling end point of 0.000050 cm. The viscosity profile for this slightly modified coating is given as curve *B* in Fig. 28-6. The successive time increments to reach complete leveling, based on the 2.0-dynes/cm^2 yield value, are listed in Table 28-1. It should be pointed out that the time increments in Table 28-1 correspond to the amplitude half-reductions that are shown on the graph (designated by successive circled points on the profile curves).

Inspection of Fig. 28-6 indicates clearly how the successive down-trending lines for the half-reduced amplitudes cut across the unsweeping viscosity profile curves. With this picture in mind it can also be recognized that an early rapid leveling action can readily be stopped short of completion by a high yield value or for that matter by a rapidly unsweeping profile curve (corresponding to ultrahigh viscosities at ultralow shear rates).

The amplitude values noted in Fig. 28-6 apply specifically to the leveling conditions assumed for the problem considered above. However, these lines and their spacing can be thought of as a fixed grid that can be moved sidewise (45° slantwise direction) to fit other conditions. By the use of this grid of fixed parallel amplitude lines and one point to fix the grid position, leveling times can be worked out graphically by this stepwise procedure for any given viscosity profile.

EFFECT OF THIXOTROPY ON LEVELING

The assumption of a constant viscosity profile during the leveling operation is generally true only for nonthixotropic coatings. This assumption does not hold for thixotropic coatings, since the viscosity profile through the lower shear rate range is presumably shifting upward as leveling proceeds. This is almost certainly the case for thixotropic coatings that have been vigorously sheared just before the start of their leveling action. This perturbing factor calls for additional experimental data that provide recovery rates for the low-shear viscosities of coatings that have been subjected to high shear. It adds one more complicating factor to the already complex problem of assessing levelability.

EFFECT OF SOLVENT LOSS ON LEVELING

If the leveling time is protracted, solvent loss is another factor contributing to leveling (in that it acts to raise the coating viscosity). This loss may be due to solvent evaporation into the atmosphere or to solvent diffusion into a porous substrate.

Solvent Evaporation

With solution coatings (binder dissolved in solvent), compaction of the shrinking coating proceeds in an orderly manner to give a soundly integrated film. In general the viscosity increase due to solvent release is gradual and free from any abrupt transition stage. This is in contrast to latex coatings, which suddenly set to a semirigid structure when the latex particles first come into contact at an early stage during the evaporation of water. The critical solids content at which this occurs can be quite close to the original solids content (say 35 to 60%). Hence leveling is much more difficult to achieve with latex coatings than with solution coatings.

Wicking

The diffusion of solvent into a porous substrate by capillary action (wicking) may also exert a controlling influence on the leveling action (8). As previously stated, latex paints are commonly formulated close to the solids content where the latex particles contact and coalesce. Even a minimal wicking action can drain off enough water to reduce the solids content to this critical semisolid stage with a resulting abrupt cessation of leveling.

Water evaporation from a latex paint at room temperature is fairly slow (about 0.2 mg/min-cm^2-unit difference in relative humidity) and does not vary appreciably from latex to latex. Conversely, the rate of water loss by wicking is quite variable, depending on the substrate and latex formulation, and may well control the setting rate of the latex coating.

Experimental evidence indicates that the achievement of a satisfactory latex application viscosity through the use of a thickener results in a lower wicking rate than if the same viscosity were obtained by an increased pigment content. In turn the nature of the pigmentation can also affect wicking. Thus calcium carbonate is reported to contribute less to wicking than does diatomaceous earth.

TEST METHODS FOR MEASURING LEVELING

Experience has shown that the evaluation of leveling by subjective judgment leads to wide disagreement. Hence considerable effort has been expended within the paint industry to establish a routine procedure that will provide a quantitative index of levelability.

Surface analyzers are commercially available that can measure the surface profiles of dry paint films to a resolution of at least 0.025 μm (0.0000025 cm) (1,9). However, this direct approach to the measurement of leveling requires a high degree of expertise and a considerable expenditure of time. Furthermore, the cost of the necessary instrumentation is presumably beyond the means of the average laboratory (10). The determination of a coating viscosity profile to predict leveling behavior (and the shifting nature of the viscosity profile with time in the case of thixotropic coatings) also calls for experienced operators, ample time, expert interpretation, and relatively expensive instrumentation (10, 11). Although these two methods for evaluating leveling are basic both to research and for a proper understanding of the mechanics of leveling, they are not feasible for providing a rapid quantitative assessment of leveling that can be adopted and used on an industry-wide bases.

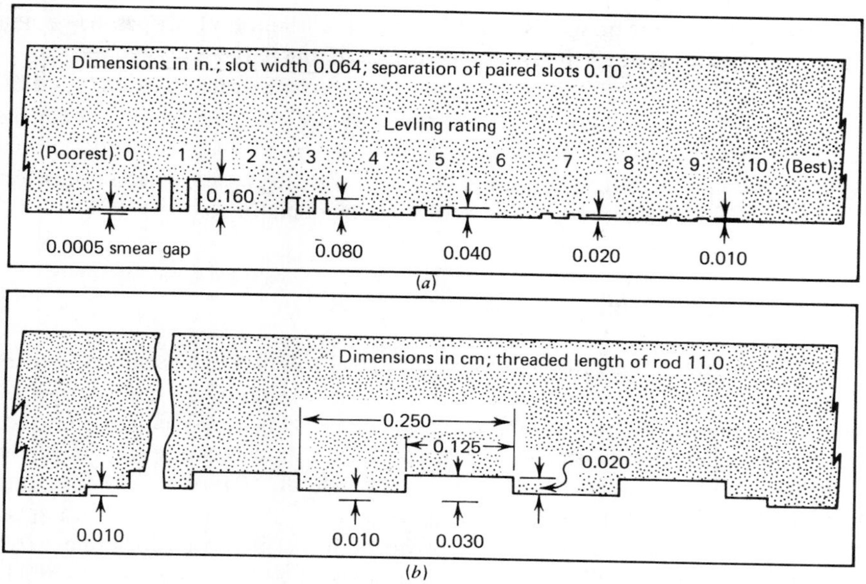

Fig. 28-7 Schematic diagrams of N.Y.P.C. and Leneta drawdown leveling blades.

At the present time several leveling test methods are being employed by the paint industry. Two depend on the use of specially designed drawndown blades, as schematically shown in Fig. 28-7.

New York Paint Club Leveling Test Blade (ASTM Method D2801)

The New York Paint Club (N.Y.P.C.) leveling blade (Fig. 28-7*a*) is machined to provide five pairs of rectangular notches, each pair of notches being spaced 0.10 in. apart (4, 12). The paired notches are designed to give progressively smaller heights (successive half-reductions from 0.016 to 0.010 in.).

A suitable quantity of presheared coating is placed in front of the blade on a drawdown chart, and the blade is drawn down at a uniform speed of about 1.0 ft/sec. Five paired stripes of decreasing thickness are formed and are allowed to dry in a horizontal position.

Levelability is determined by the ability of the paired stripes to flow together (level) to form single, merged stripes. The rating system assigns values of 0, 2, 4, 6, and 8 to the five paired stripes (from the 0.16- to the 0.010-in. height), with the leveling value for the test coating being the lowest numbered pair that fails to merge plus 1 if the merger is about half accomplished. If no stripes merge, the rating is 0 (poorest); if all the stripes merge, the rating is 10 (best).

Unfortunately the method is not applicable to unpigmented, textured, or dark coatings (10). Furthermore, the procedure does not appear suitable for comparing coatings with widely different viscosity characteristics or latex flat paints (10). Round robin testing is presently under way to decide whether this test method should be withdrawn in favor of the Leneta method.

Leneta Leveling Test Blade Method

The Leneta leveling test blade consists of a steel rod (1.1 cm in radius; 16.5 cm long) that has been machine-threaded to provide a series of alternate circumferential rectangular grooves (0.020 cm deep; 0.0125 cm wide) along its internal length, as schematically shown in Fig. 28-7*b* (4). Shoulders at each end of the rod hold the internal length of the rod a distance of 0.010 cm away from any planar surface on which the rod is placed. As a result, the effective gap clearances of the grooved and ungrooved sections of the blade on a drawdown are 0.030 and 0.010 cm, respectively. These in turn provides alternate wet coating thicknesses on a drawdown of about 0.0150 and 0.0050 cm (half the gap clearances).

The test is run by placing 8 to 10 cm^3 of the presheared test coating on a drawdown chart in front of the leveling blade and drawing down the blade at a uniform rate of about 2.0 ft/sec. This produces a striated coating with a standardized wavelength of 0.250 cm, an average thickness of about 0.010 cm, and an initial amplitude of about 0.0050 cm. After drying, a representative 3 × 5 in. section is cut out with the striations parallel to the long edge for rating the leveling.

Levelability is evaluated by comparing the cut-out section with commercially prepared levelness drawdown standards (4). These standards are numbered 1 to 9 in the order of improved leveling (see Fig. 28-3). Drawdowns poorer than the number 1 drawdown are rated 0; drawdowns better than the number 9 drawdown are rated 10 (perfect leveling).

Paint Research Association Leveling Blade

A crenellated leveling blade that spreads a coating with sets of ridges of variable width over a wet film of uniform thickness has been devised by British workers at the Paint Research Association (P.R.A.) (13). Four bands of wavelengths 0.13, 0.20, 0.26, and 0.34 cm are ground into the applicator to a depth of 0.0038 cm, and this depth is superimposed on an overall gap clearance of 0.0076 cm to give an average depth of 0.0095 cm. This average clearance results in a coating drawdown thickness of about 0.0045 cm. The design is such that brushing conditions are closely simulated (time of shear, constancy of shear rate, breakdown of coating structure), and the different sets of ridges provide different wavelengths.

Levelability is rated during the leveling process in terms of a zigzag pattern given by the reflection of a slit-light source from the surface of the applied coating. The levelness of dried glossy films (but not flat paints) can be rated by this optical technique. Agreement among the flow properties, drawdown ratings, and simple brush-outs using this procedure are reported as good for both Newtonian and non-Newtonian coatings.

Other Leveling Applicator Blades

Alternative designs have been devised for leveling applicators for research studies, such as a series of triangular slots ground into a drawdown bar (9) or even a wire-wound applicator rod (14 tight wire turns to the inch) (6).

Leveling Measured by Gloss Readings

The levelness of gloss paints (not flats) can be measured by 60° glossmeter readings taken parallel and perpendicular to the direction of brushing (10). Good leveling is indicated by a minimal difference in gloss readings between the two directions. However, the accuracy is such that only an approximate ranking among a series of test paints can be obtained.

Brush-out Leveling Test Method

A standardized test method for brushing out latex paints to evaluate leveling has been specified in detail by the New York Society for Paint Technology (10). The type of substrate (such as Leneta Form 7B paint test chart), the type of brush (2-in. nylon, tapered, 3-in. bristle, $\frac{3}{4}$ in. thick), the exact procedure (preconditioning of brush, loading of brush, brush-out technique, rate of application), and the manner of ranking the brush-outs after drying in a horizontal position are all clearly defined. Brush-out levelness standards are also commercially available for providing comparative ratings (4). These are counterparts of the drawdown levelness standards previously discussed and, like them, are numbered 1 to 9 in order of improved levelability. Brush-outs poorer than 1 are rated 0, and those better than 9 are rated 10 (perfect leveling).

Leveling Test Based on Wedge-Type Films and Comb-Produced Striations

As far as this writer is aware, no leveling test has ever taken advantage of wedge-shaped coatings where striations are introduced by a comb that is drawn across the film from the low (or zero) side to the high side of the wedge. A wedge gap film applicator (ASTM D823) commercially available in a 10-cm (4-in) width provides gap clearances of 0 to 100 μ (4 mils), 0 to 200 μ (8 mils), 100 to 500 μ, and 300 to 1500 μ (14). Ordinary combs of different tooth counts per unit length can be used to produce striations crosswise of the drawdown width immediately after the film drawdown. By this test procedure, leveling data are provided over a continuous range of film thicknesses and for different striation spacings (different combs). The timing for complete leveling can even be followed down the striation furrow, starting from the high side of the wedge (possibly instantaneous leveling) down to the low side of the wedge (infinite time for leveling for zero thickness). The use of combs with different tooth counts permits cross checks on results by using appropriate leveling equations.

In suggesting this procedure, the present writer recommends that the comb-produced striations be started at the thin side of the wedge. It is also the writer's opinion that the shape of the comb tooth is not critical, although a circular profile and a gap between the teeth equal to the diameter of this circular profile seem logical for the comb design.

INSPECTION OF LEVELING

Leveling is most effectively observed by viewing the leveled surface under oblique illumination, since this brings out the shadows cast by the surface contours. A commercial instrument has been designed to provide a standard source of oblique illumination and a correct viewing angle for examining such surface irregularities as brushmarks (Level-Luminator) (4).

Tinting white latex paints with a tinting paste (such as a water-dispersible yellow oxide) renders the brushmarks more easily discernible, especially in the middle range of leveling (it decreases the distracting effect of hiding) (10).

REFERENCES

1. Smith, N. D. P., S. E. Orchard, and A. J. Rhind-Tutt, "The Physics of Brush Marks," *JOCCA*, **44,** 618–633, September (1961).
2. Pearson, J. R. A., *J. Fluid Mech.*, **7,** 481 (1960).
3. Patton, T. C., *Paint Flow and Pigment Dispersion*, 1st ed., Wiley-Interscience, New York, 1964.
4. Catalog No. 3, "Paint Test Charts and Test Equipment," The Leneta Company, Ho-Ho-Kus, N.J. 07423, 1976.
5. Orchard, S. E., *Appl. Sci. Res.*, **A11,** 451 (1962).
6. Quach, A. and C. M. Hansen, "Evaluation of Leveling Characteristics of Some Latex Paints," *JPT*, **46,** No. 592, 40–46 (1974).
7. Sarker, N. and R. H. Lalk, "Rheological Correlation with the Application Properties of Latex Paints," *JPT*, **46,** No. 590, 29–34, March (1974).
8. Garnett, B. S., W. S. Prentiss, and J. D. Scott, "Factors Affecting the Leveling of Latex Paints," *Off. Dig.*, **31,** No. 409, 213 (1959).

9. Dodge, James S., "Quantitative Measures of Leveling," *JPT*, **44**, No. 564, 72–78, January (1972).
10. New York Society for Paint Technology, "Experiments in Measurement of Leveling of Latex Paints," *JPT*, **46**, No. 589, 31–36, February (1974).
11. Ehrlich, A., T. C. Patton, and A. Franco, "Viscosity Profiles of Solvent Based Paints: Their Measurement and Interpretation," *JPT*, **45**, No. 576, 58–67, January (1973).
12. Stieg, F. B., "The Evaluation of Leveling by a Drawdown Method," *Off. Dig.*, **32**, No. 430, 1435 (1960).
13. Camina, M. and D. M. Howell, "The Leveling of Paint Films," *JOCCA*, **55**, No. 10, 929–940, October (1972).
14. *Handbook of Paint Testing Instruments*, Paul N. Gardner Company, Fort Lauderdale, Fla. 33316.

29 *Sagging, Slumping, and Draining*

A wet coating applied to a vertical surface is subject to a downward flow (influence of gravity) that is variously referred to as running, curtaining, sagging, or draining. A certain degree of downward flow is tolerable and in fact must be accepted to attain satisfactory leveling. However, excessive sagging is inexcusable in a modern coating and fortunately is rarely seen in conventional coatings today.

SAGGING RHEOLOGY

A wet coating applied to a vertical surface can be visualized as consisting of a large number of very thin layers, each of thickness dx, piled up against each other (like pages in a book). Let one of these thin layers be located at a depth x in from the outside wet film surface, and consider a surface area of 1.0 cm^2, as schematically shown in Fig. 29-1a. At its inside location the thin layer is subject to a downward pull over its surface of 1.0 cm^2, since it is supporting the weight of the coating layers outside of it. The weight of the volume of coating (x cm thick) that is trying to slide down over this inner area is ρx, where ρ is the coating density. The downward drag over this unit surface area (equal to the shear stress τ_x) is given by Eq. 1, where g is the acceleration of gravity (980 cm/sec^2).

$$\tau_x = \rho x g \tag{1}$$

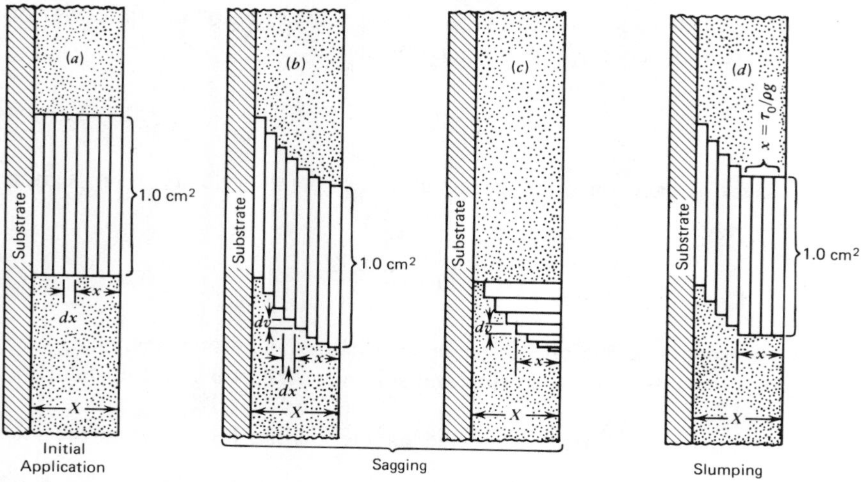

Fig. 29-1 Schematic diagrams illustrating (*a*) the geometry that applies to a coating initially applied at uniform thickness to a vertical substrate, (*b*) the geometry that applies to this coating after a period of sagging, (*c*) the incremental volumes of coating that sagged past an arbitrary level during the sag period, and (*d*) the geometry that applies to the slumping that occurs when the coating has a significant yield value.

Offsetting this downward drag is the viscous resistance of the coating, which opposes the tendency to sag. Noting that the shear rate D across the thin inner layer is $-dv/dx$, we obtain Eq. 2 expressing the sagging condition at the inner layer surface.

$$\eta = \frac{\tau}{\mathrm{D}} = \frac{\rho x g}{dv/dx} \tag{2}$$

Solving explicitly for dv yields Eq. 3.

$$dv = -\left(\frac{\rho g}{\eta}\right) x \cdot dx \tag{3}$$

Integrating Eq. 3 gives Eq. 4.

$$v = -\frac{\rho g x^2}{2} + K \tag{4}$$

The constant of integration K can be obtained by noting that, when $x = X$ (where X is the total coating thickness), the velocity $v = 0$, since the layer next to the substrate does not move (see Fig. 29-1*b*). Solving for K and substituting this value in Eq. 4 yields an expression for the velocity v_x of the thin layer of paint at depth x in a coating film of total thickness X.

$$v_x = \frac{\rho g}{2\eta} \cdot (X^2 - x^2) \qquad (5)$$

For a layer of paint on the outside surface ($x = 0$) the velocity of sag is given by Eq. 6.

$$v_0 \text{ (at surface)} = \frac{\rho g X^2}{2\eta} \qquad (6)$$

This corresponds to the rate of sag under idealized conditions (assuming such model conditions as Newtonian flow, no solvent evaporation, and no temperature change).

The maximum shear stress is exerted on the coating layer next to the substrate, as expressed by Eq. 7.

$$\tau_X \text{ (maximum)} = \rho X g \qquad (7)$$

Consider next the volume of paint V_s which sags (or runs) across a given cross section of the film in time t. For purposes of calculation, the portion of the coating that sags is considered as being sliced up into thin layers that are *perpendicular* rather than *parallel* to the vertical substrate, as schematically shown in Fig. 29-1*c*. From this standpoint it can be seen that the differential volume of coating that flows across a cross-sectional area $(1.0 \cdot X)$ in differential time dt is given by Eq. 8.

$$\frac{dV_s}{dt} = \int_{x=0}^{x=X} x \cdot dv \qquad (8)$$

From Eq. 3 dv can be expressed in terms of dx. Substituting this value for dv in Eq. 8 yields Eq. 9.

$$\frac{dV_s}{dt} = \frac{\rho g}{\eta} \int_{x=0}^{x=X} x^2 \cdot dx \qquad (9)$$

Integrating Eq. 9 over the given limits results in a relatively simple expression for the total volume of coating that sags over a given cross section of the film $(1.0 \cdot X)$ during time t.

$$V_s = \frac{X^3 \rho g t}{3\eta} \qquad (10)$$

Note the extreme influence of the coating thickness (raised to the third power) on the volume of sagging coating. This explains why a professional painter takes great care to apply a uniform coating. Thus, if a coating applied at a higher level has twice the thickness of that applied at a somewhat lower level, eight times (2^3) as much coating will sweep over the lower level as will leave it, leading to almost certain curtaining.

PROBLEM 29-1

Assume that the viscosity of an enamel at very low shear rates is 1.5 poises. If its density is 9.2 lb/gal (1.1 g/cm^3), calculate the distance h that the outside surface of the enamel will sag during a 10.0-min (600-sec) period when applied to a vertical surface at a 3.0-mil (0.0076-cm) wet film thickness. Also compute the volume of coating flowing past a horizontal line 1 cm wide during the 10.0-min period. Assume Newtonian flow and negligible solvent loss.

Solution. Substitute the given appropriate data in Eqs. 6 and 10, noting that the surface sag is given by $h = v_0 t$.

$$v_0 = \frac{1.1 \cdot 980 \cdot 0.0076^2}{2 \cdot 1.5} = 0.0207 \text{ cm/sec}$$

$$h = 0.0207 \cdot 600 = 12.4 \text{ cm}$$

$$V_s = \frac{\cdot\, 0.0076^3 \cdot 1.1 \cdot 980 \cdot 600}{3 \cdot 1.5} = 0.063 \text{ cm}^3$$

Comment. It is obvious that this enamel exhibits an intolerable degree of sagging.

In Problem 29-1 several factors were ignored that normally act to improve the sag resistance. Thus, in proportion to the rate of solvent loss, the viscosity of the wet film is raised and the film thickness is reduced. Moreover, solvent loss cools the film and thus further increases the viscosity. In the second place the low-shear-rate viscosity of 1.5 poises assumed for the enamel, although correct for a high-shear-rate application viscosity, is altogether too low for a shear rate viscosity applicable to sagging conditions. More realistic would be a viscosity 100 times as large, bringing the sagging that was exhibited into an acceptable range.

Since viscosity is invariably the major factor controlling sagging, the coating formulation must be designed to provide high viscosities at the low shear rates that apply to coating flow over vertical substrates. Presumably any commercial coating contains a colloidal flow additive in sufficient amount to raise the low-shear-rate viscosity to hundreds or thousands of poises or even to impart a yield value.

PROBLEM 29-2

A coating (density 1.5 g/cm^3) is applied to a vertical substrate to provide a uniform thickness of 5.0 mils (0.0127 cm). A point on the surface of the wet film is observed to move downward a distance of 0.30 cm during a 10.0-min (600-sec) period. Calculate the effective average viscosity of the coating during this sagging period and also the maximum shear stress and shear rate involved.

Solution. Solve for the effective average viscosity by inserting in Eq. 6 appropriate data as given in the problem.

$$v_0 = \frac{0.30}{600} = \frac{1.5 \cdot 980 \cdot 0.0127^2}{2\eta}; \qquad \eta = 237 \text{ poises}$$

The maximum shear stress (at the substrate surface) is calculated from Eq. 7.

$$\tau_X = 1.5 \cdot 0.0127 \cdot 980 = 18.7 \text{ dynes/cm}^2$$

The maximum shear rate is also present at the substrate surface, as calculated from $\mathrm{D} = \tau/\eta$ (or Eq. 2).

$$\mathrm{D} = \frac{18.7}{237} = 0.079 \text{ sec}^{-1}$$

Control of sag resistance permits the application of high-build coatings that can lead to significant savings in labor cost (a one-coat instead of a two- or three-coat application). Also wet edge times can be extended, since slower evaporating solvents can be introduced without fear of excessive running.

SLUMPING RHEOLOGY

Slumping is a form of sagging exhibited when a coating composition has a significant yield value that must be exceeded before flow down a vertical substrate occurs (schematically indicated in Fig. 29-1*d*). This special case of sagging is occasionally observed when relatively thick, pigmented coatings with high yield values are applied to vertical or inclined surfaces. Reference to Figs. 29-1*b* and *d* clarifies the difference between sagging and slumping. A coating with no yield value sags at all points throughout the applied coating; a coating possessing a yield value is characterized by an outside slump portion that moves as a relatively rigid unit with a thickness x, given by Eq. 11.

$$x = \frac{\tau_0}{\rho g} \tag{11}$$

If the yield value is made sufficiently high, as with heavily pigmented pseudoplastic compositions, the x of Eq. 11 may equal or exceed X, at which point all sag is arrested and no downward movement occurs.

SAGGING AND SLUMPING WITH NON-NEWTONIAN SYSTEMS

In the foregoing sections on sagging and slumping either Newtonian flow or an average effective viscosity was assumed. Unfortunately, pigmented coatings are rarely, if ever, Newtonian in nature. Consequently, complex equations are required to describe the sagging behavior of these non-Newtonian systems.

The problem of controlling sagging is especially critical for high-solids coatings (70 to 85%w) based on oligomeric (low-molecular-weight) resins, since these coatings lose little solvent (~5%w) during spray application and before baking, whereas conventional coatings of high molecular weight are typically 30%w solids at the spray gun and 85%w solid on the substrate before baking. Based on the assumption of "power-law" fluid systems with a yield value, equations have been developed for application to non-Newtonian systems (1). From this study it is concluded that pseudoplastic systems are most desirable for sag and slump control. Even Newtonian systems fail to provide good sprayability and good sag control at the same time, and dilatant systems are worst of all.

DRAINING RHEOLOGY

A flow situation related to sagging is draining, such as is encountered when a panel is uniformly coated and then allowed to drain in a vertical or inclined position. In this case, since no coating is located above the top of the panel, there is no replenishment of the coating material that drains down from the face of the panel. As a result the overall thickness of the applied coating is immediately reduced, with the thinning becoming most pronounced at the top, as shown by the schematic diagram of Fig. 29-2. An expression (Eq. 12) that provides a means for calculating the changing film thickness for a Newtonian system under such draining conditions has been adapted from a mathematical derivation developed by Erneto (2) and made available through the courtesy of NL Industries. Figure 29-2 illustrates a typical draining condition (β = angle of inclination) where x_0 is the initial uniform coating thickness and x_h is the film thickness at time t at a distance h down from the top of the draining film.

$$x_h = \sqrt{h/(\rho g \cos \beta t/\eta + h/x_0^2)} \tag{12}$$

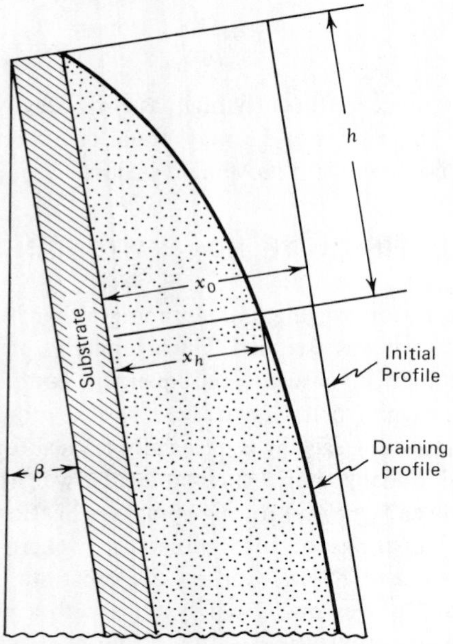

Fig. 29-2 Schematic diagram showing the geometry that applies to a uniformly deposited coating draining over the face of an inclined panel.

PROBLEM 29-3

A 3 X 5 in. test panel is uniformly coated to a thickness of 3.0 mils (0.0076 cm) with a clear varnish (density 0.90 g/cm^3; viscosity 1.4 poises) and then allowed to drain in a vertical position (3.0-in. edge at bottom). Calculate the thickness of the varnish film at the end of 10.0 min at points 1.0 and 4.0 in. down from the top of the panel.

Solution. Substitute appropriate data given in the problem in Eq. 12, and solve for the required thicknesses.

$$x \text{ (1.0 in. down)} = \sqrt{2.54/(0.9 \cdot 980 \cdot 1.0 \cdot 600/1.4 + 2.54/0.0076^2)}$$

$$= 0.00245 \text{ cm } (= 0.97 \text{ mil})$$

$$x \text{ (4.0 in. down)} = \sqrt{10.2/(0.9 \cdot 980 \cdot 1.0 \cdot 600/1.4 + 10.2/0.0076^2)}$$

$$= 0.00428 \text{ cm } (= 1.69 \text{ mils})$$

Comments. It is seen that there is a marked reduction in the thickness of this freely draining coating. Such reduction is normally overcome as required by increasing the low-shear-rate viscosity of the coating system.

SAGGING MEASUREMENT

One test procedure for rating sag or slump (Federal Test Method 4493; September 1965) that yields reproducible qualitative data is based on direct sagging observation (3). In this procedure a representative sample of the test paint is applied as a 2-in. wide strip to a flat substrate (such as a hiding power chart) at a 3.0-mil wet thickness. Immediately after the drawdown, two broad lines are rapidly and completely drawn across the wet film with the sag-liner device shown in Fig. 29-3, giving two paint-free lines of different widths for observation. A duplicate set of these lines is normally made further down on the film for check purposes. The drawdown is then immediately hung in a vertical position, and the wet coating is allowed to dry. Sag development is evaluated from inspection of the dry paint film, using the visual rating scheme shown in Fig. 29-4.

Films revealing no evidence of paint flow are rated "no sag"; films showing flow, but of a degree insufficient to cross even the narrower gap, are rated "very slight sag"; films in which the coating has run across the narrower sag line, but not the broader line, are rated "slight sag"; films where the paint has flowed

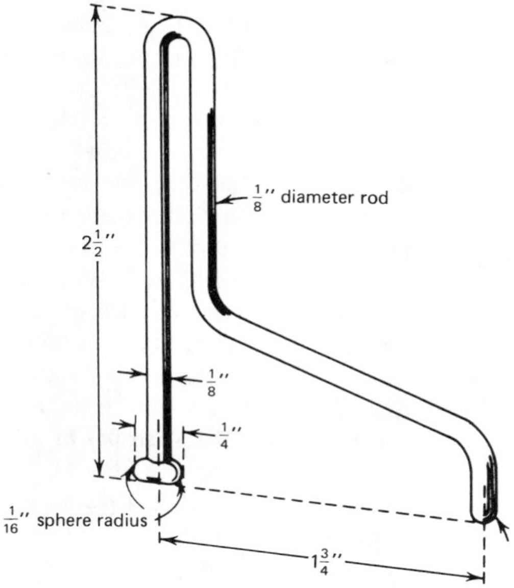

Fig. 29-3 Illustration of a sag-liner used to produced sag lines for assessing the sagging tendency of a coating applied at uniform thickness to a vertical substrate.

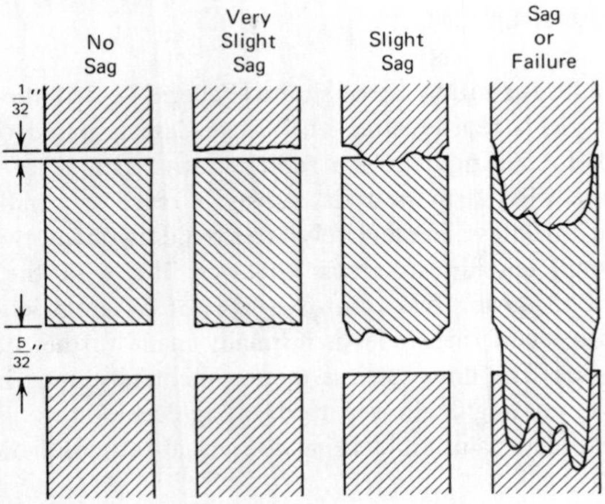

Fig. 29-4 Schematic diagram illustrating the visual rating schedule used for rating the sagging tendency of a coating applied to a vertical substrate.

across both lines are rated "sag or failure" (or "runs" for the severe sagging condition where the lines are completely obliterated by excessive flow).

A second test method quantitatively assesses the sagging tendency of a coating in terms of the maximum wet thickness of coating that can be applied to a vertical substrate with no more than a specified marginal amount of sag occurring. This criterion for judging sag is in contrast to the procedure outlined above, in which sag is rated by the flow that occurs for a wet film of uniform initial thickness.

The second test, as called for in several U.S. government specifications (4), is carried out using a uniquely designed sag-index blade (5, 6). The blade is essentially a straight-edge applicator notched with a series of flat $\frac{1}{4}$-in. wide gaps (separated from each other by a spacing of $\frac{1}{16}$ in.). Ten to eleven of these gaps may be notched over a width of about 3 in., with the gap clearances running in steps of 3 to 12 mils for a standard (architectural coating) range; 1 to 6 mils for a low clearance (production finish) range; and 14 to 60 mils for a high clearance (high-build coating) range. A drawdown on a chart with such a notched blade provides a series of parallel wet stripes of different thicknesses. It should be noted that the wet film thickness is about half the gap clearance. Immediately after the drawdown the test chart is placed in a vertical position with the stripes horizontal (thickest stripe at the bottom).

As a result of the sagging of the wet stripes, some of the uncoated intermediate $\frac{1}{16}$-in. spaces on the chart may or may not become completely covered. At the end of the dry period the chart is inspected to determine the gap clearance producing the thickest stripe that did not sag sufficiently to come into

Table 29-1 Antisag Index Ratings and Interpretation (5)

Antisag Index (Gap Clearance, mils)	Probable Wet Film Thickness (mils)	Qualitative Rating[a]	Index Range for Representative Coatings
3	1.5	Very poor	
4	2.0	Poor	
5	2.5	Poor/fair	
6	3.0	Fair	Four-hour enamel
7	3.5	Fair/good	Interior trim enamel
8	4.0	Good	
9	4.5		Interior alkyd flat
10	5.0	Very good	
11	5.5		Exterior house paint
12	6.0	Excellent	

[a]Based on observations of brush-out sag resistance.

contact with the stripe below it. This gap clearance is taken as a measure of the resistance of the coating to sagging (an antisag index). Interpolation between antisag indices (gap clearance values) is also possible by adding to the antisag index an interpolated fraction corresponding to the degree that the next lower stripe fails to merge with the stripe still lower down.

A rough relationship between antisag index and a qualitative rating scale of sagging performance is given in Table 29-1 (6).

SIMULTANEOUS SAGGING AND LEVELING

The objectives of maximum leveling and minimum sagging are mutually antagonistic. Thus a reduction in viscosity to assist leveling also promotes sagging. Or increasing film thickness to accelerate leveling also speeds up the sag rate. This mutual opposition means that some compromise must be worked out in practice to obtain satisfactory leveling without excessive sagging. Generally, the small amount of sag that is accepted to permit time for proper leveling is of no serious consequence. However, there are exceptions such as high-solids coatings (70 to 80%w). The mathematics covering the latter situation have recently been developed and reported with the conclusion that only pseudoplastic systems are capable of providing sag control, leveling, and sprayability simultaneously (7).

REFERENCES

1. Wu, Souheng, "Rheology Of High Solid Coatings: I. Analysis of Sagging and Slumping," ORPL Preprint, American Chemical Society Meeting, August 1977, pp. 315–322.

2. Erneto, Modesto, "The Rheology of Falling Films," private communication from NL Industries, Hightstown, N.J.
3. Patton, T. C., "A Simple Pigment-Settling Gage and a Simple Anti-Sag Test," *Off. Dig.*, **29,** 10–25, January (1957).
4. Federal Test Method Standard No. 141a, Method 4484: TT-E-508 Interior alkyd semigloss, TT-E-506 Alkyd gloss enamel, TT-P-1511 Interior latex gloss and semigloss, TT-I-564 Black stencil ink.
5. Schaeffer, L., "Design of an Improved Sag Tester," *Off. Dig.*, **34,** 1110–1123 (1962).
6. Catalog No. 3, "Paint Test Charts and Test Equipment," The Leneta Company, Ho-Ho-Kus, N.J. 07423, 1976.
7. Wu, Souheng, "Rheology Of High Solid Coatings: II. Analysis of Combined Sagging and Leveling," ORPL Preprint, American Chemical Society Meeting, August 1977, pp. 323–327.

30 *Film Applicators*

The precise evaluation of paint or ink properties is dependent, to a major extent, on the ability of the paint man or woman to produce as a matter of routine a wet film of controlled uniform thickness. Brushing and spraying, which are common industrial methods of coating application, do not readily lend themselves to the application of films of closely specified thickness to areas of limited dimension (test panels). Hence alternative methods of application that are characterized by simplicity and precise control have been devised for the express purpose of evaluating film properties.

STRAIGHT-EDGE APPLICATORS (1–3)

Straight-edge applicators, as a class, include all film applicators which are variously called blade applicators, doctor blades, casting knives, coating blades, and straight-edge spreaders. For any one of these applicators, there is present the same underlying principle of a straight edge smearing a thin layer of coating over a flat substrate, the thickness of the layer being controlled by machined shoulders or projections at points along the straight edge which mechanically separate the straight edge from the flat substrate underneath (control the gap clearance).

Originally, it was tacitly assumed that the thickness of the wet film laid down by the straight-edge applicator would be identical with the clearance of the spreader gap. That this was an erroneous premise is now well acknowledged, and the considerable data that have since been developed reveal that the profile of the straight edge plays the dominant role in determining the relationship between wet film thickness T and spreader clearance S.

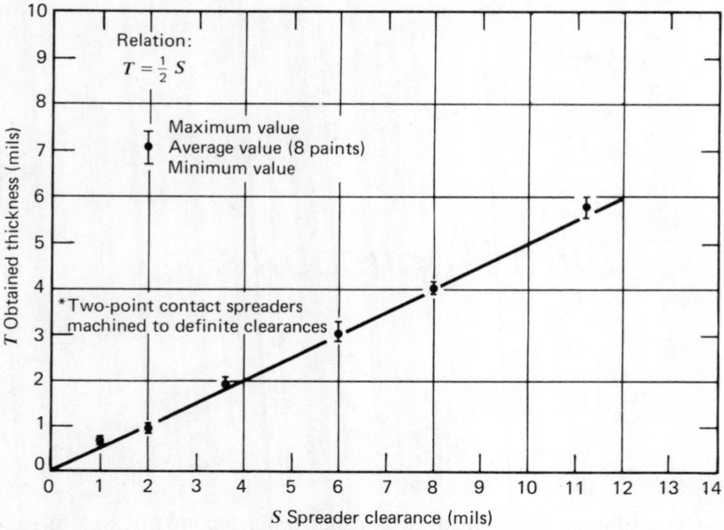

Fig. 30-1 Relationship of wet film thickness to gap clearance for a straight-edge film applicator with a flat profile. From reference 3.

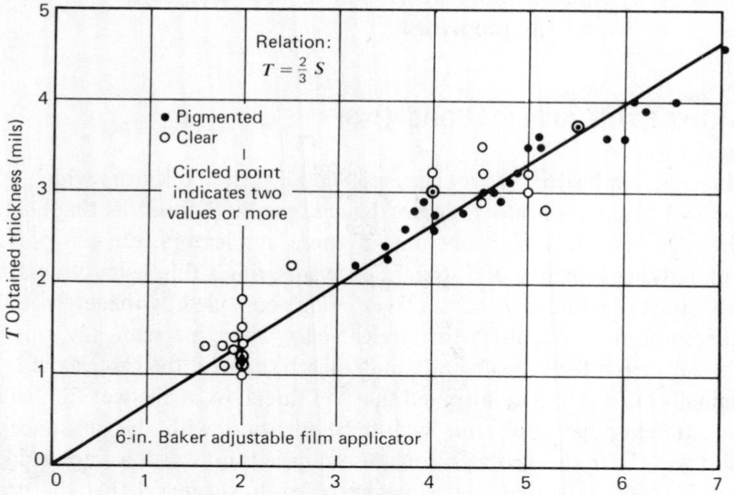

Fig. 30-2 Relationship of wet film thickness to gap clearance for a straight-edge film applicator with a circular profile. From reference 1, pp. 725, 726.

Figure 30-1 graphs experimental data relating the wet film thickness T and gap clearance S for a spreader with an essentially flat (blunt) profile; Fig. 30-2 gives this T/S relationship for a straight-edge spreader with a circular profile.

Analysis of the data developed in a more recent study (2) on the influence of such variables as binder solids content, binder density, vehicle viscosity, and vehicle density on the wet film thickness of clear vehicles (applied by a blade with an essentially flat profile) reveals somewhat higher values for the wet film thickness/gap clearance ratio (0.75 for a 2-mil gap clearance; 0.63 for a 4-mil clearance; 0.62 for a 6-mil clearance). The variables of density and vehicle viscosity were shown to be without effect on the wet film thickness. The films were prepared using the procedure specified in ASTM test method D823-53.

For a true knife-edge profile it has been shown that the wet film thickness closely approximates the actual gap clearance. The several approximate relationships as given by Eqs. 1, 2 and 3 are summarized and graphed in Fig. 30-3.

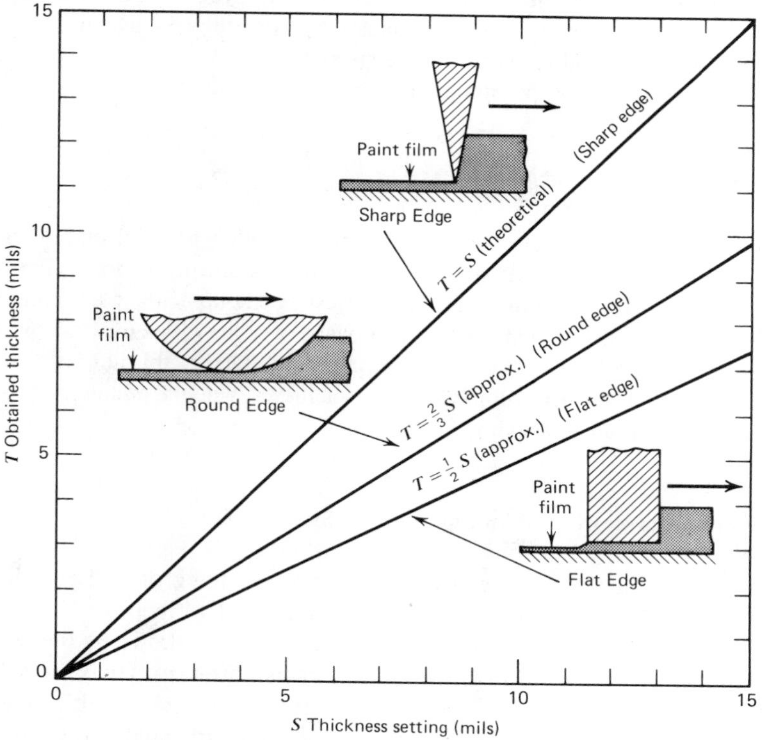

Fig. 30-3 Plot of relationships of wet film thickness to gap clearance for straight-edge film applicators with varying profiles.

Flat profile: $T = \frac{1}{2}S$ (1)

Circular profile: $T = \frac{2}{3}S$ (2)

Knife-edge profile: $T = S$ (3)

It is noteworthy that these relationships are independent of the rate of drawdown, the viscosity, or the pigment content in the case of pigmented systems. This topic is discussed further in Chapter 28 on leveling.

Many ingenious concepts have been built into these straight-edge applicators. Thus in one design a variable gap clearance is attained by the use of barrel-type micrometers. In another design this is achieved by rotating the cylindrical blade relative to locked-in end plates. In a third design pockets behind the straight edge provide a film separation so that two or three individual films can be drawn down simultaneously for side-by-side comparison.

To ensure flatness of the substrate, vacuum plates ($\frac{3}{32}$-in. holes spaced 1 in. apart to provide suction points) and magnetic chucks (a stainless steel or other nonmagnetic metal applicator must be used) are employed to guarantee a planar surface for cardboard and sheet steel substrates, respectively. Details of these and other variants of individual straight-edge applicators which are commercially available can be found in the catalogs of laboratory supply houses.

SPINNING-DISK APPLICATORS (4)

Offhand, it would seem unreasonable to expect a puddle of paint placed on the center of a spinning horizontal disk to flow out to a uniformly thick film over the disk surface. A more natural thought suggests a pile-up of the paint along the outside of the disk as the paint is slung outward over the disk surface. However, it can be shown both theoretically and experimentally that this does not occur, and that instead a uniform film thickness is achieved when a puddle of paint is spread over the center of a spinning disk.

Derivation of the Uniform Film Equation for Spinning Disks

The theoretical relationships which hold for the spreading of a viscous liquid such as a coating over a spinning disk can be derived by considering a cross section of the liquid in the form of a section of a band, as shown by the section *ABCD* of Fig. 30-4. This band is located at a radial distance r from the axis of the disk rotation. Assume the band section to have a differential thickness dr, length r, and height h. Consider the forces acting on a horizontal element of this band *ABCD*, of thickness dz, at a height z above the disk surface. This horizontal element is shown as the cross-hatched section in Fig. 30-4.

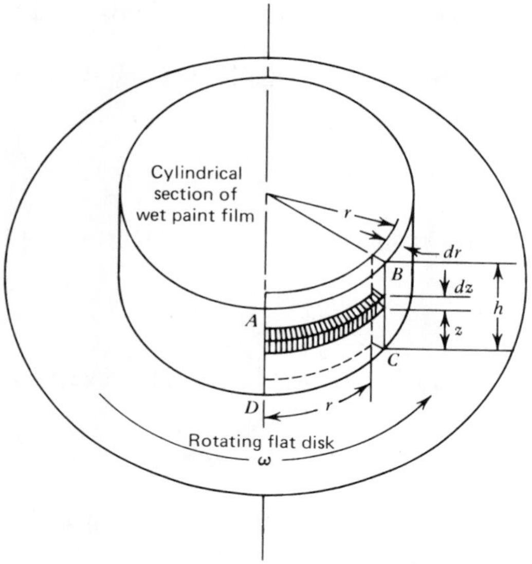

Fig. 30-4 Schematic diagram giving the space relationships for a portion of coating located on a spinning disk.

As the disk rotates, the paint directly above the element is subjected to a centrifugal force F_c which acts to move this portion of the paint outward in a radial direction. This force is given by Eq. 4, where m is the mass of paint directly above in the element *ABCD* and ω is the angular spin velocity of the rotating paint film in radians/sec.

$$F_c = m\omega^2 r \tag{4}$$

The mass of paint m can be expressed in terms of the density and the volume of the paint by Eq. 5.

$$m \text{ (in g)} = \rho\,[r\,dr(h - z)] \tag{5}$$

Hence the centrifugal force acting on the mass can be written alternatively as Eq. 6.

$$F_c = \rho\,[r\,dr(h - z)]\,\omega^2 r \tag{6}$$

Opposing the centrifugal force is the viscous resistance F_v of the paint to flow. This resisting force is obtained by applying the basic viscosity equation to the horizontal element of band *ABCD*.

$$F_v = \frac{dv}{dz}\,\eta r\,dr \tag{7}$$

Here dv is the differential velocity across differential height dz, and η is the paint viscosity.

Setting the opposing forces of centrifugal force and viscosity resistance equal to each other gives Eq. 8.

$$\rho[r\,dr(h - z)]\,\omega^2 r = \frac{dv}{dz}\,\eta r\,dr \tag{8}$$

Canceling out the common factors and rearranging terms leads to Eq. 9.

$$\frac{dv}{dz} = \frac{\rho\omega^2}{\eta}(h - z)r \tag{9}$$

For simplification let $k = \rho\omega^2/\eta$. Then Eq. 9 can be expressed as Eq. 10.

$$\frac{dv}{dz} = k(h - z)r \tag{10}$$

Integrating Eq. 10 gives velocity v for the element of paint at height z above the surface of the rotating disk as expressed by Eq. 11.

$$v = khrz - \frac{krz^2}{2} + K \tag{11}$$

When $z = 0$, the radial velocity v is also zero, since the paint at this location is lodged against the disk surface. Hence the constant of integration K is zero, and Eq. 11 can be expressed simply as Eq. 12.

$$v = kr\left(hz - \frac{z^2}{2}\right) \tag{12}$$

Attention is directed next to calculating the volume rate of flow for the paint which moves outwardly across the outer face of the band section $ABCD$. Let this flow rate be $Q = V/t$, where V is the paint volume that flows in time t. Note that the velocity of flow is maximum at the top of the band section ($z = h$) and zero at the bottom ($z = 0$). At any intermediate height the velocity will be given by Eq. 12.

Now the total outward flow rate of paint Q can be computed by summing up (integrating) the outward flow for all the horizontal elements (each of thickness dz) which are found in the band $ABCD$ ranging from $z = 0$ to $z = h$, as given by Eq. 13.

$$Q = \frac{V}{t} = \int_0^h vr\,dz = \int_0^h kr\left(hz - \frac{z^2}{2}\right) r\,dz$$

$$= kr^2 \Big|_0^h \left[\frac{hz^2}{2} - \frac{z^3}{6}\right] = \frac{kr^2h^3}{3} \tag{13}$$

As the paint flows out of the front or outer face of the band *ABCD*, other paint is simultaneously flowing into the band *ABCD* through the back or inner face. The paint flow through the front face is greater than that through the back face. Let this difference in flow through the two faces (which are separated by differential distance *dr*) be at the rate of $dQ = d(V/t)$. This differential flow can be computed by differentiating Eq. 13 for Q in terms of r.

$$\frac{dQ}{dr} = \frac{2krh^3}{3} \qquad \text{or} \qquad dQ = \frac{2kh^3}{3} r\,dr \tag{14}$$

Since there is a depletion of paint volume in the band section *ABCD* because of unequal flow rate (more goes out than goes in), the height of the band is necessarily lowered. Let the rate at which the height is reduced be dh/dt. Equation 15 then gives the rate at which paint volume is lost by the band section *ABCD*.

$$dQ = r\,dr \frac{dh}{dt} \tag{15}$$

Now two equations have been derived for dQ (Eqs. 14 and 15). These are equated to give Eq. 16.

$$\frac{2kh^3}{3} r\,dr = r\,dr \frac{dh}{dt} \tag{16}$$

Canceling out the common factor $r\,dr$ yields Eq. 17, an expression for dh/dt which is completely free of r.

$$\frac{dh}{dt} = \frac{-2kh^3}{3} \tag{17}$$

This equation tells us that the rate at which a section of paint film thins out on a spinning disk is in no way influenced by how far it is from the spin axis (a rather unexpected phenomenon). It also means that, if a film starts out with uniform thickness, spinning merely reduces the thickness in a uniform manner by spreading it out over a larger area.

In the paint laboratory these facts can be put to good use. For example, if a uniform paint film as drawn down on a panel proves to be too thick for a given purpose, placing it on a spinning turntable for a controlled period of time will uniformly reduce the thickness to the desired value. An expression for the height reduction that occurs in a given time period t will now be derived. This is done by integrating Eq. 18, which is the reciprocal of Eq. 17.

$$dt = \frac{-3}{2k} \frac{dh}{h^3} \tag{18}$$

Integrating Eq. 18 gives Eq. 19.

$$t = \frac{3}{4kh^2} + K \tag{19}$$

To establish a constant of integration, note that, at the beginning of the spinning operation ($t = 0$), the height h has a value h_0. Hence K is equal to $-3/(4kh_0^2)$. The expression for t in terms of height h and the initial height h_0 is then given by Eq. 20.

$$t = \frac{3}{4k}\left(\frac{1}{h^2} - \frac{1}{h_0^2}\right) \tag{20}$$

A more useful form of Eq. 20 is obtained by solving this equation for h in terms of t, h_0, and k, as given by Eq. 21.

$$h = \frac{h_0}{\sqrt{1 + 4kh_0^2 t/3}} \tag{21}$$

The fractional remaining height for time t and initial height h_0 is given by Eq. 22.

$$\frac{h}{h_0} = \sqrt{3/(3 + 4kth_0^2)} \tag{22}$$

Theoretically this spinning arrangement could serve to determine paint viscosity. This can be seen by solving Eq. 20 explicitly for viscosity (since $k = \rho\omega^2/\eta$) and noting that a measurement of h_0, h, t, ρ, and ω leads to the evaluation of η.

$$\eta = \frac{4t\rho\omega^2}{3(1/h^2 - 1/h_0^2)} \tag{23}$$

PROBLEM 30-1

A puddle of paint having a viscosity of 1.0 poise and a density of 1.0 g/cm^3 is placed on a flat tin panel, and the panel in turn is placed at the center of a record-playing turntable rotating at 45 rpm. Calculate the reduction in height for sections of the paint puddle which are initially 4 and 32 mils in thickness at the end of 3 min.

Solution. Substitute the given data in Eq. 22, where k by definition is equal to $\rho\omega^2/\eta$. Note that 4 mils = 0.0102 cm, and 32 mils = 0.0813 cm.

$$k = 1.0 \cdot \frac{(45 \cdot 2\pi/60)^2}{1.0} = 22.2$$

$$\frac{h}{h_0} = \sqrt{3/(3 + 4 \cdot 22.2 \cdot 180 \cdot 0.0102^2)} = 0.801; \qquad h = 3.2 \text{ mils}$$

$$\frac{h}{h_0} = \sqrt{3/(3 + 4 \cdot 22.2 \cdot 180 \cdot 0.0813^2)} = 0.166; \qquad h = 5.3 \text{ mils}$$

From these calculated data it is seen that the 4-mil section of the puddle has a thickness of 3.2 mils at the end of 3 min (a 20% reduction in thickness), whereas the 32-mil thick section has a thickness of 5.3 mils at the end of this same period (an 83% reduction in thickness).

From the results of Problem 30-1, it is clearly evident that on spinning a paint of varying film thickness leveling to a uniform thickness will occur, since thicker sections are reduced in thickness at a much faster rate than are thinner sections. From Eq. 20 it can be shown that by doubling the rotational speed the time to attain a given leveling result is reduced to one fourth, or by tripling the rotational speed the time is reduced to one ninth. It can also be demonstrated from Eq. 22 that the time required to halve a section thickness is equal to $9\eta/4\rho\omega^2 h_0^2$.

DIP APPLICATION (5–7)

The successful application of a uniformly thick film to a test panel by dip coating is largely dependent on a proper withdrawal rate. Thus for any given paint system there is a critical region of withdrawal above which the paint film thickness tends to be nonuniform, and below which approximate uniformity of thickness is obtained for the entire length of the panel. Inspection of Fig. 30-5, which delineates regions of uniform and nonuniform thickness for four typical paint systems that have been dip-coated at varying viscosities and different withdrawal rates, shows that this critical region of withdrawal, in general, lies slightly outside a withdrawal rate of 4 in./min. In line with this general experience, an industrial dip coater has been designed that provides test panel withdrawal rates of 2, 3, and 4 in./min.

It is also evident from a study of the data graphed in Fig. 30-5 that both a higher viscosity and a higher withdrawal rate contribute to increased film thickness. Conversely, decreased viscosity and decreased withdrawal rate lead to reduced film thickness. Taken in conjunction with a specified nonvolatile content of the paint system, there are then available three straightforward ways of controlling the dry film thickness of a dip coating: paint nonvolatile content, paint viscosity, and panel withdrawal rate.

Dip coating is especially useful for the application of thin coatings an area of application that presents the straight-edge applicator with difficult if not in-

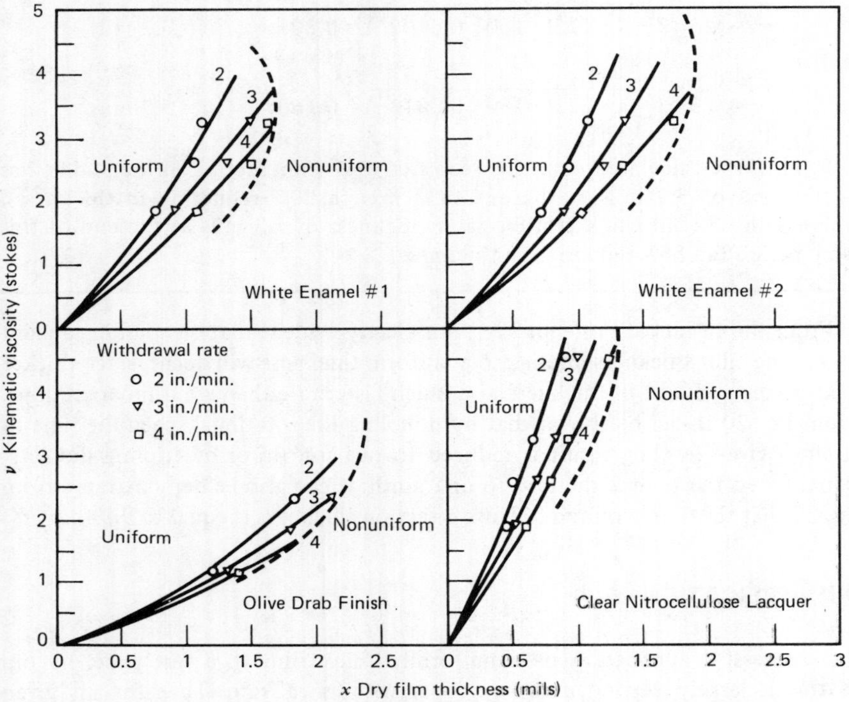

Fig. 30-5 Graphs showing the effect of viscosity and withdrawal rate on the dry film thickness of dip-coated panels for four typical paint systems.

superable mechanical difficulties. Thus by thinning the paint sufficiently (thereby simultaneously reducing the nonvolatile content and viscosity), it should be possible to routinely apply a dip coating at a thickness of 0.1 mil or less with good accuracy. Other advantages of the dip coating technique are excellent reproduction of results, no need for special skills in carrying out the dip coater operation, and adaptability to nonplanar surfaces.

The physical relationships involved in the draining of a wet film over a vertical panel are discussed in Chapter 29.

REFERENCES

1. Kaiser, E. B. and J. H. Coulitte, "Hardness, Abrasion Resistance, and Accelerated Weathering Tests on Pure Pigmented and Unpigmented Paint Vehicles," *Off. Dig.*, **23,** No. 322, 724 (1951).
2. Wetz, J. M., B. Golding, and L. C. Case, "Film Thickness Relationships of Organic Coatings," *Off. Dig.*, **31,** No. 410, 419 (1959).

3. New Jersey Zinc Company, "The Thickness of Paint Films Applied with a Straight Edge Spreader," *Leaves Paint Res. Notebook*, **1**, No. 5, October (1937).
4. Emslie, A. G., F. T. Bonner, and L. G. Peck, "Flow of Liquid on a Rotating Disk," *J. Appl. Phys.*, **29**, 858 (1958).
5. Payne, H. F., "Application of Uniform Films," *Off. Dig.*, **16**, No. 238, 400 (1944).
6. Payne, H. F., "The Dip Coater," *Ind. Eng. Chem., Anal. Ed.*, **15**, 48 (1943).
7. Gardner, H. A. and G. G. Sward, "Film Preparation and Thickness," Chapter 5 in *Physical and Chemical Examination of Paints, Varnishes, Lacquers and Colors*, 12th ed., Bethesda, Md., 1962.

31 *Floating, Flooding, Cratering, Foaming, and Spattering*

FLOATING AND FLOODING

A coating system containing more than one pigment occasionally suffers from a differential type of pigment separation that is visually recognized as floating or flooding (1-13).

Definitions

"Floating" is a term used to describe a mottled, splotchy, or streaked appearance exhibited by a paint film. "Flooding" is used to describe a more generalized alteration in the surface of a paint, wherein the surface appearance is different from the appearance of underlying layers. More specifically, "floating" denotes an uneven distribution of multiple pigments present in a paint film after application, usually in streaks or cellular delineations, whereas "flooding" denotes a uniform color change occurring in a wet paint film after application, caused by the segregation at the surface of a high ratio of one or more of the multiple pigments present. These two conditions are schematically illustrated in Fig. 31-1.

Marangoni Effect

A charming example of surface tension forces at work is afforded by the curious flow pattern that takes place in a clean wine glass half-filled with a fairly strong

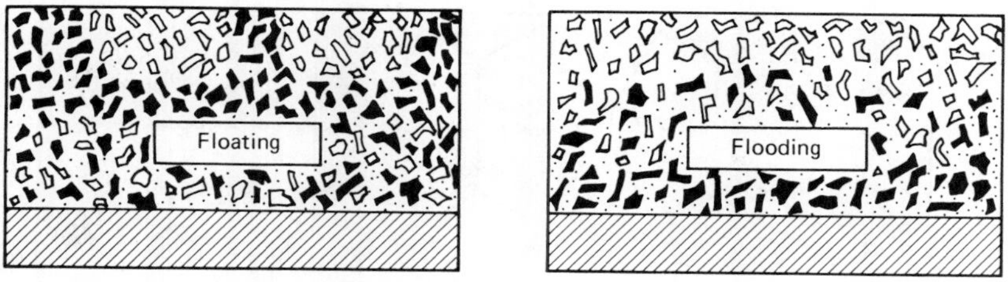

Fig. 31-1 Schematic illustration of floating and flooding conditions.

wine such as port. The liquid is observed to climb up the sides of the glass and then to gather into drops that run back into the bulk liquid. This pattern is repeated many times. This phenomenon (referred to as the Marangoni effect) was correctly explained nearly 100 years ago. The thin layer of wine crawling up the side of the glass is exposed to alcohol-free air to a greater degree than is the wine further down. Hence it loses alcohol more rapidly by evaporation. Moreover, no diffusion of alcohol in significant amount from below is available to restore the thin film to its original alcohol content. Hence the thin layer of wine at the top of the glass becomes enriched in water with a resulting increase in surface tension. This higher surface tension force pulls up more wine from below. As the process proceeds, so much wine accumulates at the top that drops (tears) are formed, which by their sheer weight, run back into the bulk liquid to repeat the cycle. A clear understanding of this effect makes much of the following discussion more comprehensible.

Physical Basis of Floating and Flooding (Benard Cells)

The process causing the differential separation is related to the vehicle circulation currents set up by solvent evaporation. Thus, when solvent volatilizes at the surface of a wet film, solvent further down in the body of the film commences to diffuse or migrate to the surface to replenish the surface supply. To expedite this movement, escape channels are often formed. Such escape routes terminate by erupting at the surface and frequently lead to the formation of geometrical patterns that are observable at the surface of a film. When observed under a microscope, especially in the case of fast-evaporating lacquers of low viscosity, the surface of a wet film is seen to be made up of miniature seething volcanoes that spew forth paint. This paint in turn spreads away from each volcanic apex to the outside of this erupting center. The circulatory or vortex motion and the type of cell structure that is built up as a result are schematically illustrated in Fig. 31-2, where both a top (plan) and a side view are given. Although the existence of such cellular currents was noted as far back as 1678, these so-called

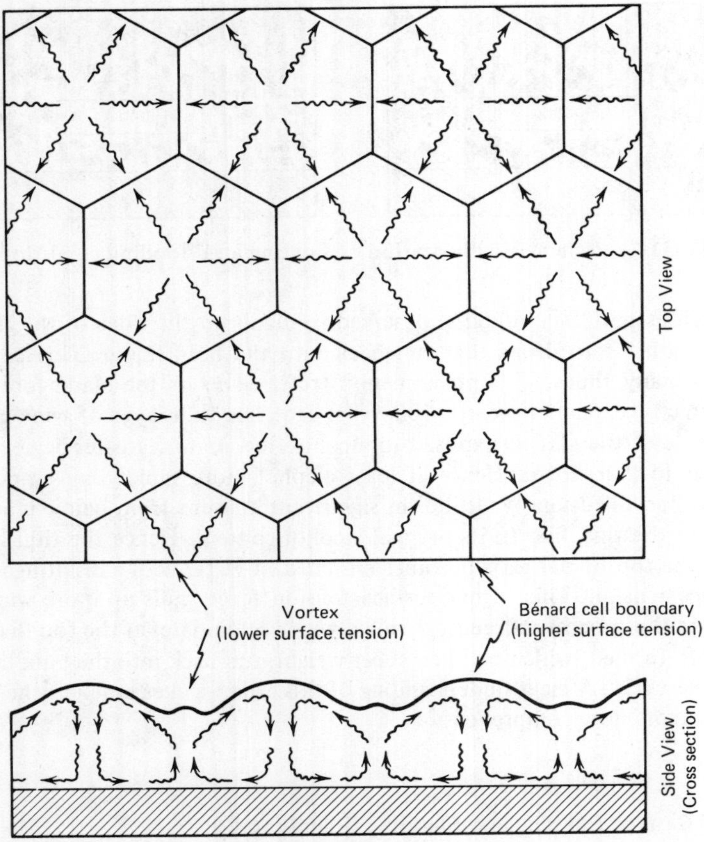

Fig. 31-2 Schematic illustration of vortex flow in a wet coating film. The Benard cells that form frequently give rise to symmetrical hexagonal configurations.

Benard cells were not scientifically studied and their action elucidated until the present century.

As solvent and associated binder stream to the surface of these cell centers, they carry with them suspended pigment. It is at this stage that a differentiation in the rate of particle transport and a corresponding differential pigment separation by type can occur. Thus finer pigments possessing greater surface will be conveyed more readily, whereas coarser particles with smaller relative surface area will be more resistant to movement. As a consequence a classification or redistribution of pigment occurs with the net result that one pigment type arrives at the surface at a higher concentration (relative to the other pigment) than when they started. This is a rather oversimplified explanation of the source and development of floating and flooding, for other factors, such as surface tension and flocculation forces, also contribute to differential pigment separation and, more often than not, are probably the controlling influences.

Figure 31-3 represents an attempt to classify the type of action that may be anticipated with pigments that differ in particle size, density, and degree of flocculation (other things being equal). For example, the blue flooding of chrome green enamels is largely due to the minute size of the iron blue pigment as opposed to the relatively larger size of the chrome yellow pigment. Also, fine-particle-size carbon blacks should normally be avoided for tinting (larger size bone, lamp, or furnace carbon blacks should be used); otherwise, floating and flooding due to particle size discrepancy are again apt to develop.

In a very real sense, flooding in paints is analogous to ore flotation, for both processes bring about a separation of disparate particle types, one being segregated in the body and one in the surface of the system in question. Since air entrap-

Relative Particle Size	Particle Flocculation	Pigment A ◊ / Pigment B ◆	Particle Densities: Equal	Particle Densities: Unequal
Equal	Both pigments deflocculated	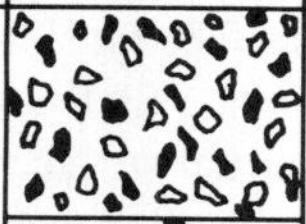	OK	Slight flooding
	One pigment flocculated		Floating	Floating
	Both pigments flocculated	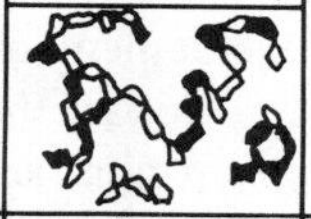	OK	OK
Unequal	Both pigments deflocculated		Flooding	Flooding
	One pigment flocculated	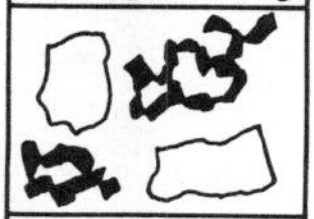	OK to floating	Floating
	Both pigments flocculated	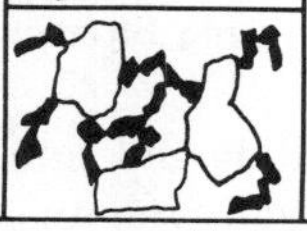	OK	OK

Fig. 31-3 Schematic illustration of the types of pigment conditions that can be anticipated for pigment mixtures that differ in particle size, density, and degree of flocculation.

ment (frothing) and surface tension effects are the major drives that enable beneficiation of the ore to be effected, it is not unreasonable to assume that the air adsorbed on paint pigment particles may also play a part in paint flooding. That the hydrophilic/hydrophobic balance is very delicate is demonstrated by the fact that with certain paint systems flooding will occur with high- but not with low-humidity conditions. In general floating and flooding become more pronounced the thicker the film (below a 2-mil wet film thickness little or no pigment differentiation is observable). In the case of a floating condition, the mottled pattern also becomes larger and more distinct with thicker wet films.

Electrical forces are believed to exert very little influence on floating or flooding. Rather, surface tension is believed to be the overriding factor that dominates the situation. An explanation offered by the present writer for vortex action due to surface tension is given in the following argument. In Fig. 31-2; the vehicle mixture that has just erupted from the body of the film at the vortex is rich in solvent. As this solvent evaporates, the vehicle mixture becomes richer in binder. The mixture is also cooled by solvent volatilization. Both of these actions normally work in the direction of increasing the surface tension. Hence the vehicle (with suspended pigment) that moves away from the center of the cell vortex is at the same time acquiring a higher surface tension. Arriving at the periphery of the cell of Fig. 31-2, the vehicle exhibits a higher surface tension than the vehicle at the center of the cell. This differential surface tension automatically stimulates and perpetuates the erupting action (pulls the center to the outside). This action continues until such time as the film surface tension forces are equalized and/or the viscosity of the mixture becomes so high that vortex motion is brought to a virtual standstill.

This explanation of the development of vortex activity by this mechanism makes it possible to account for the fact that floating and flooding usually occur in the same way whether the paint film faces up or down (although occasionally gravitational forces will dominate and reverse this observation). It also explains the finding that a marked lowering of surface tension by the use of a powerful surface tension reducing agent (say methyl silicone) invariably overcomes floating but not necessarily flooding, since the first is a surface and the second a body phenomenon.

Remedial Measures for Overcoming Floating and Flooding

Measures for overcoming floating and flooding can generally be worked out by systematically eliminating or nullifying the conditions contributing to differential pigment separation. Figure 31-3 immediately suggests that pigment fines can be source of trouble, as well as the flocculation of one pigment to the exclusion of the other (the latter effect being a matter of differential wetting). In the first instance, selection of a brand of pigment that is free of fines may clear up the difficulty. In the second instance, deflocculating the flocculated pigment with a suitable wetting agent such as lecithin (say $2\frac{1}{2}\%$ on the weight of the pigment for titanium dioxide) or the fatty acids of castor oil may alleviate the floating or flooding condition. If two deflocculated pigments are not closely matched in size, density, and shape, it may be necessary to resort to developing an overall pigment flocculated condition (say by addition of water with or without a surfactant such as soap) to prevent differential pigment separation.

A similar approach is to use additives that build up a thixotropic structure in the paint film, a sort of network that discourages pigment separation. Such additives might be high-oil-absorption pigments (5 to 25% of china clay or precipitated silica) or oxide pigments such as MgO and ZnO (about 2 and 5%, respectively, on the pigment content) that react with any residual acidity to form soaps. However, the best course of action is to employ colloidal-type thixotropes that set up an independent colloidal structure in the paint which is capable of immobilizing the vehicle without disturbing the original distribution of the pigment components. In this way pigment dispersion is maintained, and the adverse effects of pigment flocculation, such as reduced hiding, irregular and impaired color development, and degraded film integrity, are avoided. In the case of floating (primarily concerned with surface activity), it is possible, by the use of surfactants such as methyl silicone (0.01%), to reduce the surface tension forces to the point where cell circulation currents cease to exist and the development of flotation is thereby arrested. Only sufficient surfactant to just overcome the floating should be used. Any excess tends to nullify the correcting action or introduces unwanted side effects.

Differential pigment separation can also be minimized by increasing the vehicle viscosity, by replacing the faster evaporating solvents in the mixture with slower evaporating types, by insisting on a good initial grind, and by not overlooking the fact that pigment manufacturers commonly supply nonfloating (surface-treated) types of pigment that can frequently be employed to sidestep flotation and flooding problems.

Orange Peel

The development of an overall appearance of slight or faint indentations over the surface of an applied lacquer film is referred to as orange peel (because the surface appearance is similar to that of an orange skin). Orange peel is normally associated with lacquer spray application. The phenomenon arises from the formation of Benard cells generated by the circulatory currents set up by the evaporating solvent. The development of the orange peel appearance may be due to improper spray application (spray gun held either too close to substrate so that the air blast generates ripples in the surface or so distant that "dry" spray hitting the substrate cannot level properly); to incorrect (spatter) atomization; to an incompletely dispersed binder; to excessive lacquer viscosity; to an excessively drafty spray booth; or to questionable solventation. Frequently the addition of a small amount of slow-evaporating solvent (say 2% of an ether alcohol such as butoxyethanol or 1% of an alcohol such as *n*-butanol) will induce a secondary flowout that dissipates the orange peel pattern. The addition of a minute amount of a silicone agent to cut the lacquer surface tension drastically and thus radically dampen the vortex circulation currents may also prove effective in elminating orange peel.

Hammer Finish (14, 15)

In certain instances, a controlled pattern of vortex cells is deliberately developed in a paint system to create some special effect, such as a hammer finish. To produce such a consistent pattern of Benard cells, the film must be relatively thin (on the order of 1 mil when dry) and the solvent must be medium-to-fast evaporating to ensure both rapid generation of the hammer pattern and subsequent quick immobilization of the film to "freeze" the hammered effect. This film immobilization prevents erasure of the hammer appearance by secondary flow. Dominant influences that control the hammer effect are a proper spray application technique (relatively poor atomization to give a moderately wet spatter coating), a fairly high vehicle viscosity (30 to 60 sec through a Ford No. 4 cup), and a suitable type of binder (such as nitrocellulose or styrenated alkyd). Pigmentation with extrafine, nonleafing aluminum lining paste (0.3 lb/gal) enhances the hammer effect. Toluene, a commonly used solvent, represents a compromise in solvent selection (neither too slow to allow the pattern to disappear nor so fast as to render the initial hammer pattern indistinct).

CRATERING

A freshly applied wet coating represents a critical stage during which the film is susceptible to a variety of surface defects. One such serious defect is cratering, which, as the term implies, is characterized by scattered depressions in the film surface.

Surface tension forces (discussed in Chapter 9) are responsible for the development of craters. Since cratering is a liquid/liquid spreading phenomenon, all surfaces involved are considered smooth ($i = 1.00$), and contact between any two liquid surfaces is considered complete ($a = i$). The expression for one liquid spreading over a second liquid as a duplex film is given by Eq. 1 (see Eq. 23 of

Chapter 9). A duplex film is defined as a film of such thickness (several molecules thick) that the top and bottom faces of the film act independently of each other.

$$W_S \text{ (spreading coefficient)} = (\gamma_W - \gamma_I) - \gamma_L \qquad (1)$$

The convention is observed that γ_W is the larger surface tension (liquid W), γ_L is the lower surface tension (liquid L), and γ_I is their interfacial tension. A positive spreading coefficient results in spontaneous spreading.

Spontaneous spreading can also take the form of a spreading monmolecular layer rather than a duplex film.

Cratering results from the rapid spreading of a crateror (an isolated, discontinuous liquid phase such as a localized contaminant or incompletely solvated globular speck of material) over the surface of a continuous phase such as an applied coating (the crateree), as schematically shown in Fig. 31-4. In general, if

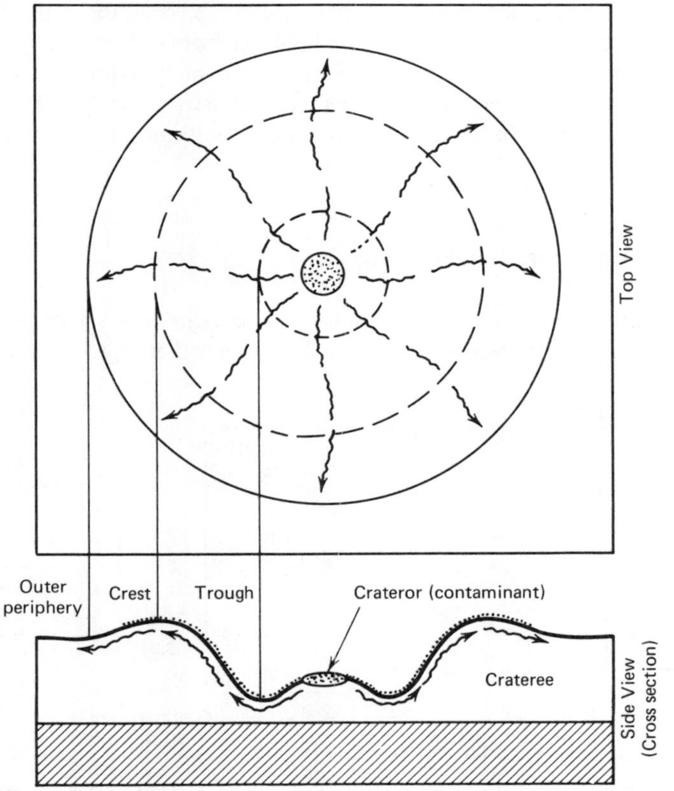

Fig. 31-4 Schematic diagrams showing top and side views of a coating crater. The crateror is characterized by a lower surface tension than that of the crateree.

the crateror (discontinuous phase) has the higher surface tension, spreading and cratering are highly unlikely. Conversely, if the crateror has the lower surface tension, spreading over the continuous phase (the crateree) with attendent cratering is very probable. The crateror source may be an airborne particle, an internal contaminant, a condensed vapor, a mist (silicone), or even a localized and temporary dislocation of surface tension forces at some spot on the surface.

Viscous Drag of Spreading Top Layer on Underlying Liquid Layers

Regardless of the source of the radially directed outward spreading action from the central crateror location, the expanding circular surface drags outward with it a substantial portion of the underlying liquid because of viscous resistance. If the film is sufficiently thick, liquid from beneath moves into the depression that is formed and this replenishment serves to obliterate (heal) the crater formation. However, if the film is thin, there is little liquid from below to fill into the depression, and, lacking liquid replenishment, a permanent crater results.

The formation of a crater is schematically illustrated in Fig. 31-5, together with representative values for typical crater dimensions and the time scale for their formation (16).

Equlibrium and Nonequilibrium Conditions

The application of even a small volume of a coating to a substrate creates an enormous expanse of fresh surface area. This means that ingredients such as sur-

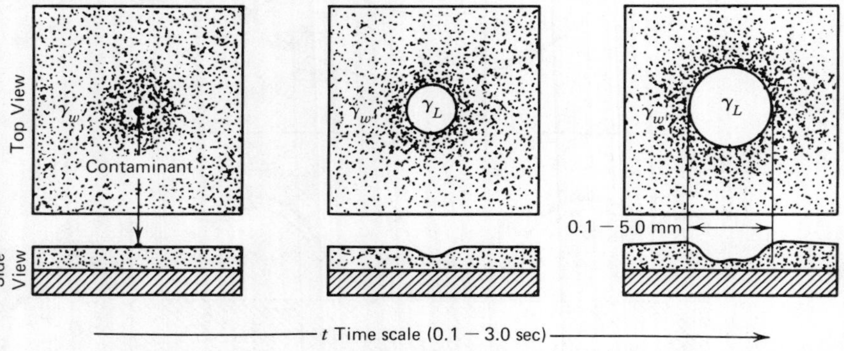

Fig. 31-5 Schematic illustration of three stages in the formation of a coating crater. The process is dramatic in that such a minute amount of crateror can cause such a large volume of liquid to be displaced in so short a time.

factants, which during storage must perforce, for the most part, remain submerged in the bulk coating (existing possibly as micelles), are now afforded the opportunity to migrate to the newly created surface and take up a more or less oriented position. Also, because of the violent physical disturbance that occurs during film application, unevenness in the concentration of these ingredients can occur, and this in turn can cause irregularities in surface tension at the surface of the applied film. As pointed out by Bierwagon, establishment of an equilibrium surface tension is highly time dependent (16). Thus differences in surface composition from point to point along the fresh surface that arise from film application can frequently create temporary surface tension gradients with resultant localized flow patterns (a physically generated crateror situation). From this it is apparent that crater formation is closely related to the method used for applying a coating to a substrate (the rate and manner of forming the fresh surface area). Systems applied slowly to ensure equilibrium may show no sign of surface tension flow, but when applied rapidly as thin films may develop visible craters. Sometimes a coating will show defects when applied by spraying but not when applied by brushing (16).

It has been observed that falling sheets of film (such as in curtain coating) that develop perforations also tend to exhibit craters in the applied film (16).

Mathematical Cratering Models

The design of a mathematical model to describe crater formation should be able to account quantitatively for (*a*) the rapid spreading action (visible as the expansion of an outside leading edge of a circle from the crateror nucleus) and (*b*) the viscous drag of the spreading top layer on underlying layers, causing a thinning out at the center and a piling up of coating at or beyond the periphery of the circular leading edge. Such models have been developed, but they yield intractable solutions. Hence approximations are introduced, thereby stripping the models of much of their practical usefulness (16). However, the expressions developed do agree with the experimental observations that smaller globular craterors are faster spreading and final crater shapes are closely related to film thickness and viscosity (as was also shown in Chapter 28 for leveling).

Experimental Demonstration of Cratering

When the center of a very thin layer of uncontaminated water in a flat tray is touched by a drop of alcohol, immediate drawback (spreading) occurs, accompanied by violent turbulence. This activity subsequently ceases as equilibrium conditions are reached (18). By placing drops of various solvents (discontinuous crateror phase) in contact with linseed oil (continuous phase) Hahn observed and reported a variety of induced cratering behavior (17). From his experimental work he concluded that the arresting of crater growth may be due in substantial part to the retraction of the spreading liquid and less so to viscosity increase, since the latter is too slow a process to exclusively control crater arrest.

Internally Created and Externally Created Cratering Systems

Hahn classifies cratering varieties into two groups, self-contained systems (cratering generated internally) and externally dependent systems (cratering generated by an external source) (17). In the first case all the ingredients for crater formation are already present, say in a closed coating system where cratering becomes manifest on application of the coating to a substrate. In the second case contact of the coating with some specific crateror during and/or after application is required to initiate the cratering action.

Practical Systems

Internally Generated Craters. Internally generated craters may appear either immediately on application or sometime thereafter, say during a baking cycle. Delayed cratering may result from selective solvent loss, which induces spontaneous spreading through a reduction in surface tension at localized spots.

Spraying films are more susceptible to cratering than are films applied by a straight-edge applicator, partly for the reason that spraying tends to deposit self-contained craterors indiscriminately (both inside and outside the applied film), whereas a straight-edge applicator tends to bury the craterors in the continuous coating phase (crateree), whence the craterors must work their way to the surface to exert their effect (17).

A polymer that exhibits no cratering when used alone may sometimes generate a cratering effect when combined as a blend with a different polymer of the same species (as occasionally exemplified by alcohol-soluble phenolic resin varnishes) (17). On the other hand, the addition of a small amount of a soluble polymer sometimes suppresses cratering (polyvinylbutyral modification of phenolic baking varnishes) (17).

Long-oil alkyds present few cratering problems, presumably because of the compatibilizing effect of their fatty acid terminal groups. This is in contrast to oil-free polyesters, which are prone to cratering defects.

Peculiarly, some troublesome systems lose their cratering tendency on aging. This is attributed to the gradual solubilization of the crateror with time within the system (e.g., octyl alcohol defoamer in latex paint) (17).

Increasing the viscosity of the continuous phase by using a thickening agent can sometimes correct or alleviate a cratering tendency.

Externally Generated Craters. Condensation of a vapor can produce particularly serious cratering problems. Airborne dust particles often serve as carriers for condensed vapors (e.g., transportation of a silicone mold-release agent from a distant location). The drifting particles may be too small for sighting, but the craters produced are highly visible.

Overspraying periodically generates cratering problems among adjacent booths. Thus a coating free from internal cratering can induce cratering on a nearby wet film of different composition. Even overspray of a polymeric system on a similar system (except that the two differ in solvent tolerance) can produce cratering (17).

When the crateror is present as a speck on the surface of a first coat, the application of a second coat may give rise to a deep void, referred to as pinholing (the residual vestiges of an oil crayon marking can produce this type of defect).

Control of Cratering

Elimination of cratering requires recognition of the forces responsible for the unbalanced surface tensions that produce this defect. If the cratering is traceable to an external crateror source, scrupulous care to avoid the surface contamination is all that is required. If this approach is not feasible or if the cratering is self-generated, then, aside from reformulation, it is necessary to nullify the spreading drive by reducing the crateree surface tension below that of the crateror. One method of accomplishing this is to introduce a surface agent that possesses, in addition to low surface tension, low vapor pressure (to avoid indiscriminate condensation) and relatively good compatibility for the system in question. As discussed in Chapter 9, surface tension and solubility parameter are closely allied.

A rise in the viscosity of a coating system has also been found effective in certain situations (it slows down the migration of a potential crateror to the surface of the coating), although a viscosity increase also acts adversely by retarding the coating flow that is necessary to fill in and eliminate a crater already in existence.

Other Unbalanced-Surface-Tension Effects

The phenomena so far discussed have a common origin (an unbalanced-surface-tension condition) that results in liquid/liquid-type spreading. Many other surface effects result from such unbalanced-surface-tension forces.

The beading observed at the edge of an applied film is initiated by the faster drying at the edge location. This gives rise to a differentially higher surface tension due to a differentially higher rate of cooling and a differentially higher solids content, which in turn induces liquid flow to the outside, resulting in a heavier edge deposit. With dark, colors this effect is called "picture framing."

Even leveling may be affected adversely. Thus solvent loss is faster from a ridge (convex surface) than from a trough (concave surface). Such a difference in evaporation rate means that the ridge, by possessing the higher surface tension (due to both an increased cooling rate and a higher vehicle solids), acts to build itself up by pulling up vehicle from the trough, which becomes correspondingly deeper. This action opposes the overall leveling drive. In this regard it is notable that agents that reduce surface tension have been successfully used to promote leveling (17), although surface tension reduction should theoretically lower the leveling action.

Hahn has proposed that wrinkling and in certain cases gloss reduction may be closely allied to differential surface tension effects during the drying or curing stage (17).

Foam instability, discussed in the next section, is undoubtedly initiated many times by the rupture of the foam wall at a crateror site.

FOAMING (19-21)

A foam consists of a mass of gas or vapor bubbles separated by thin walls of liquid film. The confining liquid film is more or less elastic, and the bubble walls themselves exhibit a unique geometry: any three walls coming together at an edge always meet at an angle of 120° (see Fig. 31-6). At least two components are necessary for foam generation (pure liquids do not foam). Since the formation of a foam involves the development of a considerably expanded surface area and since this generation of area is carried out against a liquid surface tension force, it follows that the lower the liquid surface tension the less energy is required to generate a given amount of stable foam.

Theory of Foam Formation and Collapse

Foams are inherently unstable. However, some foams are remarkably persistent. The collapse of a foam generally occurs as a result of liquid drainage from the bubble walls to the wall edges. When the film drains to a thickness of about 100 Å, molecular motion within the liquid film is normally sufficient to rupture it and destroy the bubble structure.

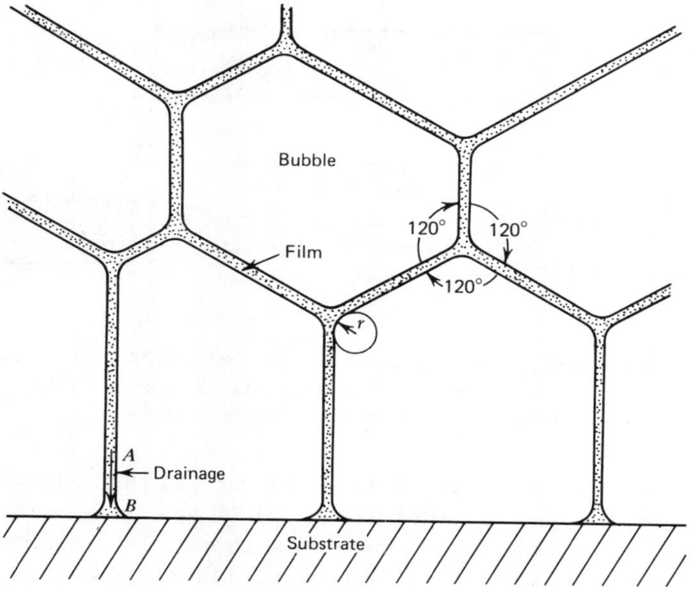

Fig. 31-6 Schematic illustration of foam structure and the mechanism of drainage to border regions.

Edge Drainage. The mechanism responsible for the drainage and thinning of liquid films is related to the peculiar draining action that occurs at the edges or borders of the film. These border regions are bounded by surface areas of sharp curvature (concave outward), as opposed to the relatively flat walls that confine the film (schematically illustrated in Fig. 31-6). Hence the liquid pressure within any border region at *B* is less than the film pressure at *A*, which is in turn equivalent to the outside atmospheric pressure. As a result of this differential pressure, the film adjacent to a border is sucked (actually pushed) from *A* to *B* into the border region and thinning outside the border edge occurs. To take the place of the film that has been drawn into the border region, film further removed from the border moves in, and in this way a thinning or pseudodrainage takes place. It is not a drainage in the sense that liquid flows within the film walls to the border region, for unless the bubble walls are moderately thick, such flow has been shown by calculation to be relatively insignificant.

Film Elasticity. Another factor as important as border drainage in determining foam stability is film elasticity. If a film consisting of liquid and surfactant is suddenly stretched at some one location, the concentration of the surfactant in

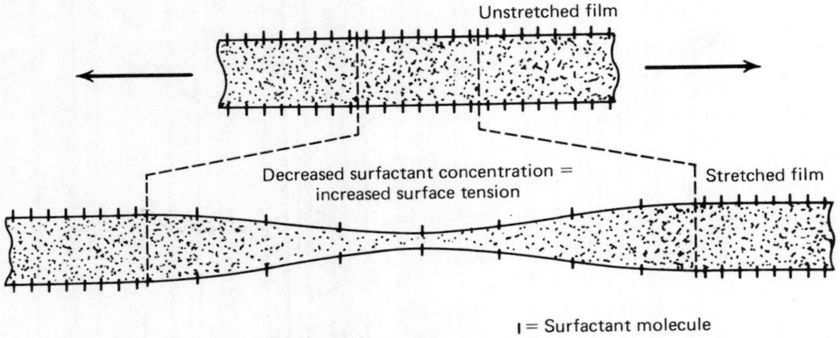

Fig. 31-7 Schematic diagrams illustrating the decrease in surface surfactant concentration that occurs when a thin film of foam is stretched. This decreased surfactant concentration gives rise to increased surface tension.

the surface is temporarily reduced and as a result the liquid surface tension at this point is correspondingly increased (see Fig. 31-7). As a result, the stretching action is then communicated to other surface areas in the immediate vicinity where the surface tension is less.This distribution of the imposed stress prevents the buildup of a stretching force sufficient to rupture the film readily at any one point. This stress distribution then provides the bubble film with a mechanism for resisting sudden breakage. The surfactant at the surface has endowed the film with the property of elasticity. Pure liquids fail to foam for the reason that they lack surface elasticity. They contain no second component that can concentrate in the surface and set up equalizing surface tension forces capable of resisting sudden stretching or contracting actions.

Other factors that are antagonistic to foam stability are gravity and surface evaporation, both of which contribute to film thinning and foam collapse.

Surface Transport. A third important concept that relates to foam stability is surface transport. This theory postulates that, as the surface of film moves away from an area of low to one of high surface tension, it drags along with it (transports) a number of underlying layers of liquid to the new location. This transport is somewhat remeniscent of talcum powder spreading over a liquid as it accompanies the surface movement from an area of low to one of high surface tension, as described in Chapter 13. Hence in the region of a stretched film (high surface tension) there is an immediate flow toward it that tends so fill it with fluid transported from regions of lower surface tension. This effects a restoration of the stretched region to its original thickness. In this regard a viscous surface may be desirable, since it entrains more liquid in rushing over to fill in the thinner section. If it drags along too much liquid, however, it may thin out its own original location to the breaking point and bubble collapse may occur for another reason.

In an effort to replenish the concentration of surfactant at the surface and thus equalize the surface tension forces, surfactant can also diffuse upward from underlying layers, rather than being brought in laterally from neighboring surface areas. Hence there are two competing processes. If the rate of surface adsorption from bulk liquid below is of the same order of magnitude or at a higher rate than the sidewise surface movement, the transport of liquid into the stretched region becomes correspondingly reduced and thinning and bubble collapse are more likely to occur.

These several physical concepts relating to foam stabilization, taken in conjunction with the more obvious factors that affect foam stability, such as mechanical and thermal shock, adsorption and diffusion of gases, and depletion of surfactant by a substrate (or by soil), provide a basis for a better understanding of the extremely complex interplay of the forces affecting and controlling foam stability and foam collapse.

Foam Stabilizers

The addition of a third component (a foam stabilizer) to the pure liquid and surfactant is sometimes effective in enhancing foam stability. Thus foam stability can be increased by selecting a third component that acts to increase the surface viscosity. It has also been demonstrated that additives that lower the critical micelle concentration tend to induce a significant degree of foam-stabilizing action.

Foam Inhibition

A reversal of the mechanisms advanced for providing foam stability is generally effective for inhibiting or preventing foam generation or for provoking foam collapse. Defoamers frequently function by chemically reacting with the foamers to neutralize or destroy their effectiveness (acids and calcium salts break up soap films). However, most defoamers or antifoamers of major industrial importance function by spreading over the foam surface, and in so doing displace and sweep before them the surfactant responsible for the foam stabilization. To do this, the surface tension that comes into play over the swept surface must be less than the surface tension of the original foamed film. If the foaming agent desorbs easily and retreats rapidly, resistance to the defoamer is minimal. However, if the foamer desorbs reluctantly and the surface viscosity is high, spreading proceeds at a retarded rate. A good defoamer should act with dispatch. It is usually a material that exhibits very low solubility in the bulk liquid and produces a marked reduction in the liquid surface tension. The more effective antifoam agents have chemical structures that enable them to lie flat when oriented at the foam interface. In aqueous systems this arrangement is promoted by chain branching, by the scattering of bulky polar groups at the center or along the length of the defoamer molecule, and by the introduction of just sufficient methylene groups to achieve water insolubility.

To facilitate the initial distribution of a defoamer, many industrial antifoamers make use of spreading solvents or emulsifying agents that act as carriers. This ensures that the defoamer is well disseminated to all parts of the surface for effective defoaming action.

Foam Evaluation

The most straightforward method for evaluating foam generation and foam stability (or instability) is simply to shake a measured amount of the surfactant solution in a graduated container under standardized conditions and note the amount and texture of the foam developed, together with its persistence on aging. A second method resorts to mechanical mixing or beating under specified routine conditions for the foam generating. Here again foam height, texture, and loss of height with time are observed.

A third widely used procedure for evaluating foamability uses the Ross-Miles apparatus. In it 200 cm^3 of surfactant solution (120 F) is poured from a height of 90 cm through a 0.29-cm diameter orifice into a 50-cm^3 puddle of the solution contained in a cylindrical container 5 cm in diameter (ASTM D1173-53). The foam height developed with this pouring technique is measured immediately after the pouring is completed and again at the end of 5 min.

Defoaming effectiveness is normally rated in terms of foaming control with no defoamer present.

Classification of Defoamers (Antifoaming Agents)

Defoamers can be conveniently classified into seven chemical types. This arbitrary breakdown has no special physical significance.

Alcohols (2-Ethylhexanol; Polyalkylene Glycols). Branched-chain alcohols and polyols fulfill the requirements of minimal water solubility and the imparting of a low surface tension, which are the major characteristics demanded of a good defoamer.

Fatty Acids and Derived Esters (Sorbitan Trioleate; Diglycol Stearate). A number of water-insoluble fatty acids and fatty acid esters have been successfully employed as defoamers. Many of them are suitable for use in food processing.

Amides (Distearoylethylenediamine). This class of defoamer is usually characterized by a moderately high molecular weight.

Phosphate Esters (Trioctyl Phosphate; Tributyl Phosphate; Sodium Octyl Phosphate). Many phosphate esters are effective at very low concentrations.

Metallic Soaps (Calcium Stearate; Magnesium Palmitate). Water-insoluble metallic soaps often provide antifoaming action when predispersed in a suitable vehicle. They have occasionally been found effective for defoaming nonaqueous foams.

Chemicals with Multiple Polar Groups (Di-terti-amylphenoxyethanol). Many effective foam inhibitors contain two or more polar groups. Castor oil is a natural product that fulfills the chemical spatial requirements for such a defoamer (distributed hydroxyl and ester groups) and has been used to control the foaming of boiler water and the foaming encountered in sugar processing operations.

Silicone Oils (Polysiloxanes). This type of defoamer constitutes a class of materials which has proved outstanding in antifoaming effectiveness for both aqueous and nonaqueous foams.

Because of the many complicating factors contributing to foam stability, it is difficult to establish simple defoamer generalizations. Thus a defoamer effective for a latex paint may be ineffective for controlling the foaming of the polymer latex from which the paint is prepared. Also an optimum defoamer may not necessarily be one that is effective at the lowest concentration. Certainly a defoamer that encourages the formation of occasional "fish-eyes" in a protective coating has questionable qualifications. Moreover, a larger quantity of a less effective, low-priced defoamer may frequently be more economical to use than a smaller quantity of a more effective, high-priced defoamer.

SPATTERING

Roller Coating Spatter

The spattering that results from roller coating application is difficult to assess and even more difficult to treat from a theoretical standpoint.

When a roller coater covering (charged with liquid coating) is rolled across a flat substrate, the innumerable fibers of the roller covering fabric compress and then expand to meter out the coating from the fiber interstices. As the trailing roller surface retreats, the fabric fibers are jerked away from the substrate and the liquid coating is forced to split almost instantaneously between the fiberous covering and the coated substrate. The result is a highly confused mechanical breakup.

Such splitting between roller covering and substrate has been photographed in great detail (22).

The last coating material to split gathers at the ends of fibers or fiber clumps that are scattered randomly over the roller covering surface. As the roller moves along, this accumulated material is quickly stretched into filaments, webs, or "legs," which almost at once are snapped apart. It is at this point that some of the stretched coating breaks away completely as droplets (spatters), rather than pulling back and returning to either the roller covering or the substrate. The retraction of the lower major portion of the broken filament or webbing to the substrate gives rise to craters or tracks that flow out (level) to a greater or smaller extent, depending on the rheological properties of the coating composition (see Chapter 28 on leveling).

The amount of spattering that is generated is a function of the nature of the roller coating covering (fabric), the nature of the substrate, the rheological properties of the applied coating, and the speed of application.

Roller Coater Covering. As discussed by Burns, the improvements and refinements in the design of roller coating fabrics represent a new technology that has grown apace with the development of latex paints (23). For example, it is now known that lambs-wool rollers are not suited to water-thinned paints, but at an earlier time it was not so apparent that water severely reduces the resilience of lambs-wool, leading to poor application. Today the cover fabric can be competently designed to provide required effects such as low bubbling or foaming (relatively low-resilience, fine-denier fibers), low lint discharge (woven fabric, secure tuft bonding), good wetting (conformability to the substrate, such as short-pile fabrics for smooth surfaces and longer pile fabrics for rougher surfaces), good leveling (nondirectional, high-density fabrics), low discharge rate (high crimp, special finish) and low spatter (a fine, irregular, and relatively dense fabric of low resilience) (23, 24).

Roller Coating Substrate. Since the nature of the substrate normally is not subject to change, resort is generally made to the other variables that can be suitably adjusted to contribute minimum spatter.

Coating Formation. The rheological behavior of the applied coating is doubtless the major factor controlling spattering. In turn the rheological behavior is dictated by the coating composition. In this regard it is intuitively deduced that moderate to high shear rates must be involved during the stretching, snapping apart, and retraction of the filaments and webs that are formed in the wake of a roller application. Laboratory work indicates that low-shear-rate rheological information is irrelevant (22).

Compositional ingredients that impart viscoelasticity to a stretched coating filament or webbing at moderate to high shear rates apparently encourage spattering. For example, although low-molecular-weight latex thickeners fail to promote spatter, in progressing to higher molecular weights spatter is observed to increase with molecular weight and this factor eventually becomes the major determinant of the spatter developed (25). Also, bulky thickeners that impart rigid colloidal structures, such as hydroxyethylcellulose (HEC), contribute far less to spatter than stringy thickeners that impart elastic colloidal structures, such as polyethylene oxide (PEO). The presence of defoamers mildly promotes spatter. On the other hand, ethylene glycol is without effect (25).

Increasing the pigment content beyond the point of incipient dilatancy results in increased spatter (25).

It is notable that the two factors of small latex particle size and high thickener hydrophilicity, which favor flowout (a low-shear-rate phenomenon), are without demonstrable effect on spattering (a moderate- to high-shear-rate phenomenon) (25).

Rate of Application. Spatter increases with increased rate of application.

Assessment of Roller Coating Spattering. An ASTM task force (ASTM Committee D.01.42–Task Group 17) is presently working on a test procedure for measuring spattering. Much of this activity centers on the design of a suitable test roller, such as a cylinder of nonporous composition fitted with notched flanges that create controlled discontinuities for breaking the filaments of the stretched coating film (26). The degree of spattering is evaluated in terms of the droplets that spatter and fall on a sheet of paper located below the vertical surface to which the coating is being applied by the roller coater. Initial tests indicate that increased spatter is given by (*a*) an increase in the number of notches, (*b*) an increase in the rolling rate, and (*c*) an increase in the coating thickness (26).

A second method for gathering spattered droplets calls for catching the droplets on an approximately 5 × 6 in. cardboard sheet mounted on the handle of a 3-in roller coater (27).

INK MISTING ("INK FLY")

Inks occasionally create very fine mists of tiny ink droplets during printing referred to as "ink fly." Certain black inks used on fast-running presses often generate this type of misting. Ink fly is a form of spatter resulting from the stretching and breakup of filaments of ink formed between parting ink rollers. The droplets are usually from 1 to 100 μ in diameter, and although not visible in the air they settle out as a fine ink haze and cover exposed surfaces in the vicinity with a thin film of ink (28).

Long inks are more susceptible to misting than short inks. Also, misting increases very rapidly above a certain threshhold roller speed. Misting can be dealt with either by reformulating the ink to suppress filament formation (such as the introduction of additives that weaken the filament strength) or by suitably handling the mist after its formation (using blowers that force the mist back onto the rollers, vents that exhaust the mist to collection areas, electrostatic devices that repel the mist, as by a taut charged wire at the exit nip) (28).

Ink fly can be measured either by judiciously placing a sheet of graph paper in a location close to the rollers and counting the ink particles that land in selected squares per given time (microscopic observation), or by drawing air from the immediate vicinity of the printing rollers through filter paper and evaluating the darkening that occurs (28). Peculiarly, ink fly can be most dense at the entering nip rather than the exit nip, since the mist formed at the exit is frequently carried around as a cloud with the rollers to collide and accumulate on the other side.

REFERENCES

1. Van Loo M., and F. E. Bartell, "Colloid Chemistry of Color Varnish–Behavior of Surface Films: I," *Ind. Eng. Chem.*, **17,** 925 (1925).
2. Bartell, F. E., and M. Van Loo, "Colloid Chemistry of Color Varnish–Pitting, Seeding, Silting, and Surface Dulling: II," *Ind. Eng. Chem.*, **17,** 1050 (1925).
3. Fisher, E. K., *Colloidal Dispersions*, Wiley, New York, 1950, pp. 138–144.
4. Ferguson, I., "Flotation in Paints–A Suggested Mechanism," *JOCCA*, **42,** 529 (1959).
5. Bell, S. H., "Pigment Flocculation in Paints: A Dynamic Phenomena," *JOCCA*, **35,** 373 (1952).
6. LeBras, L., et al., "Flooding and Floating of Titanium Dioxide-Iron Blue Enamels," *Off. Dig.*, **27,** No. 368, 607 (1955).
7. Lock, A. B., "Some Observations on the Flotation and Flooding in Paints," *JOCCA*, **43,** 859 (1960).
8. "Causes of Floating Pigments in Paints and Enamels," *Natl. Paint Varn: Lacquer Assoc. Sci. Sect. Circ.* No. 370, 1930.
9. Kramer, A. E., "Pigment Colors–A Review," *Off. Dig.*, **18,** No. 252, 2 (1946).
10. Loque, L. A., "Flooding and Flotation in Paints," *Paint Manuf.*, **31,** 5, January (1961).
11. Dewey, P. H., "Rheology Factors in the Formulating and Use of Protective Coatings," *Off. Dig.*, **18,** No. 259, 336 (1946).
12. Van Loo, M., "Physical Chemistry of Paint Coating," *Off. Dig.*, **28,** No. 383, 1126 (1956).
13. Shur, E. G., "The Flooding and Floating of Multi-Pigment Paint Systems," *Off. Dig.*, **23,** No. 311, 867 (1951).
14. Lamm, V. P., "Factors Affecting Patterns in Hammer Finishes," *Off. Dig.*, **33,** No. 432, 1408 (1961).
15. Bogin, C. D., "Formulation of the Volatiles in Nitrocellulose Lacquers," in *Paint and Varnish Technology*, (W. Von Fischer, Ed.), Reinhold, New York, 1948, p. 255.
16. Bierwagen, G. P., "Surface Dynamics of Defect Formation in Paint Films," *Prog Org. Coatings*, **3,** 101–113 (1975).
17. Hahn, Frank J., "Cratering and Related Phenomena," *JPT*, **43,** No. 562, 58–67, November (1971).

18. Boys, C. V., *Soap Bubbles; Their Colours and the Forces Which Mould Them*, rev. ed., Dover, New York, 1959.
19. Adamson, A. W., *The Physical Chemistry of Surfaces*, Interscience, New York, 1960.
20. Berkman, Sophia, and G. Egloff, *Emulsions and Foams*, Reinhold, New York, 1941.
21. Bikerman, J. J., *Foams: Theory and Industrial Applications*, Reinhold, New York, 1953.
22. Glass, J. Edward, "Dynamics of Roll Spatter and Tracking; Part I Commercial Latex Trade Paints," *JCT*, **50,** No. 640, 53–60, May (1978).
23. Burns, F. B., "Latex Semi-Gloss–Challenge for the Roller Industry," company brochure, EZPainter Corporation, 4051 South Iowa Avenue, Milwaukee, Wis. 53207.
24. Burns, F. B., "Your Complaints–Are They Due to the Paint or the Roller?" paper presented at January 1965 meeting of New York Society for Paint Technology.
25. Glass, J. Edward, "Dynamics of Roll Spatter and Tracking: Part II. Formulation Effects in Experimental Paints," *JCT*, **50,** No. 640, 61–68, May (1978).
26. Burns, F. B., Chairman "Report on Revised Spatter Test Method," ASTM Committee D.01.42–Task Group 17, June 24 1977.
27. Sward, G. G., "Architectural Paint," in *Paint Testing Manual* (G. G., Sward, Ed.), American Society for Testing and Materials, 1916 Race Street, Philadelphia, Pa. 19103, 1972.
28. Askew, F. A., (Ed.), *Printing Ink Manual*, Society of British Printing Ink Manufacturers, W. Heffer & Sons, Ltd., Cambridge, England, 1969.

Appendix

List of Major Symbols and Abbreviations

CAPITAL LETTERS

A	area; constant
A	acid number
AP	aniline point
B	breadth; width; constant
Btu	British thermal unit
C	concentration; crushing force; chord length
C	Centigrade
CED	cohesive energy density
CPVC	critical pigment volume concentration (content)
CR	contrast ratio
D	diameter
D	shear rate
D	diffusion constant
E	energy; evaporation rate or index; fraction of extender by volume in pigment mixture
F	force
F	Fahrenheit
FN	flocculation number
FP	flow point
G	constant; molar attraction constant

H	heat of vaporization; heat input
H	Hegman fineness-of-grind reading
HLB	Hydrophilic lipophilic balance
HP	hiding power
ID	inside diameter
K	constant; optical absorption coefficient
KB	kauri-butanol number
KE	kinetic energy
KU	Krebs units
L	length
LCPVC	latex critical pigment volume concentration
L.P.I.	latex porosity index
M	torque; molecular weight; ionic concentration
M_n	number-average molecular weight
M_w	weight-average molecular weight
MEK	methyl ethyl ketone
MS	mineral spirits
MW	molecular weight
N	number of molecules; Avogadro's number
NV	nonvolatile content
$\overline{\text{OA}}$	oil absorption by spatula rub-out method
$\overline{\text{OA}}_c$	oil absorption by Coleman method
OD	outside diameter
OMS	odorless mineral spirits
P	power
P	Production Club fineness of grind reading
$[P]$	parachor
P.I.	porosity index
PVC	pigment volume concentration (content)
Q	rate of volume flow
R	radius; gas constant
R_∞	optical reflectance
Re	Reynolds number
RH	relative humidity
S	saponification number
S	specific heat; optical scattering coefficient; size; gap clearance; specific surface area; volume of sand mill
SUS	Saybolt universal seconds
T	temperature; wet film thickness; tangential force; total
TS	tinting strength
U	ultimate pigment volume fraction
UPVC	ultimate pigment volume concentration

V	volume; pigment volume fraction
V_A	attraction potential
V_R	repulsion potential
VS	vehicle solids
W	work; weight
WP	wet point
X	thickness; weight percent
Z	amplitude

SMALL LETTERS

a	inner radius; apron roll transfer fraction; constant
b	outer radius; constant
b.p.	boiling point
c	concentration; center roll transfer fraction; center-to-center distance
cP	centipoise
d	differential (calculus symbol); diameter
e	binder efficiency; binder index; escaping coefficient
f	factor
g	gravitational acceleration (980 cm/sec^2)
h	height; head of liquid
i	rugosity (irregularity factor)
k	constant
log	common logarithm (to the base 10)
ln	natural logarithm (to the base 2.71828)
m	mass
n	exponent; ratio of peripheral roll velocities; number of particles; index of refraction
p	pressure; vapor pressure
p_T	total pressure
psi	pounds per square inch
r	radius
rpm	revolutions per minute
s	standoff distance
t	time
u	average velocity
v	velocity; fractional volume
w	fractional weight of pigment
x	thickness; gap clearance; fractional part
y	exponent
z	ion valence

GREEK LETTERS

Σ	sum of
Δ	increment (change in)
α	angle
γ	surface tension
δ	solubility parameter; double layer thickness
ζ	zeta potential
η	viscosity
η_0	solvent viscosity; viscosity of unpigmented vehicle
η_∞	viscosity at infinite shear rate
η_r	viscosity ratio
η_{sp}	specific viscosity
η'	plastic viscosity
$[\eta]$	intrinsic viscosity (limiting viscosity number)
θ	angle; contact angle
Λ	reduced pigment volume concentration
λ	wavelength
μ	micron; dipole moment
ν	kinematic viscosity
π	constant (3.1416)
ρ	density
ϑ	dielectric constant
τ	shear stress
τ_0	yield value
ϕ	pigment packing factor; packing coefficient; volume fraction (solution)
ω	angular velocity

Index